Ecology and Ethics

Volume 5

Series Editor

Ricardo Rozzi, University of NorthTexas, Denton, TX, USA

This series is devoted to continuing research at the interfaces of ecology and ethics (embedded in the multiple fields of philosophy and ecology) to broaden our conceptual and practical frameworks in this transdisciplinary field. Confronted with global environmental change, the academic community still labors under a tradition of strong disciplinary dissociation that hinders the integration of ecological understanding and ethical values to comprehensively address the complexities of current socio-ecological problems. During the 1990s and 2000s, a transdisciplinary integration of ecology with social disciplines, especially economics, has been institutionalized via interdisciplinary societies, research programs, and mainstream journals. Work at this interface has produced novel techniques and protocols for assessing monetary values of biodiversity and ecosystem services, as illustrated by the Millennium Ecosystem Assessment. At the beginning of the 2010s, however, an equivalent integration between ecology and philosophy still remains elusive. This series undertakes the task to develop crucial theoretical and practical linkages between ecology and ethics through interdisciplinary, international, collaborative teamwork. It aims to establish a new forum and research platform to work on this vital, but until now insufficiently researched intersection between the descriptive and normative domains. The scope of this series is to facilitate the exploration of sustainable and just ways of co-inhabitation among diverse humans, and among humans and other-than-human co-inhabitants with whom we share our heterogeneous planet. It will address topics integrating the multiple fields of philosophy and ecology such as biocultural homogenization, Planetary or Earth Stewardship.

Ricardo Rozzi • Alejandra Tauro •
Noa Avriel-Avni • T. Wright • Roy H. May Jr.
Editors

Field Environmental Philosophy

Education for Biocultural Conservation

Springer

Editors
Ricardo Rozzi
Sub-Antarctic Biocultural Conservation Program, Department of Philosophy and Religion and Department of Biological Sciences
University of North Texas
Denton, TX, USA

Alejandra Tauro
Omora Ethnobotanical Park
Institute of Ecology and Biodiversity
Puerto Williams, XII - Punta Arenas, Chile

Noa Avriel-Avni
LTSER-Europe
Dead Sea and Arava Science Center
Mitzpe Ramon, Israel

T. Wright
Philosophy and Religion
University of North Texas
Denton, TX, USA

Roy H. May Jr.
Departamento Ecuménico de Investigaciones
San José, Costa Rica

ISSN 2198-9729 ISSN 2198-9737 (electronic)
Ecology and Ethics
ISBN 978-3-031-23370-8 ISBN 978-3-031-23368-5 (eBook)
https://doi.org/10.1007/978-3-031-23368-5

Cover illustration: Cover design by deblik, Berlin.

This Springer imprint is published by the registered company Springer Nature Switzerland AG
The registered company address is: Gewerbestrasse 11, 6330 Cham, Switzerland

Foreword

Our precious globe faces extinction crises of both biological and cultural diversity. The recent "Living Planet Report 2022" (WWF 2022) documents a decline of about 70% in animal species over the last 50 years. In fact, many scientists believe we are living through the sixth mass extinction crisis on Earth (e.g., Barnosky 2011). Yet, at the same time, many cultures and less charismatic species are evermore threatened and endangered as well. This circumstance is exacerbated by what many call an *extinction of experience* with biological and cultural diversity, which means that this tremendous loss is largely invisible to large segments of the human family. This is not a simple matter, and the need for action toward biocultural conservation is more urgent than ever. While climate change has consumed much of the attention around the globe, this multifaceted extinction crisis, and the need to sustain as much biocultural diversity as possible, demands equal attention. The ethical dimension advanced so eloquently here is an essential counterpoint to the marginalization of species, landscapes, and cultures in the way many scientists look at defining and solving environmental "problems." To address these crises, this compendium provides a powerful philosophical argument, indeed a moral imperative, that cultural diversity plays a key role in valuing and understanding biological diversity, and its conservation and sustainability.

In *Field Environmental Philosophy: Education for Biocultural Conservation*, Rozzi and his colleagues give voice, and even agency, to two arenas: "other-than-human" life forms and "other-than-dominant" cultures that, together, comprise the rich biocultural world we live in. From the elegant argument for using the term "other-than-human" and the parallel call for celebrating, protecting, and learning from the "other-than-dominant" cultures, this volume provides a sound philosophical line of reasoning for why cultural and biological diversity are *both* important. This importance is, itself, complex and multifaceted. On the "peril" side, both biological and cultural diversity are severely threatened by forces of ecological and social/cultural homogenization. On the "plus" side, each dimension stands to enrich the other: rich local biota supports and inspires local cultures, while cultures are poised to explore, understand, nurture, and sustain their local biota. The book celebrates the less-than-obvious species, almost out of sight or beyond our awareness, as well as the often marginalized cultures that possess tremendous wisdom and different ways of knowing about the landscapes they inhabit. The authors recognize the tensions between valuing species and ecosystems for utilitarian purposes and valuing them because of their intrinsic worth, a significance that rises above short-term, human-only purposes. While the connection between biocultural diversity and any specific ecosystem service is for the most part implicit, they make a compelling case for conserving diverse species and cultures for their rich information content, and the diversity of possibilities they represent for the sustainability of all life on our planet in the long term.

This book builds on and expands the work started more than a decade ago, when Rozzi and his colleagues answered the siren call for a flexible model of education about the environment that brings science, culture/society, and ethical thinking together in one place. He and his colleagues were looking for something different from traditional environmental and ecological science education, often defined by national or inspirational "standards" that can result in homogenization and one-way communication from science to the public. The resulting field environmental philosophy (FEP) approach, with its deep and authentic concern and valuing of the "other-than-dominant" cultures, invites diverse people into the exploration, conversation, and conservation process. In this way, it is distinct from traditional efforts to "diversify" participation in environmental education when "underrepresented groups" are perceived as "outsiders," and it integrates dimensions of both topic and process that, at its best, builds in dynamic science, pedagogy, and community/cultural engagement. The ethical and philosophical components of the FEP approach are an important enhancement on traditional environmental education. The ethical dimension explicitly integrates rigorous biological and social science and provides cultural agency and empowerment for everyone involved. The philosophical component provides a rigorous way of accomplishing this integration by giving people both the tools and the agency to think rigorously about ethics—the moral right and wrong—to find intellectual support for a range of values beyond simple short-term utilitarian aims.

The four-step FEP model from Rozzi et al. (2010) serves as a frame of reference for discussion and comparison of the invited case studies in the book: (1) *Transdisciplinary Ecological and Philosophical Research*—participants identify a phenomenon, problem, or need and then develop an understanding that integrates the bio/ecological, social/cultural, and ethical dimensions, and explore the balance between research in the biophysical dimensions of reality, cultural, and institutional perspectives. For example, we have the analogy of *reading the book of nature* and *reading philosophical texts*. But there is also the challenge of reading the institutions (e.g., schools, protected areas, biological station, LTSER sites) that support and enable these education/research practices. (2) *Composition of metaphors and communication through narratives*—this integrative dimension focuses on invention/discovery and art/science to create narratives, metaphors, and interpretive signs, thereby stimulating *analogical thinking*. Figuring out what the story is and how to tell it personalizes learning and sense of place and can be very powerful. (3) *Field activities guided with an ecological, social/cultural, and ethical orientation*—this step ensures that learning is grounded in the real and local world. Ethics is integrated as a relevant dimension in socio-ecological studies in support of the "3Hs" conceptual framework of the biocultural ethic (Rozzi 2013): *life habits, habitats,* and *co-inhabitants* (a concept that is consistent with Leopold's 2004 framework). And (4) *Implementation of areas for in situ biocultural conservation*—here participants engage in action. This crucial step, emerging from the first three phases of the process, builds cultural agency.

Readers of this excellent book will find an incredibly rich and diverse set of examples and case studies. Throughout, the FEP model is aspirational, and most of the examples in the book provide both powerful successes and insightful, and sometimes cautionary, notes about challenges and limitations to fully realizing FEP's potential. Perhaps not surprisingly, there is no single-standard approach that stands in equal opposition to the "dominant" approach to environmental and ecological education. Instead, we learn from and celebrate localized and diverse approaches. This makes a synthesis of just a few pearls from the case studies challenging: but indeed, that is the point of diversity! Nonetheless, most examples showcase the inherent, intrinsic (rights based or deontological) value of both species and cultures. In many cases the metaphors are vivid and compelling, while in others they are revealed in the examples as the programs are considered through a retrospective FEP lens. Further, many entail amazing examples of intercultural dialogue and co-production of both ecological and philosophical knowledge via research. The integration across science, art, philosophy, ethics, and social dimensions is a hallmark of the FEP approach and is evinced in these pages across a wide range of systems and settings around the globe. And we witness the establishment of long-term relationships between scientist, citizen, philosopher, artist, and activist.

How might the FEP approach scale? There are many opportunities around our planet in urban landscapes with a blend of cultures and biodiversity where the FEP model may help us address urgent threats to biocultural diversity and sustainability. While most of the highlighted programs are local in scale or reach, let us not allow the geography of the examples to limit us to the rural and more "wild" spaces that are most evident in this collection. Indeed, there is much to be gained by considering how to apply the FEP elements, and the lessons we take from them, in working (e.g., agricultural) landscapes as well as the urban places (habitations) humans are increasingly living in. An important frontier for the application of the FEP model is the rich biocultural life of cities, where we bring our cultural heritage with us and can use that knowledge to create new urban landscapes where biocultural diversity may be revealed in unexpected ways, such as the urban landscapes in Berlin where endangered species are being found in cemeteries and other green spaces (Popkin 2022).

The authors make a compelling case that participatory, multi-sensory, and multidimensional education can be a positive force in shaping the world and the way we identify, address, and solve problems. This rich approach elevates "other-than-dominant" cultures into a more vital role in addressing the complex problems facing both "other-than-human" life forms and humans alike. Compared to the more mainstream roles of education in preparing people for work and inculcating dominant social values (implicitly, perhaps), FEP fosters and celebrates important and

Acknowledgments

This volume conveys the work of many researchers, students, and people from different regions of the world. We are grateful to our institutions in Chile and in the USA, which have supported our editorial activities at the Cape Horn International Center (CHIC) and the University of Magallanes (UMAG) in Chile, and the University of North Texas (UNT) and the Center for Environmental Philosophy (CEP) in the USA.

We appreciate the encouragement and editorial guidance provided by Springer for this fifth volume of the *Ecology and Ethics* book series. At UMAG, we are especially thankful for Francisca Massardo at UMAG editorial office of the Sub-Antarctic Biocultural Conservation Program (SBC) and Kelli Moses, International Coordinator, who provided invaluable assistance for the production and edition of this volume. At the Department of Philosophy and Religion at the University of North Texas, we would like to thank Irene Klaver for her insights and ideas in the initial stages of the development of this book project.

Part of the inspiration for developing this book came from the paper "Field environmental philosophy and biocultural conservation at the Omora Ethnobotanical Park: Methodological approaches to broaden the ways of integrating the social component ("S") in long-term socio-ecological research (LTSER) sites" by Ricardo Rozzi and collaborators published in 2010 in the *Revista Chilena de Historia Natural*, 83(1), 27-68. A decade since its publication, and more than two decades of working with the FEP methodology in the field, this book is the first to completely and reflectively present the methodology.

We thank all the people who participated in the first steps of the creation of field environmental philosophy (FEP) at the Omora Ethnobotanical Park, Chile. We are thankful for the national and international graduate students, teachers, primary and secondary students from the Donald McIntyre Griffiths school, local authorities of Puerto Williams, and all our colleagues from the sciences, philosophy, and arts who nourished the practices that led to the consolidation of FEP as it is today.

We would like to acknowledge funding from the National Agency for Research and Development of Chile (ANID CHIC - ANID PIA/BASAL (PFB210018) and the

Public Affairs Section Project (#SCI80020GR0025) of the United States Embassy in Santiago, Chile. This book is a contribution to the Cape Horn International Center (CHIC) coordinated by UMAG-Puerto Williams, Chile.

Contents

Contributors

Iris Alkaher Kibbutzim College of Education Technology and the Arts, Tel-Aviv, Israel

Nimrod Aloni Kibbutzim College of Education Technology and the Arts, Tel-Aviv, Israel

Nirit Assaf Kibbutzim College of Education Technology and the Arts, Tel-Aviv, Israel

Noa Avriel-Avni Dead Sea and Arava Science Center, Mitzpe Ramon, Israel

Netta Baryosef-Paz Kibbutzim College of Education Technology and the Arts, Tel-Aviv, Israel

Marysia Borucinska-Begg Biocitizen LA, Los Angeles, CA, USA

Hugh Breakey Institute for Ethics, Governance and Law at Griffith University, Brisbane, QLD, Australia

Noreen Breakey School of Business, The University of Queensland, St Lucia, QLD, Australia

J. Baird Callicott University Distinguished Research Professor Emeritus, University of North Texas, Denton, TX, USA

Micheline Cariño-Olvera Department of Humanities, Universidad Autónoma de Baja California Sur, La Paz, Mexico

Jesse Carmichael Biocitizen California, Los Angeles, CA, USA

Tamara Contador Sub-Antarctic Biocultural Conservation Program, Universidad de Magallanes, Punta Arenas, Chile
Wankara Laboratory, Universidad de Magallanes, Puerto Williams, Chile
Núcleo Milenio de Salmónidos Invasores, Concepción, Chile

Cape Horn International Center (CHIC), Universidad de Magallanes, Puerto Williams, Chile
Institute of Biodiversity of Antarctic and Subantarctic Ecosystems (BASE), Santiago, Chile

Peter Convey British Antarctic Survey, NERC, Cambridge, UK

Magda Costa-Carvalho Faculty of Human and Social Sciences, University of the Azores, Ponta Delgada, Portugal
Institute of Philosophy of the University of Porto, Porto, Portugal

Ramiro D. Crego Smithsonian National Zoo and Conservation Biology Institute, Conservation Ecology Center, Washington, DC, USA

Zenorina Díaz-Gómez Department of Humanities, Universidad Autónoma de Baja California Sur, La Paz, Mexico

Enrique Dussel Facultad de Filosofía, Universidad Nacional Autónoma de México, Ciudad Universitaria, México, DF, Mexico

Rob Dussler The Meadows Center for Water and the Environment, Texas State University, San Marcos, TX, USA

Adiv Gal Kibbutzim College of Education Technology and the Arts, Tel-Aviv, Israel

Melisa Gañan Wankara Laboratory, Universidad de Magallanes, Puerto Williams, Chile
Núcleo Milenio de Salmónidos Invasores, Concepción, Chile

Dafna Gan Kibbutzim College of Education Technology and the Arts, Tel-Aviv, Israel

Arran Gare Swinburne University of Technology, Melbourne, VIC, Australia

Julia González-Calderón Artesana de la Comunidad Indígena Yagán, Puerto Williams, Chile

Humberto González-Galván Department of Humanities, Universidad Autónoma de Baja California Sur, La Paz, Mexico

Lissy Goralnik Department of Community Sustainability, Michigan State University, East Lansing, MI, USA

Kurt Heidinger Biocitizen, Inc., Westhampton, MA, USA

Jennifer M. Holzer Environmental Sustainability Research Centre, Brock University, St. Catharines, ON, Canada

R. Justin Hougham Department of Youth Development, University of Wisconsin, Madison, WI, USA

Benny Jacobs-Schwartz Biocitizen LA, Los Angeles, CA, USA

María Evelinda Santiago Jiménez Tecnológico Nacional de México Instituto Tecnológico de Puebla, Puebla, Mexico

Benn Johnson Kagawa University, Kagawa, Japan

James Kennedy Department of Biological Sciences; Sub-Antarctic Biocultural Conservation Program, University of North Texas, Denton, TX, USA

Miri Lavi-Neeman Arava Institute, Ketura, Israel

Ayanda Lepheana GroundTruth Associate, Durban, South Africa

Naama Lev Kibbutzim College of Education Technology and the Arts, Tel-Aviv, Israel

Benjamin Lukey The Uehiro Academy for Philosophy and Ethics in Education, University of Hawaii, Honolulu, HI, USA

Roy Mackenzie Cape Horn International Center (CHIC), Universidad de Magallanes, Puerto Williams, Chile

Javiera Malebran Cape Horn International Center (CHIC), Universidad de Magallanes, Puerto Williams, Chile

Andrés Mansilla Laboratorio de Ecosistemas Marinos Antárticos y Subantárticos (LEMAS), Universidad de Magallanes, Punta Arenas, Chile
Cape Horn International Center (CHIC), Puerto Williams, Chile

Adva Margaliot Kibbutzim College of Education Technology and the Arts, Tel-Aviv, Israel

Guillermo Marini Facultad de Educación, Pontificia Universidad Católica de Chile, Santiago, Chile
Cape Horn International Center (CHIC), Puerto Williams, Chile

Julie Markiewicz Department of Community Sustainability, Michigan State University, East Lansing, MI, USA

Francisca Massardo Cape Horn International Center (CHIC), Universidad de Magallanes, Puerto Williams, Chile

Samuel Massey The Meadows Center for Water and the Environment, Texas State University, San Marcos, TX, USA

Roy H. May Jr Departamento Ecuménico de Investigaciones (DEI), San José, Costa Rica

Manuela Méndez-Herranz Facultad de Educación, Pontificia Universidad Católica de Chile, Cape Horn International Center (CHIC), La Florida, Chile

Brant G. Miller Department of Curriculum and Instruction, University of Idaho, Moscow, ID, USA

Mario Monteforte-Sánchez Aquaculture Academic Program, Centro de Investigaciones Biológicas del Noroeste, S.C, La Paz, Mexico

René Moreno-Terrazas-Troyo Department of Humanities, Universidad Autónoma de Baja California Sur, La Paz, Mexico

Lucas Muñoz-Petersen Master of Science Program in Biosphere Reserves Management, Eberswalde University for Sustainable Development, Eberswalde, Germany

Kari O'Connell STEM Research Center, Oregon State University, Corvallis, OR, USA

Rob O'Donoghue Environmental Learning Research Centre, Rhodes University, Grahamstown, South Africa

Alexandria K. Poole Faculty of Behavioural, Management and Social Sciences (Philosophy), University of Twente, Enschede, The Netherlands

Javier Rendoll-Cárcamo Sub-Antarctic Biocultural Conservation Program, Universidad de Magallanes, Punta Arenas, Chile
Wankara Laboratory, Universidad de Magallanes, Puerto Williams, Chile
Núcleo Milenio de Salmónidos Invasores, Concepción, Chile
Cape Horn International Center (CHIC), Universidad de Magallanes, Puerto Williams, Chile
Institute of Biodiversity of Antarctic and Subantarctic Ecosystems (BASE), Santiago, Chile

Sebastián Rosenfeld Laboratorio de Ecosistemas Marinos Antárticos y Subantárticos (LEMAS), Universidad de Magallanes, Punta Arenas, Chile
Cape Horn International Center (CHIC), Universidad de Magallanes, Puerto Williams, Chile
Centro de Investigación Gaia-Antártica, Universidad de Magallanes, Punta Arenas, Chile
Instituto Milenio Biodiversidad de Ecosistemas Antárticos y Subantárticos (BASE), Universidad de Chile, Santiago, Chile

Ricardo Rozzi Sub-Antarctic Biocultural Conservation Program, Department of Philosophy and Religion and Department of Biological Sciences, University of North Texas, Denton, TX, USA
Cape Horn International Center (CHIC), Universidad de Magallanes, Puerto Williams, Chile

Juan Carlos A. Sandoval-Rivera Instituto de Investigaciones en Educación, Universidad Veracruzana, Xalapa, Mexico

Elke Schüttler Cape Horn International Center (CHIC), Universidad de Magallanes, Puerto Williams, Chile

Taly Segal Kibbutzim College of Education Technology and the Arts, Tel-Aviv, Israel

Frederick J. Swanson US Forest Service, Pacific Northwest Research Station, Corvallis, OR, USA

Alejandra Tauro El Colegio de Puebla A.C, Puebla, Mexico
Cape Horn International Center (CHIC), Universidad de Magallanes, Puerto Williams, Chile

Jim Taylor Centre for Water Resources Research, University of KwaZulu-Natal, Durban, South Africa

Mitsuyo Toyoda Sado Island Center for Ecological Sustainability, Niigata University, Niigata, Japan

Rika Tsuji Kagawa University, Kagawa, Japan

Jaime Ojeda School of Environmental Studies, University of Victoria, Victoria, BC, Canada
Cape Horn International Center (CHIC), Universidad de Magallanes, Punta Arenas, Chile

Nora Ward School of History and Philosophy, National University of Ireland Galway, Galway, Ireland

Justin Williams Department of Philosophy, Texas State University, San Marcos, TX, USA

Egleé Zent Laboratorio de Ecología Humana, Centro de Antropología, Instituto Venezolano de Investigaciones CientÚficas (IVIC), Altos de Pipe, Miranda, Bolivarian Republic of Venezuela

Stanford Zent Laboratorio de Ecología Humana, Centro de Antropología, Instituto Venezolano de Investigaciones CientÚficas (IVIC), Altos de Pipe, Miranda, Bolivarian Republic of Venezuela

Danqiong Zhu Department of Philosophy and Religion, Subantarctic Biocultural Conservation Program, University of North Texas, Denton, TX, USA
Xidian University, Xi'an, Shaanxi, China
Cape Horn International Center (CHIC), Universidad de Magallanes, Puerto Williams, Chile

Chapter 1
Introduction to Field Environmental Philosophy: A New Methodological Approach for Biocultural Education and Conservation

Ricardo Rozzi, Roy H. May Jr, Alejandra Tauro, and Noa Avriel-Avni

Abstract Modern education has fostered universal forms of knowledge, practices, pedagogical paradigms, and infrastructure. Uniform educational models can overlook the richness of regional biological and cultural diversity, and their interactions (in short, biocultural diversity). The omission of vernacular forms of knowledge and also endemic biota and their interrelationships can indirectly drive biocultural diversity losses. At the same time, there is also a diversity of educational practices that promote the conservation of biocultural diversity. In this book, we present a novel educational methodological approach to appreciate local biocultural diversity. This methodology is called "field environmental philosophy," because it integrates sciences, humanities, arts, and ethics to learn about biophysical, symbolic-linguistic, and political dimensions of biocultural diversity. A central goal of FEP is that students and other participants not only learn about biocultural diversity but also foster respectful ways of becoming responsible co-inhabitants. This book is organized into four parts that present: (1) concepts and practices of FEP; (2) a diversity of philosophical schools of thought and concepts and methods that inform FEP, and other complementary educational models; (3) case studies that exemplify valuable

R. Rozzi (✉)
Sub-Antarctic Biocultural Conservation Program, Department of Philosophy and Religion and Department of Biological Sciences, University of North Texas, Denton, TX, USA

Cape Horn International Center (CHIC), Universidad de Magallanes, Puerto Williams, Chile
e-mail: Ricardo.Rozzi@unt.edu

R. H. May Jr
Departamento Ecuménico de Investigaciones (DEI), San José, Costa Rica

A. Tauro
Cape Horn International Center (CHIC), Universidad de Magallanes, Puerto Williams, Chile

El Colegio de Puebla A.C., Puebla, México

N. Avriel-Avni
Dead Sea and Arava Science Center, Mitzpe Ramon, Israel
e-mail: noa@adssc.org

R. Rozzi et al. (eds.), *Field Environmental Philosophy*, Ecology and Ethics 5,
https://doi.org/10.1007/978-3-031-23368-5_1

non-formal education initiatives for biocultural conservation; and (4) long-term research and education platforms that support field experiences that facilitate an understanding, valuation, and conservation of biocultural diversity. Our ultimate goal is to contribute to dynamic, intercultural, and interregional approaches to planetary stewardship initiatives. This book builds on more than 20 years of FEP activities that progressively have included (1) graduate students from different parts of the world, (ii) transdisciplinary approaches involving not only academia, but also public servants, members of indigenous and other local communities, as well as from the tourism sector, (iii) the integration of theory and practice into the creation of multiple-use protected areas, new tourism themes and activities, and biocultural conservation policy. Our goal is to offer educational concepts and practices that reconnect society with biocultural diversity for the well-being of both.

Keywords Biocultural ethic · Ecology · Experiential pedagogies · Sense of place · Social-environmental Justice

1.1 Introduction

We are immersed in a climate crisis as well as in social-environmental crisis that entails the rupture of inter-human and inter-species relations. In times of rapid social and environmental change or when unexpected contingencies, such as the COVID pandemic occur, we need to address these complex problems. This complexity demands that we examine the interactions (or lack of) that institutions and citizens of global society have with biological and cultural diversity, and their interrelationships (in short biocultural diversity). Directly or indirectly we coinhabit the various land and marine communities of the planet. However, today biocultural diversity is often underappreciated in the theory and practice of environmental sciences, education, policy, and the prevailing concepts of well-being. To overcome this omission, in this book we present educational concepts and methodologies that contribute to making biocultural diversity more visible and to foster habits of respectful (re)-connection with the vast biological and cultural diversity with which we coexist on the planet (Tauro et al. 2021).

This biocultural (re)connection is urgent because a root cause of the current socio-environmental crisis stems from the losses of biological and cultural diversity. Biocultural diversity, along with its knowledge and values, has been increasingly excluded from everyday experiences. To achieve biocultural reconnections, we propose multi-sensorial educational methodologies grounded in analogical thinking, the integration of science and philosophy, and habits of care for the multiplicity of human and other-than-human co-inhabitants and the singularity of their shared habitats.

Throughout this book, we use the term "other-than-humans" to refer to non-human living organisms, such as plants or other animals, to ecological entities such as rivers, mountains, the sea, the stars, the moon or the sun, as well as symbolic beings and mythological figures. The alternative term "non-human beings" suggests a dualistic dichotomy that maintains the *Homo sapiens* species at the center and separates it from other beings, thereby ignoring vital biophysical and symbolic-linguistic interconnections among all beings (including humans). We also consider the term "more-than-human beings," which is used with increasing frequency in the environmental ethics literature, to be problematic, because it suggests a realm of beings that appears to be beyond the human, even above the human, suggesting again a separation that denies the interconnection. The term "other-than-human beings" expresses "otherness," but it maintains humans embedded among a plurality of biological, geological, as well as symbolic beings. This ranges from subcellular beings that live within the human body, to elements such as water, air, heat, and minerals, and beyond the biophysical beings to symbolic beings that are perceived in wakefulness and in dreams, those captured with reason, the senses, intuition or surprise, and also the mystery of otherness, which deserves not only to be investigated but also respected (see Rozzi 2018, p. 37). The concept of "other-than-humans" is consistent with an education that values biocultural diversity.

Biocultural education will prepare students for future contingencies of global social-environmental change. To undertake this task, the educational methodologies introduced in this book involve ethical dimensions that demand rethinking the prevailing life habits of citizens of global society that often suffer an alienation of biocultural diversity at two levels. First, the growing concentration of urban populations generates *physical barriers* that distance people from daily contact with biocultural diversity. Second, digital society generates *conceptual barriers* that mediate perceptions to the extent that face-to-face encounters with biocultural diversity have been rapidly reduced. This double alienation, physical and conceptual, generates a lack of awareness regarding the richness of biocultural diversity that seriously imperils the sustainability of human and other-than-human life on the planet.

To address this cognitive and ethical change in the relationships of society with biocultural diversity, we offer in this book a novel methodology developed over the last two decades: *field environmental philosophy* (FEP). We complement FEP with other educational practices and pedagogies rooted in diverse cultural expressions and peoples from different world regions. These approaches integrate science and ethics and a wide variety of disciplines and forms of vernacular knowledge. We propose this integration based on three principles.

First, *integration of sciences, humanities, and other disciplines*. This integration facilitates understanding of both direct drivers (or proximate causes) and indirect drivers (or ultimate causes) of the current global socio-environmental crisis (Rozzi 2001). Such education integration can assist in a reconnection between students (and other citizens) and biocultural diversity by using *physical tools* offered by science and technology and *conceptual tools* offered by philosophy and the humanities (Rozzi et al. 2012, p. 235). This combination of approaches helps students understand and value biocultural systems and guide them toward thoughtful ways of co-inhabiting urban, rural, and remote ecosystems while respecting biological and cultural diversity.

Second, *field education* (place-based) helps in an unsurpassable way to observe biological diversity and cultural diversity *in situ, to know their interrelationships and thus value nature in its biocultural complexity*. Direct observation, meeting other cultures, and scientific understanding expand conscious awareness about biocultural diversity that stimulates ontological and epistemological transformations for understanding the natural world. In turn, this awareness triggers ethical transformations that attend to the principle of responsibility (*sensu* Jonas 1984) that reach beyond the purely human sphere (see Larrère 2019; Pommier 2019).

Third, the *conceptual framework of the "3Hs" (Habitats, Habits, co -Inhabitants) of the biocultural ethic* (*sensu* Rozzi 2012, 2013) applied to education fosters a sense of responsibility and care for biocultural diversity. The biocultural ethic values the vital links between specific *habitats* and their *co-inhabitants* with their *habits* of life. Each of these "3Hs" includes *biophysical dimensions, cultural dimensions* (involving both material and immaterial culture, i.e., symbolic—linguistic), and *institutional dimensions* that comprise social, political, infrastructural, and technological factors (Rozzi 2018). With this biocultural framework, we aspire to carefully interrelate the biophysical and symbolic-linguistic-cultural dimensions in the design of educational programs.

The "3Hs" triad of the biocultural ethic offers a systemic and contextual approach that situates the task of education in specific habitats or places. In each school and its region, there are multiple co-inhabitants with unique and interrelated lifestyles. This provides a heuristic model for visualizing, studying, and valuing the interdependencies among diverse co-inhabitants (humans and other-than-humans). Appreciation of these interdependencies generates greater awareness about the consequences of our individual and collective actions for the well-being of biocultural communities.

1.2 Field Environmental Philosophy: Education for Biocultural Conservation

In the Ecology and Ethics book series, we have examined education as an indirect driver of biocultural homogenization. In this volume, we deepen these analyses and offer complementarily solutions to this problematic facet of formal education (Rozzi et al. 2008). After introducing the methodological approach of field environmental philosophy (FEP), we address biological and cultural diversity at multiple scales (from local to global). FEP integrates diverse disciplines and cultural forms of knowledge rooted in specific localities. We discuss the potential of FEP and other approaches to biocultural education developed in different regions of the world that present as complementary practices for biocultural conservation.

In 2010, Ricardo Rozzi and colleagues published the article "Field environmental philosophy (FEP) and biocultural conservation in the Omora Ethnobotanical Park: Methodological approaches to expand the ways of integrating the social component

("S") in Long-Term Socio-Ecological Study Sites (SESELP)" to recount ten years of practical experiences and conceptual formulations around biocultural conservation (Rozzi et al. 2010). Ten years later, in 2020 we gathered together to synthesize FEP concepts and practices. This is the fifth volume in the Ecology and Ethics series. Its first volume integrated ecological sciences and environmental philosophy. The second incorporated the former disciplines into Earth Stewardship. The third exposed biocultural homogenization drivers and offered concepts and practices for biocultural conservation. The fourth documented an exemplary case of international collaborations between European women and Mahatma Gandhi's work in India, which illustrates the value of intercultural dialogue and grassroots actions. This fifth volume on FEP has three general goals:

- First, to introduce FEP as a novel educational approach to contribute to reconnecting students and other citizens of global society with biocultural diversity.
- Second, to examine how formal education, if decontextualized from local biocultural diversity, can be an indirect driver of biocultural homogenization.
- Third, to learn about formal and non-formal educational programs embedded in local communities and institutions (e.g., biological stations, botanical parks, long-term socio-ecological research sites) fostering biocultural conservation in different regions of the world.

Our book is organized into four parts. Part I introduces FEP concepts and practices illustrated with case studies developed in Chile and other countries. Part II exposes problems as well as solutions in formal education (from pre-school to higher education) and non-formal education actions to respect biocultural diversity. Part III provides case studies with contrasting approaches of biocultural education in different regions of the world. Part IV focuses on the key role that long-term social-environmental research, education and/or conservation centers can offer for the integration of diverse disciplines and cultural forms of knowledge rooted in specific localities.

1.3 Part I: FEP a New Methodological Approach to Biocultural Conservation

In Part I, we present the methodological approach of Field Environmental Philosophy (FEP), to foster an ethics rooted in specific places or habitats, and to develop a methodological approach that broadens the social component considered in socio-ecological research and education programs—which prevailingly focuses on economic values (Parr et al. 2002; Redman et al. 2004; Ohl et al. 2007; Haberl et al. 2009). Toward this aim, a group of scientists, artists, philosophers, and other professionals, both Chilean and foreign, initiated a program of biocultural conservation in 1999 that led to the creation of the Omora Ethnobotanical Park in Cape Horn at the southern tip of South America (55 o S, south of Tierra del Fuego) in 2000.

Omora Park aspires to provide a "biophysical, conceptual, and institutional space" for long-term biocultural research, education, and conservation (Rozzi et al. 2006). As a biophysical space, Omora Park aims to protect the unique and fragile biodiversity of the Magellanic sub-Antarctic region, from the Andean highlands to deciduous and evergreen forests, peatlands, and coastal ecosystems. It includes a biological reserve that protects the watershed of the Róbalo River, the source of drinking water for Puerto Williams, the world's southernmost town and capital of the Chilean Antarctic province on Navarino Island (Contador et al. 2012).

As a conceptual space, Omora Park fosters educational experiences and methodologies to integrate ecological sciences and environmental ethics into biocultural education and conservation (Rozzi et al. 2010). It includes a natural laboratory, an outdoor classroom, and a training center, whose functions involve three broad domains of action: (1) transdisciplinary research; (2) formal and non-formal education through pre-school, school, university, and training courses; and (3) biocultural conservation linked with environmental decision-making and sustainable tourism. As an institutional space, it articulates local, regional, national, and international scales in long-term socio-ecological research, education, and conservation.

FEP integrates the three domains of action of Omora Park. The adjective **field** highlights the firsthand experience that facilitates discovery for oneself of the components and processes of biocultural diversity that are often omitted or distorted in formal education. Also, in the field participants gain an integral, *in situ*, perception of biocultural diversity by engaging their senses, emotions, and rationality in the interactions with other beings. Most importantly, in the field "face-to-face" encounters with other human and other-than-human beings transform the understanding of biocultural diversity where other beings cease to be mere objects of study and acquire the status of co-inhabitant subjects. The adjective **environmental** in the title of the FEP methodological approach makes explicit the goal of overcoming the reduction of ethics to purely human affairs. It emphasizes that human existence is immersed in co-inhabitation relationships within communities of human and other-than-human co-inhabitants. Finally, the **philosophical** approach enriches field ecology because it integrates epistemological and ethical dimensions. Participants address epistemological questions to investigate not only biological and cultural diversity, but also the methods, disciplines (sciences, humanities, and arts), languages, and worldviews through which scientific and other forms of knowledge about biological and cultural diversity are forged. The goal of FEP, however, is that participants do not limit themselves to only learning about biocultural diversity, but also to explore respectful, sustainable ways of co-inhabitation (Rozzi 2013).

FEP has historically been practiced with graduate students who have formulated their theses employing this methodology with different audiences. Among the main audiences are children from kindergarten and primary school to university tourism students. There are also experiences of applying the FEP with tourists and other visitors of Omora Park. FEP has spread to other regions of Chile and to other countries such as Mexico (Moreno-Terrazas et al. 2023) and the United States (Heidinger et al. 2023).

FEP is practiced in a multi-directional four-step cycle. These steps or movements function on two levels simultaneously: to generate knowledge and conservation action, and to help people value what is often unperceived and to establish a value framework to guide them. To incorporate FEP into educational programs, we designed an interrelated four-step cycle, which we briefly summarize below.

Step 1: Transdisciplinary Ecological and Philosophical Research In-reach and not only out-reach seeks to recover dialogue between scientists and other disciplines and ways of knowing, such as philosophy and the arts. Practitioners of FEP also participate with community members and actors from different institutions, so that practical knowledge is incorporated from members of different disciplines, institutions, and sociocultural groups, who speak different languages and have different forms of ecological knowledge and practices. Comparative analyses are carried out to identify similarities and differences between different ways of knowing, valuing, and living with biocultural diversity.

Step 2: Composition of Metaphors and Narratives Graduate students and practitioners conducting FEP emphasize analogical thinking through the creation of metaphors, narratives, and artistic representations to generate knowledge. Analogical reasoning is a cognitive underpinning of the ability to notice and draw similarities across contexts (Lakoff and Johnson 1980). This is an essential ability for biocultural research and conservation practices. Metaphors constitute cognitive-linguistic figures, which are part of the fundamental cognitive structure of humans in their everyday environments as in their scientific thought (Lakoff and Johnson 1980). Metaphors are compositions inspired by the surrounding physical reality (Hermsen 2009) in order to synthesize learnings from Step I as well as to create histories, stories, and poetry that facilitate intercultural dialogues, engagement with the general public, and sharing new actions oriented to biocultural conservation.

Step 3: Design of Field Activities Guided with an Ecological and Ethical Orientation Practitioners of FEP co-create "face-to-face encounters" with diverse co-inhabitants in their habitats in such a way that all are enabled to connect with diverse co-inhabitants. By integrating emotions and concepts derived from philosophical, scientific, and vernacular knowledge into field activities, practitioners and the general public experientially apprehend the concept of co-inhabitants through direct encounters. These activities foment reciprocity. For example, participants "walk in the shoes of others" by emulating the life habits of cushion plants that grow in the mountains of Chile (Méndez-Herranz et al. 2023), or by snorkeling in the undersea forests of the Magellanic coast (Ojeda et al. 2023) or the aquatic trails of the Sea of Cortez in Baja California Mexico (Moreno-Terrazas et al. 2023).

Step 4: Implementation of In Situ Conservation Spaces The activities contemplated in the first three steps will foster a sense of responsibility as citizens who are ecologically and ethically educated, and proactively participate in caring for biocultural diversity. FEP requires graduate students to be involved in biocultural conservation actions; for example, the design and implementation of interpretive stations along trails, or areas for the protection of the inhabitants, their habits, and habitats (native habitats, species, and ecological interactions). Likewise, it is expected that others who are involved in FEP, even tourists, will engage in conservation actions.

In summary, FEP offers a methodological approach to integrate ecological sciences and environmental ethics into biocultural conservation and biocultural ethic, which are ecologically and culturally contextualized. The FEP four-step cycle helps students to gain not only an understanding of scientific and traditional ecological knowledge but foremost an in situ ethical practice, which leads to a better awareness of ecological and cultural interactions, the interrelated habits and habitats of the communities of co-inhabitants, their ecological, economic, aesthetic, ethical, and biocultural values. Based on this theoretical and practical understanding of biocultural diversity, the FEP cycle fosters research and conservation habits for respectful forms of co-inhabitation within the diverse communities of human and other-than-human co-inhabitants.

1.4 Part II: Educational Practices Between Biocultural Homogenization and Conservation

There is a delicate balance—a razor's edge—between education as a driver promoting biocultural homogenization and education promoting biocultural conservation. Part II assesses problematic concepts and methods that overlook biocultural diversity in education. Part II also introduces novel educational concepts and methodologies developed in the context of different disciplines and cultural traditions of the world. These foster biocultural conservation at local scales, as well as networking from local to global.

Philosopher Gare (2023) explains how Western schools of thought are diverse. Historical thinkers such as Vico, Herder, and others, too often are forgotten by mainstream formal education, yet they offer rich approaches to biocultural education today. By exploring the legacy of these Western thinkers, readers are invited to critically understand the importance of including Western schools of thought that are often overlooked. Understanding the diversity of Western schools of thought is essential to better assess education as a driver of biocultural homogenization, and to visualize alternatives to prevent this problem.

A central axis of the problem is that abstract ideas and ideals that shape education and sustainability often are detached from local contexts. For this reason, it is necessary to include educational approaches that emphasize singularity, local history, and personal experiences. Many such approaches have been developed in recent years and it is important to recover the materials and methods that are being used, especially in environmental education, to promote and conserve biological and cultural diversity. Traditional ecological wisdom and modern scientific knowledge can complement each other. In this sense, recovering the value of Indigenous wisdom is fundamental.

1.5 Part III: Educational Practices to Promote Biocultural Reconnections Around the World

The problems identified in Part II are addressed in Part III. Part III's main goal is to examine a diversity of ongoing initiatives that propose complementary approaches to FEP. We introduce ongoing approaches illustrated by local cases and networks that focus on local biodiversity and forms of ecological knowledge. In different parts of the world, formal and non-formal education initiatives provide guidance for reconnection among different co-inhabitants and habitats. Often these alternatives are invisible to the eyes of the prevailing structures; they are self-managed, and many are sporadic. However, they summarize the organized social effort that advocates change and social transformations (Kothari 2021).

Reconnecting means connecting again, putting back in communication, and recovering linkages; etymologically it is associated with the verb to bind. Recovering the links between humans, today mainly city dwellers, and the diversity of co-inhabitants leads us to a re-understanding of what it means to live together in a matrix of diverse cultures, living beings, and ecosystems. This reconnection resonates with the original meaning of the term religion which leads us to re-link ourselves with "the whole" or that portion of life that animates us. Following the metaphor of seeing through different eyes, ("children's eyes," "eyes of the earth," and "fraternal eyes"), we present cases of biocultural reconnection.

Experiences developed in Portugal, Japan, the United States, Venezuela, Mexico, and South Africa reconnect with the small, the silent, and the invisible in childhood and other co-inhabitants too often ignored by the prevalent view of education. In this way, these initiatives recover the sense of community involving children and adults in different initiatives through philosophical dialogue. Additionally, it is proposed to reconnect society with biocultural diversity by incorporating ecotourism as an effective practice of non-formal education. This can also generate economic alternatives based on social solidarity for local communities. Cases illustrate collaborative actions and education that are facilitated by technological tools and incorporate local development, territory defense, and traditional local knowledge. These cases offer a broader and contextual understanding of education as a tool that reconnects people with biocultural diversity. These biocultural reconnections weave together local initiatives and global movements that seek to defend and create conditions that make possible the continuance of life in its biological and cultural diversity on this planet.

1.6 Part IV: Long-Term Views to Broadening Educational Practices Into Biocultural Platforms

Part IV focuses on long-term research sites, understood as platforms to integrate science, arts, and humanities into participatory education. This section presents a variety of cases associated with biological stations, Sense of Place initiatives, botanical gardens, and other local research platforms. We give particular emphasis to Long-Term Ecological (or Socio-Ecological) Research (LTER or LTSER) sites as platforms for the integration of diverse disciplines and cultural forms of knowledge rooted in specific localities to conduct biocultural educational programs.

These platforms can play key roles in promoting biocultural conservation if they are integrated into the local cultural contexts and take seriously the biocultural world that surrounds them through novel and participatory programs. The research conducted can highlight the values of local biota, show cultural relationships between people and biota, and stimulate local conservation initiatives. For example, botanical gardens not only can be places of recreation for exhibiting the beauty of plant life but also for the conservation of endangered or culturally important species—including indigenous languages. As Christopher Dunn (2017, p. 396) affirms, "botanical gardens engaged in conservation have a unique and substantial opportunity to include cultural revitalization and awareness as a component of a larger conservation effort and messaging." Furthermore, overlooked species can be promoted and valued providing new perspectives on common biota, reconnecting human beings with the natural world. In turn, botanical gardens as models of biocultural conservation respond to different international strategies such as the Convention on Biological Diversity (CBD) that are implemented in different government policies around the world.

Even the location of the platform can promote diversity and knowledge of overlooked places. The Cape Horn Biosphere Reserve is a good example. Until 2005 there was no research site and/or long-term environmental monitoring in the subpolar latitudes between 50° and 60°S of the southern hemisphere (Rozzi et al. 2012). To fill this void, the Omora Ethnobotanical Park was established as a long-term ecological research (LTER) site in 2008. However, there is still a marked bias toward the northern hemisphere that, for example, includes 89% of the 733 sites of the International Network for Long-Term Ecological Research (ILTER) (Rozzi et al. 2020). Moreover, more than 90% of ILTER publications are from researchers from the global north (Li et al. 2015).

The chapters of Part IV illustrate the capacity of long-term research platforms to initiate participatory educational programs with local people and promote biocultural, including linguistic, conservation. This, in turn, generates resources, data, and scientific programs to better understand global environments and their connections with local ones. For these reasons, such long-term platforms provide valuable sites to foster intercultural dialogues and production of knowledge, as well as the integration of sciences, arts, and philosophy within a site-based frame of reference.

1.7 Concluding Remarks

This FEP book focuses on methodological tools solidly conceptualized and rooted in natural and social science, philosophy and ethics to re-think human practices and inspire actions toward the conservation of biocultural diversity over time. As several chapters show how education can accentuate the disconnection with biocultural diversity; yet as a transformative driver, education recovers and restores the interrelationships that characterize life on this planet. We hope the chapters of this volume contribute in practical ways to education that uplifts the richness and diversity of forms of local knowledge embedded in biological and cultural diversity that are specific to different regions of the world.

Acknowledgments Authors thank valuable comments by Kelli Moses and acknowledge the support from the Cape Horn International Center (CHIC) - ANID PIA/BASAL (PFB210018).

References

Contador T, Kennedy JH, Rozzi R (2012) The conservation status of southern South American aquatic insects in the literature. Biodivers Conserv 21(8):2095–2107

Dunn CP (2017) Biological and cultural diversity in the context of botanic garden conservation strategies. Plant Diver 39(6):396–401

Gare A (2023) Challenging the dominance of the grand narrative of economic growth in global education and culture. In: Rozzi R, Tauro A, Wright T, Avriel-Avni N, May RH Jr (eds) Field environmental philosophy: education for biocultural conservation. Ecology and ethics, vol 5. Springer, Dordrecht, pp 309–326

Haberl H, Graube V, Díaz-Delgado R, Krauze K, Neuner A, Peterseil J, Plutzar C, Singh SJ, Vadineanu A (2009) Towards an integrated model of socioeconomic biodiversity drivers, pressures and impacts: a feasibility study based on three European long-term socio-ecological research platforms. Ecol Econ 68:1797–1812

Heidinger K, Camichael J, Jacob-Schwartz B, Borucinska-Begg M (2023) Biocitizen's approach to biotic wonder, citizenship, and field environmental philosophy. In: Rozzi R, Tauro A, Wright T, Avriel-Avni N, May RH Jr (eds) Field environmental philosophy: education for biocultural conservation. Ecology and ethics, vol 5. Springer, Dordrecht, pp 131–152

Hermsen T (2009) Poetry of place: helping students write their worlds. National Council of Teachers of English

Jonas H (1984) The imperative of responsibility: in search of an ethics for the technological age. Chicago University Press, Chicago

Kothari A (2021) These Alternative Economies Are Inspirations for a Sustainable World. https://www.scientificamerican.com/article/these-alternative-economies-are-inspirations-for-a-sustainable-world/

Lakoff G, Johnson M (1980) The metaphors we live by. University of Chicago Press, Chicago

Larrère C (2019) "A Life Worthy of Being Called Human": the actuality of Hans Jonas' maxim. Environ Ethics 41(4):319–332

Li B, Parr T, Rozzi R (2015) Geographical and thematic distribution of publications generated at the international long-term ecological research network (ILTER) sites. In: Earth Stewardship. Springer, Cham, pp 195–216

Méndez-Herranz M, Marini G, Rozzi R (2023) Sub-Antarctic High Andean "Gardeners:" cultivating caring relationships. In: Rozzi R, Tauro A, Wright T, Avriel-Avni N, May RH Jr (eds) Field environmental philosophy: education for biocultural conservation. Ecology and ethics, vol 5. Springer, Dordrecht, pp 71–86

Moreno-Terrazas R, Micheline Cariño M, Díaz Z, González H, Monteforte-Sánchez M (2023) Starfishes and sky stars: a field environmental philosophy education and ecotourism experiences in Baja California, México. In: Rozzi R, Tauro A, Wright T, Avriel-Avni N, May RH Jr (eds) Field environmental philosophy: education for biocultural conservation. Ecology and ethics, vol 5. Springer, Dordrecht, pp 113–130

Ohl C, Krauze K, Grünbühl C (2007) Towards an understanding of longterm ecosystem dynamics by merging socio-economic and environmental research: criteria for long-term socio-ecological research sites selection. Ecol Econ 63:383–391

Ojeda J, Tauro A, Contador T, Mansilla A, Rosenfeld S, Ban N, Rozzi R (2023) "Pay attention, dive with your eyes wide-open:" a field environmental philosophy activity to foster reciprocity between people and nature. In: Rozzi R, Tauro A, Wright T, Avriel-Avni N, May RH Jr (eds) Field environmental philosophy: education for biocultural conservation. Ecology and ethics, vol 5. Springer, Dordrecht, pp 87–100

Parr TW, Ferretti M, Simpson IC, Forsius M, Kovács-Láng E (2002) Towards a long-term integrated monitoring programme in Europe: Network design in theory and practice. Environ Monit Assess 78:253–290

Pommier E (2019) The problem of environmental democracy: responsibility and deliberation. Environ Ethics 41(4):305–318

Redman CL, Grove JM, Kuby LH (2004) Integrating social science into the long-term ecological research (LTER) network: social dimensions of ecological change and ecological dimensions of social change. Ecosystems 7:161–171

Rozzi R (2001) Éticas ambientales latinoamericanas: raíces y ramas. In: Primack R, Rozzi R, Feinsinger P, Dirzo R, Massardo F (eds) Fundamentos de Conservación Biológica: Perspectivas Latinoamericanas. Fondo de Cultura Económica, México, pp 311–362

Rozzi R (2012) Biocultural ethics. Environ Ethics 34(1):27–50. https://doi.org/10.5840/enviroethics20123414

Rozzi R (2013) Biocultural ethics: from biocultural homogenization toward biocultural conservation. In: Rozzi R, Pickett STA, Palmer C, Armesto JJ, Callicott JB (eds) Linking ecology and ethics for a changing world: values, philosophy, and action, Ecology and ethics, vol 1. Springer, Dordrecht, pp 9–32

Rozzi R (2018) Biocultural homogenization: A wicked problem in the anthropocene. In: From biocultural homogenization to biocultural conservation. Springer, Cham, pp 21–48

Rozzi R, Massardo F, Anderson C, Heidinger K, Silander J Jr (2006) Ten principles for biocultural conservation at the southern tip of the Americas: the approach of the Omora Ethnobotanical Park. Ecol Soc 11(1):43. [online] URL: http://www.ecologyandsociety.org/vol11/iss1/art43/

Rozzi R, Arango X, Massardo F, Anderson C, Heidinger K, Moses K (2008) Field environmental philosophy and biocultural conservation: the Omora Ethnobotanical Park educational program. Environ Ethics 30(3):325–336

Rozzi R, Anderson CB, Pizarro JC, Massardo F, Medina Y, Mansilla A et al (2010) Field environmental philosophy and biocultural conservation at the Omora Ethnobotanical Park: Methodological approaches to broaden the ways of integrating the social component ("S") in long-term socio-ecological research (LTSER) sites. Revista Chilena de Historia Natural 83(1): 27–68

Rozzi R, Armesto JJ, Gutiérrez JR, Massardo F, Likens GE, Anderson CB et al (2012) Integrating ecology and environmental ethics: earth stewardship in the southern end of the Americas. BioScience 62(3):226–236

Rozzi R, Crego RD, Contador T, Schüttler E, Rosenfeld S, Mackenzie R et al (2020) Un centinela para el monitoreo del cambio climático y su impacto sobre la biodiversidad en la cumbre austral de América: la nueva red de estudios a largo plazo Cabo de Hornos. Anales del Instituto de la Patagonia 48(3):45–81

Tauro A, Ojeda J, Caviness T, Moses K, Moreno R, Wright T, Zhu D, Poole A, Massardo F, Rozzi R (2021) Field environmental philosophy: a biocultural ethic approach to education and ecotourism for sustainability. Sustainability 13:4526. https://www.mdpi.com/2071-1050/13/8/4526. https://doi.org/10.3390/su13084526

Part I
Field Environmental Philosophy: Concepts and Case Studies

Chapter 2
Introduction to Part I. Field Environmental Philosophy: Concepts and Case Studies

Ricardo Rozzi

Abstract Education plays a major role in the formation of citizens and attitudes about biological and cultural (biocultural) diversity. Today, we are increasingly disconnected from everyday interactions with our environments and the living beings that inhabit them. To reorient this trend, we present a novel methodological approach, Field Environmental Philosophy (FEP), to reconnect students and other participants with biocultural diversity. This methodological approach is called "field environmental philosophy," and not "field ecology" because it integrates epistemological, ontological, and ethical dimensions. Graduate students and other participants use methods from sciences humanities and arts. They combine analyses of philosophical, ethnographic, and ecological texts with field experiences to investigate biophysical, symbolic-linguistic, and political dimensions of biocultural diversity. The ultimate goal of FEP, however, is that participants do not limit themselves to only learning about biocultural diversity, but also propose and practice respectful, and responsible ways of co-inhabitation.

Keywords Biocultural conservation · Ecology and science education · Ecotourism · Ethics · Global change

2.1 Introduction

Purely scientific or other cognitive arguments have not been sufficient to open the "ecological eyes" of most citizens to appreciate the magnitude of the current social and environmental crisis, and the essential need to appreciate life in its cultural and biological diversity. The vast diversity of life forms and their intricate biocultural interrelationships with humans remain largely invisible to global society (Rozzi

R. Rozzi (✉)
Sub-Antarctic Biocultural Conservation Program, Department of Philosophy and Religion and Department of Biological Sciences, University of North Texas, Denton, TX, USA

Cape Horn International Center (CHIC), Universidad de Magallanes, Puerto Williams, Chile
e-mail: Ricardo.Rozzi@unt.edu

R. Rozzi et al. (eds.), *Field Environmental Philosophy*, Ecology and Ethics 5,
https://doi.org/10.1007/978-3-031-23368-5_2

2019). This represents a complex problem because the conservation of biocultural diversity is essential for the sustainability of life.

We inhabit the Anthropocene era in which we have rapidly transitioned from rural to urban society, and from direct interactions with biodiversity to technologically mediated interactions (Rozzi 2013). These changes convey that everyday direct interactions with biodiversity have been lost for a growing number of people (Miller 2005, Soga and Gaston 2016, Soga et al. 2016). To foster a reconnection between society and regional biological and cultural (biocultural) diversity, education can play a major role (Poole 2018). Part I of this book aims to contribute to education methodologies that reconnect present and future generations with biocultural diversity.

To incorporate educational concepts and practices that can cognitively and emotionally transform our understanding and appreciation of biocultural diversity, Part I introduces concepts, practices, and case studies of a novel methodological approach: "field environmental philosophy" (FEP). In 2000, I led the development of the FEP methodological approach at the Omora Ethnobotanical Park in the Cape Horn Biosphere Reserve, Chile (Rozzi et al. 2008). In this remote region, characterized by fjords, sea, mountains, glaciers, peatlands, and old-growth forest, with a group of educators, scientists, artists, and philosophers, we started a series of FEP courses linked to the University of Magallanes, Chile, and later since 2005, partnering with the University of North Texas, USA. This series of courses at Omora Park provided the basis for a systemic platform for the development of graduate theses and dissertations. Some of these FEP projects that were led by former graduate students are included in Part I of this book.

Graduate students have used the FEP methodology to facilitate preschool, primary, and secondary students, teachers, policymakers, tourists, and other citizens to understand and value biocultural diversity. To conduct these FEP experiences graduate students have integrated concepts and practices from:

(i) Ecological sciences and natural history (such as observation, drawing, and "direct encounters" with other living beings in their native habitats)
(ii) Philosophy, humanities, and the arts (such as critical reading of texts and composition of metaphors that stimulate analogical thinking)
(iii) Biocultural conservation and ethnographic practices (such as participant observation with members of indigenous and other local communities) (Rozzi et al. 2006)

For students, getting out of the classroom and having intercultural and interspecies interactions in the field facilitates a reconnection with biocultural diversity and an understanding that we co-exist in relationships of co-inhabitation with diverse human and other-than-human co-inhabitants. To understand ourselves as co-inhabitants (in contrast to a still prevailing individualistic view) who share habitats with a multitude of cultures and other living and ecological beings (such as rivers, mountains, rocks, or oceans) has deep implications in three philosophical realms: (i) epistemological, (ii) ontological, and (iii) ethical.

(i) *Epistemological*, because in the field, participants learn about multiple forms of knowledges, customs, and crafts to understand and coinhabit biocultural diversity. Historically, modern science education and the study of biodiversity tended to focus on individual categories and specimens. Under this "logic of the specimen," biological species were studied as objects in scientific collections (Neri 2011). This way of studying individuals (separated from their native habitats) has favored relationships of exploitation of objects or discrete natural resources, decontextualized from their relationships of co-inhabitation (Rozzi 2018). FEP orients students with conceptual frameworks and guided field experiences for understanding both ecological and biocultural interwoven relationships.

(ii) *Ontological*, because in the field students are surprised to discover not only the ecological and biocultural interrelationships but also to discover the similarity of the structures and functions of our bodies with those of other living beings. Through analogical thinking and guided field experiences, students and other participants can observe how plants are formed mainly by water and are interconnected by their roots and communities of fungi and bacteria. In field experiences in the sea, rivers, forests, wetlands, farms, orchards, urban parks or gardens, we appreciate how human beings and other-than-humans do not exist as isolated individual beings, but in interrelationships of cohabitation that forge our identities and well-being.

(iii) *Ethical*, because field experiences help students to realize more clearly how we share urban, rural, and remote habitats with diverse humans and living beings. In the field, students can encounter different forms of knowledge and human skills as well as unexpected behaviors displayed by other living beings. Students learn about a vast diversity of cultural expressions and living beings, most of which are excluded from educational programs and go unnoticed by citizens; in particular, small organisms with amazing life histories, morphologies, and essential ecological roles. These tiny co-inhabitants are precious in themselves and are critical to the health of humans and ecosystems. Thus, FEP methodology includes ethical reflections based on field experiences guided by biocultural concepts that stimulate an awareness of coinhabiting and sharing habitats with diverse human beings belonging to different ethnic, socio-economic, and linguistic groups and with myriads of other living and ecological beings. The understanding of biocultural relationships of co-inhabitation entails the ethical imperative of having moral considerations and relations of care that are not limited to the human sphere but that contemplate the well-being of all beings.

This methodological approach is called "field environmental philosophy," and not "field ecology" because it integrates these types of epistemological, ontological, and ethical dimensions. Systematically, graduate students and other participants address epistemological questions to investigate not only biological and cultural diversity, but also the methods, disciplines (sciences, humanities, and arts), languages, and worldviews through which scientific and other forms of knowledge about biological and cultural diversity are forged. FEP participants also conduct analyses of selected philosophical, ethnographic, and ecological texts, and investigate the biophysical, symbolic-linguistic, and political dimensions of biocultural relationships of co-inhabitation. The ultimate goal of FEP, however, is that participants do not limit themselves to only learning about biocultural diversity, but also propose and practice respectful, and responsible ways of co-inhabitation.

Toward the end of the twentieth century, the German philosopher Hans Jonas (1984) stated the ethical principle of responsibility. With the imperative "act in such a way that the effects of your action are compatible with the permanence of a genuine human life on Earth," Jonas called for responsibility for the sake of future generations and the world of life in general (Pommier 2019). Today, FEP broadens Jonas' principle of ethical responsibility to all forms of life. FEP aims to guide respectful forms of intercultural and interspecies co-inhabitation.

The FEP education methodology integrates biological and cultural diversity and their interrelationships in specific territories. In this way, FEP methodology orients students to act in a way that is, at least partially, in line with the ethical principle of responsibility for protecting the biocultural diversity that sustains life on planet Earth. Toward this aim, FEP systematically integrates sciences, humanities, arts, and ethics into biocultural education. As described in the previous chapter of this book, this integration is organized through the FEP four-step cycle. This cycle includes the following steps:

(i) Transdisciplinary ecological and philosophical research
(ii) Poetic communication through composition of metaphors and simple narratives
(iii) Design of field experiences guided with an ecological and ethical orientation
(iv) Implementation of *in situ* conservation areas (Fig. 2.1)

We begin Part I by presenting a series of chapters that explain how the FEP methodology started with field experiences involving intercultural exchanges and inter-species encounters with the participation of local and international students (Rozzi 2023). Subsequently, we include four chapters that illustrate how the four-step cycle was adapted by graduate students in their biocultural research, education, and conservation work at Omora Park. These chapters are the result of doctoral dissertations (Contador et al. 2023; Crego et al. 2023) or master theses (Méndez-Herranz et al. 2023; Ojeda et al. 2023). With the FEP methodology, graduate students not only conducted research, but also created innovative educational and ecotourism activities, such as "underwater with a hand lens: ecological sciences and environmental ethics to value freshwater biodiversity" in the world's cleanest rainwaters and rivers (Contador et al. 2023), "cultivating relations of care with sub-Antarctic high Andean gardeners" appreciating the mutualist life habits of cushion plants that grow in the southernmost mountains of the Americas (Méndez-Herranz et al. 2023), "pay attention, dive with eyes wide open" snorkeling in the kelp forests of Cape Horn in the sub-Antarctic seas (Ojeda et al. 2023), and "appreciate the eyes of the trees" in world's southernmost forests (Crego et al. 2023). These cases illustrate how students combine their scientific research and discoveries with poetic creativity and participatory conservation actions. Their work has helped to

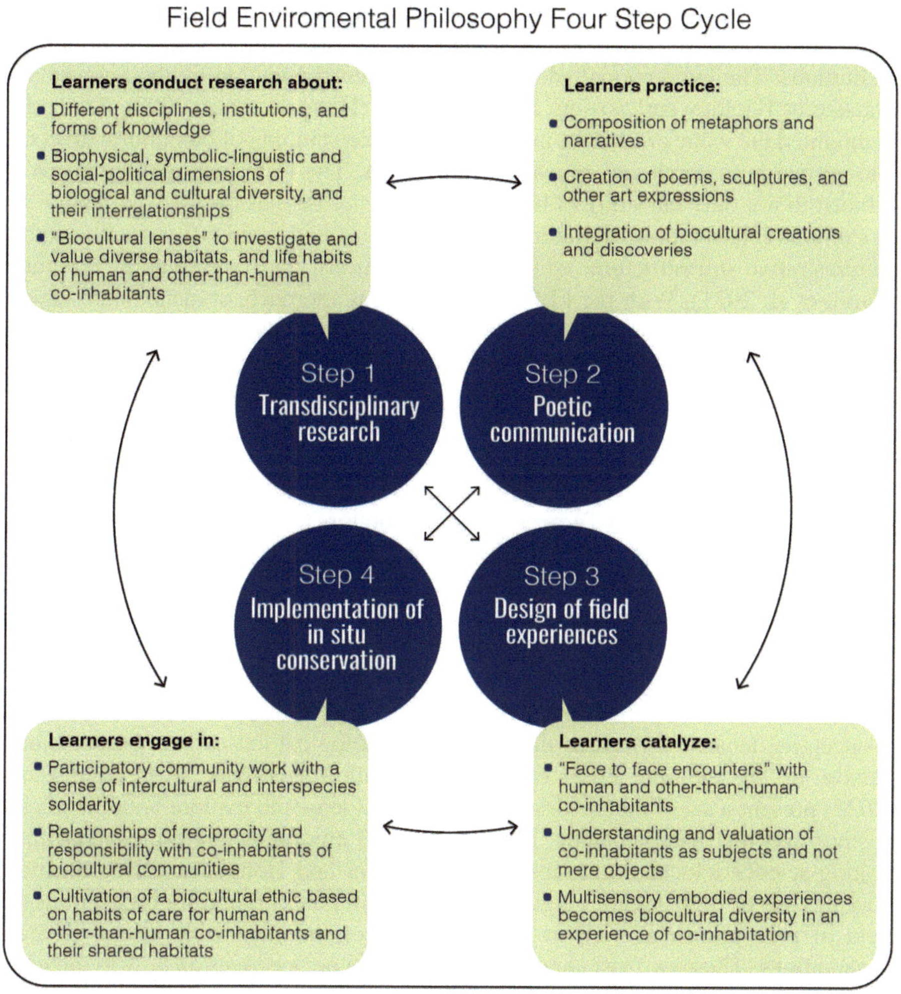

Fig. 2.1 The circles in the diagram depict each of the steps of the Field Environmental Philosophy (FEP) 4-step cycle. The boxes provide three essential methodological components that need to be considered in each of the steps. Each of these key components must be included when designing and teaching theoretical and practical concepts and activities throughout the FEP cycle. The bidirectional arrows indicate that the cycle is sequential but not one directional: each of the steps can be revisited at any moment of the cycle

make visible groups of organisms (such as freshwater invertebrates or tiny plants growing on cushion plants) as well as ecological and biocultural processes (such as the impacts of invasive exotic mammals or coevolutionary interactions between fishers and kelp forests). These organisms and processes were previously overlooked by formal education as well as decision-making processes.

Launched as a local initiative at the Omora Ethnobotanical Park, the FEP methodology has subsequently been applied at other national and international institutions. The international adoption of FEP gained attention by the Science and Practice of Ecology and Society Award in 2008 (Hargrove et al. 2008). This prize highlighted the value of FEP for integrating an understanding of biocultural diversity into academic research and educational programs. The FEP methodology has contributed to not only the study of biocultural diversity but has been also effective for biocultural conservation by interrelating the theory and praxis of research, within administrative infrastructure, educational institutions, and conservation programs (Tauro et al. 2021). With the FEP methodological approach, students learn ethical and cultural dimensions of global social-environmental change and practice participatory research methods. Two chapters of Part I illustrate how FEP's four-step cycle to integrate research, education and biocultural conservation has been adopted by graduate students in Mexico and also by educators of school-aged students in the USA.

In the coastal ecosystems of the Sea of Cortez, one of the world's most biodiverse marine sites, Rene Moreno-Terrazas and collaborators (2023) developed the activity to appreciate "starfishes and sky stars" using FEP in education and ecotourism experiences in Baja California, México. Biocitizen, Inc., an outdoor school in the USA and Chile has developed a novel approach to guide students toward biotic wonder, citizenship, using FEP in gradients of urban, rural, and wilderness habitats (Heidinger et al. 2023).

The last two chapters of Part I illustrate an essential aspect of FEP: the use of "conceptual lenses" as well as the use of "technological lenses" to observe biodiversity. In "the multiple lenses of ecotourism with a hand-lens," Rozzi and Tauro (2023) present a set of metaphorical or conceptual lenses to explore both the micro-cosmos and the macro-cosmos. In "beyond field environmental philosophy: integrating science education and technology," Miller and Hougham (2023) illustrate how the use of digital technologies in the field aids observation by broadening the field of perceptions, as well as the scope of communities with whom to share observations. These two last chapters present future perspectives on how to integrate experiences of the bodily senses in dialogue with access to knowledge mediated by new digital technologies and remote sensors.

To effectively address social-environmental problems it is essential that students can understand with clarity both direct and indirect drivers of global change. Sciences contribute mainly to identify and tackle direct drivers (or proximate causes), such as pollution, carbon emissions, deforestation, and other land-use changes. Sciences and technology help in finding solutions and defining what we can do. Philosophy contributes to understand and assess indirect drivers (or ultimate causes) of global change, including values and beliefs. The ultimate causes of global change have essential ethical dimensions rooted in values and worldviews (White 1967; Bormann and Kellert 1991; Rozzi 1999, 2001; May 2018; Taylor 2018). FEP prepares students to engage in dialogues and discussions that lead to dialectical coproduction of knowledge using scientific and technological tools as much as philosophical concepts and methods.

One of the most striking aspects of the FEP practices is that students and other participants see the "lenses" through which we perceive and value reality. For example, in the practices of ecotourism with a hand lens, the magnifying glass has multiple effects. It enables participant to observe small living beings, but it also requires the participant to slow down to observe these small and often overlooked organisms. Slow education and slow tourism experiences enable FEP participants to pay more attention to small biodiversity, their senses, and to reflect about the lenses through which they observe the world. This FEP experience provides participants with a magnifying glass that is both a physical tool as well as a conceptual one. In formal and non-formal education, FEP participants gain new critical philosophical tools as much as new poetic crafts to appreciate life in its biological and cultural diversity. FEP is a multi-sensorial experience that is guided by analogical thinking, and scientific and philosophical tools to admire the multiplicity of co-inhabitants and the singularity of their life habits linked to particular types of habitats. By practicing FEP, students and other participants are invited to cultivate the ethical practice of undertaking responsibility for caring for biocultural diversity. FEP equips students with methodological tools to explore and value biological and cultural diversity and fosters a biocultural ethic that concerns itself with human societies as much as with nature and aims for the well-being of both.

Acknowledgment I would like to thank the valuable comments by Alejandra Tauro, Kelli Moses, and Roy May, and to acknowledge the support from the Cape Horn International Center (CHIC) - ANID PIA/BASAL (PFB210018).

References

Bormann H, Kellert S (1991) Ecology, economics, ethics: the broken circle. Yale University Press, New Haven

Contador TA, Rendoll-Cárcamo J, Gañan M, Ojeda-Villarroel J, Kennedy JH, Convey P, Rozzi R (2023) Underwater with a hand lens: ecological sciences and environmental ethics to value freshwater biodiversity. In: Rozzi R, Tauro A, Wright T, Avriel-Avni N, May RH Jr (eds) Field environmental philosophy: education for biocultural conservation. Ecology and ethics, vol 5. Springer, Dordrecht, pp 53–70

Crego RD, Ward N, Rozzi R (2023) The eyes of the tree: applying field environmental philosophy to tackle conservation problems at long term socio-ecological research sites. In: Rozzi R, Tauro A, Wright T, Avriel-Avni N, May RH Jr (eds) Field environmental philosophy: education for biocultural conservation. Ecology and ethics, vol 5. Springer, Dordrecht, pp 101–112

Hargrove E, Arroyo MTK, Raven PH, Mooney H (2008) Omora ethnobotanical park and the UNESCO cape horn biosphere reserve. Ecol Soc 13(2) http://www.ecologyandsociety.org/vol13/iss2/art49/

Heidinger K, Camichael J, Jacob-Schwartz B, Borucinska-Begg M (2023) Biocitizen's approach to biotic wonder, citizenship, and field environmental philosophy. In: Rozzi R, Tauro A, Wright T, Avriel-Avni N, May RH Jr (eds) Field environmental philosophy: education for biocultural conservation. Ecology and ethics, vol 5. Springer, Dordrecht, pp 131–152

Jonas H (1984) The imperative of responsibility: In: search of an ethics for the technological age. Chicago University Press, Chicago

May RH Jr (2018) Land grabbing and violence against environmentalists. In: Rozzi R, May RH Jr, Chapin FS III, Massardo F, Gavin M, Klaver I, Pauchard A, Núñez MA, Simberloff D (eds) From biocultural homogenization to biocultural conservation. Ecology and ethics, vol 3. Springer, Dordrecht, pp 109–123

Méndez-Herranz M, Marini G, Rozzi R (2023) Sub-Antarctic high Andean "Gardeners:" cultivating caring relationships. In: Rozzi R, Tauro A, Wright T, Avriel-Avni N, May RH Jr (eds) Field environmental philosophy: education for biocultural conservation. Ecology and ethics, vol 5. Springer, Dordrecht, pp 71–86

Miller B, Hougham J (2023) Beyond field environmental philosophy: integrating science education and technology. In: Rozzi R, Tauro A, Wright T, Avriel-Avni N, May RH Jr (eds) Field environmental philosophy: education for biocultural conservation. Ecology and ethics, vol 5. Springer, Dordrecht, pp 175–191

Moreno-Terrazas R, Micheline Cariño M, Díaz Z, González H, Monteforte-Sánchez M (2023) Starfishes and sky stars: a field environmental philosophy education and ecotourism experiences in Baja California, México. In: Rozzi R, Tauro A, Wright T, Avriel-Avni N, May RH Jr (eds) Field environmental philosophy: education for biocultural conservation. Ecology and ethics, vol 5. Springer, Dordrecht, pp 113–130

Neri J (2011) The insect and the image: visualizing nature in early modern Europe, 1500-1700. University of Minnesota Press, Minneapolis

Ojeda J, Tauro A, Contador T, Mansilla A, Rosenfeld S, Ban N, Rozzi R (2023) "Pay attention, dive with your eyes wide open:" a Field Environmental Philosophy activity to foster reciprocity between people and nature. In: Rozzi R, Tauro A, Wright T, Avriel-Avni N, May RH Jr (eds) Field environmental philosophy: education for biocultural conservation. Ecology and ethics, vol 5. Springer, Dordrecht, pp 87–100

Pommier E (2019) The problem of environmental democracy: responsibility and deliberation. Environ Ethics 41(4):305–318

Poole AK (2018) The UN sustainable development goals and the biocultural heritage lacuna: where is goal number 18? In: Rozzi R, May RH Jr, Chapin FS III, Massardo F, Gavin M, Klaver I, Pauchard A, Núñez MA, Simberloff D (eds) From biocultural homogenization to biocultural conservation, ecology and ethics, vol 3. Springer, Dordrecht, pp 315–331

Rozzi R (1999) The reciprocal links between evolutionary-ecological sciences and environmental ethics. Bioscience 49(11):911–921

Rozzi R (2001) Ética ambiental: raíces y ramas latinoamericanas. In: Primack R, Rozzi R, Feinsinger P, Dirzo R, Massardo F (eds) Fundamentos de conservación biológica: Perspectivas latinoamericanas. Fondo de Cultura Económica, México, pp 311–362

Rozzi R (2013) Biocultural ethics: from biocultural homogenization toward biocultural conservation. In: Rozzi R, Pickett STA, Palmer C, Armesto JJ, Callicott JB (eds) Linking ecology and ethics for a changing world: values, philosophy, and action, ecology and ethics, vol 1. Springer, Dordrecht, pp 9–32

Rozzi R (2018) Biocultural homogenization: a wicked problem in the Anthropocene. In: From biocultural homogenization to biocultural conservation. Springer, Cham, pp 21–48

Rozzi R (2023) Inter-species and inter-cultural encounters: the education and biocultural ethics program of the Omora ethnobotanical park. In: Rozzi R, Tauro A, Wright T, Avriel-Avni N, May RH Jr (eds) Field environmental philosophy: education for biocultural conservation. Ecology and ethics, vol 5. Springer, Dordrecht, pp 153–174

Rozzi R, Tauro A (2023) The multiple lenses of ecotourism with a hand-lens: fundamental concepts and practices. In: Rozzi R, Tauro A, Wright T, Avriel-Avni N, May RH Jr (eds) Field environmental philosophy: education for biocultural conservation. Ecology and ethics, vol 5. Springer, Dordrecht, pp 27–52

Rozzi R, Massardo F, Anderson C, Heidinger K, Silander JA Jr (2006) Ten Principles for biocultural conservation at the southern tip of the Americas: the approach of the Omora Ethnobotanical Park. Ecol Soc 11(1) http://www.ecologyandsociety.org/vol11/iss1/art43

Rozzi R, Arango X, Massardo F, Anderson C, Heidinger K, Moses K (2008) Field environmental philosophy and biocultural conservation: the Omora Ethnobotanical Park educational program. Environ Ethics 30(3):325–336

Tauro A, Ojeda J, Caviness T, Moses KP, Moreno-Terrazas R, Wright T, Zhu D, Poole AK, Massardo F, Rozzi R (2021) Field environmental philosophy: a biocultural ethic approach to education and ecotourism for sustainability. Sustainability 13(8):4526. https://www.mdpi.com/2071-1050/13/8/4526

Taylor B (2018) Biostitutes and biocultural conservation: empire and irony in the motion picture avatar. In: Rozzi R, May RH Jr, Chapin FSIII, Massardo F, Gavin M, Klaver I, Pauchard A, Núñez MA, Simberloff D (eds) From biocultural homogenization to biocultural conservation, ecology and ethics, vol 3. Springer, Dordrecht, pp 71–82

White L Jr (1967) The historical roots of our ecologic crisis. Science 155:1203–1207

Rozzi R (2019) Taxonomic Chauvinism, no More! Antidotes from Hume, Darwin, and biocultural ethics. Environ Ethics 41(3):253–288

Soga M, Gaston KJ (2016) Extinction of experience: the loss of human–nature interactions. Front Ecol Environ 14:94–101

Soga M et al (2016) Urban residents' perceptions of neighbourhood nature: does the extinction of experience matter? Biol Conserv 203:143–150

Miller JR (2005) Biodiversity conservation and the extinction of experience. Trends Ecol Evol 20: 430–434

Chapter 3
The Multiple Lenses of Ecotourism with a Hand Lens: Fundamental Concepts and Practices

Ricardo Rozzi and Alejandra Tauro

Abstract Today a great diversity of living beings and human values are invisible to the prevailing culture of global society. This culture usually associates the word biodiversity with large organisms such as trees and mammals. Paradoxically, most of the animal biodiversity is made up of small organisms that remain invisible in global culture and are under-represented in philosophy, science, and education. In this chapter, we present various conceptual lenses that contribute to broadening the appreciation of biological and cultural diversity to capture the beauty and relevance of small organisms and the multiplicity of languages, forms of knowledge, and values that different cultural traditions give to biodiversity. We link these lenses with didactic practices of a new recreational activity and formal education: ecotourism with a magnifying glass. This activity integrates science, arts, humanities, and ethics. It is nourished by the tradition of naturalists but with philosophical, scientific, and technological concepts typical of the twenty-first century. Students, tourists, and other participants practice forms of analogical thinking to co-create knowledge and biocultural expressions of their own, instead of being passive recipients of information. Finally, this field environmental philosophy activity combines bodily and sensory activities with theoretical readings that provide conceptual foundations for cultivating an appreciation for the small co-inhabitants with whom we share our local habitats and the global biosphere.

Keywords Biodiversity · Biocultural ethics · Education · Field environmental philosophy · Otherness

---

R. Rozzi (✉)
Sub-Antarctic Biocultural Conservation Program, Department of Philosophy and Religion and Department of Biological Sciences, University of North Texas, Denton, TX, USA

Cape Horn International Center (CHIC), Universidad de Magallanes, Puerto Williams, Chile
e-mail: Ricardo.Rozzi@unt.edu

A. Tauro
Cape Horn International Center (CHIC), Universidad de Magallanes, Puerto Williams, Chile

El Colegio de Puebla A.C., Puebla, México

R. Rozzi et al. (eds.), *Field Environmental Philosophy*, Ecology and Ethics 5,
https://doi.org/10.1007/978-3-031-23368-5_3

3.1 Introduction

The Anthropocene is characterized by a sixth mass extinction of biodiversity interwoven with global climate change and growing problems of socio-environmental justice and degradation of human health and well-being (Primack et al. 2001; McNeill and Engelke 2016). Educators, scientists, policymakers, and other citizens are challenged today by conceptual and practical problems associated with global change, which involves complex and interrelated ecological and social dimensions (Bormann and Kellert 1991). To address these challenges educators and researchers need to undertake a kind of "Kuhnian scientific revolution" (*sensu* Kuhn 1970), which conveys new scientific paradigms that emphasize the importance of culture in research and education (Pickett et al. 1994; Worster 1994; Rozzi et al. 1998; Latour 1999).

Amidst the current massive extinction of biodiversity, it is pressing to develop educational methodologies that can reconnect society with the beauty and the multiple ecological, economic, and ethical values of the exuberant diversity of living beings with whom we co-inhabit the biosphere. Biodiversity losses have cultural drivers (Isbell et al. 2022). Hence to protect biodiversity we need cultural transformations. With the so-called "digital world" the mediatization of knowledge has been installed in formal and non-formal education, in the globalized culture. In the predominantly urban global society, communication is carried out through media communities, to the point of turning face-to-face encounters with biological and cultural diversity into an atavistic memory (Bilbeny 1997). However, communication without face-to-face encounters and sensory experiences leads to a knowledge about biological and cultural diversity that ends up not being felt. Beyond the complexities of social-environmental problems that we face today, an ethical challenge for education in the information society is to train informed but indifferent individuals. A pressing ethical task is to avoid apathy and that the senses fall asleep.

To awaken the senses, emotions, and feelings of empathy with the diversity of living beings (including small organisms that often go unnoticed), field experiences, offer a bright window for formal and non-formal education, including tourism. Field experiences allow students and citizens of today's digital world and prevailingly urban society to reconnect with the biophysical reality that pulsates in amazing urban, rural and wild habitats. To effectively orient this type of field experiences, an indispensable step is to clearly understand the interrelationships of biophysical and symbolic-linguistic dimensions involved in the observation and appreciation of biodiversity. Very often, however, in science education as well as in nature tourism the weight of words is forgotten and attention is given only to the biophysical reality (Ahl and Allen 1996).

Our goal in this chapter is to present a conceptual framework and practical activities that examine both the languages and values of culture and the biophysical dimensions of biodiversity. We develop this biocultural approach to education based on a new type of educational and slow tourism activity: *ecotourism with a hand lens*.

3.2 Ecological and Philosophical Lenses

3.2.1 An Ecological Lens: Biodiversity Components, Patterns, and Processes

The word "biodiversity," a compound term referring to biological diversity, was coined by various authors (Laura Tangley, Thomas Lovejoy, Robert L. Peters, and Walter G. Rosen) in the early 1980s (Sarkar 2021). This simultaneous and independent creation of the term underlines that scientific innovations are products of cultural and historical contexts (Rozzi 2023a). Initially, biodiversity was most commonly used to describe the number of species (Lovejoy 1980; Swingland 2001). This focus aligned with the prevailing interest inherited from modernity in the collection and enumeration of biological species, and the logic of the specimen (Neri 2011). In summary, in the early 1980s uses and measurements of biodiversity tended to narrowly focus on units or *components* of biological diversity (Worster 1994).

The meaning of the term biodiversity, however, rapidly expanded to include components at multiple scales from genes to species, and ecosystems. For example, it was defined as "the abundance, variety, and genetic constitution of native animals and plants" (Dodson et al. 1998), and later encompassed all living kingdoms (Noss 1990). In addition, in the 1990s, growing attention was given to ecological patterns and processes among biodiversity components (Rozzi 2001). For example, interest directed toward understanding and conserving the diversity of ecological interactions (Young et al. 2016) and evolutionary processes (Primack et al. 2001), which highlighted the dynamic character of biological systems. Hence, biological units are not static, like items in museum collections, but are dynamic entities with internal processes and interactions with other entities.

A systemic approach is helpful to interrelate biodiversity: components, patterns, and processes. One of these approaches was developed by forest ecologist Jerry Franklin and his collaborators (1981) who recognized three primary attributes of ecosystems: composition, structure, and function. To characterize the biodiversity of an area, Ricardo Rozzi and collaborators (2001) adapted and defined Franklin's three attributes as follows:

- *Composition* refers to the identity and variety of elements or components in a collection and includes species lists as well as measures of species diversity and genetic diversity.
- *Pattern* refers to the structure of a system, and includes the ways elements (e.g., populations of species and their gene pools) are distributed in a given habitat and the complexity of biophysical organization from the small scale of microhabitats (e.g., cracks in rocks, cavities in logs) to the large scales of landscapes (e.g., a river basin, mountainous slopes across altitudinal gradients) or planet Earth (e.g., distribution of tropical, temperate, and polar biomes).

- *Process* refers to ecological interactions and evolutionary phenomena that take place among the components in a given habitat and across habitats, and includes migration of populations, gene flows, disturbances, and nutrient cycling.

For field ecology education, this systemic has a great heuristic value. Students can practice observations and drawings in which they identify different species (composition), they understand how they are distributed in the habitat (patterns), and recognize ecological interactions (e.g., processes). Students can also clearly understand how the three attributes are interdependent. For example, a structural simplification of an ecosystem (e.g., transformation of diverse forests into a monospecific plantation) might disrupt fundamental ecological processes (e.g., pollination interactions), which lead to losses of the composition (e.g., losses of local populations or even endemic species).

For *field environmental philosophy* (FEP), however, these attributes are not sufficient. As mentioned in the Introduction to Part I (Rozzi 2023b), this methodological approach differs from *field ecology* because FEP integrates epistemological, ontological, and ethical dimensions. FEP students and other participants address epistemological questions to not only study biological diversity, but also the methods for investigating and naming biodiversity. They analyze philosophical, ethnographic, and ecological texts to investigate interrelated biophysical and cultural (both material and immaterial culture with its symbolic-linguistic features) dimensions of biodiversity. Ontological questions are critical today, because the web has become one of the main sources of biodiversity information, providing detailed information about specimens' distribution in space and time. Field experiences complemented with philosophical and biological research prepare FEP students to ask about the nature of other-than-human living beings. Are they merely objects for human use and study? Or, are they subjects with intentionality and sentiency? Epistemological and ontological questions provide a basis for ethical questions about ways humans interact with biodiversity and value other-than-human living beings. To address these philosophical questions, in the practice of FEP we have adopted the "3Hs" conceptual framework of the biocultural ethic (*sensu* Rozzi 2013), which complements the ecological framework to characterize biodiversity.

3.2.2 A Philosophical Lens: The 3Hs Triad of the Biocultural Ethic

The "3Hs" conceptual framework of the biocultural ethic values the vital links among *co-in-Habitants* and their life *Habits* that take place in shared *Habitats*, as has been explained in previous chapters. Below we concisely define these terms to facilitate its application in the practice of ecotourism with a hand lens.

The term *co-inhabitant* was coined by Ricardo Rozzi motivated by early field experiences (Rozzi 2004). Rozzi spent his childhood in the Andean mountains of central Chile, where in the mornings he was woken up by the songs of thrushes, the condors raised their chicks with dedication and some sparrows regularly came to ask him for food. It is surprising how the Scottish philosopher David Hume also paid attention to the ability of understanding and caring behavior of birds. From early field experiences and readings of Hume and other philosophers and ethologists, Rozzi conceives birds as subjects, not merely objects of study. Birds present intentional behaviors, have sentient capacity and a type of moral sentiments for the care of their progeny. In addition, with birds we share habitats, hence we are co-inhabitants.

The term co-inhabitant acquires a *descriptive* and also a *normative* sense. Descriptive, because habitat sharing is a phenomenon that has emerged from a long history of ecological and evolutionary interactions among humans, birds, and other living beings. Normative, because taking care of the habitat for the well-being of human beings, birds, and other living beings is a necessary condition for an ethic of co-inhabiting biodiverse communities. We can understand this double meaning (descriptive and normative) of the term *co-inhabitant* through an analogy with the term *companion*. The latter originally alluded to sharing bread (from the Latin, *cum* = with; *panis* = bread), and in many communities to share the bread (food) is part of rituals and an ethical imperative (Rozzi 2018). Co-inhabitant refers to sharing a habitat, and it should also become a conscious understanding and an ethical imperative for having relationships of care with other co-inhabitants and the shared habitats.

Habitats are the condition of possibility for the existence of cohabitants and their life *habits*. For example, a woodpecker can only dig holes if there are trees to dig. If the trees disappear, the habit of digging also disappears and the population of woodpeckers will probably become extinct since they will not be able to feed or reproduce in a habitat devoid of trees. These biophysical interactions are captured by diverse languages. At the southern end of the Americas, in the forests of Cape Horn, the Magellanic woodpecker (*Campephilus magellanicus*) is called by the indigenous Yahgan people: "lana." This name derives from the Yahgan word "lan," which means tongue. It alludes to the habit of the woodpecker to extending its long tongue to extract larvae from the holes it pecks in the trunks of old growth trees. The scientific name of the genus, *Campephilus*, defines the bird as a lover (*philus* in Greek) of caterpillar (*Campei* in Greek), and its specific name indicates that it inhabits the forests of Magallanes (*magellanicus* in Latin). Its English common name, Magellanic woodpecker, characterizes the identity of this bird by its habit of pecking wood in the austral woodlands. Hence, the connections between the habitats, habits, and the identity of this bird co-inhabitant can be detected in the *biophysical* existence of the forests as well as in the *cultural* meanings of the languages of indigenous populations, scientific nomenclature, and European settlers. This type of biophysical and cultural interrelationship helps us to introduce the biocultural meaning of the "hand lens" in the following section.

3.3 Four Lenses of Ecotourism with a Hand Lens

In this section, we present four types of conceptual lenses that we use in the activity of ecotourism with a hand lens: the biocultural lens, the aesthetic lens, the ethical lens, and the economic lens. These conceptual lenses orient students, tourists, and other visitors to appreciate biological and cultural diversity. For each lens, we first introduce ecological and philosophical concepts, then we propose a didactic activity, and we finish by outlining a concise reflection that integrates perspectives from the sciences, the arts, the humanities, and ethics.

3.3.1 The Biocultural Lens of Ecotourism with a Hand Lens

We, humans, participate not only in the biophysical but also in the cultural structures and processes of habitats. Our human perceptions and understanding of biological diversity are informed by the properties of other living beings as well as by attributes of our material (technology) and immaterial (symbolic and linguistic) culture. The compound term biocultural expresses this integration. We use the concept of "biocultural lenses" to indicate that any human "observer" (including scientists with their research methods, and conceptual taxonomies) interprets biodiversity informed by biophysical and cultural attributes (Fig. 3.1). In turn, the ways we humans perceive and understand biodiversity and their environment will influence the ways we coinhabit ecosystems, modifying (conserving or disrupting) their structure, processes, and composition. We will illustrate this point with the practice of ecotourism with a hand lens.

As illustrated above in the woodpecker example, organisms are named by different languages. Let us now review the case of the lichen observed in Fig. 3.1, considering four languages used by different human cultures that coinhabit the region of southern South America where this lichen grows (Fig. 3.2). First, the indigenous culture of the Mapuche people calls this lichen *kalcha-aliwen*, where *kalcha* means hair and *aliwen* means tree. This Mapuche name expresses a remarkable similarity with the lens of the 3Hs of biocultural ethics. *Kalcha-aliwen* indicates that the habit of this lichen is to grow in the form of hair and to do so on another cohabitant, a tree. Both the lichen and the tree share the habitat of the temperate forests of South America. Second, the culture of the Yagan people calls this lichen "chirlej." In the Yagan language, this is a generic name for lichens and mosses. Third, the scientific name coined by the Argentine lichenologist, Susana Calvelo, is *Protousnea magellanica*. The name of the genus expresses an evolutionary attribute proposing that it is an ancestral form (*Proto* in Greek) of the lichens of the genus *Usnea*. Its specific name indicates that its geographical distribution corresponds to the Magellanic forests. Fourth, the Arabic *ushnah* is the source for the scientific name of the genus *Usnea*. Fifth and sixth, the English and Spanish names for this lichen are *Old Man's beard* and *Barba de Viejo*, respectively. The meaning of these

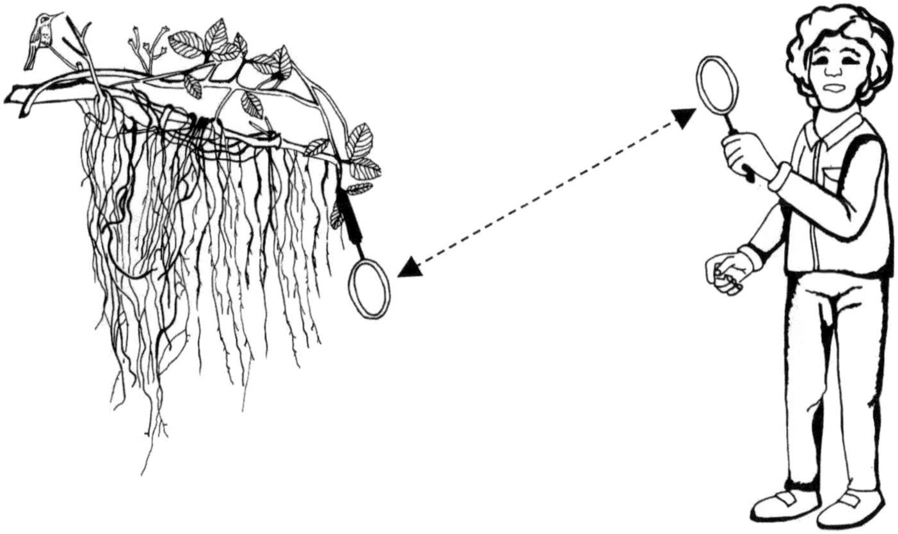

Fig. 3.1 Scheme of the use of a biocultural hand lens. The bidirectional dotted line illustrates that observing biodiversity entails dialectical relationships between the biophysical attributes of the observed living beings (e.g., a lichen) and the cultural attributes of the human observer. The latter includes material culture (e.g., technological devices such as a magnifying glass) and immaterial culture (e.g., languages that include scientific and/or vernacular names of other living beings). Illustration drawn by Mauricio Álvarez

two European languages is the same, and they point out an analogy between the shape of human beard hairs and that of lichens hanging from trees. This analogy is a clear expression of the biocultural imagination that European settlers had to name this lichen.

As mentioned above biocultural relationships involve both immaterial and material culture. Regarding the latter, some uses of the lichen by different cultures are illustrated in Fig. 3.3 (images and text below modified from Rozzi 2012). First, the indigenous Mapuche people use *kalcha-aliwen* lichens to dye wool. The tincture is extracted by boiling the lichen in water. Additionally, an infusion is prepared to purify the blood, heal ulcers and stop diarrhea. For this reason, the lichen is also called *kalcha-lawen*. Mapuche people call medicinal plants *lawen*. Second, Yahgan people use *chirlej* as fuel to light the stove, and they smoke this lichen. Third and fourth, contemporary science and traditional Arabic cultures have used this lichen for medicinal purposes. Like in the Mapuche region, in the Islamic region as well as in China, Europe, and North America, this lichen has been utilized in traditional medicine for over a thousand years. In the nineteenth century, modern science discovered that it possesses a strong antibiotic that is effective in the treatment of several bacterial diseases, and it also exhibits antiviral, antiprotozoal, antimitotic, anti-inflammatory, and analgesic activity: usnic acid. Today, this acid is included as an ingredient in medicines, toothpastes, hair shampoos, and many other products that express multifaceted biocultural, material, interrelationships between humans and

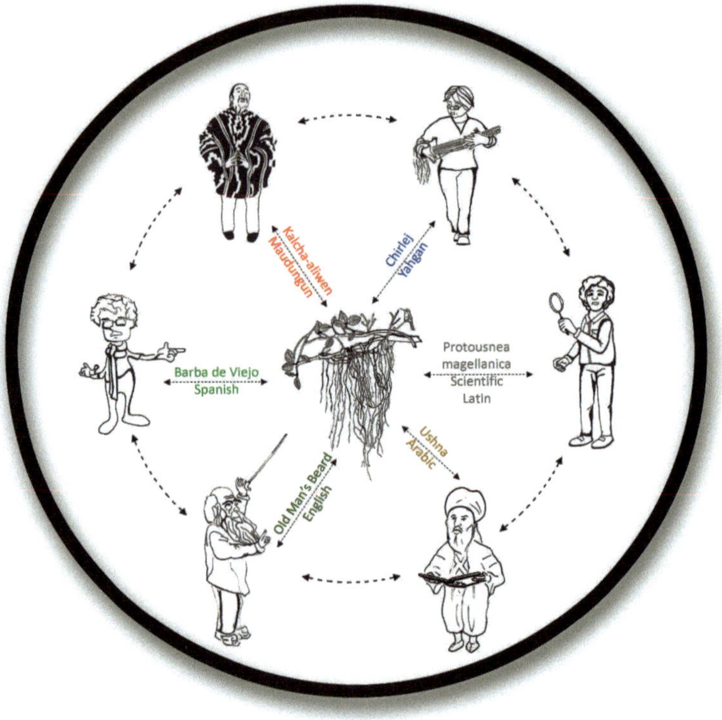

Fig. 3.2 Representation of the diversity of languages to name a species of lichen that grows in the Magellanic forests in southwestern South America. The figure illustrates the persons from whom we learned each of the names for this species of lichen. Clockwise these are Mapuche poet Lorenzo Aillapan, Yahgan handcrafter Julia González, Argentinean lichenologist Susana Calvelo, Middle Age Arabic doctor Ebubekir Muhammed bin Zekeriya Razi, British naturalist Charles Darwin, and Chilean biologist Humberto Maturana. Figure modified from Rozzi et al. (2012a, p. 139), and redrawn by Mauricio Álvarez

lichens in modern society. Fifth and sixth, English and Spanish biologists have used this lichen as a study object to investigate evolution. Naturalist Charles Darwin who extensively traveled in southern South America included examples of lichen—tree interactions to develop his theory of evolution based on natural selection. Complementarily, Chilean biologist Humberto Maturana proposed an alternative theory of evolution based on the mechanism of natural drift. These alternative theories illustrate how the same biophysical reality can inspire different scientific theories, which in turn are linked to contrasting cultural backgrounds and values (Rozzi et al. 1998; Rozzi 1999).

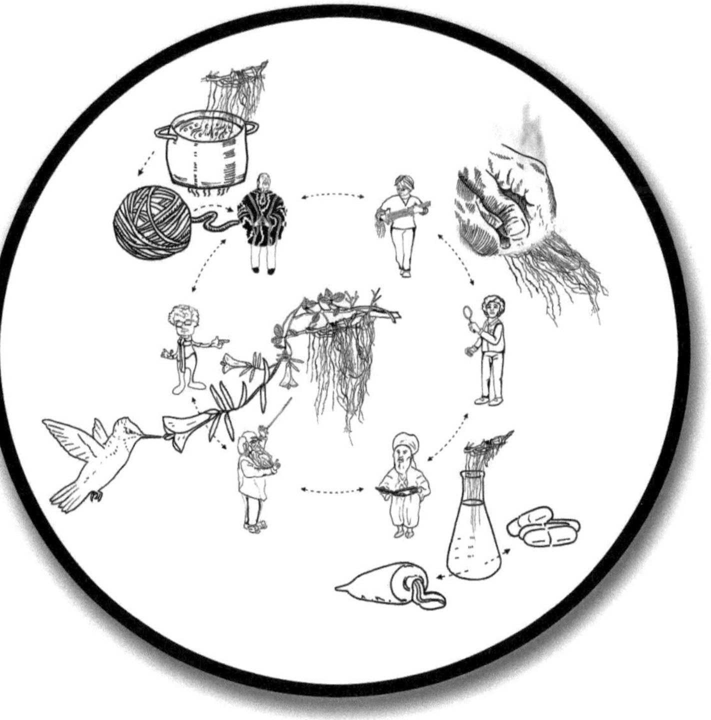

Fig. 3.3 Representation of the uses of the Old Man's beard lichens. The figure illustrates the persons from whom we learned each of the uses for this type of lichen. Clockwise these are Mapuche poet Lorenzo Aillapan, Yahgan handcrafter Julia González, Argentinean lichenologist Susana Calvelo, Middle Age Arabic doctor Ebubekir Muhammed bin Zekeriya Razi, British naturalist Charles Darwin, and Chilean biologist Humberto Maturana. Figure modified from Rozzi (2012, pp. 64–67), and redrawn by Mauricio Álvarez

3.3.1.1 A Didactic Practice of the Biocultural Lens

Biocultural relationships emerge from observations and practices that include both discovery and creation (Rozzi et al. 2010). For this reason, in the practice of ecotourism with a magnifying glass, students are invited to observe and also invent. Following the naturalist tradition, students or even tourists and other participants are invited to look for different types of lichens or small plants and draw them (Fig. 3.4). For this, they are provided with field notebooks and are asked to pay special attention to the habit of life, the habitat and microhabitat where they grow, and the co-inhabitants that they can identify and draw. This activity can be carried out in an urban habitat, such as a plaza or garden, in a rural habitat such as an orchard, or in a wild habitat such as an ancient forest.

A more structured exercise that has been particularly useful to understand the 3Hs model of the biocultural ethic, and to practice field observation, drawing and

Fig. 3.4 Representation of an ecotourism with hand lens practice to experientially understand the concept of the biocultural lens. In the field, students search for micro-habitats to observe lichens or other small organisms such as mosses or invertebrates. For example, in a city, they can find a park bushes or trees coinhabiting with snails. Students use the magnifying glass to observe them and the conceptual biocultural lens to draw them identifying life habits, co-inhabitants, and the habitat they share. Participants in this Field Environmental Philosophy experience are also invited to invent a name for each of the observed co-inhabitant. Later they are asked to compare their invented names with scientific names and vernacular names given to the co-inhabitants they depicted and named

participatory creativity has been the use of predefined tables that need to be completed by participants in ecotourism with a hand lens experiences. One of these exercises is based on the different types of non-vascular plants (mosses and liverworts) and lichens that coinhabit on the surface of rocks, trunks, or soil. Students are asked to find, observe, draw, and create a name for each type of co-inhabitant with a characteristic life habit on each of the main types of micro-habitats (Fig. 3.5). After completing their tables, participants show to the partners of this experience the co-inhabitants they found, pointing out their life habit and microhabitat. Then they show their drawings and name for each co-inhabitant. In the field, they compare their drawing and names among themselves and reflect on the similarities and differences in the attributes that caught their attention.

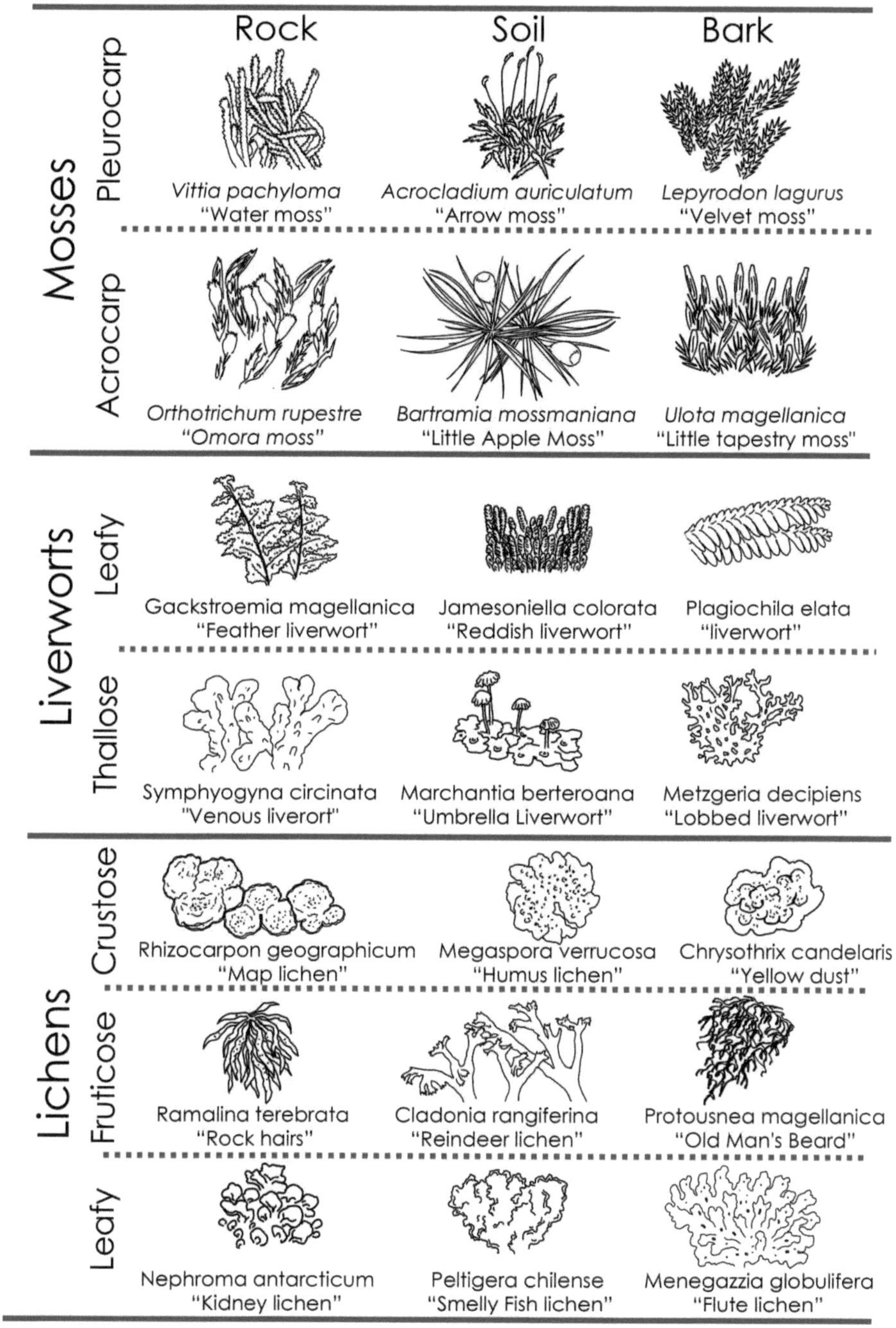

Fig. 3.5 Model of a table completed by a student, tourist, or participant in the experience of ecotourism with hand lens. To complete this table with drawings and names, participants have to first identify three main types of co-inhabitants of "miniature forests": mosses, liverworts, and lichens. For each of them, they have to search for species that have one of the main life habits:

3.3.1.2 A Reflection on the Biocultural Lens

The biocultural lens complements other scientific and technological lenses. These lenses are complementary, not exclusive. Together they allow us to appreciate that there is a great diversity not only among mosses, lichens, and other organisms but also among the perceptions and attitudes that humans have about and toward them. The interrelationships between biological diversity and cultural diversity generate biocultural diversity: we co-inhabit at the cognitive and ethical intersections that emerge between the diverse living beings and the different languages and forms of ecological knowledge through which we interact with them. Biocultural conservation aims for the conservation of both diversities and their interrelationships, which have coevolved for centuries in many places of the planet, and today offer sustainable forms of biocultural co-inhabitation. In this way, the biocultural lens also sensitizes the participants regarding the need for a socio-environmental justice that defends well-being, diverse expressions of human and other-than-human life, and a more bio-culturally diverse global society.

3.3.2 The Aesthetic Lens in Ecotourism with a Hand Lens

In the twentieth century, philosopher Ludwig Wittgenstein called attention to language and the ways of seeing the world. In this FEP practice, we highlight the relevance of the diversity of languages and disciplines to observe the world. Sciences, humanities, and arts play complementary roles in the appreciation of biodiversity. When we began teaching FEP, it was clear that students gained not only by being exposed to field experiences and scientific research, but also to artistic practices such as composing metaphors, writing poems, keeping a personal journal, or drawing. For twenty-first-century science education, it is necessary to counterbalance specialization and recover practices that were common among nineteenth-century naturalists such as Alexander von Humboldt or Ernst Haeckel.

In ecotourism with hand lens practices, we ask participants to focus on drawing as a way to understand fundamental aesthetic dimensions involved in the observations of biodiversity. We ask them to focus on ordering principles on several levels, such as the interrelationships between the type of organisms and microhabitat (e.g., drawing of organisms on contrast faces of trunks exposed to sunlight or rain versus faces protected from direct radiation or rainfall) or morphological architecture of organisms with distinct patterns of symmetry in their bodies. For example, students

---

Fig. 3.5 (continued) pleurocarp or acrocarp for mosses; leafy or thallous for liverworts, crustose, fruticose, or leafy for lichens. Finally, they have to find each of the identified co-inhabitants and their life habits growing on the three main types of substrates or micro-habitat: rock, soil, or trunk. This model table is based on real exercises that have been conducted at Omora Park since the year 2000. Drawings by Mauricio Álvarez

Fig. 3.6 Representation of an exercise applying the aesthetic lens of ecotourism with hand lens. With a hand lens in his hand, the student observes small non-vascular plants that are often overlooked. While observing them he is able to distinguish patterns of radial symmetry that characterize mosses and patterns of bilateral symmetry that characterize liverworts. These patterns of symmetries are located both in the plants that are being observed and in the mental categories of the observer that study them. The latter is a conceptual lens that we call the "aesthetic lens." Photographs by Adam Wilson. Figure drawn by Mauricio Álvarez

have to draw mosses characterized by radial symmetry and leafy liverworts characterized by bilateral symmetry (Fig. 3.6).

This exercise helps to focus their observations and provides the basis for discussions on a basic biocultural question. Are the radial and bilateral symmetries attributes of mosses and liverworts or are they attributes of mental classificatory schemes of the observer? This question opens a reflection on biocultural dialectical relationships between the discovery of geometrical forms linked to patterns of symmetry in nature and the invention of geometrical categories linked to patterns of symmetry that are projected by observers onto nature to classify different types of organisms. With a magnifying glass, small organisms can be observed and examined in the field, with the mental biocultural lens these organisms can be classified.

3.3.2.1 A Didactic Practice of the Aesthetic Lens

Drawing practices enable students to experientially understand the interrelationships between mental representations of the observer and the biophysical attributes of the observed organisms. They understand that this type of biocultural interfaces is an essential part of scientific work as much as of nature art work. Both scientists and artists invent concepts and observe patterns in nature. To put this understanding into other practices we also ask students to create small sculptures in clay (Fig. 3.7). During the twentieth century, this unity between art and science was lost due to disciplinary specialization in academia.

3.3.2.2 A Reflection on the Aesthetic Lens

With the FEP activity of ecotourism with a hand lens we aim to bring the artistic and scientific practices back together. Students work at Omora Park in teams with artists and scientists to discover and invent "lenses," which later are used to enhance the experience of tourists and other visitors who are better equipped to discover and

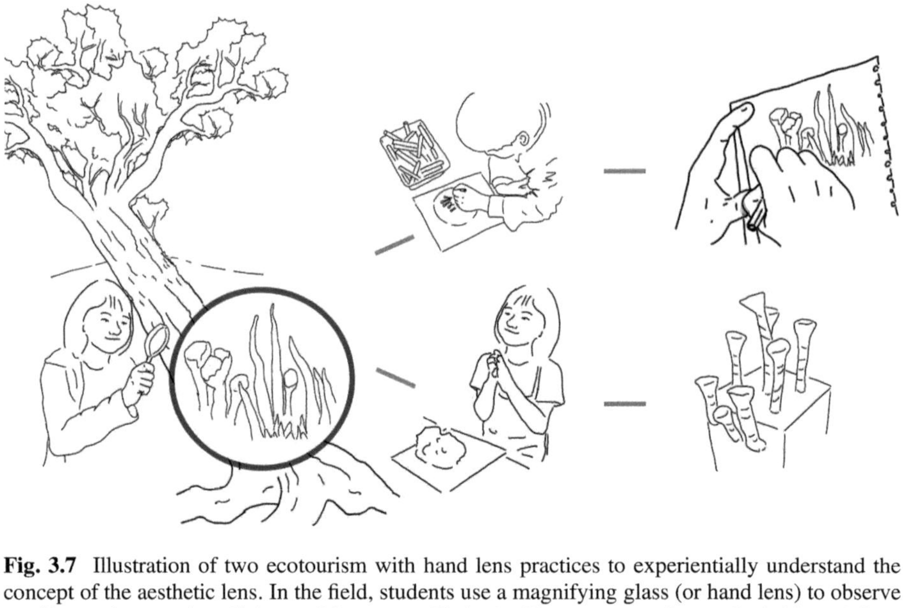

Fig. 3.7 Illustration of two ecotourism with hand lens practices to experientially understand the concept of the aesthetic lens. In the field, students use a magnifying glass (or hand lens) to observe small organisms such as lichens of the genus *Cladonia*. They pay attention to the habitats (where they grow) and habits (how they grow) and then represent their observations and understanding by drawing using colors and/or creating small sculptures in clay. Participants in this Field Environmental Philosophy experience are invited to reflect on the distinction between the present actual living beings and their representations. In particular, they are asked to reflect on the attributes that are lost (and those that are added) in the representation as compared to the actual organisms that are present in the habitat that is being explored in the practice of ecotourism with a hand lens. Drawings by Mauricio Álvarez

enjoy "the invisible world" of small organisms. The physical magnifying glass amplifies their observed organisms; the aesthetic hand lens helps visitors to learn new mental schemes and concepts through which to observe and appreciate biodiversity.

3.3.3 The Ethical Lens in Ecotourism with a Hand Lens

Nature observations and scientific discoveries may have implications for the ways we coinhabit the biosphere. Physical and conceptual hand lenses help us to perceive, understand, and value biodiversity. The smallest beings (such as mosses, insects, bacteria, and viruses) often remain invisible in our daily lives. Making visible the invisible will help us to understand our position in the world. One of the most striking results of the ecotourism with a hand lens activity FEP practices is that students and other participants are able to gain awareness about the "lenses" through which we perceive and value biodiversity.

Having a magnifying glass in the hand has multiple effects. First, it enables participants to observe small living beings. This requires participants to slow down to observe these overlooked organisms. This change of rhythm, from an accelerated life at school or cities to a slow education and slow tourism experiences, enables FEP participants to pay more attention on the one hand to small biodiversity and on the other to their own senses and breathing. FEP is a multi-sensorial experience that is guided by analogical thinking as well as by scientific and philosophical concepts. In the field, the combination of senses and concepts orients the realization of sharing habitats with plants, other animals and members of local communities. At moments like this, with the realization that we share habitats, biocultural diversity ceases to be merely a concept and begins to be an experience and awareness of co-inhabitation with diverse living beings and life histories, which regularly remain outside the experiential domain of formal education (Fig. 3.8).

By practicing FEP, students and other participants are invited to cultivate the ethical practice of undertaking responsibility for caring for other-than-human co-inhabitants and the habitats we share with them. Our connections to biodiversity are profound, and through the ethical lens, FEP participants can appreciate the *intrinsic value* of other-than-human living beings. To explore these concepts and values linked to the paradigm of the cellular, biochemical, and genetic *unity of life* we combine field experiences with a practice in the laboratory linked to the notion that humans share with all eukaryotes some basic cellular structures.

3.3.3.1 A Didactic Practice of the Ethical Lens

A first activity begins in the field where students collect a few leaves of mosses or liverworts. Then they bring the plant samples to a laboratory equipped with microscopes. In the laboratory students rub a cotton ball on their buccal mucosa, extracting

Fig. 3.8 Representation of slow observation applying the ethical lens of ecotourism with hand lens. With a hand lens in his hand, the student slows down and is able to smell the fragrances of plants, feel his breathing and encounter small plants "face to face" in a shared habitat. This moment enables the realization of being co-inhabitants with a plethora of other-than-human living beings. This realization can inspire an ethics of co-inhabitation that demands caring relationships with the shared habitats and the community of diverse co-inhabitants. Photographs by Adam Wilson. Figure drawn by Mauricio Álvarez

some cells. They then prepare slides with samples of plant leaves and other slides with the cells of their oral epithelium. Under the microscope, they observe the remarkable similarities, particularly the presence of the nucleus and mitochondria in both types of cells (Fig. 3.9). They also observe notable differences, the presence of cell wall and chloroplasts only in plant cells.

A second activity follows by returning to the field where students perform a "mindful breathing" exercise (Fig. 3.10). They exercise slow movements of air inhalation and exhalation. Integrating their senses, their biology knowledge, and their observation under the microscopes they reflect on the processes of cellular respiration that take place in the mitochondria of humans and plants. They also reflect on the photosynthesis processes that take place in the chloroplasts of plants, and generate oxygen, an essential element for human respiration.

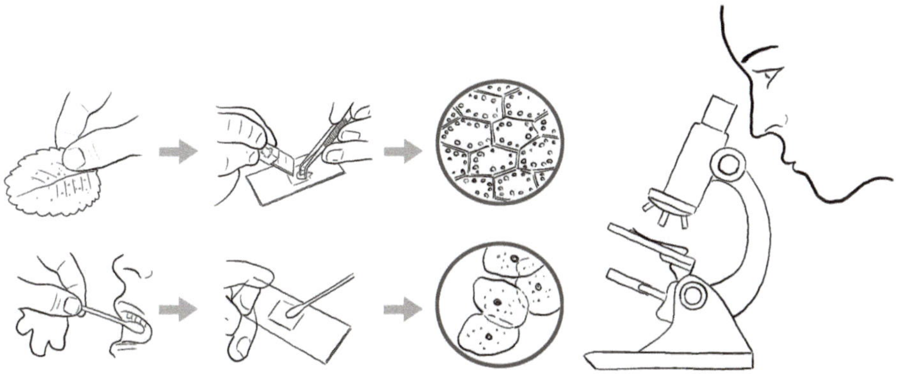

Fig. 3.9 Illustration of an ecotourism with hand lens practice in the laboratory to experientially understand the concept of the ethical lens. In the field, students collect leaves of non-vascular or vascular plants. In the laboratory, they cut thin slices of plant tissue from leaves, and they also rub cotton on their buccal mucosa, extracting some cells. Then they prepared slides with plant tissue and other with their buccal tissue. Under the microscope, they carefully observe structural similarities and differences between both types of cells. Figure drawn by Mauricio Álvarez

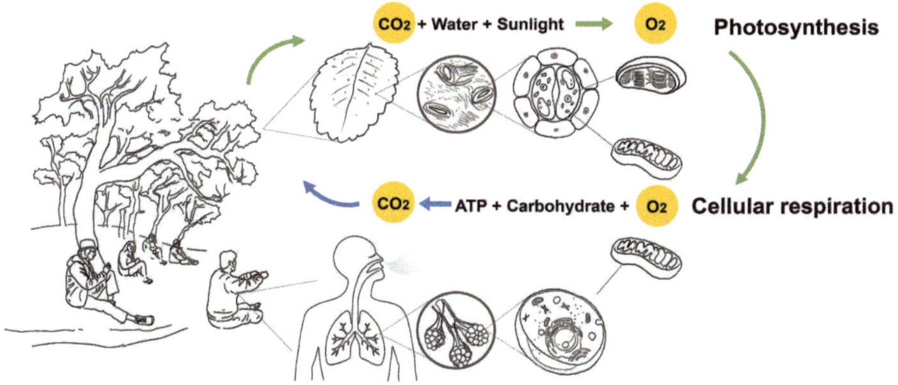

Fig. 3.10 Illustration of an ecotourism with hand lens practice in the field to experientially understand the concept of the ethical lens. Students mindfully breathe, with slow inhalations and exhalations focusing on plant fragrances, and reflecting on the cycle of air nourished by the oxygen (O2) generated in the chloroplasts of plants by the process of photosynthesis and the carbon dioxide (CO2) generated in their mitochondria (as well as the mitochondria in plant cells) by the process of cellular respiration. This mindful breathing experience integrates biological processes and senses as well as cultural understanding of the biogeochemical cycles that link human and plant co-inhabitants in particular habitat as well as in the biosphere as a whole. Moreover, this cycle links both humans and plants with the energy of the sun that is the ultimate source to produce the molecules of Adenosine triphosphate (ATP), which provide the energy required by humans and other living beings to move, grow, and reproduce. Figure drawn by Mauricio Álvarez

3.3.3.2 A Reflection on the Ethical Lens

Ecotourism with a hand lens not only amplifies the vision to see mosses, and other organisms of the miniature forests of the Cape Horn; it also gives participants an *ethical lens* that broadens our mental, perceptual and affective images about nature and our relationship with nature. Science teaches us that mosses, humans, and all living beings share the common vital pulse of cellular respiration. Through the discovery of the miniature forests of small organisms, FEP participants learn that mosses and other living beings also breathe, grow, and reproduce in a similar way to us humans. When participants return to their homes, gardens, fields, streets, or plazas, they will then be able to discover the little mosses, liverworts, and lichens living in their own everyday habitats, and they will know that they are not mere "natural resources" but, like ourselves, they are living, diverse and beautiful co-inhabitants. Through observation of the diversity and unity of life, the experience of ecotourism with a hand lens helps students, tourists, and other citizens to understand multiple and complementary ecological, aesthetic, ethical as well as economic values of biodiversity.

3.3.4 The Economic Lens in Ecotourism with a Hand Lens

The economic lens of ecotourism with a hand lens can be introduced in relation to a unique attribute of the region of Cape Horn: the world's cleanest rainwaters have been recorded at Omora Park (Rozzi et al. 2012b). Non-vascular plants play a critical ecological role in the regulation of hydrologic flow; they contribute to the avoidance of floods when high precipitation events occur, and prevent the drastic diminution of water volumes during periods of drought (Fig. 3.11). The bryophytes, lichens, and other types of vegetation also contribute to maintaining the quality of the water. On the one hand, they prevent erosion; on the other, they act as a filter that retains sediments suspended in the water, especially in peat bog habitats. The quality of the water is critical not only for drinking but also for the operation of fishery canneries established in Cape Horn. For this industry, water use with a high degree of purity is crucial to avoid chemical processes, such as the rusting of cans that are used to pack commercially valuable shellfish species, such as king crab and queen crab (Rozzi et al. 2010).

3.3.4.1 A Didactic Practice of the Economic Lens

FEP participants learn that little plants play important ecological and economic roles, acting like "natural sponges" that filter and regulate the flow and quality of water. On trees that grow in a university campus, urban park, or in old growth forests, FEP students and/or other participants (importantly, environmental policy and decision

Fig. 3.11 Representation of the water cycle regulated by mosses to appreciate the economic lens of ecotourism with hand lens. With a magnifying glass in his hand, the student can observe how carpets of bryophytes (mosses and liverworts) regulate the stem flow of rain waters. Bryophytes act like sponges that filter and regulate the flow of water (detail of a magnified moss through which water runs off). Water in the form of mist or rain runs through the treetops and down their trunks. In the sub-Antarctic Magellanic forests of the Cape Horn Biosphere Reserve, bryophytes form thick covers on the surface of trunks and capture the rainwater that flows slowly toward the ground. This enhances the quality of drinking water (exemplified in a glass of water) and prevents floods when high precipitation events occur. These "miniature forests" provide an essential ecosystem service that has high economic value for humans as well as high ecological value for ecosystem functioning. Photographs by Paola Vezzani and Adam Wilson. Figure composed by Mauricio Álvarez

makers) mount an experiment to demonstrate the ecosystem service provided by mosses that grow on trunks and regulate rainwaters' flows and quality (Fig. 3.12).

3.3.4.2 A Reflection on the Economic Lens

The economic lens is linked to the concept of ecosystem services that connects natural systems and social and economic systems (Costanza et al. 1997). During the twenty-first century, worldwide studies on the status and relevance of ecological systems for society have been carried out under the umbrella of the United Nations

Fig. 3.12 Illustration of an ecotourism with hand lens practice in the field to experimentally understand the concept of the economic lens. FEP participants select four trees (treatments) whose trunks have different degrees of inclination and have contrasting covers of mosses. The ideal setting is: Tree 1 = trunk has no inclination.no moss cover; Tree 2 = trunk with no inclination and high moss cover; Tree 3 = high inclination (> 45o) and no moss cover; Tree 4 = high inclination (> 45o) and dense moss cover. To measure stem flow volume in the field, students install a bisecting hose around the circumference of the trunk near the base of each of the selected trees. They connect this hose to a graduated rain gauge for water collection. In each trunk, students pour muddy water over the trunk and the water that runs off is captured in a rain gauge connected to the base of the trunk. The time it takes for the water to drain and the amount of water captured is measured. With school children and university students, this setting can be installed with five replicates for each treatment; twenty trees in total. Figure drawn by Mauricio Álvarez

Environment Program (UNEP) and the Millennium Ecosystem Assessment (Braat and de Groot 2012). In ecotourism with a hand lens, participants conduct simple experiments in the field to learn about the economic lens and concept of ecosystem services.

The prevailing mental image is the dominant economic model the "Mickey Mouse Model" in which the environment, society, and economy represent separate spheres (Lucena et al. 2010). In addition, economy is disproportionally more important than society and the environment (Fig. 3.13). The activities in ecotourism with a hand lens help participants to understand a novel model of sustainability in which economy is central, but it is inserted within the broader context of society and the life support systems (Fig. 3.13). The concepts and experiences with economic lens help participants to critically compare the "Mickey Mouse" and the "Bull's Eye" economic models. Under the latter, economy represents one component of society that, in turn, is immersed within the broader system that sustains life on Earth. In this way, the *economic lens* contributes to an understanding of the *instrumental value* of biodiversity, and the updated definition by the United Nations Sustainable Development Goals and Agenda 2013 of sustainable development as

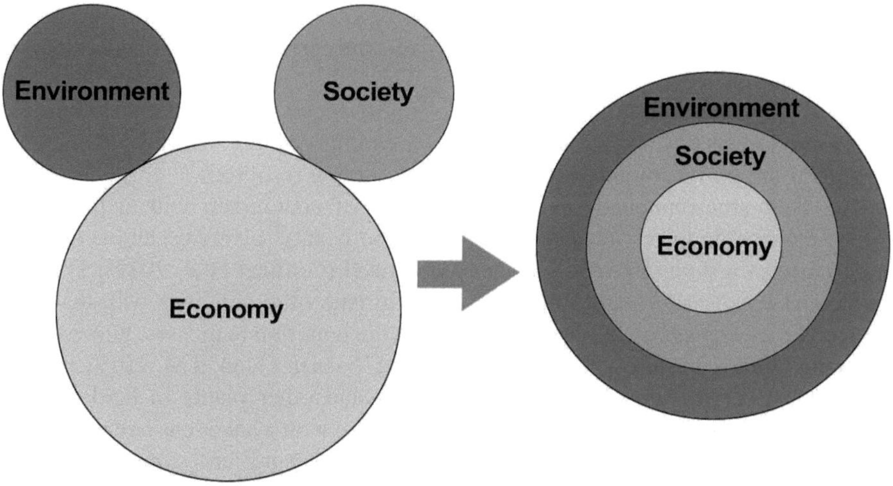

Fig. 3.13 Diagram illustrating how the economic lens aims to foster the vision of interactions between the environment, society, and the economy. A. On the left, the old *Mickey Mouse Model* shows economy as most important with society and the environment as separate, minor, side issues. B. On the right, the new *Bullseye Model* illustrates an understanding that our society is dependent on the environment and that our economy is dependent on our society. Figure modified from Lucena et al. (2010, p. 34), and drawn by Mauricio Álvarez

"one that satisfies the needs of present generations, while safeguarding the system that sustains the life of the Earth" (Griggs et al. 2013; Brandi 2015).

3.4 Implications of the Four Lenses for Biocultural Conservation: A Real Story from Cape Horn

Today a vast diversity of living beings and human values are invisible, because they do not have a place in the narrow worldview that governs global society. The limitations of a worldview—the lenses through which we view the world—become evident when confronting otherness. As an example, global standard biodiversity assessments based on vascular plants had considered floristic diversity of southwestern South America to be poor. However, long-term botanical fieldwork in the region disclosed its floristic otherness; non-vascular plants had a greater diversity than vascular plant species. Moreover, the sub—Antarctic Magellanic ecoregion hosts >5% of the world's species of non-vascular plants. This stimulated the research team at Omora Park to "change the lenses to assess biodiversity richness" and focus on non-vascular plants for defining conservation priorities in high-latitude ecoregions (Rozzi et al. 2008). This change made visible an idiosyncratic biodiversity and provided an argument to create the UNESCO Cape Horn Biosphere Reserve in 2005.

For the first time in the world, a biosphere reserve was designated based on the diversity of mosses. These tiny organisms had rarely been perceived and valued in the international conservation community. This led to a change in the language referring to mosses, and to an awareness of inter-species co-inhabitation. Thus, the "miniature forests of Cape Horn" metaphor was composed to communicate that little organisms are co-inhabitants rather than mere "natural resources."

The field environmental philosophy activity of ecotourism with a hand lens helped citizens and decision-makers discover the beauty, diversity, and ecological importance of a flora that regularly goes unnoticed (Goffinet et al. 2012). Through the ethical lens, visitors are able to observe floristic otherness in a way that they recover the awareness of cellular processes that are common to mosses, humans, and all living beings. In this type of "forest bathing" (*sensu* Guan et al. 2017; Hansen et al. 2017), the direct encounter with mosses and other plants in field and the orientation provided by the four lenses of ecotourism with a hand lens help visitors to experience the vital pulse of life. Ecotourism with a hand lens summons ethical, aesthetical, and ecological values that broaden the perspectives that prevail in the relationship of global society with nature, and reconnect citizens with the wonders of biocultural diversity.

3.5 Concluding Remarks

The Portuguese philosopher Magda Costa-Carvalho (2023) affirms that philosophy contributes to deconstructing and reconstructing ideas in education with provocative concepts such as small, silent, and (in)significant. Ecotourism with a hand lens uses physical or technological tools, such as the optical lenses of a magnifying glass, or a microscope. Ecotourism with a hand lens also uses conceptual or symbolic lenses, such as languages, values, and concepts, that guide us in the observation of the cosmos. These conceptual lenses allow us to reflect on how we view the biodiversity that we research and use. These lenses allow us to access the microcosm of biodiversity and discover that what is small is beautiful, essential, and diverse.

Biocultural tourism, as proposed by Breakey and Breakey (2015, 2023), provides education within natural and/or cultural heritage experiences through the broadening of ethical consideration and care to new ecological and/or cultural phenomena. The practices of ecotourism with a hand lens have a transformative effect on students, tourists, and other members of the global urban and digital society who are offered new ways to experience co-inhabiting with biological and cultural diversity. For example, through face-to-face encounters with members of other cultures and with small living beings. These encounters revitalize the understanding of the sense of belonging to biocultural communities. This sense of co-inhabitation in biocultural communities compensates for the disconnection exacerbated by the rationalistic individualism that dominates the urbanized and digitized global society.

In summary, with the four lenses of ecotourism with a hand lens, we propose: (1) to recover the connection with biocultural diversity through direct and multisensory experiences, (2) to generate creative educational activities based on the singularities of the places, (3) to inspire invention of new activities as diverse as educational and problematic situations exist. For education, these practices present a balance and integration between the biophysical dimensions and the cultural dimensions (material and symbolic) of biodiversity. These last dimensions are expressed in the diversity of languages and recover multiple ways of "understanding and expressing the world" through analogical thinking and poetic craftsmanship. These practices seek to guide forms of intercultural co-inhabitation and inter-species relationships.

Acknowledgments Authors thank Kelli Moses for valuable comments and acknowledge the support from the Cape Horn International Center (CHIC) - ANID PIA/BASAL (PFB210018).

References

Ahl V, Allen TFH (1996) Hierarchy theory: a vision, vocabulary and epistemology. University of Columbia Press, New York

Bilbeny N (1997) La revolución en la ética: Hábitos y Creencias en la Sociedad Digital. Anagrama, Barcelona

Bormann H, Kellert S (1991) Ecology, economics, ethics: the broken circle. Yale University Press, New Haven

Braat LC, De Groot R (2012) The ecosystem services agenda: bridging the worlds of natural science and economics, conservation and development, and public and private policy. Ecosyst Serv 1(1):4–15

Brandi C (2015) Safeguarding the earth system as a priority for sustainable development and global ethics: the need for an earth system SDG. J Glob Ethics 11(1):32–36

Breakey N, Breakey H (2015) Tourism and Aldo Leopold's 'Cultural Harvest': creating virtuous tourists as agents of sustainability. J Sustain Tour 23(1):85–103

Breakey H, Breakey N (2023) Leopold's cultural harvest, biocultural tourism, and field environmental philosophy. In: Rozzi R, Tauro A, Wright T, Avriel-Avni N, May RH Jr (eds) Field environmental philosophy: education for biocultural conservation. Ecology and ethics, vol 5. Springer, Dordrecht, pp 281–294

Costa-Carvalho C (2023) Small, silent and (in)significant: childhood as a minoritarian experience of education. In: Rozzi R, Tauro A, Wright T, Avriel-Avni N, May RH Jr (eds) Field environmental philosophy: education for biocultural conservation. Ecology and ethics, vol 5. Springer, Dordrecht, pp 339–358

Costanza R, d'Arge R, De Groot R, Farber S, Grasso M, Hannon B et al (1997) The value of the world's ecosystem services and natural capital. Nature 387(6630):253–260

Dodson SI, Allen TFH, Carpenter SR, Ives AR, Jeanne RL, Kitchell JF, Langston NE, Turner MG (1998) Ecology. Oxford University Press, New York

Franklin JF (1981) Ecological characteristics of old-growth Douglas-fir forests, vol. 118. US Department of Agriculture, Forest Service, Pacific Northwest Forest and Range Experiment Station

Goffinet B, Rozzi R, Lewis L et al (2012) The Miniature Forests of Cape Horn: Eco-Tourism with a Hand-lens ("Los Bosques en Miniatura del Cabo de Hornos: Ecoturismo con Lupa"), Bilingual English-Spanish edition. UNT Press–Ediciones Universidad de Magallanes, Denton/Punta Arenas

Griggs D, Stafford-Smith M, Gaffney O, Rockström J, Öhman MC, Shyamsundar P, Steffen W, Glaser G, Kanie N, Noble I (2013) Sustainable development goals for people and planet. Nature 495:305–307
Guan H, Wei H, He X, Ren Z, An B (2017) The tree-species-specific effect of forest bathing on perceived anxiety alleviation of young-adults in urban forests. Annals Forest Res 60(2):327–341
Hansen MM, Jones R, Tocchini K (2017) Shinrin-yoku (forest bathing) and nature therapy: a state-of-the-art review. Int J Environ Res Public Health 14(8):851
Isbell F, Balvanera P, Mori AS, He JS, Bullock JM, Regmi GR et al (2022) Expert perspectives on global biodiversity loss and its drivers and impacts on people. Front Ecol Environ. https://doi.org/10.1002/fee.2536
Kuhn TS (1970) The structure of scientific revolutions. University of Chicago Press, Chicago
Latour B (1999) Pandora's Hope. Essays on the reality of science studies. Harvard University Press, Cambridge
Lovejoy TE (1980) In: Barney GO (ed) The Global 2000 Report to the President The Technical Report, vol. 2. Penguin, pp 327–332
Lucena J, Schneider J, Leydens JA (2010) Engineering and sustainable community development. Synthesis Lect Eng Technol Soc 5(1):1–230
McNeill JR, Engelke P (2016) The great acceleration: an environmental history of the Anthropocene since 1945. Harvard University Press
Neri J (2011) The Insect and the image: visualizing nature in early modern Europe, 1500-1700. University of Minnesota Press, Minneapolis
Noss RF (1990) Indicators for monitoring biodiversity: a hierarchical approach. Conserv Biol 4: 355–364
Pickett STA, Kolasa J, Jones CG (1994) Ecological understanding: the nature of theory and the theory of nature. Academic Press, Orlando, xiii + 206 pp
Primack R, Rozzi R, Feinsinger P, Dirzo R, Massardo F (2001) Elementos de conservación biológica: perspectivas Latinoamericanas. Fondo de Cultura Económica, Ciudad de México, México
Rozzi R (1999) The reciprocal links between evolutionary-ecological sciences and environmental ethics. Bioscience 49(11):911–921
Rozzi R (2001) Ética ambiental: raíces y ramas latinoamericanas. In: Primack R, Rozzi R, Feinsinger P, Dirzo R, Massardo F (eds) Fundamentos de conservación biológica: Perspectivas latinoamericanas. Fondo de Cultura Económica, México, pp 311–362
Rozzi R (2004) Implicaciones éticas de narrativas yaganes y mapuches sobre las aves de los bosques templados de Sudamérica austral. Ornitología Neotropical 15(suppl.):435–444
Rozzi R (2012) Introduction to ecotourism with a hand lens. In: Goffinet B, Rozzi R, Lewis L, Buck W, Massardo F (eds) The miniature forests of Cape Horn: eco-tourism with a hand-lens, Bilingual English-Spanish edn. University of North Texas Press, Denton, pp 28–79
Rozzi R (2013) Biocultural ethics: from biocultural homogenization toward biocultural conservation. In: Rozzi R, Pickett STA, Palmer C, Armesto JJ, Callicott JB (eds) Linking ecology and ethics for a changing world: values, philosophy, and action. Ecology and Ethics book series, vol 1. Springer, Dordrecht, pp 9–32
Rozzi R (2018) Biocultural homogenization: a wicked problem in the Anthropocene. In: From biocultural homogenization to biocultural conservation. Springer, Cham, pp 21–48
Rozzi R (2023a) Introduction Part I. Field environmental philosophy: concepts and case studies. In: Rozzi R, Tauro A, Wright T, Avriel-Avni N, May RH Jr (eds) Field environmental philosophy: education for biocultural conservation. Ecology and ethics, vol 5. Springer, Dordrecht, pp 17–26
Rozzi R (2023b) Inter-species and inter-cultural encounters: the education and biocultural ethics program of the Omora Ethnobotanical Park. In: Rozzi R, Tauro A, Wright T, Avriel-Avni N, May RH Jr (eds) Field environmental philosophy: education for biocultural conservation. Ecology and ethics, vol 5. Springer, Dordrecht, pp 153–174
Rozzi R, Hargrove E, Armesto JJ, Pickett STA, Silander J Jr (1998) "Natural drift" as a post-modern metaphor. Revista Chilena de Historia Natural 71:9–21

Rozzi R, Armesto J, Goffinet B et al (2008) Changing lenses to assess biodiversity: patterns of species richness in sub-Antarctic plants and implications for global conservation. Front Ecol Environ 6:131–137

Rozzi R, Anderson CB, Pizarro JC, Massardo F, Medina Y, Mansilla A et al (2010) Field environmental philosophy and biocultural conservation at the Omora Ethnobotanical Park: methodological approaches to broaden the ways of integrating the social component ("S") in long-term socio-ecological research (LTSER) sites. Revista Chilena de Historia Natural 83(1): 27–68

Rozzi R, Lewis L, Massardo F, Medina Y, Moses K, Méndez M, Sancho L, Vezzani P, Russell S, Goffinet B (2012a) Ecotourism with a Hand-Lens at Omora Park. Ediciones Universidad de Magallanes, Punta Arenas, Chile

Rozzi R, Armesto JJ, Gutiérrez J, Massardo F, Likens G, Anderson CB, Poole A, Moses K, Hargrove G, Mansilla A, Kennedy JH, Willson M, Jax K, Jones C, Callicott JB, Kalin MT (2012b) Integrating ecology and environmental ethics: Earth stewardship in the southern end of the Americas. BioScience 62(3):226–236

Sarkar S (2021) Origin of the term biodiversity. BioScience 71(9):893–893

Swingland IR (2001) Biodiversity, definition of. Encycl Biodivers 1:377–391

Worster D (1994) Nature's economy: a history of ecological ideas. Cambridge University Press, Nueva York

Young HS, McCauley DJ, Galetti M, Dirzo R (2016) Patterns, causes, and consequences of Anthropocene defaunation. Ann Rev Ecol Evol Syst 47(1):333–358

Chapter 4
Underwater with a Hand Lens: Ecological Sciences and Environmental Ethics to Value Freshwater Biodiversity

Tamara Contador, Javier Rendoll-Cárcamo, Melisa Gañan, Jaime Ojeda, James Kennedy, Peter Convey, and Ricardo Rozzi

Abstract Despite their impressive diversity and ecosystem relevance, insects are undervalued and rarely considered in conservation efforts, except for those that are medically or economically important. In terms of funding and effort, insect conservation research lags far behind vertebrate research, hampering the development of methodologies to better understand their conservation needs. This taxonomic bias has a severe limitation in that it only gives moral consideration to our closest

---

T. Contador (✉) · J. Rendoll-Cárcamo · M. Gañan
Sub-Antarctic Biocultural Conservation Program, Universidad de Magallanes, Punta Arenas, Chile

Wankara Laboratory, Universidad de Magallanes, Puerto Williams, Chile

Núcleo Milenio de Salmónidos Invasores, Concepción, Chile

Cape Horn International Center (CHIC), Universidad de Magallanes, Puerto Williams, Chile

Institute of Biodiversity of Antarctic and Subantarctic Ecosystems (BASE), Santiago, Chile
e-mail: tamara.contador@umag.cl; javier.rendoll@umag.cl

J. Ojeda
School of Environmental Studies, University of Victoria, Victoria, BC, Canada

Cape Horn International Center (CHIC), Universidad de Magallanes, Punta Arenas, Chile

J. Kennedy
Wankara Laboratory, Universidad de Magallanes, Puerto Williams, Chile

Cape Horn International Center (CHIC), Universidad de Magallanes, Puerto Williams, Chile

Department of Biological Sciences, University of North Texas, Denton, TX, USA

Sub-Antarctic Biocultural Conservation Program, University of North Texas, Denton, TX, USA
e-mail: kennedy@unt.edu

P. Convey
Cape Horn International Center (CHIC), Universidad de Magallanes, Puerto Williams, Chile

British Antarctic Survey, NERC, Cambridge, UK

Institute of Biodiversity of Antarctic and Subantarctic Ecosystems (BASE), Santiago, Chile
e-mail: pcon@bas.ac.uk

R. Rozzi et al. (eds.), *Field Environmental Philosophy*, Ecology and Ethics 5,
https://doi.org/10.1007/978-3-031-23368-5_4

evolutionary relatives, excluding the vast majority of other life forms on our planet. In this regard, the Field Environmental Philosophy (FEP) methodology provides a platform for the development of interdisciplinary approaches to biocultural conservation that combine ecological research and environmental ethics. We integrated ecological research, education, and environmental ethics over the last 15 years through the FEP to promote and foster freshwater insect awareness, conservation, and value in the Magellanic sub-Antarctic ecoregion of southern Chile. The FEP practice has enabled us to reach a wide range of people and raise awareness for these under-appreciated co-inhabitants. This methodology also promotes educational practices that encourage direct encounters with the inhabitants, habitats, and life habits, making them "visible" once more. Our work aims to awaken appreciation and valuation of the small co-inhabitants with whom we share the planet by providing a platform of scientific training and thinking from the earliest stages of education, with a novel combination of disciplines aiming to reveal the intrinsic value of freshwater insects.

Keywords Biocultural ethics · Field environmental philosophy · Freshwater insects · Magellanic sub-Antarctic · Intrinsic value · Instrumental value · Taxonomic chauvinism

4.1 Introduction

An under-appreciated and seldom considered aspect of global change is its significant impact on invertebrate organisms, mainly insects (Sánchez-Bayo and Wyckhuys 2019). Even though insects account for more than 90% of the planet's terrestrial faunal diversity and powerfully influence the overall state of biodiversity, researchers and conservation biologists pay only limited attention to their modern extinction (Fonseca 2009; Briggs 2017). In agreement with biodiversity losses reported in other animal taxa, declines are particularly serious among aquatic insects (Ricciardi and Rasmussen 1999). Although we have insufficient knowledge available for most countries, we know that 33% of aquatic insects are threatened compared to 28% of terrestrial insects (Sánchez-Bayo and Wyckhuys 2019; Dudgeon 2019; Cardoso et al. 2020). In addition to the impact of global warming, the quantity and quality of water have been drastically affected by aggressive agricultural practices, degradation, and fragmentation of freshwater ecosystems worldwide (Foley et al. 2005).

R. Rozzi
Sub-Antarctic Biocultural Conservation Program, Department of Philosophy and Religion and Department of Biological Sciences, University of North Texas, Denton, TX, USA

Cape Horn International Center (CHIC), Universidad de Magallanes, Puerto Williams, Chile
e-mail: Ricardo.Rozzi@unt.edu

There is extensive evidence about the drastic insect decline; however, only 10% of conservation funding worldwide is directed toward their conservation (Cardoso et al. 2011; Samways et al. 2020). There is a clear taxonomic bias in conservation research, formal education, animal rights, and conservation initiatives, with vertebrates enormously over-represented compared to invertebrates (Clark and May 2002; New and Samways 2014; Rozzi 2019). This scenario is even more alarming in the freshwater realm, particularly in southern South America, where research on the ecology and conservation of terrestrial and aquatic insects is minimal compared to other regions of the world (Contador et al. 2012; Ramírez and Gutiérrez-Fonseca 2014; Nieto et al. 2017). Research in conservation biology is dominated by vertebrates, with work on birds and mammals receiving a high proportion of funding and effort (Fazey et al. 2005; Di Marco et al. 2017). In particular, the limited amount of conservation research on invertebrates is likely to affect the development of appropriate research methodologies to understand their conservation needs, which cannot be easily adapted from the better-studied vertebrate groups (Pawar 2003; Di Marco et al. 2017). This taxonomic bias creates a further severe limitation because it gives moral consideration solely to our closest evolutionary relatives, thus excluding the vast majority of other life forms on our planet (Rozzi 2019).

What reasons underlie these inequalities and biases favoring vertebrates? Attitudes toward insects have a philosophical and ethical basis, even though these may not be recognized (Mather 2011). Fukano and Soga (2021) have found that insects and other terrestrial arthropods are among the most disliked and least valued taxonomic groups of living organisms globally, especially in developed countries. The expansion of urban societies deepens the disconnection between most humans and the most diverse groups of animals and nature as a whole (Soga and Gaston 2016). The rapid growth of cities alienates societies from natural spaces, particularly causing a lack of awareness and negative perceptions toward insects (Basset and Lamarre 2019; Fukano and Soga 2021). Fukano and Soga (2021) propose the "urbanization-disgust hypothesis," which states that urbanization is a significant driver of the prevalence of negative attitudes and perceptions toward insects. Urbanization changes ecosystems and the diversity of habitats available for insects and other invertebrates. Consequently, it reduces the diversity of insects and direct human interactions or "face-to-face encounters" with them. Many of the most abundant insects in urban settings are associated with the transmission of diseases; hence, a psychological explanation might underly pathogen-avoidance behaviors and negative attitudes toward invertebrates. Moreover, people experience insects more frequently indoors than outdoors, thus augmenting disgust toward them (Fukano and Soga 2021).

Why do we hold less consideration for insects? Why is it possible for us to kill them as if their lives did not matter? Animal rights have focused their attention to sentient vertebrates; mostly mammals (see Regan 2004; Sunstein and Nussbaum 2004). People tend to care and positively value animals closely related to us and those who can feel pain (Mather 2011). Whether insects and other invertebrates are capable of experiencing pain is heatedly debated (Lockwood 1988; Watkins et al. 2010; Andrews 2011; Elwood 2011; Magee and Elwood 2013; Klein and Barron

2016; Adamo 2016a; Burrell 2017). The utilitarian school ascribes moral consideration only to animals capable of feeling pain. Most ethicists still assume that insects cannot suffer and have subjective experiences; therefore, there is no need for ethical (moral) consideration for them (Rozzi 2019).

The utilitarian argument is based on the distinction between an insect's ability to respond to potentially damaging stimuli (nociception) and the capacity to have subjective experiences and pain (Adamo 2016b). Psychologist Shelley Adamo (2016a) examines four areas of research that touch on whether insects feel pain: philosophy, insect neurobiology and behavior, artificial intelligence and robotics, and evolution. These areas have provided new evidence that invertebrates have forms of learning, memory capacity, and consciousness (Nityananda 2016; Perry et al. 2017). However, some contest this conclusion and the capacity of insects for suffering (Sherwin 2001). Insects have small nervous systems distributed throughout their bodies in the form of ganglia (Gullan and Cranston 2014; Adamo 2016a). This disaggregation of the nervous system is thought to limit the capacity for advanced information processing, although counter to this their principal brains contain complex neuroanatomical features, which could be equivalent to reward circuits in vertebrates (Adamo 2016a). In summary, the nervous systems of insects do not constitute a limitation for feeling pain nor memory and sophisticated and complex behaviors (Giurfa 2013).

How can we address the ethical challenges discussed above? First, we propose that insects should be valued and treated with respect, even if only limited evidence is available on their ability to feel pain or consciousness. Second, to support our proposal, we provide an overview of our long-term efforts to contribute to developing knowledge, valuing, and addressing conservation challenges relating to freshwater invertebrates in the Magellanic sub-Antarctic ecoregion of southern Chile (Contador et al. 2012, 2014a, 2015a). During the last fifteen years, we have integrated ecological sciences, education, and environmental ethics through the praxis of the Field Environmental Philosophy (FEP) methodology to contribute to the valuing and conservation of freshwater insects, their habitats, and life habits (Contador et al. 2018; Rendoll Cárcamo et al. 2020). Third, we discuss how taking this approach contributes to developing positive perceptions and values toward insects, encouraging conservation actions in the long term by communities across local, national, and international scales.

4.2 Praxis of Field Environmental Philosophy: Ethical Valuing of Freshwater Ecosystems and Their Co-Inhabitants

For over twenty years, the scientific team at Omora Park in the Cape Horn Biosphere Reserve (CHBR), Chile, has developed the Field Environmental Philosophy (FEP) methodological approach, which integrates ecological sciences, arts, and environmental ethics (Tauro et al. 2021). In freshwater ecology, this approach was initially developed as part of a doctoral dissertation (Contador 2011), and it was consolidated with a master thesis (Rendoll-Cárcamo 2018). Furthermore, to contribute toward an appreciation of insects, we adapted FEP's Four-Step cycle, which includes: (1) transdisciplinary research, (2) poetic communication through the composition of metaphors, (3) design of field activities with an ethical and ecological orientation, and (4) *in situ* conservation, to contribute to biocultural conservation (Contador et al. 2018).

4.2.1 Step 1. Transdisciplinary Research

Our transdisciplinary research has focused on the integration of ecological sciences and environmental philosophy:

(a) Ecological Framework Our ecological work has focused on the adaptations and biodiversity of freshwater and terrestrial insects to global environmental change. Our primary setting is the Róbalo River watershed protected by Omora Park (Rozzi et al. 2006). The Róbalo River runs through a steep altitudinal gradient and provides drinking water to Puerto Williams (Navarino Island, 55°S), and is managed as a natural laboratory (Contador et al. 2015a; Rendoll Cárcamo et al. 2019). In January 2008, we began studies on the biodiversity, distribution, ecophysiology, and phenology of aquatic insects along the altitudinal gradient. Our research goals are to (i) monitor climatic characteristics of Magellanic sub-Antarctic freshwater ecosystems; (ii) study the phenological responses of freshwater insects along altitudinal gradients; (iii) identify species that can play the role of "sentinels of climate change," and (iv) develop ethical research methodologies to tackle the above objectives in the long-term.

(b) Philosophical Framework The philosophical basis of our long-term multidisciplinary work includes the analysis and practice of *biocultural ethics*. It contributes a new eco-philosophical paradigm to value the vital links that sustain the life habits of living beings interacting with the other co-inhabitants with whom they share and co-constitute habitats (Rozzi 2012, 2018). These are the "3Hs" of biocultural ethics: co-in-Habitants, with their Habits forged in shared Habitats. The "3Hs" model recovers the understanding that we are part of communities of co-inhabitants, and the valuation of caring relationships, complementarity, and reciprocity, which emerge in sharing in general and in sharing habitats in particular (Rozzi 2019). Applying this philosophical approach contributes to the generation of inter-specific relationships between humans and the other living beings with whom we co-inhabit (Rozzi 2013). We primarily focus on overcoming the conceptual omission of the *intrinsic value* of freshwater invertebrates, their habits, and habitats. To recognize and ethically value the community of diverse life forms as co-inhabitants, we must overcome the dualistic view between the *subject* (human) and *object* (non-human). This dualism between *subject* and *object* has been pervasive, deeply embedded in Western thought. It is at the root of various interrelated dualisms, such as activity (or agency) versus passivity or resonating in culture versus nature (Klaver 2013). This way of perceiving the world implies a preconceived sense of superiority over all non-human beings, resulting in an implied legitimization of their use, domination, and exploitation (Klaver 2013). In the science realm, a "logic of the specimen" has prevailed studying animals and plants as individuals separated from their habitats. They are considered *objects* of study, which can be used and exploited to fulfill research objectives (Rozzi 2019). To overcome this dualism in freshwater ecology, the result of our long-term research has been the creation of a new methodology to observe, study, and value freshwater invertebrates in the Cape Horn Biosphere Reserve. This methodology can be applied in other places of the world (Box 4.1).

Box 4.1 Can We Justify the Collection and Killing of Hundreds of Individual Insects in Entomological Research?

In this box, we present a novel ethical research methodology for ecological studies in freshwater ecosystems. Entomologists use a variety of methods to collect a large number of insects to help answer a variety of ecological questions. This is not to say that entomologists do not value insects, as most follow informal and formal codes of conduct as well as collection regulations such as not indiscriminately collecting rare insects if this could lead to population decline or extinction (Didham et al. 2019). However, these codes only consider individual insects as general exemplars of their species and tend to focus on their instrumental value within ecosystems (Didham et al. 2019; Fischer and Larson 2019). This common practice in scientific research often considers individuals, populations, communities, and ecosystems as mere *objects* of study. From the perspective of biocultural ethics, insects are considered *subjects*, co-inhabitants that have an intrinsic value. Most ethical debate fails to propose actions to overcome dualities and ethical considerations in entomological research. We agree with Bob Fischer and Brendon Larson (2019) that "entomologists should aspire to study insects without killing them." For this reason, we developed a methodology based on the 3H approach to avoid causing pain and death in insects (and other invertebrates) by both researching and respecting their habits and habitats (Contador et al. 2018). This methodology includes the following phases:

1. **Rigorous training *in-situ* and in the laboratory.** Researchers undergo an in-depth training process to identify each species in the field. After the training, researchers are able to identify *in situ* each species within its habitat of origin. Scientists conduct all observations regarding the species of interest within its habitat (Fig. 4.1).
2. **Habitat characterization, sampling, and identification of co-inhabitants.** Researchers take samples using conventional methods

(continued)

Box 4.1 (continued)

(e.g., Surber samplers or D-frame nets) to ensure the quantitative rigor of sampling techniques. In the field, researchers conduct standard habitat assessments (e.g., physico-chemical habitat characterization) and sample-targeted species. Identification and quantification are performed *in situ*. Identification of invertebrates may be difficult depending on the season of the year. For example, insects in early developmental stages cannot be identified with the naked eye, so they must be observed with a stereoscope at a laboratory. In such cases, specimens are reared in the laboratory under controlled temperature and physico-chemical water conditions to ensure their survival (Fig. 4.1).

3. **Analyses and return to the habitat.** When organisms are transported to the laboratory for identification or to conduct phenological studies, organisms must be carefully measured. In this case, invertebrates are transported alive to the laboratory, where they are kept in oxygenated aquariums to be photographed and measured with stereoscopic magnifiers. Once all observations and measurements are made, the invertebrates are carefully and respectfully returned to their habitat of origin (Fig. 4.1). In most cases, however, we conduct our observations in the field. In these cases, organisms are delicately removed from their micro-habitats (e.g., rocks, aquatic plants) to be counted and identified. Afterward, they are immediately returned to their exact original place in the stream.

4.2.2 *Step 2. Composition of Metaphors:* The River as a Community of Life *and* Aquatic Insects, Invisible Sentinels

Metaphors are considered *cultural messengers,* which not only constitute a purely linguistic expression and are also a fundamental part of the cognitive structure of human beings (Rozzi 1999). The Greek root of the word "metaphor" means "to transfer or carry" and implies communication between the biophysical and cultural domains (Larson 2006). To help overcome the conceptual omission of the intrinsic value of freshwater invertebrates, their habits, and habitats, we collectively composed the metaphor "*The River as a Community of Life*." We composed this metaphor with students and tourism guides from Puerto Williams through a series of workshops conducted between 2010 and 2013. We also collaborated with artists, scientists, and anthropologists associated with Omora Park (Contador et al. 2018). The composition of metaphors took place during field activities in which participants learned to identify, recognize, and understand essential ecological concepts about freshwater invertebrates (Contador et al. 2018). These activities were conducted

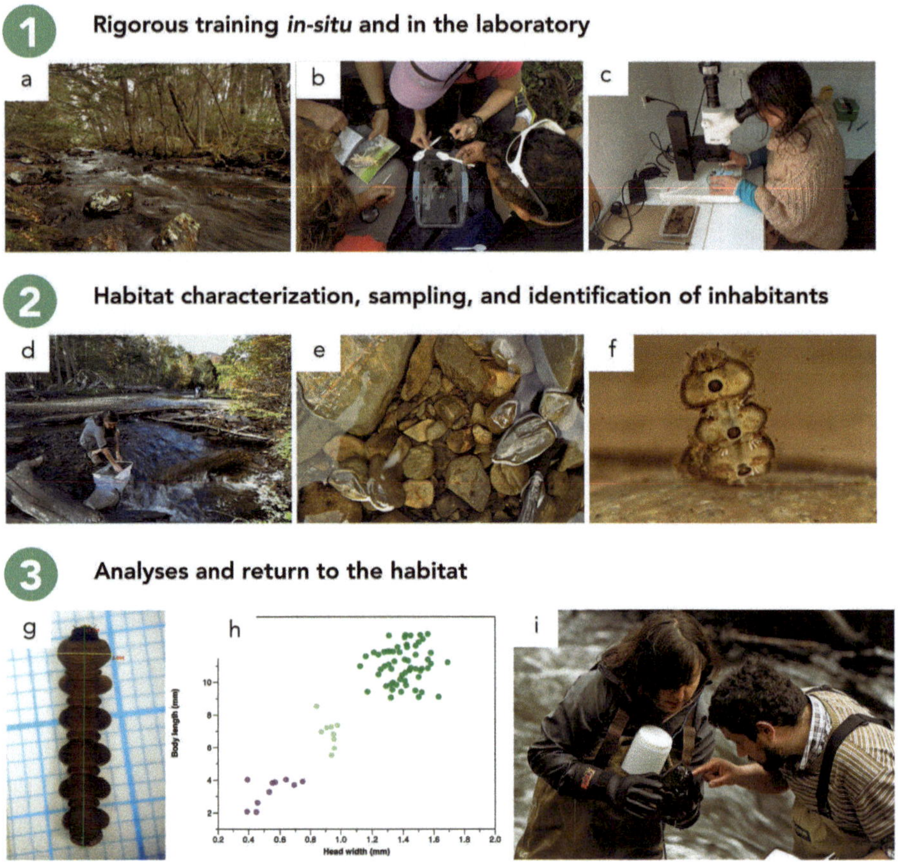

Fig. 4.1 Illustration of a long-term study on the phenology of *Edwardsina dispar* (Insecta: Diptera), a freshwater insect adapted to cold, running waters. This study exemplifies the three phases carried out in the field and in the laboratory for an ethical treatment of aquatic insects in ecological research. Phase 1: Researchers are trained in the field as well as in the laboratory to identify insects and other invertebrates (**a**, **b**, **c**). Phase 2: Habitat characterization, sampling, and identification of the co-inhabitants occur in the field, when possible (**d**, **e**, **f**). Phase 3: If necessary, invertebrates are transported to the laboratory, where they are respectfully maintained in oxygenated containers with elements from their habitat to be measured and analyzed (**g**, **h**). When all measurements are finished, the invertebrates are returned to their habitat (**i**). Photographs by Gonzalo Arriagada (**a**, **d**, **e**, **f**, and **i**); Tamara Contador (**b**); Javier Rendoll (**c** and **g**)

within the conceptual framework of *biocultural ethics* and were based on encouraging:

(a) Face-to-face encounters with insects through guided field experiences
(b) Contemplation and meditation about the experiences and perceptions generated during *face-to-face* encounters
(c) Ethical valuing of insects, their habitats, and habits

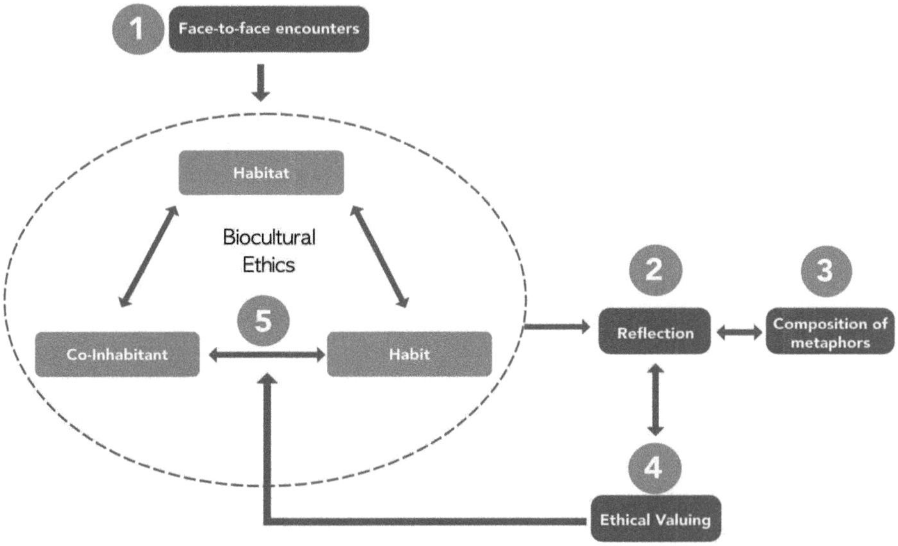

Fig. 4.2 Methodological sequence of five phases to contribute to biocultural conservation through the composition and implementation of metaphors in communicational processes about the various values of living beings. Figure adapted from Rozzi (2013), and Contador et al. (2018).

"Direct encounters" stimulate reflections on the various values that we assign to insects and other invertebrates. As participants reflected upon the importance of insects for the overall health of the river, metaphors contributed to the understanding of key ecological concepts such as nutrient cycling and phenology. Participants described aquatic insect communities as "insect villages" and found that these "villages" resembled the communities in which they lived. The river was not "just water and stones," but a "place full of life," much like Puerto Williams. In this way, with a process that combines participatory fieldwork, readings, and reflections, the metaphor *The River as a Community of Life* was created (Fig. 4.2)

Another metaphor was composed: *Aquatic Insects, Invisible Sentinels.* This metaphor highlights the *instrumental value* or usefulness freshwater insects as sentinels of a persistent problem: global environmental change. The adjective "invisible" denotes that insects (and other invertebrates), as well as their habitats and life habits, are little known and undervalued in both conservation strategies and everyday culture (Rendoll Cárcamo et al. 2020). The scientific literature focused on the usefulness of invertebrates as bioindicators exceed by orders of magnitude the literature dedicated to their state of conservation (Contador et al. 2012; Ramírez and Gutiérrez-Fonseca 2014). Hence, the term "invisible" catalyzes a breakdown of scientific and conservation paradigms, inviting students and other citizens to join the action of observing, inquiring, and knowing. Finally, through metaphors and *face-to-face* encounters with the Róbalo River and its co-inhabitants, participants recognized them as living *subjects*. When people experience a face-to-face encounter with these tiny organisms, they have an opportunity to live an ethical transformation: they stop defining them as *objects* and begin to appreciate and accept them as other subjects with whom we share habitats and some life habits. In this way, the "utility" of the insect (for example, as an indicator of water quality or "sentinel" of climate change) ceases to be the only value of these organisms, and participants recognize their intrinsic value and dignity.

4.2.3 *Step 3. Design of Field Activities with an Ecological and Ethical Orientation:* Underwater with a Hand Lens

To counteract the phenomenon of the *extinction of experience*, and to share new knowledge and discoveries generated by research on invertebrates in sub-Antarctic freshwater ecosystems, we designed the activity, *Underwater with a Hand Lens.* This activity was designed with school and university students in 2009. The "Omora Workshop on the Environment" at the local school included studies and weekly field activities on the Róbalo and other rivers near Puerto Williams. Afterward, we extended the practice of *Underwater with a Hand Lens* from formal education into scientific tourism. This novel educational and ecotourism activity continues to be practiced today, and it helps visitors discover, appreciate, and ethically value the underwater inhabitants, habits, and habitats. It is organized in the following three phases (Fig. 4.3):

First Phase: *Habitat Observation* Participants observe and identify river habitats and learn to recognize the connections with riparian ecosystems. Through observation, they understand how terrestrial and aquatic ecosystems interact, noting how tree leaves and branches fall into the river and become food for insects and other invertebrates. Familiarization with both micro and macro habitats is essential to understand the role played by the different co-inhabitants that make up the "community of life." For example, "invertebrate inhabitants" live between aquatic and terrestrial habitats, moving from one habitat to another during the different stages of their life cycles. Macro-habitat refers to different ecosystems at a landscape scale (e.g., high Andean, riverine, lacustrine habitats). Micro-habitat refers to smaller-scale structures within an ecosystem (e.g., submerged plants, woody debris, leaf packs, or rocks within a particular portion of a river or lake).

Second Phase: *Face-to-Face Encounters* Using a net, a hand lens, spoons, and/or soft-touch forceps, participants collect a small part of a micro-habitat, recording the exact sampling location. The micro-habitat sample is placed in a container with water, and participants watch as invertebrates, or "underwater co-inhabitants," begin moving in the container. Then, with the help of a magnifying glass, they carefully observe the diversity of morphologies of our submerged co-inhabitants and their life

Fig. 4.3 Underwater with a Hand Lens in the rivers of the Cape Horn Biosphere Reserve, Chile. The activity is performed in three phases. (1) Observing and recognizing macro- and micro-habitats; (2) Face-to-face encounter with co-inhabitants; (3) valuing and respecting co-inhabitants, their life habits, and habitats. Adapted from Contador et al. (2018). Photographs by Gonzalo Arriagada. Design and composition of illustrations by Felipe Portilla and Silvia Lazzarino

habits (e.g., eating behaviors and movements), relating them to the micro- and macro-habitats identified in the previous step.

Third Phase: *Respecting the Habitats, Habits, and Co-Inhabitants of the River* After observing the invertebrates, the participants return each micro-habitat sample to the exact place where they found it. They pay special attention to include all of its original insect co-inhabitants. This process is not trivial. During the field experiences, students from Puerto Williams discovered that the richness and abundance of invertebrates varied according to the type of micro-habitat and that there was a greater diversity and abundance of invertebrates on the rocky bottoms of the rivers. Through this exploration, the students concluded that most invertebrates live in rocky micro-habitats. Consequently, students proposed valuing and respecting their home or "habitats" by returning the rocks to precisely the same place where they were found. This exercise is a concrete action that integrates cognition and emotion, triggering an ethical transformation: the stone ceased to be *just a rock*. The rock now is perceived as the home of other-than-human co-inhabitants, which should be respected and valued. Participants stop seeing the freshwater insects as mere *objects* and begin to appreciate them as *subjects* and co-inhabitants, who unfold their life habits within habitats in which we all co-inhabit.

4.2.4 Step 4. In-situ *Conservation:* Underwater Inhabitants of Cape Horn Interpretative Trail

The Magellanic sub-Antarctic ecoregion is one of the last pristine areas of the planet. Nevertheless, it is under increasing pressures due to climate change, invasive species, and biocultural homogenization. If well directed, tourism can contribute positively to this region's socio-economic well-being and biocultural conservation (Tauro et al. 2021). In this context, we designed, built, and implemented the *Underwater inhabitants of Cape Horn* interpretative trail at Omora Park. This trail provides an *in-situ* conservation platform to share research results by using the metaphor *The River as a Community of Life* and the praxis of the activity *Underwater with a Hand Lens* (Fig. 4.4). An essential part of the narrative along this trail and its interpretive stations signal the appreciation for:

(a) The small and large patterns in the landscape (habitats)
(b) The diverse co-inhabitants of the Róbalo River watershed
(c) The life habits associated with ecological interactions among co-inhabitants and processes that link terrestrial with aquatic ecosystems

Visitors are guided throughout five interpretive stations which illustrate contrasting facets of the habitats of the Róbalo River valley (see Contador et al. 2018). Visitors appreciate how different life stage and habits of insects are associated with different plants and types of freshwater and terrestrial habitats (see Contador and Kennedy 2014).

Fig. 4.4 Underwater co-Inhabitants of Cape Horn Interpretative Trail at Omora Park. Platforms enable respectful observation of the invertebrate co-inhabitants within their river habitats (**a**, **b**). FEP participants and other visitors are guided to have "direct encounters" with invertebrates and to identify them in situ, often using field guides and other support materials (**c**). *Meridiolaris chiloensis,* a mayfly that is one of the most common co-inhabitants in the Róbalo River (**d**). Adapted from Contador et al. (2014b). Photographs by Gonzalo Arriagada

4.3 Concluding Remarks

There is an urgent need to rekindle the excitement and fascination for nature that is inborn within us (Wilson 1984; Simaika and Samways 2010). From 2010 to 2020, the interdisciplinary research and the metaphors described above have been shared with over 3000 visitors participating in various activities at Omora Park. Visitors include students, teachers, tourists, local people, public authorities, and researchers from around the world. We have disseminated the knowledge and concepts through over 200 publications ranging from social media to scientific publications as well as books that provide information for school education, tourism operators, and citizens who wish to learn about freshwater invertebrates in the Cape Horn Biosphere Reserve (Contador et al. 2014b; Contador and Kennedy 2014; Contador et al. 2015b). Interaction with a broad array of participants is essential to achieve biocultural conservation (Rozzi et al. 2006). The protection of the habitat is indispensable for conserving the community of freshwater co-inhabitants and the continuation of formal and non-formal educational activities. Through FEP, students learn that for achieving biocultural conservation scientific research is necessary but not sufficient; it requires continuous long-term transdisciplinary work. To strengthen the conservation of the Róbalo River, we are currently working on a proposal for legally considering it as a subject of biocultural rights, as it has been done for rivers in New Zealand, India, and Colombia (Macpherson and Ospina 2015; O'Donnell 2018).

The praxis of FEP as it applies to freshwater insects in the CHBR has allowed us to reach a wide cross-section of people and increase awareness and value for these under-perceived co-inhabitants. However, the inclusion of insects in formal education remains limited. This is particularly problematic because early childhood experiences are central in fostering pro-active support of environmental causes (Lemelin et al. 2017). The FEP methodology supports a diversity of educational practices that promote discovery, understanding, and valuation of the inherent uniqueness of habitats, life habits, and diversity of co-inhabitants in specific habitats. Direct exposure to natural habitats, its beauty, and its diversity of co-inhabitants, has become an increasingly rare experience in formal education as well as in everyday life (Feinsinger et al. 1997; Leopold 2004; Smith 2004; Louv 2008). For FEP participants, the experience of direct "face-to-face" encounters with co-inhabitants and their habitats has been essential toward obtaining a biocultural understanding (Rozzi et al. 2006). We extended the original concept of "face-to-face" encounters developed by philosopher Emanuel Lévinas within the realm of human ethics (Davy 2007) to include encounters with other-than-human co-inhabitants. These face-to-face encounters have enabled FEP participants to stop understanding biocultural diversity merely as a concept and engage in experiences of co-inhabitation with diverse living beings with their life histories that regularly remain outside the scopes considered in formal education and decision-making processes (Rozzi et al. 2008).

Educational activities that include insects in their purview provide training and scientific thinking platforms from preschool to higher education, society, industry, and policymakers. Direct encounters with these small co-inhabitants, their habitats, and habits have amazed visitors and encouraged an appreciation for aquatic insects, making them "visible" (Contador et al. 2018; Rendoll Cárcamo et al. 2020). Such activities also have provided platforms for training and developing scientific thinking from the earliest stages of education by integrating disciplines into novel approaches that combine a scientific, aesthetic, and ethical appreciation of invertebrates. In this way, we make a valuable contribution from the extreme south of the Americas that has global application for the appreciation and valuation of the small co-inhabitants with whom we share the planet.

Acknowledgment Authors Contador, Ojeda, and Rozzi received basal support from the Cape Horn International Center (CHIC) - ANID PIA/BASAL (PFB210018).

References

Adamo S (2016a) Do insects feel pain? A question at the intersection of animal behaviour, philosophy and robotics. Anim Behav 118:75–79

Adamo S (2016b) Subjective experience in insects: definitions and other difficulties. Animal Sentience 1:15
Andrews PLR (2011) Introduction: laboratory invertebrates: only spineless, or spineless and painless? ILAR J 52:121–125
Basset Y, Lamarre G (2019) Toward a world that values insects. Science 364:1230–1231
Briggs JC (2017) Emergence of a sixth mass extinction? Biol J Linnean Soc 122:243–248
Burrell BD (2017) Comparative biology of pain: What invertebrates can tell us about how nociception works. J Neurophysiol 117:1461–1473
Cardoso P, Erwin T, Borges P, New T (2011) The seven impediments in invertebrate conservation and how to overcome them. Biol Conserv 144:2647–2655
Cardoso P, Barton PS, Birkhofer K et al (2020) Scientists' warning to humanity on insect extinctions. Biol Conserv 242:108426
Clark JA, May RM (2002) Taxonomic bias in conservation research. Science 297:191–192
Contador T, Kennedy J (2014) Underwater Inhabitants of the rivers of Cape Horn. Universidad de Magallanes
Contador T, Kennedy J, Ojeda J et al (2014a) Life cycles of freshwater invertebrates and global climate change in the sub-Antarctic Magellanic ecoregion: long-term ecological research at the Omora .... Bosque
Contador T, Méndez M, Medina Y et al (2014b) Río Róbalo Comunidad de Vida
Contador T, Kennedy JH, Rozzi R, Villarroel JO (2015a) Sharp altitudinal gradients in magellanic sub-antarctic streams: patterns along a fluvial system in the cape horn biosphere reserve (55°S). Polar Biol 38:1853–1866
Contador T, Rosenfeld S, Ojeda J, Kennedy J (2015b) Historia natural de los invertebrados acuáticos del Cabo de Hornos. Universidad de Magallanes
Contador T, Rozzi R, Kennedy J et al (2018) Sumergidos con lupa en los ríos del cabo de hornos: Valoración ética de los ecosistemas dulceacuícolas y sus co-habitantes. Magallania (Punta Arenas) 46:183–206
Contador TA (2011) Benthic macroinvertebrates of temperate, sub-Antarctic streams: the effects of altitudinal zoning and temperature on the phenology of aquatic insects associated to the Róbalo river, Navarino Island (55 S), Chile. University of North Texas. Denton, Texas, USA. Doctorate Dissertation University of North Texas
Contador TA, Kennedy JH, Rozzi R (2012) The conservation status of southern South American aquatic insects in the literature. Biodiv Conserv 21(8):2095–2107
Davy BJ (2007) An other face of ethics in Levinas. Ethics Environ 12:39–65
Di Marco M, Chapman S, Althor G et al (2017) Changing trends and persisting biases in three decades of conservation science. Glob Ecol Conserv 10:32–42
Didham R, Leather S, Basset Y (2019) Ethics in entomology. Antenna 43:124–125
Dudgeon D (2019) Multiple threats imperil freshwater biodiversity in the Anthropocene. Curr Biol 29:R960–R967
Elwood RW (2011) Pain and suffering in invertebrates? ILAR J 52:175–184
Fazey I, Fischer J, Lindenmayer DB (2005) What do conservation biologists publish? Biol Conserv 124:63–73
Feinsinger P, Margutti L, Oviedo RD (1997) School yards and nature trails: ecology education outside the university. Trends Ecol Evol 12:115–120
Fischer B, Larson BMH (2019) Collecting insects to conserve them: a call for ethical caution. Insect Conserv Divers 12:173–182
Foley JA, Defries R, Asner GP et al (2005) Global consequences of land use. Science 309:570–574
Fonseca CR (2009) The silent mass extinction of insect herbivores in biodiversity hotspots. Conserv Biol 23:1507–1515
Fukano Y, Soga M (2021) Why do so many modern people hate insects? The urbanization–disgust hypothesis. Sci Total Environ 777:146229
Giurfa M (2013) Cognition with few neurons: higher-order learning in insects. Trends Neurosci 36: 285–294

Gullan PJ, Cranston PS (2014) The insects: an outline of entomology. John Wiley & Sons
Klaver I (2013) Environment imagination situation. In: Ricardo Rozzi STA, Pickett CP, Armesto JJ, Baird Callicott J (eds) Linking ecology and ethics for a changing world, values, philosophy and actions. Springer, Dordrecht, pp 85–108
Klein C, Barron AB (2016) Insects have the capacity for subjective experience. Anim Sentience 1:1
Larson BMH (2006) The social resonance of competitive and progressive evolutionary metaphors. Bioscience 56:997
Lemelin RH, Dampier J, Harper R et al (2017) Perceptions of insects: a visual analysis. Soc Anim 25:553–572
Leopold AC (2004) Living with the land ethic. Bioscience 54:149–154
Lockwood JA (1988) Not to harm a fly: our ethical obligations to insects. Between Species Online J Study Philos Anim 4:. https://doi.org/10.15368/bts.1988v4n3.10
Louv R (2008) Last child in the woods: saving our children from nature-deficit disorder. Algonquin Books
Macpherson E, Ospina F (2015) The pluralism of river rights in Aotearoa, New Zealand and Colombia. J Water Law 25:283–293
Magee B, Elwood RW (2013) Shock avoidance by discrimination learning in the shore crab (Carcinus maenas) is consistent with a key criterion for pain. J Exp Biol 216:353–358
Mather JA (2011) Philosophical background of attitudes toward and treatment of invertebrates. ILAR J 52:205–212
New TR, Samways MJ (2014) Insect conservation in the southern temperate zones: an overview. Aust Entomol 53:26–31
Nieto C, Ovando XMC, Loyola R et al (2017) The role of macroinvertebrates for conservation of freshwater systems. Ecol Evol 7:5502–5513
Nityananda V (2016) Attention-like processes in insects. Proc Biol Sci 283. https://doi.org/10.1098/rspb.2016.1986
O'Donnell E (2018) Legal rights for rivers: Competition, collaboration and water governance. Routledge
Pawar S (2003) Taxonomic Chauvinism and the methodologically challenged. Bioscience 53:861–864
Perry CJ, Barron AB, Chittka L (2017) The frontiers of insect cognition. Curr Opin Behav Sci 16: 111–118
Ramírez A, Gutiérrez-Fonseca PE (2014) Estudios sobre macroinvertebrados acuáticos en América Latina: avances recientes y direcciones futuras. Revista de Biología Tropical 62:9–20
Regan T (2004) The case for animal rights. University of California Press, Berkley
Rendoll Cárcamo J, Contador T, Gañán M et al (2019) Altitudinal gradients in Magellanic sub-Antarctic lagoons: the effect of elevation on freshwater macroinvertebrate diversity and distribution. PeerJ 7:e7128
Rendoll Cárcamo J, Contador T, Gañán M (2020) Filosofía Ambiental de Campo: Educación e investigación para la valoración ecológica y ética de los insectos dulceacuícolas. Magallania (Punta)
Rendoll-Cárcamo J (2018) Variación estacional y altitudinal en la diversidad, distribución y tolerancias térmicas de macroinvertebrados dulceacuícolas sub-Antárticos: en la búsqueda de bioindicadores térmicos para el cambio climático global. Tesis de Magíster. Punta Arenas, Chile: Departamento de Recursos Naturales ...
Ricciardi A, Rasmussen JB (1999) Extinction rates of North American Freshwater Fauna. Conserv Biol 13:1220–1222
Rozzi R (1999) The reciprocal links between evolutionary-ecological sciences and environmental ethics. Bioscience 49:911–921
Rozzi R (2012) Biocultural ethics: recovering the vital links between the inhabitants, their habits, and habitats. Environ Ethics 34(1):27–50

Rozzi R (2018) Biocultural conservation and biocultural ethics. In: et al. From biocultural homogenization to biocultural conservation. Ecology and ethics, vol 3. Springer, Cham. https://doi.org/10.1007/978-3-319-99513-7_19

Rozzi R (2019) Taxonomic Chauvinism, No More!: antidotes from hume, darwin, and biocultural ethics. Environ Ethics 41:249–282

Rozzi R, Massardo F, Anderson CB et al (2006) Ten principles for biocultural conservation at the southern tip of the Americas: the approach of the Omora Ethnobotanical Park. Ecol Soc 11

Rozzi R, Arango X, Massardo F et al (2008) Field environmental philosophy and biocultural conservation: the Omora Ethnobotanical Park educational program. Environ Ethics 30:325–336

Rozzi R, Pickett ST, Palmer C et al (2013) Linking ecology and ethics for a changing world: values, philosophy, and action. Springer, Dordrecht

Samways MJ, Barton PS, Birkhofer K et al (2020) Solutions for humanity on how to conserve insects. Biol Conserv 242:108427

Sánchez-Bayo F, Wyckhuys KAG (2019) Worldwide decline of the entomofauna: a review of its drivers. Biol Conserv 232:8–27

Sherwin CM (2001) Can invertebrates suffer? Or, how robust is argument-by-analogy? Anim Welf 10:103–118

Simaika JP, Samways MJ (2010) Biophilia as a universal ethic for conserving biodiversity. Conserv Biol 24:903–906

Smith D (2004) Issues and trends in higher education biology fieldwork. J Biol Educ 39:6–10

Soga M, Gaston KJ (2016) Extinction of experience: the loss of human-nature interactions. Front Ecol Environ 14:94–101

Sunstein CR, Nussbaum MC (2004) Animal rights: current debates and new directions. Oxford University Press

Tauro A, Ojeda J, Caviness T et al (2021) Field environmental philosophy: a biocultural ethic approach to education and ecotourism for sustainability. Sustain Sci Pract Policy 13:4526

Watkins AJ, Goldstein DA, Lee LC et al (2010) Lobster attack induces sensitization in the sea hare, Aplysia californica. J Neurosci 30:11028–11031

Wilson EO (1984) Biophilia. Harvard University Press

Chapter 5
Sub-Antarctic High Andean "Gardeners": Cultivating Caring Relationships

Manuela Méndez-Herranz, Guillermo Marini, and Ricardo Rozzi

Abstract This chapter presents a case study of the Field Environmental Philosophy four-step cycle methodology in the sub-Antarctic high Andean zone. First, based on ecological and philosophical research, we problematize the use of the term "Andean Desert" for naming this ecosystem. Second, we propose the metaphor "High Andean sub-Antarctic Gardeners" to transform our mental image from a desert toward a plant world dominated by habits that promote cooperation and care among co-inhabitants in a specific geographical place. Third, we describe a guided activity with an ecological and ethical orientation, "Co-inhabiting as high Andean cushion plants." Through this experience, we understand that, in nature, just as in human society, there is not only competition but also much collaboration and symbiosis to co-inhabit and flourish. Four, we contribute to habilitate areas for *in situ* conservation in order to protect habitats, species, and their interactions. This case illustrates how to put into practice a pedagogical aesthetic experience that considers bodies, emotions, and minds of the participants while learning about ecological relationships and their ethical implications in a pre-reflexive way. This activity also reminds us about our indissoluble union with the world and its ethical implications. The "gardeners" teach us through their life habits the relevance of cooperation for flourishing together in difficult contexts. In that sense, through our life habits, we can modify habitats in such ways that allow (or not) co-inhabitants to flourish.

M. Méndez-Herranz (✉) · G. Marini
Facultad de Educación, Pontificia Universidad Católica de Chile, Cape Horn International Center (CHIC), La Florida, Chile
e-mail: momendez@uc.cl; gmarini@uc.cl

R. Rozzi
Sub-Antarctic Biocultural Conservation Program, Department of Philosophy and Religion and Department of Biological Sciences, University of North Texas, Denton, TX, USA

Cape Horn International Center (CHIC), Universidad de Magallanes, Puerto Williams, Chile
e-mail: Ricardo.Rozzi@unt.edu

R. Rozzi et al. (eds.), *Field Environmental Philosophy*, Ecology and Ethics 5,
https://doi.org/10.1007/978-3-031-23368-5_5

Keywords Aesthetic experience · Collaboration · Ecology · Ethics · Metaphors

5.1 Introduction

High-altitude and high-latitude plants are exposed to harsh environments. Cooperation among plants and other co-inhabitants is essential for flourishing in these environments. Inspired by these ecological interactions of intra- and interspecific collaboration, we present a case study of the Field Environmental Philosophy (FEP) four-step cycle methodology (Tauro et al. 2021). This example illustrates the value of analogical thinking for interrelating the life habits of plants, humans, and other organisms. We examine problematic discrepancies between the biophysical attributes of ecosystems and organisms and the ways scientists, educators, and other citizens name them. Naming organisms and habitats is not trivial, because names have cultural and political implications. Another important contribution of this case study is the design of a novel ethical and ecologically guided activity called "co-inhabiting as High Andean plants." This activity stimulates participants to learn through their bodies, emotions, and minds in order to feel themselves as embodied members of both human and natural communities. This biocultural experience reconnects participants with the natural world.

This research was conducted in the high Andean zone above the tree line (approximately 500 m above sea level) in the mountains of the Omora Ethnobotanical Park. The park is located in the northern sector of Navarino Island (55°S, 67°W) and is part of the Cape Horn Biosphere Reserve (Rozzi et al. 2006). This chapter problematizes common mental images and valuations of the High Andean flora. Often the habitats found above the tree line in the sub-Antarctic ecoregion of Magallanes are referred to as "high Andean desert." As we will discuss below, this term is problematic because it simplifies the ecological and ethical comprehension of the region. As an alternative, we propose the metaphor of "sub-Antarctic High Andean Gardeners" as a way of naming and understanding these habitats, valuing their biophysical and biocultural complexity. Additionally, we designed the field experience "Co-inhabiting as High Andean Plants" to help learn about the vital relevance of grouping together to co-inhabit cold and windy environments. By grouping together plants form true "islands of life and care" in these conditions of adversity. The final section explores pedagogical implications, emphasizing the contribution of designing experiences from a biocultural ethics perspective that involves not only reason but also bodies and emotions.

5.1.1 Field Environmental Philosophy (FEP)

Our research on the high Andean flora encompassed biophysical, symbolic-linguistic, and institutional dimensions. This multi-dimensionality is necessary to understand biocultural conservation in general and the conservation and valuation of high mountain flora and ecosystems in particular. The three dimensions are part of the conceptual framework of biocultural ethics and its methodology associated with FEP (Rozzi 2013). In this section, we concisely present the work developed in each of the four steps of the FEP.

5.1.1.1 Step 1: Biophysical and Symbolic-Linguistic Attributes of High Andean Flora

FEP's first step integrates ecological and philosophical research (Rozzi et al. 2010). The ecological study focused on positive interactions generated by cushion plants, which have a unique life habit that aggregates hundreds or thousands of conspecific individuals forming compact mats (Figs. 5.1 and 5.2). Cushion plants are particularly common in high altitude and high latitude habitats where they facilitate the growth of other plant species and organisms. Facilitation is an interaction most prevalent in stressful environments (Bertness and Callaway 1994), where living next to neighbors enhances the survival, growth, or reproduction of a species (Callaway 2007). In these habitats, neighbors help meet basic needs such as food or shelter (Crotty and Bertness 2015). In the high Andean habitats of southern Chile and Argentina, previous research suggested that cushion plants act as "nursery plants" or facilitators since numerous species grow in association with them. These nursery plants generate microhabitat conditions that are more favorable than

Fig. 5.1 High Andean habitat where we studied the abundance and diversity of plants on two types of microhabitats: (i) cushion plants, and (ii) stony floor separated by at least 5 meters. In the background of the pictures are the Dientes de Navarino Mountains in the highlands of Omora Park, Cape Horn Biosphere Reserve, Chile. Photograph by Manuela Méndez

Fig. 5.2 Close-up of a *Bolax gummifera* cushion plant. Notice the contrast between the diversity of plants growing on the cushion plant and the lack of plants on the exposed stony surface. Photograph by Manuela Méndez

those that are exposed to a severe environment (Nuñez et al. 1999; Cavieres et al. 2002; Arroyo et al. 2003; Brancaleoni et al. 2003).

The ecological research of this study is based on the master's thesis by Manuela Méndez (2013), who investigated the spatial associations of vascular plants and mosses with cushion plants of *Bolax gummifera* in high Andean habitats in the Cape Horn Biosphere Reserve. On slopes of polar and equatorial aspect with contrasting conditions of solar radiation, temperature, snow cover, and wind exposure, we compared the abundance and diversity of plant species growing on and around cushion plants (Fig. 5.1). In total, 52 plant species were identified. A first finding was that the majority (34 species) corresponded to mosses (non-vascular plants) and only one-third (18 species) to vascular plants (Méndez et al. 2013, 2018). The higher diversity of non-vascular plants is remarkable because it contrasts with the flora of the rest of the world where vascular plants are ten times more diverse (Rozzi et al. 2008). A second finding was that most species grow on cushion plants (Fig. 5.2).

Cushion-forming plants were very abundant and belonged mainly to three species: *Bolax gummifera*, *Azorella selago,* and *A. lycopodiodes* (Méndez et al. 2013, 2018). As mentioned above, this growth form plays a key ecological role in severe high mountain environments (Körner 2003). They are particularly abundant in high Andean ecosystems throughout South America, where they generate microhabitat conditions (Arroyo et al. 2003, Cavieres et al. 2002) that facilitate the growth of other plants that only grow on these "vegetation islands" (Fig. 5.2). Recent observations by Omora Ethnobotanical Park ornithologist Omar Barroso suggest that

Fig. 5.3 High Andean bird species associated with cushion plants in the Cape Horn Biosphere Reserve, Chile. **a**. Adult and juvenile of Yellow-bridled Finch (*Melanodera xanthogramma*) on cushion plants and herbs of *Nassauvia latissima*. **b**. White-bellied Seedsnipe (*Attagis malouinus)* among grasses on cushion plants and grasses; in the background is the Beagle Channel and mountains of Tierra del Fuego. **c**. A family of Rufous-chested Dotterel (*Charadrius modestus*); an adult sheltering chick beside a cushion plant. Photographs Omar Barroso, Omora Park Archive

these "islands" may also provide critical habitat for nesting and survival of some high Andean birds, including endemic species (Fig. 5.3).

Our philosophical research focused on a hermeneutical analysis of botanical texts of the Magellanic sub-Antarctic ecoregion. Emphasis was on the terminology used to name the habitats of the high Andean zone (Moore 1983, Pisano 1980). Additionally, three philosophical texts were analyzed to generate a conceptual framework for environmental ethics: (i) Aldo Leopold's "The Land Ethic" (1949), which presents the image of the biotic community; (ii) Lynn White's "The Historical Roots of Our Ecological Crisis" (1967), which argues that how we think about nature decisively influences how we relate to nature; and (iii) Ricardo Rozzi's "Biocultural ethics" (2012), which emphasizes the value of the vital links among co-inhabitants and their life habits in a shared habitat.

Regarding botanical texts, we analyzed the work of the Magellanic explorer and botanist Edmundo Pisano (1980), an expert on the regional mountain flora. Pisano had already noticed the spatial pattern detected in our ecological study (Méndez et al. 2013). Pisano (1980, p. 218) noted that: "Only on rare occasions, plants form a continuous cover, but they are often grouped together, especially where cushion plants of *Azorella selago* and *Bolax gummifera* are present." However, when naming this environment, Pisano uses a term that invisibilizes the biodiversity that he recognized: "The summits of the mountains, as well as all altitudes above 500 m [above the tree line], are characterized by the most typical expression of the *Andean Desert*" (Pisano 1980, p. 218; italics added). The contrast between the symbolic-linguistic meaning of the word "desert" and the biophysical reality recorded by science in this high-altitude region motivated us to research the discrepancy between language and ecological reality.

First, the immediate meaning of desert is that of an "unpopulated, solitary, or uninhabited" space (RAE 2021). This meaning predominates in colloquial language where a desert is usually associated with a place where "there is nothing or nothing happens." Interestingly, such an assessment entails a tacit comprehension of "nothingness," "emptiness," or the expectation of some type of presence that is unmet. In this sense, although a hasty traveler might judge the high Andean habitats in this way, a more observant traveler (as FEP stimulates) will notice the ecological richness of the area, and will hardly find it to be an empty, lonely, or uninteresting environment.

Second, the etymological meaning of desert offers a subtler contrasting perspective. Long before referring to an uninhabited space, desert evoked the action of "cutting ties" or "abandoning." This was the expression used by Caesar and Cicero to describe soldiers who deserted from the army (thereby cutting off all relationships with their people), spouses who separate, or parents who abandon their children (Lewis and Short 1879). From this point of view, the high Andean habitat does not correspond to a desert because it exhibits fertile networks of relationships among different species that have been associated with sheltering cushion plants.

Third, in the Judeo-Christian tradition, the desert has a symbolic-spiritual character as the place from which life can be reoriented (Lewis and Short 1879). For the linguists Charlton Lewis and Charles Short (1879), the fundamental example of this understanding of desert is the encounter between Yahweh and Moses in the Sinai desert (Numbers 1:1). Ever since, the desert has provided monks and prophets with a habitat that is free from worldly concerns and that inspires contemplation and prayer (Bratton 1988). Thus, deserts are places for retreat and contemplation of the meaning of life. From this viewpoint, visiting and studying the high Andean desert would facilitate the transit from the inhospitable to the fertile, from severed relationships to reconnections, from an impression of absences and silences to the confirmation of concrete and supportive presences.

These three meanings of desert encourage us to integrate philosophical and ecological research and to apply it to the next FEP's steps: composition of metaphors, and field activities. These help us to interpret and experience the high Andean habitat and its vegetation islands. Before doing this, however, we want to point out the importance of the links among *co-in-Habitants* and their life *Habits* and shared *Habitats* (Rozzi 2012). Under this "3Hs" conceptual framework of the biocultural ethic, "the cushion *life habit* generates a type of *habitat* that facilitates life in the high mountains for plant and animal *co-inhabitants*. Therefore, it is necessary to conserve each of the elements" (Rozzi et al. 2014). The life habit of cushion plants is characterized by cooperation between individuals of the same species and other species (e.g., co-inhabitants) in harsh environments, which has implications for an environmental ethic (Méndez-Herranz and Rozzi 2017).

5.1.1.2 Step 2: Sub-Antarctic High Andean Gardeners

FEP's second step involves the poetic practice of composing metaphors to change the mental image of the "Andean desert" by communicating the values and results obtained in FEP's Step 1. The term "Andean desert" contradicts what Pisano recorded on the habitats of the southern mountains, and what Manuela Méndez and collaborators found about the life habits of high mountain plants. This "Andean desert" is not devoid of life but is inhabited by numerous cushion plants that generate microhabitat conditions hosting numerous species of plants and even birds. In this harsh environment, cooperative intra- and interspecific interactions are relevant to plant growth habits. Cushion plants surprise us as islands of vegetation or real gardens with a variety of shapes, colors, and textures in a prevailing stony environment.

Metaphorically, we could say that cushion plants "care" for their neighbors by providing nutrients—trapped in their leaf litter—and a stable substrate—thanks to their strong roots that resist the harsh winds that lash the Andean peaks. These ecological attributes of cushion plants inspired our metaphor "jardineras," a Spanish word that means both "container gardens" and "gardeners." The first evokes a special garden or habitat that brings together diverse life forms such as flowering plants and insects in a confined area such as a pot. From this garden aromas, colors and other sensations are emitted. Additionally, container gardens are usually maintained carefully. The second meaning, "gardener," refers to the person who actively tends a garden, caring for the needs of those diverse lives. Interpreted from the 3Hs model of biocultural ethics, the metaphor "jardinera" refers both to the habitat and to the habit of care. In this sense, people are invited to assume caring relationships with plants and other beings, including humans.

We added the adjective "sub-Antarctic" to the metaphor to denote the geographical location of these cushion plants. They grow in the Magellanic sub-Antarctic ecoregion, a unique but still little-known area of the world (Rozzi et al. 2006, 2012).

For this reason, it is relevant to distinguish it from the general term "subpolar," which is often used for high latitudes. Sub-Antarctic is more specific than subpolar and it helps to mark the differences between the largely continental subpolar areas in the northern hemisphere and the largely maritime areas in the southern hemisphere (Arroyo et al. 1995; Mark et al. 2001).

We also added a second adjective to the metaphor: "high Andean." This adjective corrects the Eurocentric projection of the more commonly used term "alpine" (Körner 2003). The use of the term alpine for the high mountain floras of mountain ranges such as the Himalayas in Asia, the Rocky Mountains in North America, and the Andes in South America is problematic because it projects a European cultural reality onto other regions. "Alpine" invisibilizes the uniqueness of these other mountain ranges by subordinating them under this colonial heritage term. Therefore, "high Andean" resituates in language and culture those places that are above the tree line, highlighting the uniqueness of the high altitude habitat of the mountain range that runs throughout South America: the Andes.

In summary, the metaphor "High Andean sub-Antarctic Gardeners" transforms our mental image from a desert toward a plant world dominated by habits that promote cooperation and care among co-inhabitants. This metaphor refers explicitly to the southernmost mountain peaks of the Andean Cordillera.

5.1.1.3 Step 3: "Co-Inhabiting as High Andean Cushion Plants"

FEP's third step involves the design of guided activities with an ecological and ethical orientation. In this research, we incorporated corporal and emotional dimensions to contribute experientially to an integral understanding of the cognitive transformation proposed by the metaphor "High Andean sub-Antarctic Gardeners." Step 1 challenged and enriched the understanding of Andean desert, and Step 2 coined a logical synthesis between habitats–habits–co-inhabitants. Step 3 addresses the experiential component. In other words, to paraphrase the philosopher Maurice Merleau-Ponty (1964), it is necessary to question ourselves symbolically, metaphorically, perceptually, and ethically until we can recognize that what we had called desert is really the web of life with which we have not yet connected. The aim of designing our ecologically and ethically guided field activity is to offer an invitation to connect ourselves with the high Andean world.

To address the challenges posed, in this step we promoted "direct encounters" with the co-inhabitants of the sub-Antarctic mountains. Inspired by the question "How do cushion plants and other species manage to live in this harsh environment?" we designed an activity that we called "Co-inhabiting as high Andean plants." This activity has been carried out with people from the local community of Navarino Island, researchers, and students from different universities in Chile and the world. Participants represent disciplines often far removed from ecology and philosophy, such as economics and business administration.

The activity begins with a leisurely walk of approximately 2 hours, along an accessible and well-defined trail that crosses different types of forests until reaching

Fig. 5.4 Group hiking the trail that goes from the Beagle Channel to the summit with high Andean habitats. A. Along the trail participants cross through different forest types, such as the evergreen beech and the high-deciduous beech. Photograph by Manuela Méndez

the high Andean plateau (Fig. 5.4). During this hike, we observe how the landscape and the species that accompany us change as we go up the slopes. We observe how the trees are decreasing in size, until we finally cross the tree line and arrive at the high Andean open habitats. The experience in the high Andean zone begins with an invitation to slow down and observe with all our being. We suggest that participants contemplate the vast view from the summit, feel the climate on their skins, listen, and smell. We invite them to get close to the ground, feel the rocks, and see from other perspectives. We encourage them to observe from ground level and pay attention to the biodiversity and its ways of co-inhabiting.

After this initial experience, we perform a series of slow movements that guide the observation. First, we invite participants to stand and extend their arms exposing as much as possible their body surfaces, "as if they were a tall tree," and to pay attention to the bodily sensations this generates. Next, we invite them to lie down on the ground, "as if they were a cushion plant." We ask them about their different physical sensations (e.g., in relation to the wind or the thermal sensation of their bodies) when standing and when laying in contact with the rocks (Fig. 5.5). Next, we invite them to stand again, but this time, closely grouped in contact with their companions. We ask them to describe their sensations of inhabiting individually as compared to co-inhabiting closely together with their companions.

Fig. 5.5 Group of participants (**a**) standing apart from each other with their arms extended, and (**b**) laying on ground. Photographs by Manuela Méndez

Finally, we invite them to remain together and crouch at ground level. Through this exercise we learn with our bodies, emotions, and mind in an analogical way to experientially understand the life habits of high Andean cushion plants that are formed by numerous individuals growing together (Fig. 5.6). We also invite them

Fig. 5.6 Group of participants crouching together at ground level. Photograph Cristián Valle

to reflect on how these cushion plants shelter a diversity of life forms. "From this experience, we understand that, in nature, just as in human society, there is not only competition but also much collaboration and symbiosis to co-inhabit and flourish, as oases in deserts, even in the harsh environments of the sub-Antarctic high Andean summits" (Méndez-Herranz and Rozzi 2017, pp. 27–28).

5.1.1.4 Step 4. *In situ* Conservation

FEP's fourth step is related to actions that promote the conservation and/or restoration of biocultural diversity, such as the design and implementation of interpretive trails or areas for the protection of habitats, their co-inhabitants, and life habits. This step commits participants to assume responsibility, as citizens and as co-inhabitants, to care for the diversity of life and the integrity of ecosystems (Tauro et al. 2021).

Chile is characterized by a beautiful and wild mountain range in most of its extension, especially at its southern extreme. Numerous people from all over the world travel here to camp, climb, and trek. However, visitors often only contemplate the grand landscapes and open spaces, omitting the opportunity to slow down and perceive the marvelous and diverse communities of small co-inhabitants, which inhabit the "previously called high Andean desert." Even less frequently, they observe the ways in which these co-inhabitants interact positively in these harsh environments.

The great heterogeneity of microhabitats (generated by the variety of substrates, slopes, exposures, and altitudinal gradients) explains the high species diversity found in the Chilean mountainous regions (Arroyo et al. 1981; Rozzi et al. 1997; Körner 2021). Each time we ascend mountain peaks temperature decreases, snow and wind often increase, the height of trees decreases, and plant species grow more and more closely to ground level. To truly learn about high mountain biodiversity and its specialized life strategies, we must get close to the ground, stop, and take in our surroundings.

In order for mountain biodiversity to continue to exist, it is not enough to change the mental image from "Andean desert" to "high Andean sub-Antarctic gardeners," nor is it enough to implement "direct encounters" or ecologically and ethically guided field experiences. It is also necessary to habilitate areas for *in situ* conservation in order to protect habitats, species, and their interactions. With this intention, we worked at Omora Park, which is crossed by the southernmost section of the Chile Trail ("Sendero de Chile"). The Chile Trail is an environmental education and ecotourism project created in 2001 by the National Environmental Commission (now the Ministry of the Environment). The purpose of the trail is to link the Andean Cordillera from north to south so that it can be traversed by foot, by horseback, or by non-motorized vehicles (Fundación Sendero de Chile 2021).

In Chile, tourist visitation in the mountains has rapidly increased in recent decades, generating opportunities but also threats, such as the increase of exotic plants growing in association with the hiking trails (Liedtke et al. 2020). The conservation of these habitats is necessary in order to allow cushion plants and all native flora with their diverse life forms to continue to flourish. Working in the field to conserve these habitats and flora teaches us to interact collaboratively with our co-inhabitants in harsh environments.

5.2 Cultivating Caring Relationships

From the composition of the metaphor "High Andean Sub-Antarctic Gardeners" and the field experience "Cohabiting as High Andean Cushion Plants," we derive valuable lessons for co-inhabiting in "deserts" or harsh environments; and, hence, to cultivate caring relationships with our diverse neighbors. The "gardeners" resist and transform the deserts of misunderstanding, abandonment, and loneliness. They teach us through their life habits the relevance of cooperation for flourishing together when the context is difficult. Paraphrasing Merleau-Ponty (1964) again, they also remind us about our indissoluble union with the world. Through our life habits, we can modify habitats in such ways that allow (or not) co-inhabitants to flourish.

In 1967, Lynn White wrote a provocative essay asserting that the ecological crisis we are currently experiencing derives from problematic ideas about the human–nature relationship, particularly those historically promoted by Baconian science and Western Christianity. Both emphasize the transcendence and domination of humanity over nature. This view on Christianity was quickly criticized and shown to be more diverse and complex than White's argument (Whitney 1993, 2013, 2015; Warde 2011).

Decisively, in 2015 Pope Francis addressed this problem through the publication of the Encyclical Letter *Laudato Si'. On Care for Our Common Home*. In a passage in which he quotes from a homily by Benedict XVI, Francis (2015, p. 141) argues that if ""the external deserts in the world are growing, because the internal deserts have become so vast'... the ecological crisis is also a summons to profound interior conversion." He recognizes that the current ecological crisis requires the ecological conversion of all humanity. This ecological conversion, he explains, "also entails a loving awareness that we are not disconnected from the rest of creatures, but joined in a splendid universal communion" (Francis 2015, p. 143). These ideas go to the deep root of today's crisis.

Along this line, education is often posited as the solution that will transform our ideas about the human–nature relationship or bring about "ecological conversion." However, we frequently lack methodologies or practical activities to help us experience what these mean in our embodied lives. We can live in deserts of individualism, selfishness, excessive greed, or we can cultivate gardens creating cooperative ties with those we live with. The metaphor "High Andean Sub-Antarctic Gardeners" and the activity "Co-inhabiting as High Andean Cushion Plants" help us to stop, to pay attention to the environment, to perceive and to feel the shared habitat integrally by putting ourselves in the "skin" of the cushion plants. Feeling like the plant, we (re-)encounter it physically and emotionally. Moreover, from this new perspective, we begin to (re-)discover many co-inhabitants whom we often overlooked. We begin to feel the habits that generate communities in the "deserts."

This case illustrates how to put into practice a pedagogical experience that considers the bodies, emotions, and minds of the participants. This is relevant for working on deep aspects related to ways of comprehensively understanding our values and relationships with the world. However, it is worth noticing that the "paradox of perception" pervades formal education: although perception is always present in every learning process, it tends to go unnoticed (Marini 2021). This fact can be explained because we have not been educated in the practice of stopping and paying attention to what we are perceiving, and because of this, we fail to let ourselves be surprised by the way our perception shapes the senses and purposes of education (Marini 2018). For example, formal school curriculum still retains a strong preference for literacy-numeracy, over other ways of understanding and organizing experience such as aesthetic sensibility. It is as if the awareness of our own perceptions, in dialogue with our emotions and symbolic projections, were less relevant than the skills of naming and operating mathematically and textually in the world. Because of this, the case study based on the "High Andean Sub-Antarctic Gardeners" provides an important contribution to formal education.

In this sense, the ecologically and ethically guided activity, "co-inhabiting as high Andean cushion plants" can be understood as an aesthetic experience. Aesthetic experiences connect how the environment is felt through the senses and how what is perceived is felt affectively, thus uniting cognitive and emotional aspects. Aesthetic experiences have also been characterized as pre-reflexive and pre-language ways of knowing that involve the body and movement. The aesthetic reclaims the indivisibility of mind, body, and world (Merleau-Ponty 1962). The richness of this exercise is that, by involving the body in a pre-reflexive encounter with the high Andean habitat and its co-inhabitants, feelings of empathy, recognition, and surprise are awakened and move us emotionally. Even though perception is an important but scarcely addressed topic of research in environmental education (Ghisloti-Iared et al. 2016, 2017), it has much to contribute to education that promotes sensitivity to co-inhabitants by cultivating the common ties that bind human beings to all other beings.

What is more, aesthetics, specifically everyday aesthetics, which focus on the habitual activities of our daily lives such as cooking and personal hygiene, emphasizes that educating our perceptions is a moral necessity (Saito 2010, 2017, 2018), since many of our decisions are oriented—or determined—by aspects such as beauty, taste, or comfort. Therefore, promoting "care for our common home," as Pope Francis urges, implies revisiting, without inhibitions, the ecological impact of such aesthetic preferences. For example, when visiting high mountains we may feel more comfort or softness when walking on cushion plants or other vegetation rather than on rocks. However, we must question how this aesthetic preference affects the lives of these co-inhabitants and the state of the habitat.

Finally, high Andean sub-Antarctic gardeners and everyday aesthetics lead us to ask how education contributes (or not) to turning our experiences into deserts. For example, how does formal education promote bonds among human co-inhabitants and between human and other-than-human co-inhabitants? What sensory stimuli does my day-to-day life provide and what does it teach me about how to relate to other co-inhabitants of our common home? The sub-Antarctic High Andean Gardeners can inspire us to cultivate a truly diverse garden in the midst of harsh environments thus making us gardeners in the service of caring for the diversity of life.

Acknowledgments Authors Marini and Rozzi receive basal support from the Cape Horn International Center (CHIC) - ANID PIA/BASAL (PFB210018).

References

Arroyo MTK, Armesto JJ, Villagrán C (1981) Plant phenological patterns in the high Andean Cordillera of Central Chile. J. Ecol. 69:205–223

Arroyo MTK, Cavieres L, Marticorena C, Muñoz M (1995) Convergence in the mediterranean floras of central Chile and California: insights from comparative biogeography. In: Arroyo MTK, Fox M, Zedler P (eds) Ecology and biogeography of Mediterranean ecosystems in Chile, California and Australia. SpringerVerlag, New York, pp 43–88

Arroyo MTK, Cavieres LA, Peñaloza A, Arroyo-Kalin MA (2003) Positive associations between the cushion plant Azorella monantha (Apiaceae) and alpine plant species in the Chilean Patagonian Andes. Plant Ecol 169:121–129

Bertness MD, Callaway R (1994) Positive interactions in communities. Trends Ecol Evolut 9(5): 191–193

Brancaleoni L, Strelin J, Gerdol R (2003) Relationships between geomorphology and vegetation patterns in subantarctic Andean tundra of Tierra del Fuego. Polar Biol 26:404–410
Bratton SP (1988) The original desert solitaire: early Christian monasticism and wilderness. Environ Ethics 10(2):31–53
Callaway RM (2007) Positive interactions and interdependence in plant communities. Springer, Berlin
Cavieres L, Arroyo MTK, Peñaloza A, Molina-Montenegro M, Torres C (2002) Nurse effect of Bolax gummifera cushion plants in the alpine vegetation of the Chilean Patagonian Andes. J Veg Sci 13:547–554
Crotty S, Bertness M (2015) Positive interactions expand habitat use and the realized niches of sypatric species. Ecology 96(10):2575–2582. https://doi.org/10.1890/15-0240.1
Francis (2015) Laudato Si'. On Care for Our Common Home. Encyclical Letter. Our Sunday Visitor Publising Division, Huntington, Indiana
Fundacion Sendero de Chile (2021) Nuestra historia. Available via: https://www.fundacionsenderodechile.org/historia
Ghisloti-Iared V, Torres de Oliveira H, Payne PG (2016) The aesthetic experience of nature and hermeneutic phenomenology. J Environ Educ 47(3):191–201
Ghisloti-Iared V, Torres de Oliveira H, Reid A (2017) Aesthetic experiences in the Cerrado (Brazilian savanna): contributions to environmental education practice and research. Environ Educ Res 23(9):1273–1290
Körner C (2003) Alpine plant life, 2nd edn. Springer, Berlin
Körner C (2021) Alpine plant life: functional plant ecology of high mountain ecosystems. Springer, Switzerland
Leopold A (1949) The land ethic. A sand county almanac and sketches here and there. Oxford University Press, New York
Lewis C, Short C (1879) A latin dictionary. Clarendon Press, Oxford
Liedtke R et al (2020) Hiking trails as conduits for the spread of non-native species in mountain areas. Biol Invasions 22(3):1121–1134
Marini G (2018) Three theses on the relationships between aesthetics and education: reintervening school experience through perception. In: Lecciones de Clase (ed) Francisco Schwember. Ediciones UC, Santiago de Chile, pp 74–97
Marini G (2021) An introduction to everyday aesthetics in education. Studies in Philosophy and Education. https://doi.org/10.1007/s11217-020-09740-x
Mark AF, Dickinson KJM, Allen J, Smith R, West CJ (2001) Vegetation patterns, plant distribution and life forms across the alpine zone in southern Tierra del Fuego Argentina. Austral Ecol 26(4): 423–440
Méndez M, Rozzi R, Cavieres L (2013) Flora vascular y musgos en la zona altoandina de la Isla Navarino (55°S) Reserva de la Biosfera Cabo de Hornos Chile. Gayana Botánica 70(2):338–344. https://doi.org/10.4067/S0717-66432013000200011
Méndez MO, Cavieres L, Rozzi R (2018) Jardineras subantárticas: Conocimiento y valoración de la flora altoandina. Magallania (Punta Arenas) 46(1):125–135. https://doi.org/10.4067/S0718-22442018000100125
Méndez-Herranz M, Rozzi R (2017) Filosofía Ambiental de Campo. Jardineras subantárticas altoandinas en el Parque Etnobotánico Omora. Revista del Jardín Botánico Chagual XV(15):23–29
Merleau-Ponty M (1962) The phenomenology of perception. Routledge, London
Merleau-Ponty M (1964) El Ojo y el ESpíritu. Éditions Gallimard, España
Moore DM (1983) Flora de Tierra del Fuego. Oswestry, UK & Missouri Botanical Garden, St. Louis, USA
Nuñez CI, Aizen MA, Ezcurra C (1999) Species associations and nurse plant effects in patches of high-Andean vegetation. J Vegetation Sci 10:357–364
Pisano E (1980) Distribución y características de la vegetación del archipiélago del Cabo de Hornos. Anales del Instituto de la Patagonia 11:191–224

Real Academia Española [RAE] (2021). https://dle.rae.es/desierto. Retrieved 4/19/2021
Rozzi R (2012) Biocultural ethics: the vital links between the inhabitants, their habits and regional habitats. Environ Ethics 3:27–50
Rozzi R, Kalin MT, Armesto JJ (1997) Factors affecting gene flow between populations of Anarthrophyllum cumingii (Papilionaceae) growing on equatorial and polar facing slopes in the Andes of Central Chile. Plant Ecol 132(2):171–179
Rozzi R, Massardo F, Anderson C, Heidinger K, Silander JA Jr (2006) Ten principles for biocultural conservation at the southern tip of the Americas: the approach of the Omora Ethnobotanical Park. Ecol Soc 11(1):43. https://doi.org/10.5751/ES-01709-110143
Rozzi R et al (2008) Changing lenses to assess biodiversity: patterns of species richness in sub-Antarctic plants and implications for global conservation. Front Ecol Environ 6(3):131–137
Rozzi R, Anderson C, Pizarro JC, Massardo F, Medina Y, Mansilla AO, Kennedy JH et al (2010) Filosofía ambiental de campo y conservación biocultural en el Parque Etnobotánico Omora: Aproximaciones metodológicas para ampliar los modos de integrar el componente social ("S") en Sitios de Estudios Socio-Ecológicos a Largo Plazo (SESELP). Revista Chilena de Historia Natural 83:27–68
Rozzi R, Armesto JJ, Gutiérrez J, Massardo F, Likens G, Anderson C, Poole A et al (2012) Integrating ecology and environmental ethics: earth stewardship in the southern end of the Americas. BioScience 62(3):226–236
Rozzi R, Massardo F, Contador T, Crego R, Méndez M, Rijal R, Cavieres L, Jiménez JE (2014) Filosofía ambiental de campo: ecología y ética en las redes LTER-Chile e ILTER. Bosque 35(3): 439–447
Saito Y (2010) Everyday aesthetics. Oxford University Press, New York
Saito Y (2017) Aesthetics of the familiar: everyday life and world-making. Oxford University Press, New York
Saito Y (2018) Consumer aesthetics and environmental ethics: problems and possibilities. J Aesthetics Art Crit 76(4):429–439
Tauro A et al (2021) Field environmental philosophy: a biocultural ethic approach to education and ecotourism for sustainability. Sustainability 13:4526
Warde E (2011) Christianity and the environment: the Lynn White controversy. Ex Post Facto XX:45–62
White L (1967) The historical roots of our ecological crisis. Science 155:1203–1207
Whitney E (1993) Lynn White, ecotheology, and history. Environ Ethics 15(2):151–169
Whitney E (2013) The Lynn White thesis: reception and legacy. Environ Ethics 35(3):313–331
Whitney E (2015) Lynn White Jr.'s 'The Historical Roots of Our Ecologic Crisis' after 50 years. Hist Compass 13(8):396–410

Chapter 6
"Pay Attention, Dive with Eyes Wide Open": A Field Environmental Philosophy Activity to Foster Reciprocity Between People and Nature

Jaime Ojeda, Alejandra Tauro, Tamara Contador, Julia González-Calderón, Javiera Malebran, Sebastián Rosenfeld, Francisca Massardo, Andrés Mansilla, and Ricardo Rozzi

Abstract This chapter presents a case study of a diving experience co-designed with the participation of Yaghan fishers, scientists, and philosophers. It is an adaptation of the field environmental philosophy (FEP) methodology, which is based on the conceptual framework of the biocultural ethic. By experiencing interrelationships with macroalgae and various invertebrates, the activity integrates the ecological knowledge of the different participants to forge a "biocultural lens" that guides the participants to understand the concept of co-inhabitant which is basic to

J. Ojeda (✉)
School of Environmental Studies, University of Victoria, Victoria, BC, Canada

Cape Horn International Center (CHIC), Universidad de Magallanes, Punta Arenas, Chile
e-mail: jaime.ojeda@umag.cl

A. Tauro
Cape Horn International Center (CHIC), Universidad de Magallanes, Puerto Williams, Chile

El Colegio de Puebla A.C., Puebla, Mexico
e-mail: alejandra.tauro@colpue.edu.mx

T. Contador
Sub-Antarctic Biocultural Conservation Program, Universidad de Magallanes, Punta Arenas, Chile

Wankara Laboratory, Universidad de Magallanes, Puerto Williams, Chile

Núcleo Milenio de Salmónidos Invasores, Concepción, Chile

Cape Horn International Center (CHIC), Universidad de Magallanes, Puerto Williams, Chile

Institute of Biodiversity of Antarctic and Subantarctic Ecosystems (BASE), Santiago, Chile
e-mail: tamara.contador@umag.cl

J. González-Calderón
Artesana de la Comunidad Indígena Yagán, Puerto Williams, Chile

J. Malebran · F. Massardo
Cape Horn International Center (CHIC), Universidad de Magallanes, Puerto Williams, Chile

R. Rozzi et al. (eds.), *Field Environmental Philosophy*, Ecology and Ethics 5,
https://doi.org/10.1007/978-3-031-23368-5_6

biocultural ethics. Through “face-to-face” encounters with the *otherness* of algae, mollusks, and other submerged invertebrates in the kelp forests of Cape Horn, this FEP activity helps participants to experience ancestral and novel ways of seeing, feeling, and co-inhabiting with marine biodiversity.

Keywords Biocultural conservation · Education under water · Ethics · Traditional ecological knowledge · Reciprocity

6.1 Introduction

Water affords multiple possibilities for recreation, enjoyment, productive activities such as fishing, or education about marine biodiversity and biocultural diversity. Submerging oneself and diving in water is always an exciting experience. Swimming alongside large animals such as dolphins, whales, sharks, or rays (Lawrence et al. 2016) or diving among coral reefs and enjoying other underwater excursions (Takata Experience 2022; Moreno-Terrazas et al. 2022) are impactful, especially for tourists. Many of these diving and other marine experiences are led by tourism organizations, explorer groups, and conservationists, together with school or university educational programs. These initiatives aim to generate an experience in the aquatic habitat and with its co-inhabitants, thus motivating curiosity about biodiversity and marine conservation, and self-reflection (Lucrezi et al. 2013). For others, immersing oneself in water can be a source of inspiration that leads to the creation of artistic projects. Practicing these underwater experiences sharpens observation skills, enhances the imagination, and provides the experience of co-inhabitation in a habitat

S. Rosenfeld
Laboratorio de Ecosistemas Marinos Antárticos y Subantárticos (LEMAS), Universidad de Magallanes, Punta Arenas, Chile

Cape Horn International Center (CHIC), Universidad de Magallanes, Puerto Williams, Chile

Centro de Investigación Gaia-Antártica, Universidad de Magallanes, Punta Arenas, Chile

Instituto Milenio Biodiversidad de Ecosistemas Antárticos y Subantárticos (BASE), Universidad de Chile, Santiago, Chile
e-mail: sebastian.rosenfeld@umag.cl

A. Mansilla
Laboratorio de Ecosistemas Marinos Antárticos y Subantárticos (LEMAS), Universidad de Magallanes, Punta Arenas, Chile

Cape Horn International Center (CHIC), Universidad de Magallanes, Puerto Williams, Chile
e-mail: andres.mansilla@umag.cl

R. Rozzi
Sub-Antarctic Biocultural Conservation Program, Department of Philosophy and Religion and Department of Biological Sciences, University of North Texas, Denton, TX, USA

Cape Horn International Center (CHIC), Universidad de Magallanes, Puerto Williams, Chile
e-mail: Ricardo.Rozzi@unt.edu

that as human beings we should respect and observe. In this sense, within the framework of the Field Environmental Philosophy (FEP) methodological approach, we have developed an underwater educational experience called *Ojo, Bucea con Ojo*, or in English—Pay Attention, Dive with Eyes Wide Open. Its purpose is to connect participants with the unknown and usually unvisited marine environment and therefore literally open their eyes to a whole new world.

The Field Environmental Philosophy (FEP) methodology implemented for more than 20 years in Omora Park in Cape Horn, Chile and other places in Chile and the world, consists of four steps (Rozzi et al. 2010). Together these steps guide students to learn from lived experiences and broadening these experiences by making visible aspects of the habitat and co-inhabitants that are often unnoticed, as occurs with the underwater world. FEP aims to provoke reflection about where we are, how we observe, what we feel, and how we feel before, during, and after encounters with other-than-human beings. FEP steps (see Rozzi et al. 2022 in this volume) are non-sequential and can be implemented in any order depending on the characteristics and problems in each habitat, the groups of collaborators, and the needs for change that are detected. The steps can be iterative and interchangeable depending on the work dynamics over time. In this way, the methodology allows implementing and accompanying educational and environmental management processes in the long term. That is, FEP promotes a sense of constant adaptation to social and ecological systems.

In this chapter we present the activity "pay attention, dive with eyes wide open" (Ojeda et al. 2018). This activity is an adaptation of the FEP methodology, based on concepts of biocultural ethics, that integrates: (i) cognitive and emotional experiences; (ii) intercultural dialogues; and (iii) interspecific cohabitation experiences. With this triple integration, we co-designed a diving experience with the participation of Yaghan fishers, students, scientists, and philosophers (Ojeda et al. 2018). This activity integrates "Western" ecological, scientific, and philosophical knowledge with that of Indigenous peoples to forge a "biocultural lens" that guides participants to understand an essential concept of biocultural ethics: the co-inhabitant (Rozzi 2019). This concept allows us to consider biodiversity not as a mere natural resource but rather as a community of living beings that cohabit by sharing habitats and co-evolutionary histories with human communities and other beings. Through "face-to-face" encounters with the *otherness* of oysters and other submerged invertebrates in the kelp forests of Cape Horn, co-inhabitation is experienced (Fig. 6.1).

This diving experience makes visible not only the small animals but also the ecological knowledge of local communities of artisanal fishers and Indigenous peoples, such as the Yaghan. Diving and collecting invertebrates in these cold waters has been a vital custom or "habit" of Yaghan women since time immemorial (Ojeda et al. 2018). In the material culture of the Yaghan, bivalve shells have provided the foundation for the construction of their *akar* huts which formed the intimate "habitat" of the Yaghan family. In the symbolic culture and the Yaghan worldview, mollusks are perceived as sentient beings that sanction selfish habits, but honor

Fig. 6.1 Artistic rendition of the field environmental philosophy activity "pay attention, dive with eyes wide open" (proposed by Ojeda 2013). The uniqueness of this activity is that the participants are equipped with physical lenses (such as their diving masks or filming cameras) and also with a "conceptual lens" (the "3Hs" lens of the biocultural ethic, *sensu* Rozzi 2013). Swimming among kelp forests, participants appreciate the (1) *Habitats* where invertebrates, seaweeds, and humans forge their (2) life *Habits* in (3) inter-species interactions with other *co-in-Habitants*. By engaging emotions and concepts derived from philosophical, scientific, and vernacular knowledge into the observation of surprising and unexpected behaviors of local cultures, such as artisanal fishers or understudied groups of organisms such as micro-invertebrates, students and other participants experientially learn the concept of co-inhabitants. Art by Mauricio Álvarez, Omora Park Archive

habits of collaboration and respect, as the artisan Julia González relates.[1] Both the material and immaterial culture of the original Yaghan people in the extreme south of South America include values and practices consistent with the co-inhabitant concept of biocultural ethics (Rozzi 2019). Multiple relational values are attributed to these invertebrates, including instrumental and intrinsic values. With this worldview, humans and mollusks have co-inhabited these archipelagos for 7000 years (REF Pilar Rivas).

By participating in "pay attention, dive with eyes wide open," participants dive not only into multidimensional biophysical worlds but also symbolic-linguistic worlds. This multidimensional biocultural understanding helps participants to overcome both biological (taxonomic) chauvinism and cultural (racist) chauvinisms, which often derive from the arrogance of hegemonic forms of knowledge. This

[1]See the scene with the Yaghan artisan Julia González and the researcher Jaime Ojeda in the documentary *El Regreso a la Madriguera*, co-directed by Jaime Sepúlveda and Ricardo Rozzi: https://vimeo.com/31905600, minutes: 18:53–19:07 (accessed October 17, 2018).

arrogance implies exclusion and oppression of the knowledge of indigenous and other local communities (Grosfoguel 2007). In turn, this epistemic injustice (*sensu* Fricker 2007) that leads to the direct oppression of local communities also leads to indirect oppression of the biodiversity with which these human communities co-inhabit (Rozzi 2019). For this reason, we aim to foster both inter-cultural and inter-species forms of justice.

Epistemic injustice occurs when the knowledge of some people or communities is arbitrarily ignored or oppressed (Fricker 2007). The hegemony of the concept of "natural resources" and of unidirectional relationships with nature sustained by a global economic model whereby humans exploit biodiversity, implies few, if any, responsibilities, or accountability to sustain biological and cultural diversity (Ojeda et al. 2022).

6.2 Applying FEP Step by Step. A Guide to Implementing Biocultural Ethics

In a practical sense, FEP provided a roadmap to put into practice the activity "pay attention, dive with eyes wide open." This activity is based on an interdisciplinary approach, where teachers, students, and scientists generated reciprocal experiences outside the classroom. The central axes of the activity are empathy, mutual aid from a biocultural framework, and situating oneself as a person in a territory that has both a biological and historical context. We work on the coastal edge of Robalo Bay located in the *Onashaga Channel* (currently called the Beagle Channel), Cape Horn Biosphere Reserve, Chile. We carried out experiences under intertidal waters and walking on the intertidal coast, discovering a habitat so old in the evolution of life that it has become "inhospitable" for the vast majority of human beings who live on earth (Fig. 6.2).

6.2.1 Steps in the Diving Experience, "Pay Attention, Dive with Eyes Wide Open"

Step 1: Transdisciplinary Research: Learning from Coastal Biocultural Diversity

In this step, we integrate three complementary "conceptual lenses:" ecological, ethno-ecological, and philosophical. Initial ecological research focused on macroalgae and mollusk diversity, assessing their annual phenology. We found 49 macroalgae taxa and 34 mollusk taxa. The latter maintains a relatively stable abundance throughout the year (Ojeda et al. 2014). However, due to seasonal changes in the day length, wet biomass of seaweeds is three times higher in summer than in winter (Ojeda et al. 2018).

Fig. 6.2 Children in an FEP activity exploring the coastal and intertidal zones to discover the diversity of co-inhabitants. Photograph by Jaime Ojeda

Ethno-ecological research combined bibliographic review with participant observation in activities with Yaghan fishers and handcrafters. We learned that, by being abundant throughout the year and rich in minerals, mollusks have been a staple food for Yaghan indigenous people (Ojeda et al. 2014, 2018; Orquera 1999). Mollusks have played a role analogous to bread for the Mediterranean peoples or maize in Mesoamerica. Yet, as indicated above, mollusks played another key role for the Yaghan culture: their shells provided the basic material to build the foundations for their huts or *akar* (Orquera 1999). Shells on the hut's floors created good drainage conditions in rainy environments. The mollusk's long-term use and Yaghan traditional management created a shell mound that served as protection against strong winds. Today thousands of shell middens are found along the coastal habitats of Cape Horn.

Macroalgae were also essential for traditional uses for Yaghan people. For example, they used giant kelp to build wigwams (semi-permanent domed dwellings [Lothrop 1928]) and hunted sea lions that use giant kelp forests to avoid predators (Gusinde 1986). The Yaghan also created fishing lines from giant kelp to catch rockcod (*Nothotenia tessellate*), hake (*Merluccius spp.*), and kingklip (*Genypterus*

Fig. 6.3 Aerial imagen of terrestrial and marine forests in the Sub-Antarctic Channels. This imagen represents the analogy of ecological and cultural continuity that a forest (under or over water) can embrace for nature–people relationships. Photograph by Jaime Ojeda

chilensis) (see Hyades and Deniker 1891). The Yaghan repaired canoes with seaweeds by preparing a gelatinous mix with mosses and seaweeds to cover cracks in canoes (Gusinde 1986).

Step 2. Composition of Metaphors and Narratives: The Underwater Forests Analogical thinking orients the poetic practice of creating metaphors through which FEP participants have represented submerged seaweeds as submerged forests. Analogous to terrestrial forests that present a marked seasonality in Cape Horn, seaweeds increase their biomass in synchrony with the sprouting of leaves on the trees of terrestrial forests during Spring and Summer (Fig. 6.3). Based on this ecological analogy, the metaphor of "underwater forests of Cape Horn" facilitates an understanding about the ecological interactions occurring underwater and enhances visualization through a mental image that is complementary to the more evident one of terrestrial forests (Ojeda et al. 2018).

The "underwater forests of Cape Horn" metaphor also facilitates an understanding of the co-inhabitation relationships of the Yahgan people with marine living beings that have close kinship relationships with humans. Hence, for the Yahgan, mollusks have the rights to occupy their habitats; for example, a limpet can be the owner of a rock. Consequently, Yahgan people believe that they ought to share intertidal habitats with mollusks and other marine organisms (Ojeda et al. 2018). The "underwater forests" metaphor offers a "biocultural lens" to understand not only ecological relationships but also ethical interspecific relationships that have been oriented by ethical values and traditional ecological knowledge that originated in the remote region of Cape Horn, probably before the emergence of ancient Greek philosophy (Rozzi 2019). Today, this "biocultural lens" inspires students, visitors, and other people to cultivate respectful forms of co-inhabitation with other-than-human organisms.

Step 3. Co-creation of Ethical and Ecological Field Activities: "Pay Attention, Dive with Eyes Wide Open"
Educational programs at Omora Park emphasize direct encounters between human and other-than-human beings based on the value of reciprocity. To orient direct encounters, field activities are co-created with members of the local community. With the participation of Yaghan handcrafters, fishers, teachers, scientists, and philosophers, the activity "pay attention, dive with eyes wide open" was created. In Spanish *ojo* means both "eye" and "pay close attention." It aims to reconnect humans to marine landscapes of Cape Horn (Chile) by helping participants literally to see and to pay close attention to this usually unperceived and too often alien marine environment. The activity includes both diving or snorkeling and walking along the intertidal zone. The experience brings together biological and cultural diversity, and their interrelations, and helps participants to experience ancestral and novel ways of seeing, feeling, and co-inhabiting with marine biodiversity.

"Pay attention, dive with eyes wide open" orients participants in two ways: (i) to have "face-to-face" encounters with invertebrates, seaweeds, and other organisms in the subtidal zone during high and low tides (Fig. 6.4); (ii) to motivate reciprocity

Fig. 6.4 Links between metaphor (underwater forests) and educational activity (pay attention, dive with eyes wide open) are pictorially illustrated. The terrestrial and marine forest analogy for creating reciprocal values among students, teachers, researchers, and local biodiversity is essential to learning these values. Photographs by Jaime Ojeda and Sebastian Rosenfeld

with underwater biodiversity, in which empathy is the cultural bridge to create this connection. In other words, empathy promotes how participants learn from others through embodied encounters such as facial expressions, emotional reactions, and touch.

This activity equips participants with two types of lenses: physical lenses, such as diving masks, cameras, or hand lenses, and a conceptual lens, the "3Hs biocultural lens." By swimming among kelp forests or walking the intertidal zone participants appreciate the **H**abitats where invertebrates forge their life **H**abits in interaction with other co-in-**H**abitants (the 3Hs). This biocultural lens orients participants to understand the concept of co-inhabitants through "face- to-face" encounters with the otherness of mollusks and other invertebrates in the submerged kelp forests or tidal edges of Cape Horn. By diving or exploring the intertidal zone, participants experience a life habit that is essential for the culture of the Yaghan people and other fishers, as well as for invertebrates that co-inhabit in underwater forests and edges of the sea of Cape Horn. By integrating emotions and concepts derived from indigenous, scientific, and philosophical knowledge into the observation of surprising and unexpected behaviors of invertebrates, students and other participants experientially apprehend the concept of co-inhabitants. The "pay attention, dive with eyes wide open" activity has triggered artistic inspiration and inspired novel environmental education programs, and today offers a model for new community initiatives for underwater eco-tourism.

Step 4. Conservation *in situ*: A New Marine Park in Cape Horn
The creation of marine protected areas was stimulated with the discovery of the close relationships between marine organisms and humans in the Cape Horn archipelagoes. Protecting marine ecosystems is essential to conserving experiences of co-inhabiting with algae, mollusks, and other beings (human and other-than-human beings). For this reason, in 2015 Omora Park researchers launched a complex participatory policy process to create the Diego Ramírez - Drake Passage Marine Park. Consistent with UNESCO's Man and Biosphere program that demands compatibility between social needs and conservation, the proposal maintained ancestral artisanal fishing practices and certified industrial fishing and ecotourism in ways that are harmonious with the zoning scheme of the Cape Horn Biosphere Reserve (Rozzi et al. 2017). The process was not free of tension but multi-stakeholder negotiations, including joint workshops with the Chilean Ministry of Economy and artisanal fishers, as well as with industrial fishery led to agreements about the design of the protected area, and in 2019, the Chilean government officially created the Diego Ramírez - Drake Passage Marine Park (Massardo 2020). This marine park protects critical nesting sites for albatrosses, penguins, and other endangered marine species, as well as intertidal and subtidal ecosystems with abundant macroalgae and invertebrates that reach their southernmost distribution (Rozzi et al. 2017). Today, multi-stakeholder negotiations are in the process of generating an implementation plan for this large marine protected area at the southern end of the Americas.

6.2.2 Steps of a Guided Educational School Experience Inspired by "Pay Attention. Dive with Eyes Wide Open"

Point 1. Identify a Marine Area to Visit - Step 4: Conservation *in situ*

In situ conservation (Step 4) makes it possible to have healthy habitats where we can enjoy, learn, and live together with a diversity of co-inhabitants. Having public access to the coast or beaches in many countries is limited; only certain areas not taken over by mass tourism, extractive industries, or upscale beach homes can be accessed. However, open access is fundamental for developing school and other learning activities. This means that access should be possible from where one lives and that the conditions that make life possible are maintained so that one can learn from the place itself (Fig. 6.5(1)). In our case, we promote collective public access (care) to coastal areas with respect and responsibilities to maintain marine life. To develop this activity, with prior permission, during the summer we moved groups of students to different points of the intertidal zone. We identified the places to take students based on the diversity of habitats on land and under water. These decisions were made based on knowledge of the biocultural characteristics of the coast.

Point 2. Situate Ourselves in the Place as Historical Subjects - Step 1: Transdisciplinary Research

Once in the place, we proposed the listening and narration of stories about the relationships between human beings and the sea, especially stories associated with the visited site (Fig. 6.5(2)). The stories can be diverse and encompass different dimensions of these relationships, from ecological, cultural, historical, or socioeconomic. In this way, we situate ourselves as human beings with a shared history, which makes us brothers and sisters of the fishers and the fish, of the harvesters and the mollusks, of the navigators and the waters. This point feeds into Step 1 of FEP where we identify a shared story that emerges from transdisciplinary inquiry. By inquiring about the habitat, habits, and co-inhabitants from different knowledge and narratives, we can find those unique stories that help us situate ourselves, to recognize ourselves as part of a place, of a local history and a longer history.

Point 3. Dive into the Water, Traverse the Intertidal - Step 1: Transdisciplinary Research

At this point, we recreate FEP Step 1. That is, from active knowledge and from our body and sensations, we conduct transdisciplinary research. We dive into the water or traverse the intertidal (Fig. 6.5(3)). These actions are based on the capabilities and desires of the group: if they already know how and want to dive or if they prefer to walk along the intertidal zone. To facilitate seeing, we provide "glasses": for diving we use a mask; for intertidal walking a hand lens. The diving mask clears underwater vision so that the diver can observe and appreciate the submerged forests and other forms of life. The hand lens literally magnifies the small animals, and objects that populate the intertidal zone. In this sense, lenses provide a new "eye," another organ through which to see, through which to amplify the gaze. Realizing *how* (the way

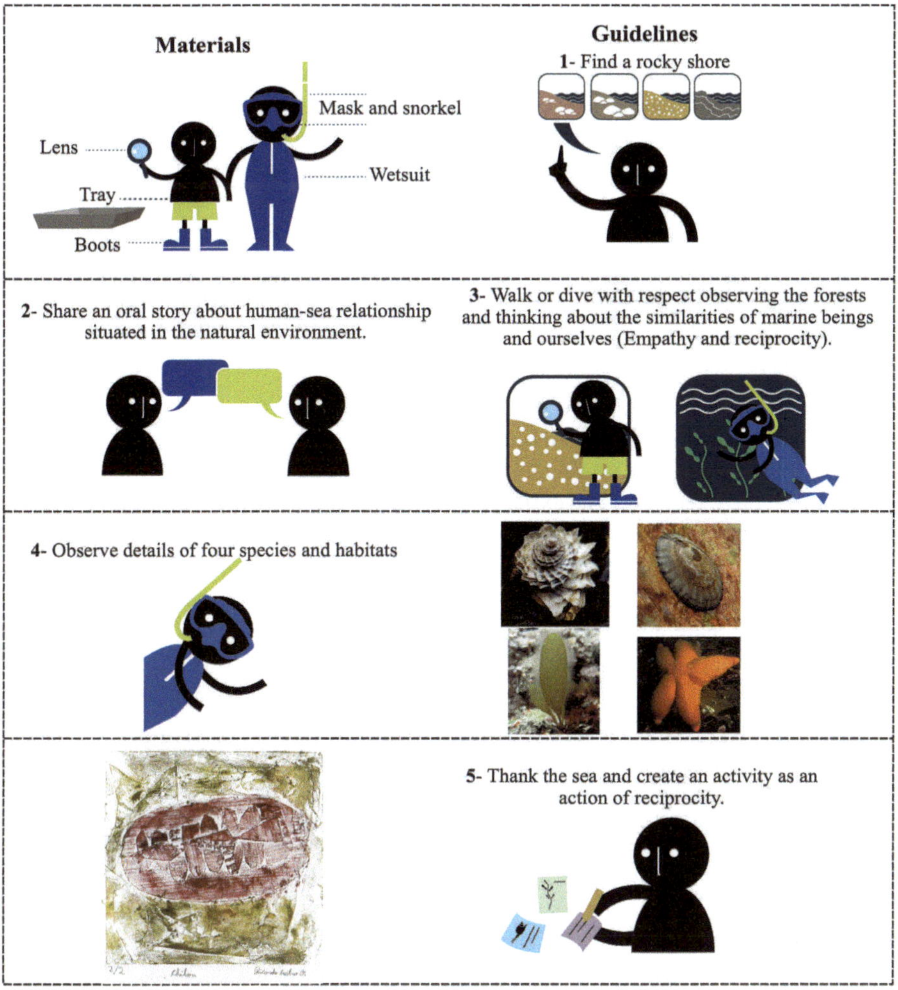

Fig. 6.5 Practical guidelines for the educational activity of "Pay Attention, Dive with Eyes Wide Open." Design and composition of illustrations by Felipe Portilla and Silvia Lazzarino. Photographs by Jaime Ojeda

we) look is learning at this point. So, we move on to the next point, where we notice *what we are* looking at.

Point 4. We Learn to Look Carefully - Steps 2 and 3: Composition of Metaphors and Narratives, Co-creation of Activities

"Take a look. What are you looking at?" Focusing on observation would appear to be familiar to sight-governed hominin species (Fig. 6.5(4)). But being fully attentive and present in the observation can take some effort for some people in the current times. Stopping to observe is learning to observe, to make visible what is in front of us. We do not see until we take another attitude, so we put on some lenses that make it easier for us to fine-tune this capacity. This lens, enriched with place-based stories, guides us as feeling observers toward the different co-inhabitants that surround us. This becomes a habit as we identify co-inhabitants and notice and feel their habits (as they are, as they are expressed in colors or movements). For example, in summer we can see the deployment of algae forests, and identify that there are "submerged forests." We begin to discover that there is an "underwater forest," or "communities between the rocks." In this way, as we become familiar with new observations and new learnings, we recreate Steps 2 and 3, and the activity is conceptualized in a metaphor. By using analogical reasoning to think in shapes and images and bring our inventiveness to name and include what we now see, we expand our perception.

Point 5. We Give Back from What Has Been Learned - Step 3: Co-creation of New Activities (and New Metaphors)
Finally, at this point, we conclude the activity by making learning conscious. We note that we have received this experience from the sea and the coast, and we want to thank, repay in some way what we received by saying "goodbye to the sea" (Fig. 6.5 (5)). To close, we propose new creations and actions inspired by Step 3. Here we co-create, but from a new experience lived in the first person by the subject and the group. We co-create and iterate the FEP cycle, motivating significant learning that adds to broader processes with the local and global territory. As a result of the experiences with different groups, we have seen that some young students create stories, drawings, and poems, and share them with their families and the school community. The adolescent students often propose actions aimed at environmental improvements, such as cleaning the coast. These actions can be as diverse and creative as the participating groups.

6.3 Conclusion

The activity "Pay attention. Dive with Eyes Wide Open" enables participants through "face-to-face" encounters to discover themselves as co-inhabitants of algae, oysters, and other invertebrates submerged in the kelp forests and intertidal zones of Cape Horn from the perspective of *otherness*. The experience of immersing oneself in the marine habitat invites one to discover an underwater forest and the intertidal communities among the rocks of the shoreline. Coupled with listening to the stories of the first inhabitants of the coasts and the detailed observation of the diversity of co-inhabitants (their shapes, colors, movements), people experience themselves as being in the habitat with other humans and other-than-humans. Furthermore, when walking along the rocky intertidal zone they discover small beings that go unnoticed. These variations of the activity (submerging or walking) include "seeing through glasses"—the diving mask or the hand lens—that provide a new "eye" through which to see differently or to amplify the gaze. In this way, participants focus on what they look at in greater detail and they also focus on how they look, what they look through (the filters they use, both physical [the diving mask, the hand lens] and mental [knowledge, beliefs, perceptions]). Being aware of the way they look helps them to be fully attentive and present in the observation. Thus, they begin to discover that there is an "underwater forest" or "communities between the rocks." This FEP activity helps participants to experience ancestral and novel ways of seeing, feeling, and co-inhabiting with marine biodiversity, fostering intercultural forms of justice as well as between species.

These learnings are possible to achieve from the design and implementation of guided educational activities within the framework of the FEP. This step-by-step guided framework supports educational activity and explicitly makes these activities bioculturally ethical. In addition to presenting an interpretive guide prepared for schoolchildren and other participants, and the detailed explanation of the activity "pay attention, dive with eyes wide open," we provide a diversity of learning that derives from the philosophical scaffolding of this methodology based on biocultural ethics. The explicit contribution of biocultural ethics in the practice of FEP allows access to deep learning such as reciprocity, otherness, epistemic justice, and co-habitation.

Acknowledgments Authors thank Roy May for valuable comments and acknowledge the support from the Cape Horn International Center (CHIC) - ANID PIA/BASAL (PFB210018).

References

Fricker F (2007) Epistemic injustice, power and the ethics of knowing. Oxford University Press, New York

Grosfoguel R (2007) The epistemic decolonial turn: beyond political-economy paradigms. Cult Stud 21:211–223

Gusinde M (1986) Los indios de Tierra del Fuego. Vol II, Los Yámanas. CAEA, Buenos Aires

Hyades P, Deniker J (1891) Anthropologie et ethnographie. In: Mission scientifique du Cap Horn (1882-1883), vol 7. Gauthier Villars et Fils, Paris

Lawrence AJ, Budziak A, Campbell I, Cornish A, Ender I, Jeffries B, Kanstinger P, Macdonald C, Marston J, Stevens G, Ward-Paige CA (2016) Turismo Responsable con Tiburones y Rayas: Una Guía para Mejores Prácticas. WWF, Gland, Switzderland and Rancho Santa Margarita, Project AWARE USA and Manta Trust, Dorset, UK

Lothrop S (1928) The Indians of Tierra del Fuego. Museum of the American Indian, Heye Foundation, New York

Lucrezi S, Saayman M, van der Merwe P (2013) Managing diving impacts on reef ecosystems: analysis of putative influences of motivations, marine life preferences and experience on divers' environmental perceptions. Ocean Coast Manag 76:52–63

Massardo F (2020) Reserva de la Biosfera Cabo de Hornos y Parque Marino Islas Diego Ramírez-Paso Drake. Anales del Instituto de la Patagonia 48(3):39–44

Moreno-Terrazas R, Micheline Cariño M, Díaz Z, González H, Monteforte-Sánchez M (2022) Starfishes and sky stars: a field environmental philosophy education and ecotourism experiences in Baja California, México. In: Rozzi R, Tauro A, Wright T, Avriel-Avni N, May RH Jr (eds) Field environmental philosophy: education for biocultural conservation. Ecology and ethics, vol 5. Springer, Dordrecht, pp nnn

Ojeda J (2013) Dinámica estacional de macroalgas y moluscos intermareales y su relación con el conocimiento tradicional ecológico yagán, en canales subantárticos del Cabo de Hornos: Una aproximación biocultural desde la filosofía ambiental de campo. Master in Sciences Program, Universidad de Magallanes, Chile

Ojeda J, Rosenfeld S, Marambio J, Rozzi R, Mansilla A (2014) Patrones estacionales y espaciales de la diversidad de moluscos intermareales de Bahía Róbalo, Canal Beagle, Reserva de la biosfera Cabo de Hornos, Chile. Rev Biol Mar Oceanogr 49:493–509

Ojeda J, Rozzi R, Rosenfeld S, Contador T, Massardo F, Malebrán J, González-Calderón J, Mansilla A (2018) Interacciones bioculturales del pueblo yagán con las macroalgas y moluscos: una aproximación desde la filosofía ambiental de campo. Magallania 46(1):155–181

Ojeda J, Salomon AK, Rowe JK, Ban NC (2022) Reciprocal contributions between people and nature: a conceptual intervention. BioScience 72:952–962. https://doi.org/10.1093/biosci/biac053

Orquera LA (1999) El consumo de moluscos por los canoeros del extremo sur. Relac Soc Argent Antropol 24:307–327

Rozzi R (2013) Biocultural ethics: from biocultural homogenization toward biocultural conservation. In: Linking ecology and ethics for a changing world. Springer, Dordrecht, pp 9–32

Rozzi R (2019) Taxonomic Chauvinism, No More! Environ Ethics 41:249–282

Rozzi R, Anderson CB, Cristobal Pizarro J et al (2010) Field environmental philosophy and biocultural conservation at the Omora Ethnobotanical Park: methodological approaches to broaden the ways of integrating the social component ("S") in Long-Term Socio-Ecological Research (LTSER) Sites. Rev Chil Hist Nat 83:27–68

Rozzi R, Massardo F, Mansilla A, Squeo FA, Barros E, Contador T, Frangopulos M, Poulin E, Rosenfeld S, Goffinet B et al (2017) Parque Marino Cabo de Hornos—Diego Ramírez; Universidad de Magallanes: Punta Arenas, Chile

Rozzi R, May RH Jr, Tauro A, Avriel-Avni N (2022) Introduction to field environmental philosophy: a new methodological approach for biocultural education and conservation. In: Rozzi R, Tauro A, Wright T, Avriel-Avni N, May RH Jr (eds) Field environmental philosophy: education for biocultural conservation. Ecology and ethics, vol 5. Springer, Dordrecht, pp nnn

Takata Experience Dive Center (2022) Eco-diving programs. Available via: takataexperience.com/eco-diving-programs/

Chapter 7
The Eyes of the Tree: Applying Field Environmental Philosophy to Tackle Conservation Problems at Long-Term Socio-ecological Research Sites

Ramiro D. Crego, Nora Ward, and Ricardo Rozzi

Abstract The Anthropocene is marked by a process of biocultural homogenization that includes the loss of biological and cultural diversity. The International Long-term Ecological Research (ILTER) network is intended to monitor, and eventually orient actions to counteract global environmental change. However, the focus of most studies on biophysical research at the expense of cultural dimensions and the lack of direct applications in management and policy limit the scope of the ILTER network. In this study, we implemented a Field Environmental Philosophy methodology in an ILTER site in southern Chile to demonstrate how the integration of ecological and philosophical studies can help overcome these limitations. The work integrates the cultural and ethical interrelations among the habitat, the community of co-inhabitants, and their life habits, to overcome ethical dilemmas in invasive ecology and find conservation solutions to the problem of invasive species. We created a metaphor, "the eyes of the tree," that draws together the ecological, technological, and cultural dimensions of the study, providing a conceptual, ethical, and technological lens that helps people perceive and understand the sub-Antarctic biota of Magallanes and monitor and control invasive species. It facilitates a conservation discourse with an eco-centric view in order to counteract the problem of biocultural homogenization.

R. D. Crego (✉)
Smithsonian National Zoo and Conservation Biology Institute, Conservation Ecology Center, Front Royal, VA, USA

N. Ward
School of History and Philosophy, National University of Ireland Galway, Galway, Ireland
e-mail: nora.ward@nuigalway.ie

R. Rozzi
Sub-Antarctic Biocultural Conservation Program, Department of Philosophy and Religion and Department of Biological Sciences, University of North Texas, Denton, TX, USA

Cape Horn International Center (CHIC), Universidad de Magallanes, Puerto Williams, Chile
e-mail: Ricardo.Rozzi@unt.edu

R. Rozzi et al. (eds.), *Field Environmental Philosophy*, Ecology and Ethics 5,
https://doi.org/10.1007/978-3-031-23368-5_7

Keywords Biocultural conservation · Biological invasions · Field environmental philosophy · ILTER · Metaphor

Large and rapid global changes are occurring in the Anthropocene, a period characterized by the prevalence of planetary scale transformations caused by human actions (Steffen et al. 2011). Among the plethora of changes occurring, reduction of biodiversity stands out due to habitat loss, climate change, and the spread of invasive species (Vitousek et al. 1997; Bellard et al. 2016). One of the most important institutions conceived to study and monitor the intricate relationship between biodiversity and human activities, and eventually guide the necessary socio-environmental global change to avoid the collapse of ecosystems, is the International Long-Term Ecological Research (ILTER) network (Li et al. 2015). The ILTER consists of a worldwide network of sites in which scientists engage in long-term ecological and socio-ecological research. The ILTER network also includes Long-Term Socio-Ecological Research (LTSER) networks and sites (Armesto et al. 2014).

The success of the ILTER and LTSER to provide global solutions to the biodiversity crisis, however, is being jeopardized by studies that focus almost exclusively on the biophysical domain, frequently omitting cultural and ethical dimensions, and that are biased toward theoretical studies, frequently preventing ILTER sites from being involved in transdisciplinary conservation actions (Rozzi et al. 2014). Additionally, more than 90% of the ILTER sites are located in the northern hemisphere, leaving the southern hemisphere forced to make conservation decisions that are based mostly on results from studies conducted on completely different socio-ecological systems (Rozzi 2012; Rozzi et al. 2014; Li et al. 2015).

A conceptual framework to address the lack of ethics in research is found in biocultural conservation and its proposed methodology, the *Field Environmental Philosophy* (FEP). Conceived in the southernmost ILTER and LTSER-Chile, the Omora Ethnobotanical Park (from now on: Omora Park; Rozzi et al. 2010), FEP offers a transdisciplinary methodological approach to education and conservation that integrates biophysical, cultural, and institutional dimensions for the complex problems we face in the Anthropocene. FEP methodology helps to assess the intricate relationships between ecosystems and people's ethical and social values, thereby increasing the possibilities of successful conservation actions (Rozzi et al. 2010, 2014). Since 2008, FEP has been used to develop a diversity of interdisciplinary studies that have generated novel conservation, education, and ecotourism activities (Rozzi et al. 2008, 2010, 2014; Berchez et al. 2015).

In this chapter, we examine how the integration of applied ecology and philosophy using the FEP can contribute to understanding the biological and ethical complexities of socio-ecosystems and help expand the scope of conservation programs in the ILTER and LTSER networks. We focused on the conservation problems associated with the introduction and rapid expansion of invasive mammalian species at the UNESCO Cape Horn Biosphere Reserve (CHBR). The integration of science and ethics in invasive species studies is especially relevant because the management of invasive species frequently involves values related to the removal of individuals and populations that limit management and policy (Simberloff 2012; Crowley et al. 2017).

7.1 Implementing Field Environmental Philosophy

The field environmental philosophy methodology uses a cycle of four steps: (1) transdisciplinary ecological and philosophical research; (2) poetic communication through composition of metaphors and narratives; (3) field activities guided with an ecological and ethical orientation; and (4) conservation *in-situ*. Following these four steps, we researched the problem of invasive mammal species at the CHBR. We then composed a metaphor and implemented field activities for conservation.

Step 1: Transdisciplinary Research on Invasive Species and Biocultural Homogenization
Invasive species are species that have been recently introduced by humans in areas where they have been historically absent. After the introduction, these species expand their range and abundances, generating large ecological and/or economic impacts (Catford et al. 2009). Invasive species can affect the biology, abundance, and behavior of native species and the structure and functioning of entire ecosystems (Ehrenfeld 2010; Vilà et al. 2011). The relocation of species around the world by human activities causes a worldwide phenomenon known as "biotic homogenization" (McKinney and Lockwood 1999; Simberloff 2013), in which the continuing degradation of the environment ends in homogenized ecosystems dominated by the same group of plants and animals (Simberloff 2013).

Additionally, Rozzi (2018) describes a complementary process, the biocultural homogenization, that is a wicked problem that results from synergistic feedbacks between biotic and cultural homogenization. Life habits, customs, and worldviews of people, conceived in deep connection with their native habitats and their other-than-human co-inhabitants, are supplanted by habits of a dominant, global capitalistic culture. This results in the loss of original cultures (Amerindians or equivalent), loss of languages, and the concentration of people in cities (Rozzi 2012). Although these two processes of homogenization have been studied independently by ecologists and sociologists, Rozzi (2012) proposed that the loss of biological and cultural diversity at local, regional, and global scales are linked in one single process that he named "biocultural homogenization." Therefore, to tackle biocultural homogenization, a biocultural approach is needed, embracing both dimensions of the problem.

The biocultural ethic provides a conceptual framework that advocates for the conservation of *habitats* and access to them by the community of *co-inhabitants* (humans and other-than-humans) as a condition of possibility for the continuation of their *habits* of life and well-being (Rozzi 2012). These "3H" of the biocultural ethic allow a greater appreciation of the plethora of forms of co-inhabiting responsibly and sustainably with the biotic community in different ecosystems (Rozzi 2012, 2015).

With this conceptual framework, the biocultural ethic integrates the biophysical and cultural domains in conjunction with the socio-political, institutional, and technological domains. The biophysical domain is common to humans and other-than-humans (that is, all the biotic and abiotic components of the biosphere). The symbolic-linguistic (cultural) domain is human, and critical for the way that many cultures value and interact with other-than-human organisms. Such socio-ecological relationships take place through institutions that also include cognitive, normative, and technological dimensions.

The sub-Antarctic Magellanic region, located in southernmost South America, constitutes an example of the biocultural homogenization process. These temperate forests of the south of South America are isolated from other forests by at least 1500–2000 km of xeric environments and oceans. The long geographical isolation has allowed populations of different species to diverge into new species (Mayr 1942), resulting in several cases of endemism (Armesto et al. 1998). The cultural habits of European settlers in South America, however, promoted the introduction of multiple alien species that came from the sub-Arctic region in the northern hemisphere. Species with fur value, such as the American beaver (*Castor canadensis*), the muskrat (*Ondatra zibethicus*), and the American mink (*Neovison vison*), were introduced in Argentina and Chile with the idea of establishing a profitable business related to the fur industry (Jaksic et al. 2002).

Nowadays, these three species that naturally interact in their native range in North America have formed a functional ensemble in the south, with severe negative effects on the native ecosystems. Beavers, which are known for their role as ecosystem engineers (Rosell et al. 2005; Nummi and Hahtola 2008), affect river flow and native vegetation in the sub-Antarctic Magellanic region, changing forests into wetlands with abundant vegetation, creating suitable habitats for muskrats (Crego et al. 2016). Muskrats, in turn, have become the main prey item for American mink populations living inland (Crego et al. 2016). The latter is a ferocious predator, preying opportunistically on a myriad of native species, especially native birds and small rodents (Schüttler et al. 2008; Jiménez et al. 2014; Crego et al. 2016). Importantly, the native biota evolved in the absence of terrestrial mammalian predators, making them potentially naïve to mink predation risk (Crego et al. 2018b). The invasive species project at Omora Park established a transdisciplinary work in conjunction with governmental institutions such as the Agricultural and Livestock Bureau (SAG) and the National Forestry Service (CONAF) of Chile (e.g., see Schüttler et al. 2019) that is essential for controlling these species.

The introduction and expansion of beavers, muskrats, and mink promoted the biological homogenization of two distant ecosystems, in which an assemblage of sub-Artic co-inhabitants, reconfigured the structure and evolutionary course of the co-inhabitants, habits, and habitats of the sub-Antarctic ecosystem. The invasion can lead to local extinctions, loss of local diversity and alter evolutionary processes and future drift (Mooney and Cleland 2001; Yoshida et al. 2003; Lambrinos 2014; Sarrazin and Lecomte 2016).

Similar to the unique biodiversity, Amerindian worldviews in subpolar regions are also unique as the result of thousands of years of co-inhabiting certain habitats in close relationship with co-inhabitants. According to the Yaghan cosmogony—people who have lived in the CHBR region for millennia—mammals, birds, and other animals descend from humans. This pre-Columbian evolutionary view implies a kinship relationship between humans and the other co-inhabitants of sub-Antarctic ecosystems (Rozzi 2004). Another unique cosmogony can be found in the Koyukon people, who have inhabited the geographic pole opposite the sub-Antarctic forests on the banks of the Koyukuk and Yukon rivers of Alaska, in the sub-Arctic. In the Koyukon cosmogony, there is a *Distant Time* in which animals and plants were humans. At some point in that *Distant Time*, certain humans died and became animals or plants. Therefore, for the Koyukon, animals and plants are literally their relatives, their ancestors, brothers, and sisters (Nelson 1984). This cosmogony has ecologically and ethically guided the Koyukon way of living. Koyukon people have coexisted for thousands of years in harmony with their surroundings, living in synchrony with the seasonality, ecological processes, and co-inhabitants of sub-Arctic forest ecosystems, practicing moderation and respect for the environment (Nelson 1984). The changes in culture that followed the colonization of the Americas have also reconfigured the lives of the Yagan and Koyukon people and promoted a cultural homogenization process. Yagan and Koyukon people today share much more than a sense of kinship and respect for the habitat.

Step 2: Composition of Metaphors and Communication Through Narratives: "The Eyes of the Tree"
The composition of metaphors is an essential step of the FEP methodology. Metaphors communicate concepts to the public by integrating ecological and philosophical studies through analogies that synthesize values and concepts in biocultural conservation and biocultural ethics (Rozzi et al. 2010). To communicate the biocultural problem of invasive species in the CHBR and the conservation view to protect the region, we composed a metaphor that combines biophysical, cultural, and technological views and proposes a holistic understanding of native ecosystems.

Unlike forest birds, which can easily be seen or heard, terrestrial predators, such as the American mink, tend to be elusive species that are difficult to observe. A technological innovation, the camera trap (Fig. 7.1), provides a novel and effective tool with which researchers can document, study, and communicate the activity of this type of species (O'Connell et al. 2011). These cameras are usually attached to tree trunks. The Magellanic sub-Antarctic forests are dominated by the genus *Nothofagus*. Some individuals of *Nothofagus betuloides* located at Omora Park in the CHBR are between 202 and 350 years old (Llancabure 2011; Lombardi et al. 2011). These individuals are "living witnesses" to the history of the area, among them, the arrival of Europeans to Cape Horn and the establishment of Western practices and beliefs, including the introduction of domestic species and the fur industry.

According to the Koyukon cosmovision, trees are always observing and judging the behavior of hunters and every member of the Koyukon community (Nelson 1984). A Koyukon never walks alone in the forest: it is "as if the trees had eyes." By attaching camera traps ("technological eyes") to the trees to study the American

Fig. 7.1 The photo shows the interpretative station "the eyes of the tree" located at the Omora Ethnobotanical Park, in southern Chile, and a close look to a camera trap. Camera traps are utilized to monitor invasive species in the most austral ILTER and LTSER site in the world. The metaphor "the eyes of the tree" illustrates how a "technological eye" allows visitors to expand the "conceptual and biocultural" lenses by generating empathy with a century-old tree that has witnessed the arrival of invasive species and the processes of biocultural homogenization in the region. Photos by Ramiro D Crego

mink on Navarino Island (Crego et al. 2018a), different methods of perception may connect us to the Koyukon worldview in some small way. The biocultural ethic asserts that the life habits of the co-inhabitants of a community are interrelated with the habitats where they evolve. In this case, the "co-inhabitant tree" has been equipped with a "techno-eye" that allows researchers to document the arrival of a new co-inhabitant that creates the ethical dilemma of how to co-inhabit and judge human behavior in this new relationship of co-inhabiting. From these three sources, scientific, Amerindian, and ethical, the metaphor "the eyes of the tree" was composed.

Step 3. Ecologically and Ethically Guided Field Activities to Connect the Eyes to the Tree.
The Omora Park education program emphasizes the importance of direct encounters with nature to stimulate the perception and appreciation of both biological and cultural diversity (Rozzi et al. 2010). To understand and communicate the metaphor of the "eyes of the tree," we designed and implemented an interpretative station along the interpretative trail "The southernmost forests of the world" at Omora Park.

The interpretative station introduces the park visitor to the "technological lenses," the camera trap (Fig. 7.1). During an ecological and ethically guided visit, visitors inquire about the "technological lens," thus learning about the instrument and the ecology of animals it registers in photographs. The photographic record reveals the existence of the invasive American mink, whose presence remains otherwise invisible to visitors. The biotic interactions and impact of the mink on the native biota are also made visible. In addition, the guided tour includes a "conceptual lens" that allows a clearer view of the co-inhabitants of the forest, their habits of life, their delicate balance, and the fragility of this ecosystem equilibrium against the arrival of invasive species. An inquiry into the "biocultural lenses" introduces the Koyukon narratives that inspired the metaphor. This is complemented by Yaghan's narratives that take place during the visits to Omora Park. In this way, "technological lenses" together with "conceptual and biocultural lenses" allow the general public, authorities, teachers, and other members of the community to deconstruct the complexity of the entire ecosystem.

Understanding the complex socio-ecological processes that involved the arrival of invasive species to the region and the ecological impacts that these species have on the delicate ecological equilibrium allows to reconsider ethics. Our actions as a society can favor the introduction and disruption of the complex equilibrium of ecosystems that jeopardize the ecological processes that allow all co-inhabitants to assess the habitat and conduct their habits of life. The ecological and ethically guided visit to the "eyes of the tree" invites all visitors to see the surrounding world with new lenses.

Step 4. In-situ Conservation: Biocultural Conservation Implications of "The Eyes of the Tree"

The "eyes of the tree" station at Omora Park illustrates, in a poetic manner, a long-term monitoring program for the conservation of native biological and cultural diversity. The station has been incorporated into the system of interpretive trails of Omora Park, complementing other educational and ecotourism activities designed using the FEP methodological approach (Rozzi et al. 2008, 2010, 2014).

Attached to a century-old tree, the station "the eyes of tree" contributes to conservation programs that transcend the application of ecological knowledge into a management program to control invasive species (Crego 2018). The "eyes of the tree" metaphor invites visitors to extend and supplement their own individual vision and move toward an "ecosystemic" and holistic vision of the forest. Such an extension connects not only with traditional indigenous perceptions of land but also relates to normative frameworks within the academic environmental ethics movement. In the land ethic, for example, Aldo Leopold encourages "thinking like a mountain," referring to modes of perception that are long-range, community-oriented, and non-anthropocentric (Leopold 1949). Likewise, Arne Naess' theory of "Self-Realization" refers to a process of enlightenment in which the ontology of the atomistic self is replaced with that of the *ecological self*—an understanding of self-identity that is holistic and relational (Naess 1987).

A holistic understanding of the forest aided by this metaphor may serve to mitigate conflicts that have emerged as a result of the management and control of invasive species. Campaigns to control and reduce the negative effects of invasive species have been historically repudiated by animal rights movements (Simberloff 2012). This movement, largely led by philosophers Peter Singer in Australia (Singer 1975) and Tom Regan in the United States (Regan 1975), argues for the welfare of individual animals on account of their sentience. For both Singer and Regan, sentience—the ability to feel pleasure and pain—is the only relevant criterion for moral consideration. Such a framework tends to differ from that of environmental ethics, which usually extends moral consideration not only toward sentient animals but also ecosystems in a broader sense. Thus, ecological relationships among species are viewed as equally important to that of individual welfare. This approach is categorized as eco-centric holism or eco-centrism (Palmer 2013).

The individualistic versus holistic distinction is of great relevance for the management of invasive species and the conservation of ecosystems. The animal rights' emphasis on reducing the suffering of individual sentient animals, conflicts with conservation practices that involve the control (e.g., culling) of individuals for the benefit of the ecosystem as a whole (Hutchins and Wemmer 1986). While the animal rights framework provides an important contribution to the process of extending moral consideration beyond humans, which has resulted in important steps toward the reduction of animal cruelty and inhumane treatment globally, we contend that it is incomplete when dealing with ecologic crises in a broader sense. Because the animal rights framework draws a firm line between individuals worthy of moral consideration—those capable of sentience—and those not worthy, it leaves large swaths of the ecosystem unprotected and unaccounted for in a moral sense. Such a position is eventually self-defeating as all sentient beings ultimately depend upon the health and protection of non-sentient beings. The "eyes of the tree" metaphor, therefore, invites visitors to move beyond a purely individualistic understanding of nature and toward a holistic or ecosystemic comprehension.

In the middle of the forest, the visitor experiences face-to-face relationships with the environment. The metaphor of "tree eyes" through a "technological eye" helps people to "see" like a tree that has lived for hundreds of years and has witnessed the arrival and impact of the invasive species in the sub-Antarctic Magellanic forest. When people "see" as a tree, they expand their "conceptual and biocultural" lens. Empathy with the tree generates an ethic toward the integrity of the ecosystem. An ethic of respect for co-inhabitants that considers not only the charismatic mammals, but also small birds, mice, plants, mosses, and all interactions occurring in the community of co-inhabitants and the ecosystem in general. An ethic that values biological and cultural diversity, countering biocultural homogenization by avoiding the introduction of species by a hegemonic culture, species which are quickly adopted by people that live in places as remote as Navarino Island.

7.2 Resolving the Limitations of the ILTER Network

In this chapter, we show how the use of biocultural ethics and the FEP as a working methodology with the use of a metaphor that integrates technological, ecological, philosophical, and Amerindian components, allow us to expand the scope of ILTER networks for conservation. The "eyes of the tree" contributes to expanding beyond the biophysical domain of the study (i.e., the ecological impact of a small group of mammalian invasive species and its implications for management), integrating the cultural and ethical interrelations among the habitat, the community of co-inhabitants, and their life habits. This helps people reflect on the ecosystem as a whole and embrace complex ethical dilemmas.

In the metaphor of the "eyes of the tree," the "technological lens" allows us to expand our "conceptual and biocultural" lenses by generating empathy with a tree that is an old living witness to the processes of biocultural homogenization. The technology used (camera trap) is integrated with an Amerindian worldview (Koyukon culture) that conceives the ability of the trees to "observe." Through experience, emotions are triggered, permitting a better understanding of the real problem that, in the process of biocultural homogenization, invasive species generate on local biodiversity. It is through empathy with the tree that "observes" the evolution of forest co-inhabitants, (humans and other-than-humans), that we can understand the importance of conserving the biological, evolutionary, and cultural wealth of a unique place. Today, in order to counteract biocultural homogenization, the FEP from the south of the world and the metaphor of "eyes of the tree" provide a conceptual, ethical, and technological lens to perceive, monitor, and understand the sub-Antarctic biota of Magallanes.

Acknowledgments Author R Rozzi receives basal support from the Cape Horn International Center (CHIC) - ANID PIA/BASAL (PFB210018). We thank Roy May for his valuable comments on the manuscript.

References

Armesto JJ, Rozzi R, Smith-Ramírez C, Arroyo MTK (1998) Effective conservation targets in South American temperate forests. Science 282:1271–1272

Armesto JJ, Núñez-Ávila M, Donoso P, Rozzi R (2014) Introducción a la Sección Especial de Revista Bosque: Avances de una red de sitios de estudios ecológicos a largo plazo en el suroeste de Sudamérica. Bosque (Valdivia) 35:413–414. https://doi.org/10.4067/S0717-92002014000300015

Bellard C, Cassey P, Blackburn TM (2016) Alien species as a driver of recent extinctions. Biol Lett 12:20150623

Berchez F, Mansilla A, Ghilardi-Lopes NP, Schwindt E et al (2015) Ecology and education in marine protected areas: insights from Brazil and South America. In: Earth Stewardship: linking ecology and ethics in theory and practice. pp 351–366

Catford JA, Jansson R, Nilsson C (2009) Reducing redundancy in invasion ecology by integrating hypotheses into a single theoretical framework. Divers Distrib 15:22–40

Crego RD (2018) Recomendaciones de manejo para el visón norteamericano basado en los últimos acances sobre el estudio de la invasión en isla Navarino. Reporte tecnico
Crego RDD, Jiménez JEJE, Rozzi R (2016) A synergistic trio of invasive mammals? Facilitative interactions among beavers, muskrats, and mink at the southern end of the Americas. Biol Invasions 18:1923–1938. https://doi.org/10.1007/s10530-016-1135-0
Crego RD, Jiménez JE, Rozzi R (2018a) Potential niche expansion of the American mink invading a remote island free of nativepredatory mammals. PLoS One 13. https://doi.org/10.1371/journal.pone.0194745
Crego RD, Jiménez JE, Rozzi R (2018b) Macro- and micro-habitat selection of small rodents and their predation risk perseprion under a novel invasive predator at the southern end of Americas. Mammal Res 103:9–15. https://doi.org/10.1016/j.radphyschem.2014.05.010
Crowley SL, Hinchliffe S, Mcdonald RA (2017) Conflict in invasive species management. https://doi.org/10.1002/fee.1471
Ehrenfeld JG (2010) Ecosystem consequences of biological invasions. Annu Rev Ecol Evol Syst 41:59–80
Hutchins M, Wemmer C (1986) Wildlife conservation and animal rights: are they compatible? In: Fox MW, Mickley LD (eds) The Humane Society of the United States, Washington, DC, pp 111–137
Jaksic FM, Iriarte JA, Jiménez JE, Martínez DR (2002) Invaders without frontiers: cross-border invasions of exotic mammals. Biol Invasions 4:157–173
Jiménez JEJE, Crego RDRD, Soto GEGE et al (2014) Potential impact of the alien American mink (Neovison vison) on Magellanic woodpeckers (Campephilus magellanicus) in Navarino Island, Southern Chile. Biol Invasions 16:961–966. https://doi.org/10.1007/s10530-013-0549-1
Lambrinos J (2014) How interactions between ecology and evolution influence contemporary invasion dynamics. Ecology 85:2061–2070
Leopold A (1949) A Sand County Almanac. Oxford University Press, New York
Li B, Parr T, Rozzi R (2015) Geographical and thematic distribution of publications generated at the International Long-Term Ecological Research Network (ILTER) Sites. In: Rozzi R, Chapin FS III, Callicott JB et al (eds) Earth stewardship. Linking ecology and ethics in theory and practice. Springer, Dordrecht, pp 351–366
Llancabure JC (2011) Relaciones entre el crecimiento de Nothofagus betuloides y el clima local y de gran escala en bosques subantárticos de la Isla Navarino
Lombardi F, Cocozza C, Lasserre B et al (2011) Dendrochronological assessment of the time since death of dead wood in an old growth Magellan's beech forest, Navarino Island (Chile). Austral Ecol 36:329–340. https://doi.org/10.1111/j.1442-9993.2010.02154.x
Mayr E (1942) Systematics and the origin of species. Columbia University Press, New York
McKinney ML, Lockwood JL (1999) Biotic homogenization: a few winners replacing many losers in the next mass extinction. Trends Ecol Evol 14:450–453
Mooney HA, Cleland EE (2001) The evolutionary impact of invasive species. PNAS 98:5446–5451
Naess A (1987) Self-realization: an ecological approach to being in the world. The Trumpeter 4:35–41
Nelson RK (1984) Make prayers to the raven. A Koyukon view of the northern forest. Chicago, USA
Nummi P, Hahtola A (2008) The beaver as an ecosystem engineer facilitates teal breeding. Ecography (Cop) 31:519–524
O'Connell AF, Nichols JD, Karanth KU (2011) Camera traps in animal ecology. Methods and analyses. Springer, Tokyo
Palmer C (2013) Contested frameworks in environmental ethics. In: Rozzi R, Pickett S, Palmer C et al (eds) Linking ecology and ethics for a changing world: values, philosophy, and action, ecology and ethics 1. Springer, pp 191–206
Regan TOM (1975) The moral basis of vegetarianism. Can J Philos 5:181–214
Rosell F, Bozsér O, Collen P, Parker H (2005) Ecological impact of beavers Castor fiber and Castor canadensis and their ability to modify ecosystems. Mamm Rev 35:248–276

Rozzi R (2004) Implicaciones Éticas de Narrativas Yaganes y Mapuches Sobre las Aves de los Bosques Templados de Sudamerica Austral. Ornitol Neotrop 15:435–444
Rozzi R (2012) Biocultural ethics: recovering the vital links between the inhabitants, their habits, and habitats. Environ Ethics 43:27–50
Rozzi R (2015) Implications of the biocultural ethic for Earth Stewardship. In: Rozzi R, Chapin FS III, Callicott JB et al (eds) Earth Stewardship. Linking ecology and ethics in theory and practice. Ecology and ethics 2. Springer, pp 113–136
Rozzi R (2018) Biocultural homogenization: a wicked problem in the Anthropocene. In: Rozzi R, May RH Jr, Chapin FS III, Massardo F, Gavin M, Klaver I, Pauchard A, Nuñez MA, Simberloff D (eds) From biocultural homogenization to biocultural conservation, Ecology and ethics, vol 3. Springer, Dordrecht, pp 21–47
Rozzi R, Arango X, Massardo F et al (2008) Field environmental philosophy and biocultural conservation: the Omora Ethnobotanical Park educational program. Environ Ethics 30:325–336
Rozzi R, Anderson CB, Pizarro JC et al (2010) Field environmental philosophy and biocultural conservation at the Omora Ethnobotanical Park: methodological approaches to broaden the ways of ("S") in Long-Term Socio-Ecological Research (LTSER) Sites. Rev Chil Hist Nat 83: 27–68
Rozzi R, Massardo F, Contador T et al (2014) Filosofía ambiental de campo: ecología y ética en las redes LTER-Chile e ILTER. Bosque 35:439–447. http://www.scielo.cl/scielo.php?script=sci_arttext&pid=S0717-92002014000300019&lng=en&nrm=iso&tlng=en
Sarrazin F, Lecomte J (2016) Evolution in the Anthropocene. Science 351:922–923. https://doi.org/10.1126/science.aad6756
Schüttler E, Cárcamo J, Rozzi R (2008) Diet of the American mink Mustela vison and its potential impact on the native fauna of Navarino Island, Cape Horn Biosphere Reserve, Chile. Rev Chil Hist Nat 81:585–598
Schüttler E, Crego RD, Saavedra-Aracena L et al (2019) New records of invasive mammals from the sub-Antarctic Cape Horn Archipelago. Polar Biol 42:1093–1105. https://doi.org/10.1007/s00300-019-02497-1
Simberloff D (2012) Nature, natives, nativism, and management: worlviews underlying controversies in invasion biology. Environ Ethics 34:5–25
Simberloff D (2013) Introduced species, homogenizing biotas and cultures. In: Rozzi R, Pickett STA, Palmer C et al (eds) Linking ecology and ethics for a changing world: values, philosophy, and action. Ecology and ethics 1. Springer, Dordrecht, pp 33–48
Singer P (1975) Animal liberation. Random House, New York
Steffen W, Persson Å, Deutsch L et al (2011) The anthropocene: from global change to planetary stewardship. Ambio 40:739–761. https://doi.org/10.1007/s13280-011-0185-x
Vilà M, Espinar JL, Hejda M et al (2011) Ecological impacts of invasive alien plants: a meta-analysis of their effects on species, communities and ecosystems. Ecol Lett 14:702–708
Vitousek PPM, D'Antonio C, Loope LL et al (1997) Introduced species: a significant component of human-caused global change. New Zeal J Zool 21:1–16
Yoshida T, Jones LE, Ellner SP et al (2003) Rapid evolution drives ecological dynamics in a predator-prey system. Nature 424:303–306

Chapter 8
Starfish and Sky Stars: Field Environmental Philosophy Education and Ecotourism Experiences in Baja California Sur, Mexico

René Moreno-Terrazas-Troyo, Zenorina Díaz-Gómez, Humberto González-Galván, Micheline Cariño-Olvera, and Mario Monteforte-Sánchez

Abstract The starfish of the Gulf of California (GoC) are part of the rich biodiversity of Baja California Sur, Mexico. In this sea, starfish have been widely studied from the natural sciences, but not from human disciplines. In this interdisciplinary chapter, from marine biology, environmental history, and philosophy we describe biocultural educational and conservation processes that have occurred in the coastal and marine areas of Baja California Sur. Although we do not have current records of living native languages, there are still fishermen and communities with a long tradition that maintain an *ethos* different from the developmentalist paradigm that entails a unidimensional approach in formal education. Based on the conceptual framework of biocultural ethics—which links the well-being and life habits of co-inhabitants with the habitats they share—we reflect on the socio-environmental changes of recent years. In this territory, there are many specialists dedicated to the sea, environmental organizations, and inhabitants that are working to defend the health of the sea, the variety of life forms that co-inhabit the ecosystems of the GoC, and, therefore, the coastal societies that depend on them. Another purpose is to contribute to social change for *buen vivir* or "living well" from an ethical environmental education and Field Environmental Philosophy methodological approach. With this approach, we designed a low-impact ecotourism alternative. We created the metaphor "from the starfish to the stars in the sky" and the field activities "starfish viewer" and "marine philosophical hiking," which facilitate face-to-face encounters with marine co-inhabitants that shine in the ineffable vermilion landscape of submerged ecosystems.

R. Moreno-Terrazas-Troyo (✉) · Z. Díaz-Gómez · H. González-Galván · M. Cariño-Olvera
Departament of Humanities, Universidad Autónoma de Baja California Sur, La Paz, Mexico
e-mail: rtroyo@uabcs.mx; dgomez@uabcs.mx; hgonz@uabcs.mx; m.carino@uabcs.mx

M. Monteforte-Sánchez
2 Aquaculture Academic Program, Centro de Investigaciones Biológicas del Noroeste, S.C., La Paz, Mexico
e-mail: montefor04@cibnor.mx

R. Rozzi et al. (eds.), *Field Environmental Philosophy*, Ecology and Ethics 5,
https://doi.org/10.1007/978-3-031-23368-5_8

Keywords Environmental education · Environmental ethics · Environmental history · *Homo viator* · Tourism

8.1 Introduction

The environmental impact generated by mass tourism on the coastal and marine ecosystems is among the main problems we face in Baja California Sur, Mexico, especially because tourism is one of the region's main sources of income (Box 8.1). This tourism presents great challenges to the conservation of the ecoregion (Arizpe and Covarrubias 2010; Cisneros-Montemayor et al. 2020). A motivation for writing this text is to propose alternatives to massive sun and beach tourism. Another approach is needed, one that is more respectful and cordial and that generates environmental culture (cf. Rozzi et al. 2010a). The alternative proposes to change the ways of approaching marine biodiversity and to broaden the spectrum of the valuable work carried out within protected areas (López-Espinosa 2002). This proposal for biocultural education and tourism seeks to lessen the deterioration and impacts on the rocky reefs caused by motorized tourism and industrial boats; dragging of industrial fishing nets; and anchors of the boats that take tourists fishing, whale-watching, snorkeling, or diving.

Box 8.1 Biocultural Diversity and Ecotourism in the Gulf of California
The Gulf of California is located in the far northwest of Mexico (Fig. 8.1). One of the most biodiverse seas on the planet, it is home to 36 species of marine mammals (Urbán 2010); 31 species of cetaceans; five of the seven species of sea turtles in the world (Seminoff 2010); more than 700 species of fish, including sharks and their surprising diversity of internal parasites (Caira and Jensen 2014); 210 varieties of birds, and just over 6000 species of macroinvertebrates (WWF 2021, Fig. 8.2). Although it has protected areas, its biodiversity is threatened by the two million tourists that visit it each year.

The Gulf of California is the only inland sea in the world whose territory belongs to a single nation. It has 49% of Mexico's coastline and nearly 50% of the country's island territory. It is one of the five marine ecosystems with the highest productivity and biodiversity on the planet (CONANP 2000). It is longer than 1300 km, has 3000 km of coastline, and has 17,000 islands and insular incidents. Due to its long north-south orientation, it includes two zoogeographic regions, one tropical and one temperate. This gives it a dynamic and complex system of currents, upwellings, and movements of large bodies of water (Ledsma-Vázquez et al. 2009). Brusca and Wallerstein (1979) introduced the term subtropical to define the biogeography of its marine environment. The Gulf is characterized by large fluctuations in

(continued)

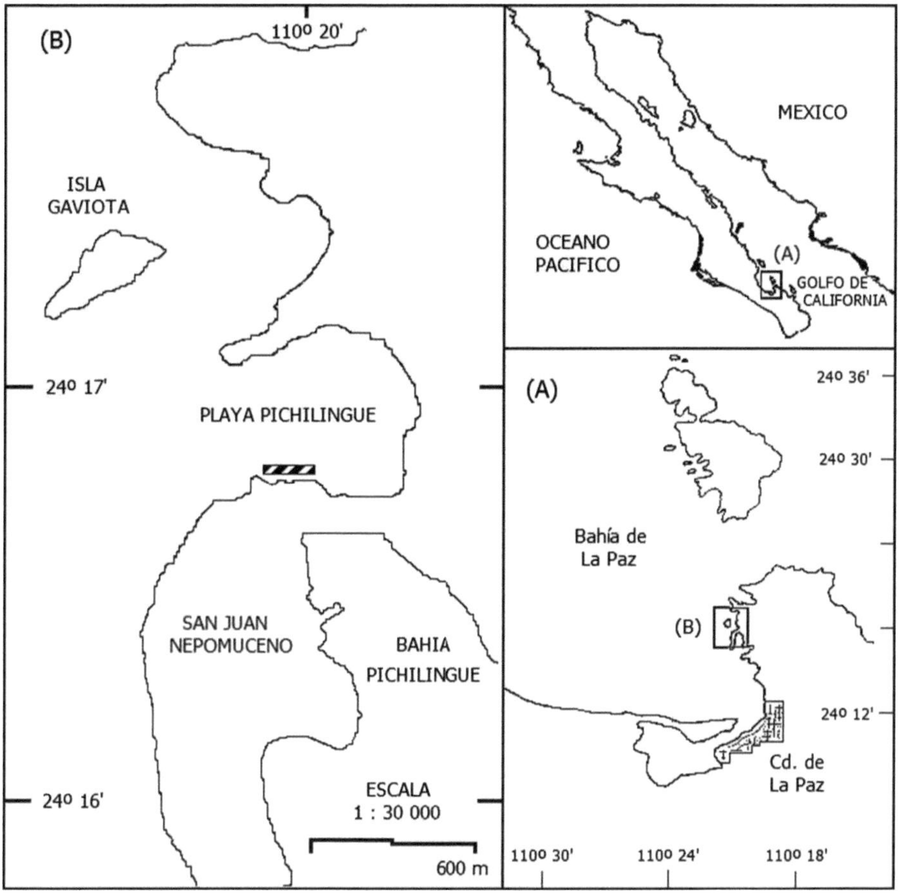

Fig. 8.1 Location of the Gulf of California, indicating (**a**) the Bahía de La Paz sector and (**b**) Gaviota Island where the main educational and Field Environmental Philosophy activities in this study are being carried out. Map prepared by José Luis Sánchez Lizaso (2011)

Box 8.1 (continued)

temperature (in winter it fluctuates between 10 and 12 °C and in summer between 30 and 32 °C) and salinity due to the dynamics of water masses, high insolation, and scarce freshwater supply (due to lack of rainfall and rivers). Its depth is highly variable. North of Tiburón and Ángel de la Guarda islands, about 50% of the seafloor is 200 m or less, but in the rest, depths can reach greater than 3500 m. As Cariño and Monteforte (2018, p. 17) explain, "The upwelling phenomena that generate these depths, coupled with the numerous

(continued)

Fig. 8.2 Invertebrate biodiversity on a rocky reef in the Gulf of California that includes taxonomic groups as diverse as starfish and corals as well as vertebrates and algae. Photograph by Francisco A. Solís Marín

Box 8.1 (continued)

estuaries and coastal lagoons, explain its high biological productivity and consequent marine biodiversity. However, any slight change in the water composition and balance of this exceptional ecosystem can have catastrophic effects on the life cycle of natural communities."

To avoid the adverse effects of mass tourism, numerous actors from different fields of knowledge are developing tourism initiatives based on more sustainable alternatives such as hiking, kayaking, and snorkeling that require little or no motorized support. In this chapter, we adapt the methodological approach of Field Environmental Philosophy (FEP) (Rozzi et al. 2010b) as a complementary model to guide us in this search for sustainable tourism. FEP is a novel approach developed at the Omora Ethnobotanical Park in southern Chile. It offers a unique holistic alternative for sustainable tourism to test as a model for Baja California, Mexico. Currently, there are no other instances of its application in the region. FEP addresses not only anthropogenic impacts on ecosystems, with an eminently ethical concern, but also promotes other life habits that are immersed in diverse ways of coinhabiting shared habitats, or what Ricardo Rozzi (2013) has coined as the 3Hs of a biocultural ethic that underpin the FEP approach. Furthermore, FEP fosters interdisciplinary work, emotiveness, empathy, and invites participants to change how they perceive biodiversity and the environment through historical, cultural, biological, poetic, and philosophical lenses (Tauro et al. 2021).

Specifically, our chapter addresses the environmental problems of the Gulf of California through five "lenses" or perspectiives, each with different approaches rooted in the disciplines of history, philosophy, and biological sciences. In this collaborative chapter, each author develops a section from his or her experience and knowledge: René Moreno-Terrazas from environmental philosophy and the starfish on Gaviota Island; Humberto González-Galván with the perspective of *Homo viator* in Cabo Pulmo; Zenorina Díaz-Gómez from the Heideggerian philosophy of inhabiting; Micheline Cariño-Olvera based on environmental history; and Mario Monteforte-Sánchez on marine sciences. This multivocality enriches the interpretation of the ecoregion.

8.2 Starry Sea and Sky of the Gulf of California

Not only does the sky have stars, but also the sea is a landscape full of them. Unlike the sky, however, in the sea, the stars are within our reach. In many parts of marine space, they shine in their own way. Asteroids and comets also flash through the sea. Immersed in the sea, the microcosm is perceived in a drop of seawater or in the eye of an octopus but then, once on the surface, looking up to the macrocosm, we travel to the celestial vault.

All cultures have had symbolic representations of the sea and the sky, each one showing a unique biocultural richness. Although the original cultures of Baja California Sur have mostly disappeared (Schmal 2007), the stories about them allow us to imagine how they may have perceived their environment.

According to several myths and legends from the southern part of the peninsula, there was a mountain cult, associated with creation by the god known as Niparajá, who visited the top of a mountain from one of the stars of Orion's Belt (Reyes 2017, pp. 27–30). According to the myth, this god created thorny vegetation, animals, and the first man and woman in the region. Niparajá is also considered the creator of the sky, the Earth, and the sea. In ceremonies, the most common offerings to Niparajá were fruits and seeds such as pitahaya (*Stenocereus thurberi*) and medesá (*Cercidium microphyllum*) (Fujita and Bulhusen 2019, p. 91). Another god, Purutahuí, is also credited with creating the stars, which, in the conception of the areas' original inhabitants, the Pericú, were made of metal (Clavijero 1852, pp. 28–29).

The Popol Vuh, the sacred Maya-Quiché text, describes that, in the time before creation, there was only the calm sea and the extensive sky (Popol Vuh 1998, p. 92). In Nahuatl mythology, the salty waters of the seas were joined to the celestial water or *ilhuicatl* (Morante 2000). Ilhuicatl, translated also as aquatic sky, divine or marvelous water, reaches out and communicates with the sky, extending into the celestial vault and blending them together, at least at night (Alcina 1997, p. 351).

Regarding the Aztec or Mexica culture, in 2018, Leonardo López Luján and collaborators (2018) identified invertebrate remains in the archaeological zone of Templo Mayor (Mexico-Tenochtitlan). To their surprise, they found six species of starfish. This finding demonstrates the symbolic importance that invertebrates had for pre-Hispanic cultures (López et al. 2018). It confirms that the Mexica offered starfish to their gods, in this case to the Mexica goddess Tlaltecuhtli. She is the goddess of the Earth who sustains creation through her own sacrifice. Tlaltecuhtli generates the cosmos and time since she is the principle that promotes the multiple creations of the world, according to this ancient myth (Magaloni 2011).

With these examples from the Maya-Quiché and Aztec cultures, we know that knowledge about the sea and the sky was important for the biocultural imaginary of these pre-Columbian traditions. With the arrival of the Jesuits, an intense process of biocultural extermination of the socio-environmental knowledge of these communities took place (Ortega 2015). Today these pre-Columbian cosmovisions have been forgotten and little is known about them. This oblivion is even more drastic with respect to the pre-Columbian cultures of the GoC, where only a few people remember them. There are some compilations of Baja California legends, but they have not been fully incorporated into today's prevailing culture. The dissemination of these stories of the ancient Californians has been limited. Including these narratives in ecotourism and experiences of FEP would be a valuable way to introduce visitors, both local and from other regions of Mexico and the world, to this ancestral biocultural knowledge.

8.2.1 Starfish in the Marine Constellation of the Gulf of California

Bahía de La Paz, on the southern Gulf coast, is home to habitats such as rocky reefs, estuaries, and mangroves. Among the co-inhabitants we can find bluish-brown or tan starfish (*Phataria unifascialis*) and the yellow-spotted or pyramid starfish (*Phataria pyramidata*) (Fig. 8.3). Oscar Holguín-Quiñones, malacologist specializing in the GoC, explains that the names change according to the area where the group of fishermen or divers live. Both species are dried and traded in the seashell market. In addition to illegal extraction, overfishing is a constant threat to the Gulf of California (Johnson et al. 2017).

Fig. 8.3 Gulf of California starfish. Left, *Phataria pyramidata* and right, *Phataria unifascialis*. Photograph Francisco A. Solís Marín

Starfish, *acitlalmichitli* in Nahuatl, can survive for several months without feeding and still conserve 80% of their body mass. Since they are opportunistic predators, they eat everything in their path. They are animals with bilateral symmetry and deuterostome embryonic development, just like humans!

Researching these organisms, scientists found what could be the common ancestor of humans and starfish. Dubbed "Saccorhytus," it dates back more than 500 million years (Han et al. 2017). The bilateral symmetry of humans is combined with the radial symmetry of starfish, and this synthesis is so important that the starry pentagon is one of the main sacred geometric figures. It is a source of inspiration for imagining the stars in the sky and symbolizes the divine proportion of the human body. This knowledge is lost in abyssal time, where esoteric, magic, and spiritual knowledges are mixed. The geometric representation is a source of inspiration in cultures all over the world and surely the pentradial stars continue to nourish their biocultural imagery (on sacred geometry see Skinner 2006). Similarly, as described by Rozzi (2019), chauvinistic boundaries and hierarchies between humans and other animals, including those small ones submerged in the sea, dissolve. In their anatomical and physiological plan, humans and other animals have eyes, mouths, arms, feet, nervous systems, and hormones such as oxytocin that stimulate desires and passions such as feeding and courtship in reproduction.

8.2.2 Field Environmental Philosophy at Gaviota Island

Gaviota Island is located within the Bay of La Paz (24°17′N; 110°20′W), on the southeastern coast of the Baja California Peninsula (Fig. 8.1). Most of the perimeter of Gaviota Island has cliffs of up to 10 m, with only one beach in the southeastern part, approximately 50 m long and with a low slope (Carmona 1993, p. 9). The island is located within the Balandra Natural Protected Area (Balandra NPA) (Colectivo Balandra 2021). This region was designated a Mexican Natural Protected Area in 2012 because of its geographic, environmental, and socioeconomic importance. It is also considered a restricted-use subzone, meaning that certain kinds of development are prohibited (Fig. 8.4).

Brown pelicans (*Pelecanus occidentalis*), cormorants (*Phalacrocorax auritus* and *Phalacrocorax penicillatus*), magnificent frigatebird (*Fregata magnificens*), and boobies (*Sula leucogaster* and *Sula nebouxii*) often roost on the island (Carmona 1993, p. 54). Fish that frequently accompany starfish include parrotfish (of the genera Calotomus, Nicholsina, and Scarus), scissortail damselfish (*Chromis Atrilobata*), damselfish of the Sea of Cortez (*Stegastes rectifraenum*), royal angel (*Holacanthus passer*), rainbow damselfish (*Thalassoma lucasanum*), Mexican chestnut (*Chromis limbaughi*), grouper (*Paranthias colonus*), copeetona or cockerel (*Bodianus diplotaenia*), banderita (*Abudefduf troschelii*), tamboril bonito or botete (*Canthigaster punctatissima*), and barber butterfly (*Johnrandallia nigrirostris*). All cohabit the rocky reefs of *Porites californica, Pocillopora elegans,* and *Pocillopora damicornis*. Others that coinhabitant with the starfish are polychaetes, bivalve

Fig. 8.4 Aerial view of Gaviota Island, where we conducted field practice on August 30, 2019. Photograph by Israel Sánchez Alcántara (2014)

mollusks, crustaceans, sponges, calcareous algae, red algae, brittle stars, and sea urchins.

An advantage of having a coastal educational experience, *in situ,* is witnessing the vast expanse of the sea and the sky coming together on the horizon in a blue continuum. The coast allows us to expand the vision of the *oikos* or home: it is one of the few places where you have an open panorama to imagine the origin and infinity of the universe. The contemplation of this immense space easily awakens the poetic gaze.

8.2.2.1 Ecotourism with a Viewfinder: Face-to-Face Encounter with the Stars

To awaken environmental awareness, ethical valuation of the marine ecosystem, and knowledge about biological evolution, we propose "face-to-face" encounters with diverse human beings and other organisms (Rozzi et al. 2006) by adapting the practice of FEP (Rozzi et al. 2010) to the biocultural locality of Gaviota Island. We have created the metaphor "viewfinder for the stars," to describe this low-impact ecotourism strategy.

The "viewfinder for the stars" experience begins with kayaking from Pichilingue beach located near the city of La Paz, to Gaviota Island (Fig. 8.5), where visitors snorkel to observe the cohabitants of the rocky reef patches (Fig. 8.2). Visitors are encouraged to appreciate the colors, shapes, and similarities, and to learn to identify other forms of life. When we return to the beach, we discuss what we have observed, make drawings, and create metaphors. Finally, we wait for the night to see the starry

Fig. 8.5 Philosophy students, sailing to Gaviota Island on August 30, 2019. Photograph by our guide and environmental interpretation instructor, Mario Escalera Gómez

sky. We notice how the horizon disappears, and the sea and the sky become one. "We are stardust," as Carl Sagan used to say... not only stellar, but also marine stardust.

8.2.2.2 Cabo Pulmo and the Experience of Newness

"Philosophical hiking" is a university elective course for humanities majors (philosophy, history, modern languages, literature), and for the degree in alternative tourism. The objective of this course is ambitious: to explore the philosophical idea of *Homo viator*, which generally means to view travel and traveling as substantive to being human. Philosophical hiking is a community pedagogy that articulates, in a peripatetic way (in the sense of Aristotle who taught while walking with his disciples, as well as in *Homo viator*), thought with emotion, and nature with personal culture. These juxtaposed terms describe a holistic experience.

As a group, we shared an experience comparable to FEP (Rozzi 2013). We snorkeled in "Los Arbolitos" bay (little tree) located near the town of Cabo Pulmo, observing and admiring some of the starfish species that René Moreno describes in the previous section. We then walked the trail to "La Hornilla" (little oven) near El Cardonal guided by Gabriela Castro (a graduate in alternative tourism and native resident of La Hornilla). I also had a personal opportunity to talk with her extensively about the various activities that the community of Cabo Pulmo implements in response to the predatory development threats to the site. Several days later, back in the classroom, we reflected on this experience from the philosophical perspective of *Homo viator* that we had been exploring since the beginning of the semester.

In this sense, the experience was carried out according to *Homo viator*, but in an unpredictable way:

> the acquisition of experiences requires certain attitudes. Man cannot force them, but he can expose himself to those situations which are the only ones in which he can have certain kinds of experiences. Let us recall the original meaning of the word [*Homo viator*] is "man on a journey" (*auf der Fahrt*), one who risks being in foreign lands because his own environment is too narrow for him. Indeed, real experience is not possible in the protected world of one's home and familiar surroundings. That is why those who have never left the narrow circle of the "belfry" are despised. No experience can be gained from a secure position; not even those that come from an elaborated scientific problem. On the contrary, to have experiences we must commit ourselves, expose ourselves to the unexpected that presents itself. We can only have real experiences when we are open to to the unexpected. But this requires daring and a previous disposition, for the acquisition of experiences is painful and involves dangers. (Bollnow 1976, p. 151)

In this sense, experiences are symbolic of "being exposed." I still have a scar on my forehead left by a branch of a Brazil tree that I walked into when I took the wrong path toward La Hornilla, which I did not even reach. That is what the *entity that walks outside* (*ex-peri-entiae*) risks: not arriving. To a "not arriving" that, nevertheless, always arrives. In this case, we reached the summit of an ancient place, a Guaycura pantheon,—named for the original inhabitants of the region—which we had not planned to "discover."

My expectations of the experience were also fulfilled in a different sense than expected. I had planned (in fact, it was what I planned the most) a bonfire at which Gabriela Castro and her brother Juan (a local service provider), would tell us about the resistance work that the community of Cabo Pulmo sustains to confront the onslaught of predatory development. This kind of development destroys everything by transforming it into merchandise and profit for individuals, without any real collective benefit.

The bonfire materialized in an unthinkable way. Each of the participants (the twelve students, plus the driver, Miguel) would offer their own description of what happened (even I have different versions of what happened). However, against the background of that Apollonian fire, I preferred to present myself as a moderator of the Dionysian spirit; a faint light in a pile of stones; an altar built by Daniel in honor of the sacred balance.[1] We were accompanied by a pair of splendid dogs, lying at the foot of the bonfire. All this led to a free-wheeling conversation to the point of exhaustion, and for me, had the personal reward of a solitary and silent return to a transitory home in a cabin at two o'clock in the morning, walking the only street of a very dark and quiet village of former fishermen, at that moment, so full of stars. *The new* is paradoxical. It is an experience of otherness, sometimes arising from the most commonplace, when one is in a state of *Homo viator*. It is more an attitude and willingness of Being, than a fixed goal. *Homo viator* journeys step-by-step into *the new* with ontological authenticity. Ernst Bloch (2007, p. 425. Italics mine) tells us:

> The same things kill slowly every day. *To crave something new: the pleasure of the journey helps this. The pleasure of the journey not only renews the expectation, before embarking on the road, but also does so in the joy of looking.* Desires fall by the wayside and can no longer be remedied. . . . The moldy also falls by the wayside, which may have adhered not only to everyday life, always the same, but also to desires that have been dragged along for too long. The dreams of desire can, in fact, fall to such an extent outside the time that is proper to them, that they can never again be realized. Whoever desired a Kodak [camera] in his youth and did not get it then, will never find the Kodak of his desires. Even if, as a man, he can buy the best of them.

Long before embarking on this trip, I had already worked up my enthusiasm with rich anticipations of perfection and limits. The concept of double anticipation (perfection and limit) circumscribes the conditions of hermeneutic possibility to each given situation. It comes indirectly from the philosophers Hans-Georg Gadamer, Martin Heidegger, Ernst Bloch, and Otto Bollnow, although none of them systematize it in any specific text. So, full of new expectations, this philosophical hiking trip to Cabo Pulmo was gestating in me. Were my expectations fulfilled? My narrative, up to this point, indicates a simultaneous yes and no, as it had to be. "Where everything falls apart, the way to the strange is not missing or has not been missing" (Bloch 2007, p. 454. Paraphrase mine).

8.3 On the Poetics of Inhabiting

In the twentieth century, Hannah Arendt (2005) expressed her concern for the triumph of humans, the *animal laborans*, over the world, as opposed to the cultivation of the active life that was appreciated by ancient Greek culture. One of her central theses is that the fateful experience of early twentieth-century totalitarianisms was conditioned by the growth processes of economic productivity. This in turn led to the rupture of community ties, and thus, the loneliness and isolation of people (Arendt 2005). According to Arendt, *labor*, *work*, and *action* are the three fundamental ways in which human beings develop their lives. The first refers to the satisfaction of biological needs, the second to the creation of satisfiers and objects

---

[1]These are alusions to myths of the Greek gods Apollo, Dionnysus, and the Book of Daniel in the Hebrew Bible.

of utility, while the last is the way in which the human transcends need and becomes a free being (Arendt 2005, p. 35). Of these three ways in which humans inhabit the earth, it is action that allows them to transcend their finite condition. As the philosopher explains, it is this possibility that the ancient people recognized in the so-called active life. The *life of pleasures*, *life in the service of the city*, and *contemplation* were recognized as conditions freed from necessity and therefore linked to beauty. Thus, in a universe where everything manifested an eternal character, *active life* was the way in which mortals could participate in eternity.

Paradoxically, this stalking of necessity, long endured by humanity, has conjured up new dangers in the modern era through scientific-technical development and an overabundance of satisfiers generated by the *animal laborans*. Far from providing security, and far from allowing longevity of life, it has worn it out. Modern life, in this sense, does not express enjoyment and fullfilledness, but weariness. The desire or the greed for novelty never ceases because it obeys the irrational demand to constantly accumulate capital. In the crudest sense, life is reduced to being consumed and discarded. It is thoroughly corroded. How can this corrosion be stopped? How should we "inhabit" or "dwell?"

The primordial unity that links all living beings is fractured. Due to their reflective capacity, instead of instinctive action, human beings must evaluate and choose among the multiple possibilities that are presented to them. They have become free beings. Their decisions will determine their lives, their own and that of their cohabitants. Therefore, their dwelling is not only a physical dwelling but a spiritual one. It is a meaningful dwelling that edifies and is carefully inhabited. In this chapter, this is the meaning of *ethos*; the biocultural *ethos,* because it contemplates not only caring for biophysical space but also caring for the symbolic-cultural and socio-political dimensions (Rozzi 2012).

The best-known meaning of the word *ethos*, which is character or way of being, appears in Plato's texts when "knowledge" had become "philosophy" and "philosophy" was something taught in schools. As Heidegger rightly expressed, this does not mean that before its formal appearance there was no sense of *ethos*. As proof of this, the philosopher recovers the ancient meaning of the word as expressed by Heraclitus in the collected tragedies of Sophocles: "The dwelling is for man, the ordinary place for the presentation of God" (Heidegger 2001, p. 78). *Ethos* is, then, the place of dwelling or inhabitating. From this reference, Heidegger continues searching until he finds that in old German the word *buan* also means "to inhabit," but not only that, in its derivation, *bauen* also means "to take care, to cultivate." Then in Wunian Gothic, "to inhabit" means "to care for, to preserve, to leave free from harm." After tracing this etymological history, he concludes that human "inhabiting" is "to take care." In clear opposition to the humanist tradition, which had placed humans as the foundation and center of the world, Heidegger answers that humanity is the shepherd of being, and that it is up to them to care for it. The "quadrant" that synthesizes this place of human belonging, is the coming together of earth and sky, divine and mortal. Heidegger called this "the quaternity" (Heidegger 1994, p. 132).

The recovery of wisdom stored in language helps us understand that the meaning of inhabiting is to take care of the physical and spiritual place of inhabiting or one's dwelling. This implies precisely safeguarding memory, the ancient wisdom that through the years has helped us to cultivate physical and spiritual spaces and relationships with those with whom we coinhabit. In this sense, caring is not only preserving, it is building because, as Heidegger warned (citing a poem by Friedrich Hölderlin), poetically humanity inhabits, or poetically humanity should inhabit.

> Is God unknown? Is he manifest as the heavens? This is what I believe. The measure of man is this: full of merit, yet man poetically inhabits this earth. But the shadow of night is not purer with the stars. If I can say this, as a man, who is called image of divinity: Is there on earth something that can be done? There is nothing. (Hölderlin 1983, p. 117)

But how to recover this memory sedimented in ancient language? How to listen again to the meaning of *inhabit*? Heidegger says that to inhabit "being," it will be necessary to be silent (2001, p. 20). Silence is the predisposition for encountering the meaning of being in the world. Silence separates us from the frenzy of modern haste, it allows a parenthesis in time for the senses to sharpen and appreciate beauty and, as in the rapture of an eternal instant, the "I" overcomes the fracture and again feels part of the whole, as cohabitant. That kind of experience is what is known as aesthetic experience, so vital, so purifying, so necessary to remember that the good life is beyond utility. Because "one *has* a world if there are useless things there, one *inhabits* a world if one does things beyond that which is useful. That which is beyond the usefulness of common things is the beauty of those things. In that sense, beauty is connected to freedom" (Heller 1999, p. 312. Italics mine). The meaning of life, therefore, is more than satisfying basic needs; it is more than a parasitic living. It is the delight of beautiful things, the freedom to experience that delight, and to live poetically, reproducing that beauty in doing and in caring for and building the dwelling place. For this reason, Trías (1997, p. 46) expressed that, while the ascending ecstasy produced in the contemplation of beautiful things, is mystical, descending from ecstasy becomes civic because it is coming back and communicating through the creative or artistic act.

Practices aimed at producing an ethical-aesthetic experience, such as those conducted by the FEP, are significant because they allow us to recognize ourselves as co-inhabitants and, therefore, responsible for dwelling sustained by physical and cultural diversity. This learning constitutes an authentic educational practice. It is formative in a profound sense if we consider that it aims to educate one's own sensibility. In this regard, let us recall the importance that the idea of "good taste" had for so long in the Western tradition. It was the foundation of civilized life, since it expressed a refinement of the senses that made it possible to distinguish the good from the bad, the worthy from the unworthy. And, as Hans Gadamer (2012, p. 72) explains, before circumscribing this quality to the mere field of art, it served as the foundation of community life. Having good taste, or good sense, meant living in community with justice. Thus, the ethics of moderation, so dear to the ancient Greeks, consisted of an ethics of good taste, founded on the contemplative experience that allowed them to conceive the idea of order and beauty, and from that recognition, the possibility of transcending a life subordinated to mere necessity. Therefore, the challenge of this reflection continues to be the one expressed by

Friedrich Schiller (1990, p. 181): "by means of beauty to escape the sense of lostness in which we find ourselves."

With aesthetics and poetry, we can create metaphors, as proposed by FEP, to recover the meanings of words (that have been lost or hidden by a dualistic, egocentric, anthropocentric ontology) that signify careful cohabitation in the same habitat. Recovering these meanings will redirect us along the path toward reconnections, to remembering our origin, and to our rootedness with the others. Doing this broadens the poetic gaze that allows us to imagine relational ontologies that will reconnect/entangle/root us with the whole.

8.4 Environmental History of the Gulf of California

For the past two decades, the Mexican government has made great efforts in environmental policy planning to take advantage of the GoC's resources and conserve its high ecological value. It has created 15 natural protected areas and the Ordenamiento Ecológico Marino del GdC (OEMGC)—a specific legal policy governing GoC habitats—was the first of its kind to be decreed (in 2006) at the national level (SEGOB 2006). In 2005, the nearly 900 islands of the Gulf were recognized by UNESCO as a Natural World Heritage Site (UNESCO 2021), and it is home to numerous coastal wetlands recognized as Ramsar Sites (RAMSAR 2009). In the national context, it is of great economic importance due to both its rich fisheries (70% of national production) and its scenic beauty, which gives rise to a diverse tourism industry. In this section, we examine concepts that contribute to a sustainable tourism industry and that provide an integral experience of appreciation of the biological and cultural diversity of the GoC.

The exploration and colonization processes in the GoC began during the colonial period. In 1533, the first Spanish ship arrived and reported its pearl wealth. Two years later, Hernán Cortés, owner of that ship, arrived at the Baja California peninsula with the intention of settling and founding a colony (Cariño and Monteforte 2012). But the arid climate of the region delayed that purpose for seventeen decades. However, during the sixteenth and seventeenth centuries, the Gulf was crossed by numerous pearling expeditions as well as others for exploratory purposes. Their reports also brought news of the diverse native peoples that inhabited both coasts of the Gulf.

The peninsula was once home to hunter-gatherers who survived less than a century after colonization by Jesuit missionaries in 1697. The original inhabitants of the northern part of the peninsula and the other coastal areas survive to this day, but through the centuries their territory and population have decreased drastically. The Cucapa live in the vast area of the Colorado River delta. The Comcáac (Seris) inhabit the north coast of the current state of Sonora and on Tiburón Island. On the southern coast of this state are the Yoremes (Yaquis) and in the north of Sinaloa, the Yoremes-Mayo. Finally, on the coast of Nayarit, the Nayeri (Coras) and Wirraritari (Huicholes) have sacred sites and ancestral territories. These peoples have resisted colonization and dispossession, live in a state of socioeconomic and political marginalization, and face serious socio-environmental conflicts (Boege 2008; Luque 2012). Nevertheless, their knowledge and culture of nature compose the rich biocultural heritage of the GoC.

Throughout history, the GoC has gone from being seen as an area separating the northwestern states—to the west Baja California and Baja California Sur, and to the east Sonora, Sinaloa, and Nayarit—to being designated an integration and administration zone based on environmental criteria (Doode and Wong 2001). It should also be noted that the Gulf and its coasts are a region in dispute because the conceptions of nature and the economic interests of different actors and social sectors (industrial fishing, coastal fishing, sport fishing, aquaculture, tourism, environmentalists, and academics, as well as indigenous peoples) are in conflict. This situation has generated a variety and growing number of environmental conflicts among those that hold certain decision-making power, as well as ecological-distributive conflicts between them and the local population.

This region has been extensively studied by the natural sciences; there are tens of thousands of international publications that address multitude aspects related to ecology, biodiversity, and threats to both. However, from the social sciences the region has been little studied, in spite of the fact that the intensive use of the natural resources of this marine-coastal space (with the corresponding affectation of its inhabitants) dates to colonial times. This absence is even more noticeable in the case of interdisciplinary studies, especially those with a critical perspective.

From the sixteenth century to the present, the GoC, due to the richness of its biodiversity, the greed for its resources, and the beauty of its landscapes, has awakened fascination and amazement. Unfortunately, an asymmetrical appropriation of power by external actors has resulted in subalternizing and ignoring its inhabitants. Their aspirations and expectations have not been considered in the decision-making processes that affect their territory and culture. This has generated processes of exclusion associated with a wave of biocultural homogenization, and, in the era of rapid socio-environmental change that we are living, it is essential to raise awareness of this situation and to overcome it through a new biocultural ethic (Rozzi 2018).

The inhabitants of these southern Californian regions are capable of weaving networks of knowledge and practices that can culturally and environmentally restore their dwelling space. This is evident in their habits of promoting activities that contribute to the restoration of the habitat and in their many environmental institutions and civic organizations. The FEP adds to this Baja California culture with its methodological approach involving transdisciplinary research, concern for environmental health, and enrichment of environmental education. The need for collaboration among diverse stakeholders is a necessary condition for implementing appropriate policies promoting forms of ecotourism that bring equitable benefits to local communities while incentivizing sustainable practices and biodiversity conservation (Johnson et al. 2019). To this end, our adaptation of the FEP structures a multi-vocal collaboration of specialists from the marine sciences with specialists from the social and human sciences, triggering change-generating synergies.

To overcome the coloniality of power, being, knowledge, and nature that has existed in the GoC since the sixteenth century, it is essential that the rural and urban communities inhabiting its coasts be made aware of the importance of their territory for their existence, and consequently, the importance and urgency of its defense. The GoC does not have a natural vocation for tourism, as real estate interests and large corporations have emphasized (Cariño et al. 2021). Nor should it be considered a reserve for fishing or mining resources (Cariño and Domínguez 2021). It is the biocultural heritage of the people who inhabit it and depend on it for their livelihoods that should be the focus of development policies. Being part of the global ocean system, the GoC is also a source of life for humanity. The only way in which this extraordinary marine and coastal space can continue to provide its vital function is by defending it against the violation of its ecological integrity and the biocultural right to life.

So far, these brief sentimental reflections, lovingly and sincerely, are committed to a relational ontology. From the coasts of Southern California, we share our eco-aesthetic gazes from the starfish to the stars of the sky, trying to expand our scales, beats, pulses, and rhythms as a vital step to improve our cohabitation with all the biodiversity of the Gulf of California and the biosphere.

Acknowledgment We appreciate the support of Roy May and Ricardo Rozzi in editing and translating this chapter. RMT thanks the support from the Cape Horn International Center (CHIC)—ANID PIA/BASAL (PFB210018).

References

Alcina J (1997) El agua primordial entre los mexicas. In: Manrique L, Castillo N (eds) Homenaje al doctor Ignacio Bernal. INAH, México

Arendt H (2005) La condición humana. Paidós, Barcelona

Arizpe O, Covarrubias M (2010) Sustainable tourism planning for the only coral reef in the Gulf of California: Cabo Pulmo national park. WIT Trans Ecol Environ 139:263–273

Bloch E (2007) El principio esperanza I. Trotta, Madrid

Boege E (2008) El patrimonio biocultural de los pueblos indígenas de México. INAH/CDI, México

Bollnow O (1976) Introducción a la filosofía del conocimiento. La comprensión previa y la experiencia de lo nuevo. Amorrortu, Buenos Aires

Brusca RC, Wallerstein BR (1979) The marine isopod crustaceans of the Gulf of California. II. Idoteidae: new genus, new species, new records and comments on the morphology, taxonomy and evolution within the family. Proc Biol Soc Wash 92(2):253–271

Caira JN, Jensen K (2014) A digest of elasmobranch tapeworms. J Parasitol 100(4):373–391

Cariño M, Domínguez W (eds) (2021) Extractivismo industrial y comercial del golfo de California (1830-2017). Volumen II de la Colección: Nuestro mar: Historia ambiental crítica del golfo de California (siglos XVI-XXI). Comares, Granada (in press)

Cariño M, Monteforte M (2012) Episodes of environmental history in the Gulf of California: fisheries, commerce and aquaculture of Nacre and Pearls. In: Boyer C (ed) A land between waters. Environmental histories of modern Mexico. The University of Arizona Press, Tucson, pp 245–276

Cariño M, Monteforte M (2018) La Historiografía Ambiental Sudcaliforniana y La Sustentabilidad Local. Historia Ambiental Latinoamericana y Caribeña (HALAC) revista de la Solcha 8(2): 14–41. https://doi.org/10.32991/2237-2717.2018v8i2

Cariño M, Valiente C, Ortega A (eds) (2021) El golfo de California: un espacio excepcional (1900-2017). Volumen III de la Colección: Nuestro mar: Historia ambiental crítica del golfo de California (siglos XVI-XXI). Comares, Granada (in press)

Carmona R (1993) Reproducción y crecimiento de dos especies de gaviota Larus livens y L. Occidentalis anidantes en Baja California Sur. Tesis de Maestría. CICIMAR-IPN

Cisneros-Montemayor AM, Townsel A, Gonzales CM, Haas AR, Navarro-Holm EE, Salorio-Zuñiga T, Johnson AF (2020) Nature-based marine tourism in the Gulf of California and Baja California Peninsula: economic benefits and key species. Nat Res Forum 44(2):111–128. https://doi.org/10.1111/1477-8947.12193

Clavijero F (1852) Historia de la Antigua ó Baja California. Colección Digital UANL. Available via: http://cdigital.dgb.uanl.mx/la/1080013153/1080013153_08.pdf. Accessed 26 May 2020

Colectivo Balandra (2021) Balandra Área Natural Protegida. Available via: http://balandraesnuestra.org/balandra-area-natural-protegida/. Accessed 18 Apr 2021

CONANP (2000) Programa de Manejo Área de Protección de Flora y Fauna Islas del Golfo de California. SEMARNAP, México

Doode S, Wong P (2001) El Golfo de California: surgimiento de nuevos actores sociales, ambientalismo y región. Estudios Sociales 11:25–56

Fujita H, Bulhusen K (2019) Sitios pictográficos en la Sierra de las Cacachilas, Baja California Sur, México. In: Lara-Galicia A (coord) Las manifestaciones rupestres en México. Técnica, iconografía y paisaje. AcerVos y Universidad Pablo de Olavide, Sevilla, pp 89–113

Gadamer H (2012) Verdad y método II. Sigueme, Salamanca

Han J, Morris S, Ou Q et al (2017) Meiofaunal deuterostomes from the basal Cambrian of Shaanxi. Nature 542:228–231. https://doi.org/10.1038/nature21072

Heidegger M (1994) Construir, habitar, pensar. In Conferencias y artículos. Serbal, Madrid

Heidegger M (2001) Carta sobre el humanismo. Alianza, Madrid

Heller Á (1999) Una filosofía de la historia en fragmentos. Gedisa, Barcelona

Hölderlin F (1983) Las grandes elegías. Versión de J. Talens. Hiperión, España

Johnson AF, Moreno-Báez M, Giron-Nava A, Corominas J, Erisman B, Ezcurra E, Aburto-Oropeza O (2017) A spatial method to calculate small-scale fisheries effort in data poor scenarios. PLoS One. https://doi.org/10.1371/journal.pone.0174064

Johnson AF, Gonzales C, Townsel A, Cisneros-Montemayor AM (2019) Marine ecotourism in the Gulf of California and the Baja California Peninsula: research trends and information gaps. Scientia Marina 83(2):177–185

Ledsma-Vázquez J, Johnson ME, Gonzalez-Yajimovich O (2009) Gulf of California geography, geological origins, oceanography, and sedimentation patterns. In: Atlas of coastal ecosystems in the Gulf of California: past and present. The University of Arizona Press, Tempe, pp 1–10

López L, Solís F, Zúñiga-Arellano B, Ochoa A, Vargas CAC, Cao-Romero CM, Méndez IE (2018) Del Océano al Altiplano. Las estrellas marinas del Templo Mayor de Tenochtitlan. Arqueología Mexicana 150:68–76

López-Espinosa de los Monteros R (2002) Evaluating ecotourism in natural protected areas of La Paz Bay, Baja California Sur, México: Ecotourism or nature-based tourism? Biodivers Conserv 11:1539–1550. https://doi.org/10.1023/A:1016887417263

Luque D (2012) Del mar y del desierto: gastronomía comcáac (seri). Ecoturismo y pueblos indígenas. Centro de Investigación en Alimentación y Desarrollo, A.C. Hermosillo, Sonora, México

Magaloni D (2011) El origen mítico de las ciudades: la iconografía de la creación. Arqueología Mexicana 107:29–33

Morante RB (2000) El universo mesoamericano: Conceptos integradores. Desacatos 5:31–44. Available via: https://desacatos.ciesas.edu.mx/index.php/Desacatos/article/view/1220. Accessed 28 May 2020

Ortega A (2015) Diálogo de saberes ambientales entre Europa-América. Agroecosistemas oasianos en Baja California Sur s. XVIII-XX. Asclepio 67(1):76. https://doi.org/10.3989/asclepio.2015.02
Popol Vuh (1998) Lecturas históricas mexicanas, 2a. ed. selección, prefacio, notas y tablas cronológicas de Ernesto de la Torre Villar. UNAM-Instituto de Investigaciones Históricas, México
RAMSAR (2009) Mexico Names Marine Site on Gulf of California. Available via: Mexico names marine site on Gulf of California | Ramsar
Reyes L (2017) Mitos, leyendas y tradiciones sudcalifornianas. ISC, La Paz, B.C.S.
Rozzi R (2012) Biocultural ethics: the vital links between the inhabitants, their habits and regional habitats. Environ Ethics 34:27–50
Rozzi R (2013) Biocultural ethics: from biocultural homogenization toward biocultural conservation. In: Linking ecology and ethics for a changing world. Springer, Dordrecht, pp 9–32
Rozzi R (2018) Biocultural homogenization: a wicked problem in the anthropocene. In: From biocultural homogenization to biocultural conservation. Springer, Cham, pp 21–48
Rozzi R (2019) Taxonomic Chauvinism, No More! Antidotes from Hume, Darwin, and biocultural ethics. Environ Ethics 41:249–282
Rozzi R, Massardo F, Anderson C, Heidinger K, Silander J Jr (2006) Ten principles for biocultural conservation at the southern tip of the Americas: the approach of the Omora Ethnobotanical Park. Ecol Soc 11(1):43 [online] URL: http://www.ecologyandsociety.org/vol11/iss1/art43/
Rozzi R, Massardo F, Cruz F, Grenier C, Muñoz A, Mueller E (2010a) Galapagos and Cape Horn: ecotourism or greenwashing in two emblematic Latin American archipelagoes? Environ Philos 7(2):1–32
Rozzi R, Anderson CB, Pizarro JC, Massardo F, Medina Y, Mansilla A, Kalin MT (2010b) Field environmental philosophy and biocultural conservation at the Omora Ethnobotanical Park: methodological approaches to broaden the ways of integrating the social component ("S") in long-term socio-ecological research (LTSER) sites. Revista Chilena de Historia Natural 83(1): 27–68
Schiller F (1990) Carta sobre la educación estética del hombre. Anthropos, Barcelona
Schmal JP (2007) Indigenous Baja. History of Mexico. Available via: History of Mexico - Indigenous Baja (houstonculture.org)
SEGOB (2006) Diario Oficial de la Federación. Available via: DOF - Diario Oficial de la Federación
Seminoff JA (2010) Sea turtles of the Gulf of California: biology, culture, and conservation. In: Brusca RC (ed) The Gulf of California: biodiversity and conservation. ASDM studies in natural history. Arizona-Sonora Desert Museum Press and University of Arizona Press, Tucson, pp 135–167
Skinner S (2006) Scared geometry. Sterling, New York
Tauro A, Ojeda J, Caviness T, Moses K, Moreno R, Wright T, Zhu D, Poole A, Massardo F, Rozzi R (2021) Field environmental philosophy: a biocultural ethic approach to education and ecotourism for sustainability. Sustainability 13:4526. https://doi.org/10.3390/su13084526
Trías E (1997) El artista y la ciudad. Anagrama, Barcelona
UNESCO (2021) Islands and protected areas of the Gulf of California. Available via: Islands and Protected Areas of the Gulf of California - UNESCO World Heritage Centre
Urbán J (2010) Marine mammals of the Gulf of California: an overview of diversity and conservation status. En: R. C. Brusca (ed) The Gulf of California: biodiversity and conservation. ASDM studies in natural history. Arizona-Sonora Desert Museum Press & University of Arizona Press, pp 188-209
WWF (2021) Golfo de California. Available via: https://www.wwf.org.mx/que hacemos/programas/golfo california/. Accessed 21 Apr 2021

Chapter 9
Biocitizen's Approach to Biotic Wonder, Citizenship, and Field Environmental Philosophy

Kurt Heidinger, Jesse Carmichael, Benny Jacobs-Schwartz, and Marysia Borucinska-Begg

Abstract The extinction crisis precipitates rapid socioeconomic changes on regional, national, and global levels. Institutional educational systems struggle to keep up with these changes, creating opportunities for start-ups to provide novel educational services that address them, such as Field Environmental Philosophy (FEP). Since 2009, Biocitizen, Inc. has provided FEP "unschooling" and eco-literacy instruction services for students K-12 in Northampton, Massachusetts, Los Angeles, California, New York City, New York, and Valparaiso, Chile. This essay details why and how Biocitizen Inc. enhances institutional educational programming with FEP.

Keywords Biocultural · Ecobiotic · Extinction crisis · Education · Pedagogy

9.1 What Biocitizen Is

Biocitizen, Inc. (hereafter Biocitizen) is a non-profit business that provides eco-literacy instruction to nurture, shape, and unleash "biotic citizens" as proposed by Aldo Leopold. Since 2009, Biocitizen has developed "unschooling" programs for K-12 students modelled after Aristotle's Lyceum, as a "walking" school (see Box 9.1, and Moreno-Terrazas et al. 2022 in this volume) that promulgates the

---

K. Heidinger (✉)
Biocitizen, Inc., Westhampton, MA, USA
e-mail: kheidinger@biocitizen.org

J. Carmichael
Biocitizen California, Los Angeles, CA, USA
e-mail: jesse@biocitizen.org

B. Jacobs-Schwartz · M. Borucinska-Begg
Biocitizen LA, Los Angeles, CA, USA
e-mail: benny@biocitizen.org; info_la@biocitizen.org

R. Rozzi et al. (eds.), *Field Environmental Philosophy*, Ecology and Ethics 5,
https://doi.org/10.1007/978-3-031-23368-5_9

Leopoldian "Land Ethic" (1949) through Field Environmental Philosophy (FEP) programming.

In the "Land Ethic," his final essay of the book *A Sand County Almanac*, Leopold stated: "a thing is right when it tends to preserve the integrity, stability, and beauty of the biotic community. It is wrong when it tends otherwise" (Leopold 1949, pp. 224–225). His land ethic "changes the role of *Homo sapiens* from conqueror of the land-community to plain member and citizen of it" (Leopold 1949, p. 223). It is important to note first that Leopold speaks of humans on a species', and not only a cultural, level, and second that living the land ethic requires us to live mindfully on the species', and not strictly cultural, level.

When Leopold presented "man the conqueror versus man the biotic citizen" (Leopold 1949, p. 223) as antipodes, and adjudged the latter as existentially consonant with the land ethic, he provided the name of our *unschool* and its mission, which aims to usher students into a vibrant awareness of their *ecobiotic selves* so that they may live and behave accordingly.

Sixty years after Leopold's essay, the Ecological Society of America's "Earth Stewardship" Initiative has called for "a new ethic of environmental citizenship on the part of individuals, businesses, and governments" (ESA 2021). Biocitizen responds to its call with programming that cultivates the biotic citizenship ethos in our students, and their families and local community. Earth Stewardship requires considering the planetary scale and the biophysical, cultural, socio-political, and institutional diversity it includes (Rozzi et al. 2015). Having emerged both in North and South America, Biocitizen's mission is transnational and cosmopolitan, while also being locally focused because its FEP curriculum is executed where its students and their families live. It builds on fruitful collaborations with the interdisciplinary team of the Omora Ethnobotanical Park that is developing Field Environmental Philosophy (FEP) praxis at the southern tip of Chile (Box 9.1). Furthermore, it builds upon a long personal and institutional history of environmental action (Box 9.2).

Box 9.1 Biocitizen Germinated in Dialogue with the Omora Ethnobotanical Park, Cape Horn Biosphere Reserve, Chile

> The lichen becomes the symbol of symbiosis. We, too, live together. And we carry our awareness of that togetherness of life and living, of our own ability to change lenses and share new experiences with others, of the importance of even the little things, with us for the rest of our lives.—Britt Holbrook (2012, p. 421)

Biocitizen began as a series of interdisciplinary research and teaching experiments conducted between 1998–2004 at the University of Connecticut (UConn), by then graduate students Kurt Heidinger and Ricardo Rozzi, from the English and the Ecology and Evolutionary Biology departments, respectively. Heidinger, who was specializing in North American nature writing, was hired by UConn's Neag School of Education to create an

(continued)

Box 9.1 (continued)
experiential-learning summer course for "gifted and talented" high school students called "Words in the Woods." This course brought students outdoors to revel in the observations and inspirations of the Transcendentalists.

Two years later, Heidinger met Rozzi when they participated in an "Interdisciplinary Working Group on Environmental Issues" that defined and launched the Edwin Way Teale "Nature and the Environment" Lecture Series. When the advisors disbanded the group the two continued critical dialogues and cultural exchanges, Rozzi joining Heidinger's "Words in the Woods" and wandering the hills, wading the rivers together with students pondering how we fit into the biophysical designs of our cosmos.

The experience of walking while teaching students ecology, literature, history, and philosophy was thrilling and it inspired them to apply for and receive funding to teach a semester-long experimental course, investigating the environmental philosophies of Leopold and Darwin. Rozzi and Heidinger used the opportunity to read deeply into the works of these authors. These readings and experiences inspired for the work conducted by Francisca Massardo at the University of Magallanes (UMAG) in Chile, leading to the formal creation of the Omora Ethnobotanical Park in 2000 near Puerto Williams, which is the provincial capital of the Chilean Antarctic Province, including Cape Horn County (Rozzi et al. 2006).

Heidinger, Rozzi and Massardo continued collaborating and conceived a first Field Environmental Philosophy (FEP) course at the Omora Ethnobotanical Park (OEP) in 2000 with students from UMAG, the US and Germany (Tauro et al. 2021). Later, in 2006, FEP courses were integrated into the curricula of UMAG and the University of North Texas. The first version of this course at UNT became the study abroad field course "Tracing Darwin's Path" co-organized with freshwater ecologist James Kennedy. It entailed three complementarity activities. First, the Omora Park team focused on creating the walking-interpretive trails as an outdoor classroom, with particular attention to small organisms such as insects, mosses, and lichens (Goffinet et al. 2012). Second, Kennedy conducted stream invertebrate research. Third, Heidinger taught a "pedagogy of the sublime" in a three-week FEP itinerary that took the students from the Beagle Channel up to the place where the Robalo River begins its journey at the highest ridges of the *Dientes de Navarino* mountains that loom above Puerto Williams. Heidinger's goal was to harness for heuristic purposes the culture shock, the urge to adventure, the immersion in such a remote, unindustrialized place where the day lasts 20 h, and the biome's biotic vivacity exults in original, wild exuberance (Rozzi et al. 2012). A twenty-four year old Darwin had the epiphany of "the descent of man" in the Cape Horn archipelago. Later, when he wrote the *Descent of Man*, Darwin profusely

(continued)

Box 9.1 (continued)
traced and included emotion-filled memories of what he experienced on the Beagle Channel (Rozzi et al. 2018).

Tracing Darwin's Path was designed to bring students to the cusp of their comprehensions, and create and sustain a state of self-consciousness right on that paradigm-shifting edge. Like Darwin, students could step outside their prior enculturation and play with giant ideas to re-evaluate the meaning of things in a place peopled by the Yaghan and other Fuegian people, whose biotic worldview pre-dates and out-lasts the industrial worldview. The Yahgan story of Omora—the hero who was both hummingbird and human—gives the name to and signals the ethos of OEP (Rozzi et al. 2008).

Biocitizen began in the Omora Park when Heidinger, out in the field with students practicing the fine art of getting lost (and found), concluded that what he was doing there with FEP involved multiple interwoven dimensions: de-industrializing character, catalyzing the psychosomatic experience of the sublime, exploring ecological literacy, teaching biocultural history in place, and practicing the land ethic. Heidinger concluded that this could and must be done, in and of itself, as a non-academic program of unschooling. Inspired by the perspective afforded at OEP, Heidinger turned to the task of preparing people for the end of one world and the beginning of another. There in the Cape Horn forests so far from the North American forests one easily witnessed the artificiality of their industrially constructed character, and whipped by the howling sleets of Cape Horn, felt the embodied emotions and the "animal instincts" that Leopold (1949, p. 203) said were the basis of the character of the biocitizen. Feeling sure it was possible to catalyze these experiences elsewhere where they could benefit others who would never be able to visit OEP, Heidinger revised the FEP pedagogy he developed there and introduced it to the suburbs and cities of the USA and Chile. In keeping with Leopold's socio-evolutionary goal of sharing and building ecobiotic community spirit, Biocitizen invites us to walk together "reading the land" to find our place in its story: locally, nationally and internationally.

Box 9.2 Biocitizen Germinated from Realizing Culture-Transforming Capabilities of Education

> Sentiment without action is the ruin of the soul. One brave deed is worth a hundred books.—Edward Abbey (1989, p. 40)

Environmental action is at the foundation of Biocitizen. In 1999, inspired by the land ethic, Heidinger began a 15-year political engagement with UConn and the state of Connecticut to conserve the largest public drinking water system in the state. As the centerpiece of a three billion dollar public-private development project called "UConn 2000," UConn's administration proposed a biotech industrial park headlined by Pfizer Pharmaceutical, then the largest, wealthiest company of its kind in the world. Unfortunately, the "park" was going to be built on "Class 1 public drinking watershed land" which by state law absolutely prohibits development. Motivated by his PhD research, Heidinger had become an elected Inland Wetlands official. Knowing the water laws, he notified relevant state agencies of UConn's impending Class 1 lands violation, while contracting lawyers to prepare a lawsuit to prevent it from using the land for any purpose except water protection. Attorney General (now Senator) Richard Blumenthal intervened, promising that if the suit was dropped, legislation would be passed to protect the Class 1 land. Heidinger agreed. The legislation was raised without the intention of passing it, however, and was summarily tabled, never getting a vote. Notwithstanding, the biotech park never appeared on Class 1 land, and the incalculable value of the water source in question became common knowledge to the people of the region. Moreover, UConn's hazardous waste transfer facility, which had been operating on this Class 1 land for years, was moved adjacent to the industrial research schools that generated the waste. The process of challenging his university to conserve its water supply was catalyzed Heidinger's determination for he realized culture-transforming capabilities as a citizen, scholar and teacher, broadcasting the land ethic, and basic watershed management principles, through academic and governmental communities and the public at large, while acting politically to enforce drinking watershed laws (Fig. 9.1).

Fig. 9.1 "A few years ago, Kurt Heidinger, then a UConn grad student and president of the Naubesatuck Watershed Council, warned UConn that, in its current growth mode, it was heading toward pumping the Fenton River too low. Tom Callahan, a special assistant of the president of the university, dismissed Heidinger's complaints as 'egregious, reckless and erroneous' in a 2002 Northeast story entitled 'Big Bad Neighbor.' More than a week ago, a stretch of the Fenton near the university's wells went dry. Heidinger, now a visiting professor at the Center for Environmental Philosophy at the University of North Texas, revisited the area and had to dig 18 inches to find a sign of water" (Morse 2005). Photo by Shana Sureck

Biocitizen (biocitizen.org) offers FEP programming at four locations—Northampton, Massachusetts, Los Angeles, California, and New York City, New York in the USA and Valparaiso, Chile (Biocitizen 2021)—and serves approximately 2000 students per year in sessions lasting one to five days. Its "Free Ranger" sessions are scheduled once a week for a month, and its afterschool sessions, and "Family Bird Walks" follow a similar a la carte model. Biocitizen's premier offering, "Our Place Summer School," is a five-day program designed to give students an immersive experience of biocultural diversity in the biome they call home. "Claws" wilderness programs last from four to seven days and take post-adolescent students out of their biomes and into terrestrial and marine environments that lack industrial infrastructures. "Now Voyager" expeditions take two to three weeks to complete, require air travel and bring adults to biomes and cultures they are not familiar with. In the area of educational enhancement, our schools (particularly LA) collaborate with public and charter schools to provide their students with FEP programming, including growing native-plant permaculture gardens and instituting cyclical "acts of care" for the habitats the schools operate in. Biocitizen's most ambitious FEP offering is the "Living Rivers School" that teams established conservation biologists and graduate assistants with students to conduct research that provides the basis for scientific reports of use to regulators, policy-makers, journalists, and the public. "Coming soon to the biome[1] you live in!": we plan on opening more schools around the planet in the future.

9.2 Why Biocitizen Exists

[M]an is, in fact, only a member of a biotic team (Leopold 1949, p. 205).

Global society plays a significant role in fomenting the extinction crisis, concludes conservation biologist Gerardo Ceballos and his collaborators:

> a sample of 27,600 vertebrate species, and a more detailed analysis documenting the population extinctions between 1900 and 2015 in 177 mammal species ... [shows that] the rate of population loss in terrestrial vertebrates is extremely high—even in "species of low concern." Dwindling population sizes and range shrinkages amount to a massive anthropogenic erosion of biodiversity and of the ecosystem services essential to civilization. This "biological annihilation" underlines the seriousness for humanity of Earth's ongoing sixth mass extinction event. (Ceballos et al. 2017, p. E6089).

They urge us to act now:

> [T]he sixth mass extinction is already here and the window for effective action is very short, probably two or three decades at most. All signs point to ever more powerful assaults on biodiversity in the next two decades, painting a dismal picture of the future of life, including human life. (Ceballos et al. 2017, p. E6095).

The challenge we face is trying to survive. Like Leopold, Ceballos and collaborators speak about humans at the species level, with less emphasis on examining cultural diversity—because the crisis is pancultural and, putatively, post-national. While the crisis worsens, each Biocitizen *unschool* is a local expression of a biocultural evolutionary process; we provide "land ethic" concepts and experiences that impel students to care for their biomes, and to adapt to changing environmental conditions, so we and future generations will survive.

Today, spurred by COVID-19 and climate change, the collective—and diverse—consciousnesses of the world's mass populations are shifting uneasily from "industrial" to "biotic" worldviews, and the general awareness is increasing of our status as *ecobiotic* beings constituted by the biome that we co-inhabit with other ecobiotic beings. Our *unschools* are expressions of this "paradigm shift";[2] we unschool students of their industrial schooling, while schooling them in the land ethic. Also expressing the paradigm shift are the global-warming protest movements like

[1] At Biocitizen we use the term "biome" loosely with a meaning that is close to Frederick Clements' definition of "biotic community" and Heinrich Walter of "biotope" (cfr Jax 1998). We avoid the more abstract term "ecosystem," because the term "biome" conveys an understanding that is closer to "home of life."

[2] We elaborate on the concept of paradigm defined by Thomas Kuhn (1970), and use it broadly to mean *a cultural construct that defines a worldview and the behaviors it rationalizes.*

"Extinction Rebellion," "Sunrise Movement," and school-skipping "Fridays for the Future" organized by Greta Thunberg. Also, over 400 scientists including delegates of the Intergovernmental Panel on Climate Change (IPCC) observed that:

> Protestors have been successful in getting the public, the media, and policymakers to pay attention to climate change. They have identified and sought to delegitimise actors (industrial, financial and political) that are responsible for driving greenhouse gas emissions and ecosystem destruction. .... But around the world today, those who put their voices and bodies on the line to raise the alarm are being threatened and silenced by the very countries they seek to protect. We are gravely concerned about the increasing criminalisation and targeting of climate protestors around the world. (Open Letter 2021)

These movements and documents illustrate the ongoing shift toward a biotic worldview. From the land ethic perspective, the biotic worldview is antagonistic to the prevailing industrial worldview, because for hundreds of years the industrial behaviors of mass populations are "wrong" because they harm "the integrity, stability, and beauty of the biotic community." (Leopold 1949, pp. 224–225). This is why Biocitizen *unschools*.

Kuhn (1970, p. 92) averred that "scientific revolutions are ... episodes in which an older paradigm is replaced in whole or in part by an incompatible new one," within a larger history of science and culture. Kuhn thus reveals how the transition between paradigms is disorganized and rife with conflict—exactly as we experience it now. For at least twenty years we have witnessed, and perhaps participated in, the political conflicts that can be interpreted as conflicts between industrialist "conquerors" and "biotic citizens." Addressing the United Nations Security Council recently, natural historian David Attenborough described the epistemological, and political, chaos that epitomizes the paradigm shift:

> We have left the stable and secure climatic period that gave birth to our civilizations. There is no going back. .... Perhaps the most significant lesson brought by these last 12 months [of COVID during 2020-2021] has been that we are no longer separate nations, each best served by looking after its own needs and security. We are a single, truly global species, whose greatest threats are shared and whose security must ultimately come from acting together in the interests of us all. (Global Citizen 2021)

Attenborough articulates the extinction crisis as being, in part, that of the industrial-to-biotic paradigm shift, and more identifies us as "a global species." "There is no going back" to the environmental conditions that industrialism obviated, he says. However, environmental damage does not present the full extent of the crisis. Another, more bewildering, part is that the industrial paradigm refuses to budge, and continues to worsen conditions and our chances of survival. Attenborough beseeches the UN to accelerate our shift from an industrial to a biotic paradigm by adopting the Leopoldian premise that *Homo sapiens* is "part of a biotic team" that includes all other species.

Attenborough's appeal is laudable and Biocitizen fully supports it. However, we cannot expect governments or corporations to shift the industrial paradigm promptly. Hence, Biocitizen educates students to be prepared to cope with the collapse of industrially-based infrastructures and ways of life.

Biocitizen addresses the cultural dimension of the extinction crisis by creating and executing Leopoldian FEP curricula that teach students about the "collective organism" that is their biome and how they are a part of it, biologically, culturally, and historically. We provide students with physical and conceptual activities that allow them to *actually* step out of industrialism and into bioticism, which is a paradigm constructed upon ecological science and land ethic values. Biocitizen acts humbly as a biocultural evolutionary agent to advance our species through and out of the extinction crisis, and to coach students to be "part of a biotic team" in an existential contest we must win.

9.3 How Biocitizen Practices FEP

9.3.1 What We Do Physically

The essential practice of our version of FEP involves walking, a psychosomatic exercise that requires students' vigorous: physicality, sensual acuity, and emotional and cognitive activity.[3] These aspects of our characters are united within us, of course, but walking delivers a different balance between them than sitting does; walkers think physically, not just mentally, as they coordinate their muscles and react to the stimulations elicited by moving through environments. When we walk, we teach the "psycho" (self) that is "logical" (i.e., *logos*- or symbol-oriented) *and* the "soma" (body) that breathes, hungers, and is sensual and emotional. Teaching the whole student, as opposed to just their brain, gives us a great advantage when teaching the land ethic, for as Leopold indicated: "We can be ethical only in relation to something we can see, feel, understand, love, or otherwise have faith in" (Leopold 1949, p. 214). Among many other things, walking stimulates ethical cogitation.

Our teachers walk students into the biotic vivacities of our biomes so they sense, feel, understand, and love them. Teachers sustain that state of cognition so the lessons in ecology and biocultural history are registered not just logically as useful information but also existentially as clues for solving mysteries of their own lives and of life itself. We teach Field Environmental "Feel-Awe-Sophy"—for from their feelings of wonder (i.e., awe) about the world and the nascent abilities they have to understand the world as it presents itself as "nature" and "culture," students derive wisdom (i.e., sophy), gain self-confidence, and develop the kind of character that is prepared to successfully make the paradigm shift and undertake earth stewardship (Fig. 9.2).

As we walk, students have no difficulty feeling somatically and thinking logically that they are members of an animal species. This animal feeling is a first step toward grasping Leopold's wise injunction that we are a part of a "biotic team" cooperating

[3] For, as Leopold noted, "The evolution of a land ethic is an intellectual as well as emotional process" (Leopold 1949, p. 225).

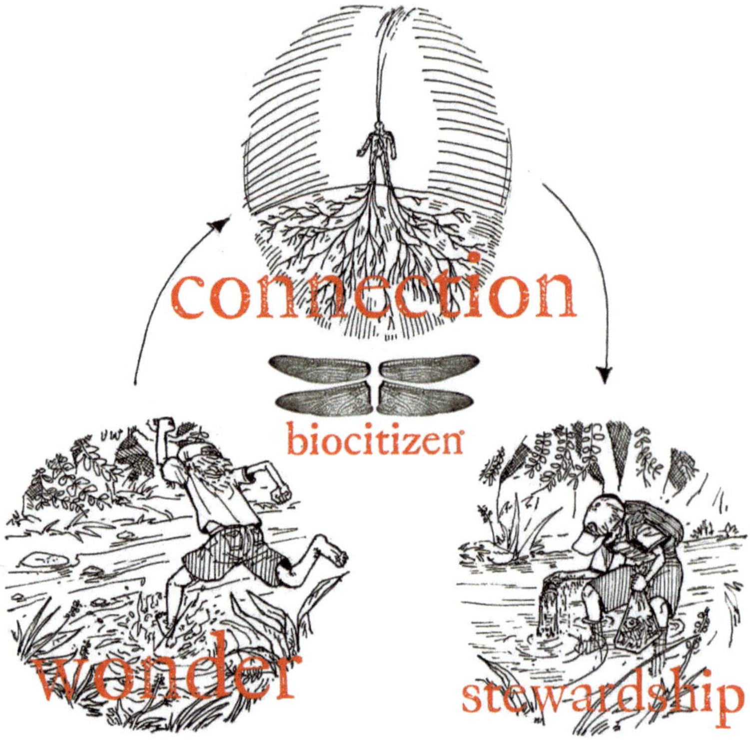

Fig. 9.2 Biocitizen transforms Field Environmental Philosophy into Field Environmental "Feel-Awe-Sophy." Students feel wonder (i.e., awe) which leads them to access and experience wisdom (i.e., sophy) that motivates them to undertake earth stewardship. Figure designed by Cormac Carmichael Pickering

within the "collective organism." Students enjoy becoming conscious of and gaining control over the animal they feel they are.[4] "Off the screen and into the green" is Biocitizen's trademarked maxim, and it is received enthusiastically by parents who worry their kids spend too much time sitting in front of computers. They are well aware of the dangers of nature deficit disorder and want their kids to recognize and exercise the animal-biotic self that complements and corrects their cyber-industrial self. In this way, the global paradigm shift occurs in their characters, and we guide and nurture that process of personal behavioral change.

[4]Another way of putting this is we teach them they are "human," which means "of the humus." Corollary to this Latin linkage (and epistemology) is the Hebraic "adam" which means "of the ground" and the Greek "autochthon" which means "sprung from the soil." We do not define "animal" as Cartesians and Social Darwinists have, as something "beneath us" that must be "dominated"; instead we define "animal" as Leopold did: as a co-operative "member of a biotic team."

Leopold explained that walking outdoors summons psychosomatic faculties he called "animal instincts," and he said that letting these instincts guide us through new "ecological situations" was a first step in living the land ethic:

> An ethic may be regarded as a mode of guidance for meeting ecological situations so new ... that the path ... is not discernable to the average individual. Animal instincts are modes of guidance for the individual in meeting such situations. Ethics are possibly a kind of community instinct in-the-making. (Leopold 1949, p. 203)

To summarize Leopold's course of biocitizen character development and apply it to our present context, we identify three complementary moments:

(i) Walking into biotic vivacities awakens our "animal instincts".
(ii) Sharing the psychosomatic experience socially generates a "community instinct."
(iii) Confronting the "new ecological situation" of the extinction crisis helps us to understand ourselves as a species amongst species that co-inhabit a biome.

Meeting the challenges of the crisis means imagining how we will survive, and what kind of world we want to build to replace the one that cannot be sustained. Walking elevates mood and "boosts creative ideation in real time and shortly after" (Opprezzo and Schwartz 2014, p. 1142). When students recognize their animal nature while walking, they also recognize the hopeful and motivated part of their character they need to address and overcome the challenges of the extinction crisis.

When we are walking we cannot dismiss the biophysical world. It is self-evident that we are in it, living it, breathing it, feeling it; it is us. For this reason, at Biocitizen we say "where we are is who we are"—the creed of the ecobiotic self. We also note that when we read about the world, we can dismiss it, since when merely re-presented as words, it is disembodied, and just an idea. Our teachers help students to revel in that dissolution, because it is liberating and enjoyable to do so. They use the revelation that we are "a part and parcel of Nature" (Thoreau 1862) as the science-based epistemological foundation upon which to re-construct their characters as "biocitizen."

As we walk, our teachers foment and sustain a state of heightened psychosomatic cognition, channeling its delightful creative energy to train them in ecosemiotics to read the world as a biocultural history of the "biotic interactions between people and the land" (Leopold 1949, p. 205). Thus, through FEP, they increasingly comprehend how "the human species together with its culture is entirely a part of the ecosystem" (Kull 1998, p. 346). This might sound difficult to do, but it is as easy as following water from mountaintop to ocean and witnessing how we use that water to constitute ourselves.

9.3.2 *What We Do Ideologically*

A more difficult, long-term culture-building task is that of helping students articulate verbally the species-based subjectivity they are developing, so they can share it socially and have a positive, reforming effect on the communities they are part of. As Leopold noted, "the land ethic [is] a product of social evolution" (Leopold 1949, p. 225). To help them express their ecobiotic content, we follow Leopold's example of creating and teaching metaphorical constructs that convey ecological concepts. He observed that:

> An ethic to supplement and guide the economic relation to land presupposes the existence of some mental image of land as a biotic mechanism. . . . The image commonly employed in conservation education is "the balance of nature." For reasons too lengthy to detail here, this figure of speech fails to describe accurately what little we know about the land mechanism. A much truer image is the one employed in ecology: the biotic pyramid. (Leopold 1949, p. 214)

The "mental image" of the "land pyramid" does not convey the ecobiotic fact that our bodies are consanguineous with our biome nor does it evoke the species-based subjectivity we are inculcating. "Leopold's 'collective organism' construct is better suited to this task, but it does not connote consanguinity, unlike the metaphor that Biocitizen floats of the 'long-body'—a 'mental image' that connotes the bios we share with all other life forms." It is scientifically based, and it conveys the constructs of species-based subjectivity and consanguinity that are core components of the biotic paradigm. So, what is the long-body, and how do we use it in tandem with our physical FEP pedagogy to create biocitizens?

Leopold's and Attenborough's assertions that we are members of an animal species that is subject to the processes of evolution by natural selection are based on Darwin's conclusion, articulated in the *Descent of Man*, "that man is descended from some less highly organised form" (Darwin 1873, p. 606). Like Leopold, Darwin posited a kind of human identity that he did not fully explain the contents or boundaries of. In both cases, they and their expositors left us with work to do. Biocitizen teachers are tasked with educating biocitizens. The metaphor of the long-body stimulates and directs our imaginations when we are in the field, generates an ecological way of (re-)cognizing the world, and lets us finish the work of biotic character building that Darwin, Leopold, and others left for us to do. Based on its provenance in the *Descent*, the metaphor of the long-body has three contexts that summon the character of the biocitizen: biological, temporal, and environmental.

Within the biological context, the long-body metaphor is grounded in our consanguineous symbiosis with other species. To give one example: we have about a liter of *Lactobacillus acidophilus* bacteria in our digestive systems who[5] allow us to derive nutrition from the food we eat, and who form an essential part of our immune system. These bio-existential facts are largely unknown by most people, even though they are advertised in "probiotic" yogurt commercials. Students are asked to imagine their own bodies as being not actually their own, but instead, as something that is shared by other organisms whose functions are absolutely necessary for survival of

[5] "Who" is used here instead of "that" because within the long-body all living creatures are subjects, not objects. Though it might seem like semantic foolishness, it is through small acts of epistemological freeplay like these that paradigms are ultimately shifted.

both. The metaphor of the long-body gives us the defamiliarizing perspective we need to (re-)cognize our bodies are not our own; we share them.

Within the temporal context, the long-body metaphor allows us to imagine the dimensions of geological time that Darwin found essential to his envisioning of evolution by natural selection. We can lead our students backward in time following the species path of *Lactobacillus acidophilus* to earlier versions of itself, perhaps over three billion years. Students contemplate how, inside of us, we have members of our biotic team who share a body with their ancestors that has been passed down in a continuum that is older than the ground we stand on, the mountains we climb, rivers we ford and streets we cross. This fact de-familiarizes the very definition of what human is and means, and our teachers mention it every now and then to install biotic character.

Within the environmental context, the long-body metaphor invites students to imagine the ways they are constituted by the things that surround them. This is not so hard to do, because we are exposed to the weather and ground conditions. These conditions determine who we are at any given moment—hot or cold, wet or dry, at ease or anxious. An approaching thunderstorm, or street crossing, determines our lives and cannot be dismissed as being separate from us; we share the bodies of storms or streets when they enclose us. The "animal instincts" are called upon to negotiate obstacles as we walk, and students are trained to be alert to the subtle fluctuations in climate and atmosphere. They learn by feeling what microclimates are, so in any environment we pass through they know where they can find warmth or coolness, or humidity or aridity. With practice they look upon the environment as a form of clothing; there are places that are "jackets" because they are cold and places that are "t-shirts" because they are warm. The somatic sensitivities generated by this form of FEP are expressions of the shift to an ecobiotic self-consciousness; and since they are based in feeling, this psychosomatic way of reading (and comprehending) the environment can be applied anywhere.

Our fundamental FEP itinerary follows rivers because they provide the best example of the long-body as a symbolic metaphor *and* biotic reality (Box 9.3). The river connects bios within a watershed, and on our walks students casually investigate how hydrology determines the biocultural character of ecobiotic communities and industrial infrastructures. Water is essential to life. Thinking like a river, and perhaps letting rivers think for us, stimulates our students to imagine how this aspect of themselves—water—permeates everything around them. Through practice, by following water, they recognize the life they share with all other lifeforms.

Box 9.3 A Week of FEP with Biocitizen Los Angeles

What is LA? This simple question launches, and recurs during, our every quest.

Los Angeles (LA) is a coastal desert city with a pleasant Mediterranean climate that sprawls over vast alluvial flats between stunning snow capped mountains and the Pacific Ocean. The Tongva people co-inhabited this biome long before the EuroAmerican invasions, and let the waters of the Los Angeles River run freely from sky to ocean. Today we find our river confined by an immense concrete canal system designed to prevent it from flooding our city and its industries (Park et al. 2009; Arroyo 2010; Pincetl 2013; Klaver and Aaron Frith 2014; Griffis 2021). Like other modern cities, LA stymies the revivifying circulations of the long-body, reducing the quality of life for all creatures, and imperilling its own sustainability. Consequently, it suffers a slow, steady, unappealing morbidity. Parents know this and send their children to us, because we get them "off the screen and into the green" where they psychosomatically experience the complex biophysical and cultural dynamics that industrialism cannot entirely control, engineer, or thwart. While having fun and making friends, Biocitizen students are taught to read what constitutes "our place" in a city, which means the industrial infrastructures that comprise it.

As a class walks through the humanly-constructed urban environment, students are cued to the traces of original ecosystems, for these are places where surficial waters still flow and pool, and where plants and animals gather and grow "wildly" (Griffis 2021). Students sense how temporary and artificial their urban environment is. They compare streets to streams and realize how steadfast the long-body is. The LA River expresses the vitality of the biome, for despite all attempts to confine it, it reclaims many spaces industrialists try to deny it (Klaver 2018). Wherever it can, it returns, and when it does the biome expresses its diverse, unique, and resilient character and serves again as home to an astounding community of endemic and "invasive" species of plants and animals.

Indeed degraded, abused, and over-developed, the biotic aspects of our city persevere (Klaver 2013). When students, who hours earlier were staring at screens, come into contact with the aromatic salvia of the coastal sage scrub, and graze their arms and legs upon the spiky impenetrable chaparral, they become aware of a part of themselves that technology cannot summon. They recover an appreciation of their animal bodies, their mammalian selves. They think kinaesthetically, drawing on the capabilities inherent in their whole psychosomatic self. Now informed by the intelligence of their senses and muscles, they *feel* themselves placed in an environment that is alive and that houses our species with other species in a biotic community, as co-inhabitants in a shared biome; and inspired by this state of heightened existential

(continued)

Box 9.3 (continued)

awareness, begin to understand and express a biocultural ethics (*sensu* Rozzi et al. 2008; Rozzi 2018). When they return home, they carry their FEP experience and a bit of that biotic self with them. Over time and recursively, the urban and wild selves complement each other, for this identity extends into both worlds, unifying them, and ending their binary opposition. The city is mighty, but the long-body precedes, survives, and outlasts it.

We follow water because it is life—and our ideal five-day FEP itinerary walks the Arroyo Seco and her sister, the Los Angeles River from sky to ocean. The Arroyo Seco headwaters begin at the peaks of the San Gabriel Mountains. The mountain portion of this watercourse is known as the Upper Arroyo, the suburban portion known as the Lower Arroyo. Where the Seco meets the LA River (just northwest of the site of what was Yaanga, the largest Tongva or Kizh village) is the historic birthplace of the city. Our teachers lead the students on this itinerary so they can witness and feel how the mountain river changes character as it passes through suburban then urban environments. In the Upper Arroyo, the river leaps down riparian canyons, cooled by canopies of alder, then enters and dries up in the corridors of concrete of the Lower Arroyo in Pasadena and then the LA River in downtown Los Angeles. Ultimately, the LA River terminates in the estuaries that border the busy, toxic Port of Long Beach. Each section tells the story of how industrialism reshapes biotic communities to serve a cultural purpose of generating monetary wealth.

Day One: begins just below the headwaters of the Arroyo Seco at Switzer Falls in the San Gabriel Mountains. With the city far below, the river cuts through the ancient walls of granite and creates waterfalls and swimming holes which catalyze an experience of deep biotic immersion that is both empowering and inspiring. In the winter of 2021 a student was so awestruck by what he saw that he turned to his friend and exclaimed "This is paradise! Do you see this!? This is amazing!" While experiencing wonder, students are cued to ponder how river systems once were, and are, a source of life and energy for the canyon's biotic communities (Fig. 9.3).

Day Two: finds us in the upper middle section of the Arroyo Seco. Known as the JPL Trail, named for its famous military-industrial-complex neighbor, NASA's Jet Propulsion Laboratory, this 8 mile walk culminates at 60 foot concrete waterfall (Fig. 9.4).

The falls provide a thrill for the students, and a firm reminder that this little river is considered a commodity by those who live far away. Most of the surrounding habitat of the Arroyo Seco in this section is defined by wild Mule Fat shrub, River Willow, and Alder, but our students scramble over crumbling dams, collapsing walls and other derelict infrastructures left behind by industrialists. These physical and symbolic signs are read together, and tell the story of how this canyon was engineered and the river contained to serve cultural

(continued)

Box 9.3 (continued)
goals until lack of interest, local tectonics, and extreme flooding made it what it is today.

Day Three: is spent exploring the confluence where Lower Arroyo joins the LA River. Reverberating the white noise of the highways that cross over and parallel it, the Lower Arroyo is dry here, and littered with shopping carts and the discards of our "single-use" consumer lifestyle (Fig. 9.5).

Under these conditions, teachers don't have to remind students how clean and beautiful the river was at the headwaters, and how, as it nears the ocean, industrial culture morbidifies it. This is the world they know as their home and yet they don't know it, because they have never followed water, and seen it from the perspective of the long body. Parents prefer to have their kids romping in pristine and wild places, but this day allows the students to fold their love for the river into their witness of its tragic neglect and abuse. Because of their earlier walks, they can envision the long body of the river. Having much to think about and feel, this is the day they "connect" to the LA river (Fig. 9.6).

On Day Four: we ascend to a vantage point in the Elysian Park hills up above the confluence of the two rivers, the lowest portion of the Arroyo Seco and the LA River. Here we are in the heart of North East Los Angeles where the urban and the suburban, and native and the exotic, intersect.

The hike starts on a "fire road" leading out of the neighborhood known as Chavez Ravine. This neighborhood is the site of the Battle of Chavez Ravine, which begins in 1951, and ends with the city forcefully evicting 300 Mexican American families under guise of urban redevelopment, supported by the Housing Act of 1949 which provides federal funds to build public housing. The fire road eventually becomes a single track trail, skirting the edge of the park, providing the perch for one of the most content-filled views of our city. We then ascend Point Grand View Park, the tail end of the Santa Monica mountains. Here we gain a view of Dodger Stadium and the downtown skyline, as well as our familiar transverse ranges: the Verdugo- and San Gabriel- Mountains. From this vantage point we observe the more industrialized versions of Arroyo Seco and LA River. Exercising our ecosemiotic powers, we see how concrete covers up a once lush and fertile alluvial valley created by thousands of years of annual floods. We identify the intersection of the Arroyo and LA Rivers as being where the migrating steelhead trout would turn as they made their way back to their cold clear mountain streams to spawn. Now there are hardly any trees to shade and cool the water, and insufficient sediment remains to host trout spawning. Instead of such habitat, we have concrete river floors, and vertical concrete walls to direct the flow out and away from the buildings and homes (Fig. 9.7).

(continued)

Box 9.3 (continued)

On Day Five: our exploration of the Arroyo Seco/LA River watershed ends at the Pacific Ocean and the Port of Long Beach, where the terrestrial long body connects with its marine self. We stop first where the Queen Mary is beached in an industrial estuary. We talk about the commerce of stuff that turns the wheels of our country's busiest harbor, where containers of rubber ducks and tupperware are lifted from the decks of ships coming from all over the world. Our students look around for the more comforting markers of nature, the softer forms of a native plant community, or the peaks of the San Gabriels, but there are none here. Seeking the beauty of life and living systems, we push on and make our way along the Palos Verdes Peninsula, as the Tongva did/do, to Whites Point tide pools and cove. Here we look out across the channel to Santa Catalina Island and reflect on the journey of the San Gabriel snow melt along the Arroyo Seco and the LA River to this entirely different world, and how they are connected. As we wade and explore the intertidal zone of the Whites Point, our students reflect back to that some of the mountain river water is still around us, mixing with the salty waters of the Catalina Channel and the Pacific.

Following water from the sky to the ocean, our students immerse themselves in the wonder of each place, perceiving and connecting the signs and symbols each place reveals into stories that, combined, express their version of the biocultural history of our biome: as we put it, "place is the story we live in." They see how previous human co-inhabitants have broken the integrity of our living systems down, and how we continue to do so . . . and they also begin to understand that they are active participants in this process, and can choose to write a new story of place, one in which they become the stewards for a new generation: Biocitizens. As we walk and explore that which surrounds us, we witness the majesty of unique and awesome places—this we call wonder. Wonder opens our senses to the life we share with other creatures—the long body—and promotes deep connection: connection to the earth, each other, and ourselves. Connection invites us to further deepen our relationship to our place and our co-inhabitants. From that caring relationship we feel compelled to act as stewards—not from a desire to follow the rules nor out of a sense of shame, but motivated by wonder and connection. We are inspired to restore, reforest, preserve, and protect life and living systems, out of love. This is the essence of biophilia: love of life and living systems. Biocitizens follow the path of Wonder—Connection—Stewardship to restore our relationship to our place and its inhabitants. That's why we walk.

Fig. 9.3 Upper Arroyo Seco River at Switzer Falls. Photo by Vicente Diez Aguirre

Fig. 9.4 Arroyo Seco, JPL Trail. Photo by Benjamin Jacobs Schwartz

Fig. 9.5 Los Angeles River, Glendale Freeway. Photo by Vicente Diez Aguirre

Fig. 9.6 Point Grand View, Elysian Park. Photo by Benjamin Jacobs Schwartz

Fig. 9.7 Whites Point Tide Pools and Cove. Photo by Vicente Diez Aguirre

Recognizing their symbioses, biocitizens know that whatever they do to the long-body they do to themselves. To save themselves, they save the long-body. Self-interest is sublimated when the self is shared. When the self is the long-body, taking care of oneself means taking care of everything. A sublimated biotic self-interest guides the biocitizen to protect, conserve, and rehabilitate familial species and the habitats we share, and are.

In conclusion, Biocitizen practices FEP to cultivate ecobiotic consciousness and produce trans-species, transgenerational selves, because through their behaviors the biotic paradigm is lived—i.e., "walked," not just "talked."

Acknowledgment We appreciate the valuable editions by Roy May and Ricardo Rozzi on an early version of the manuscript.

References

Abbey E (1989) A voice crying in the wilderness. St. Martin's Press, New York

Arroyo J (2010) Culture in concrete: art and the re-imagination of the Los Angeles River as civic space. M.C.P. thesis, Massachusetts Institute of Technology. Available via: www.arroyoseco.org/arroyojcthesis.pdf

Biocitizen (2021) About Biocitizen. Available via: https://biocitizen.org/about/

Ceballos G, Ehrlich PR, Dirzo R (2017) Biological annihilation via the ongoing sixth mass extinction signaled by vertebrate population losses and declines. Proc Natl Acad Sci 114(30): E6089–E6096

Darwin C (1873) The descent of man. John Murray, London
ESA (2021) Earth Stewardship: sustaining and enhancing Earth's life-support systems. Ecological Society of America. Available via: https://www.esa.org/programs/science-engagement/earth-stewardship/
Global Citizen (2021) 'It's Too Late': 5 Key Things to Know From David Attenborough's UN Climate Change Speech. Available via: https://www.globalcitizen.org/en/content/david-attenborough-climate-change-unsc-cop26/
Goffinet B, Rozzi R, Lewis L, Buck W, Massardo F (2012) The miniature forests of Cape Horn: eco-tourism with a hand-lens. Bilingual English-Spanish edition. University of North Texas Press - Ediciones Universidad de Magallanes, Punta Arenas, Chile
Griffis M (2021) The Los Angeles River's overlooked anglers. High Country News 53(5). Available via: https://www.hcn.org/issues/53.5/south-people-places-the-los-angeles-rivers-overlooked-anglers
Holbrook B (2012) Afterword on seeing-as. In: Goffinet B, Rozzi R, Lewis L, Buck W, Massardo F (eds) The miniature forests of Cape Horn: eco-tourism with a hand-lens. Bilingual English-Spanish edition. University of North Texas Press - Ediciones Universidad de Magallanes, Punta Arenas, Chile, pp 417–421
Jax K (1998) Holocoen and ecosystem: on the origin and historical consequences of two concepts. J Hist Biol:113–142
Klaver IJ (2013) Environment imagination situation. In: Rozzi R, Pickett ST, Palmer C, Armesto JJ, Callicott JB (eds) Linking ecology and ethics for a changing world: values, philosophy, and action. Ecology and ethics, vol 1. Springer, Dordrecht, pp 85–105
Klaver I (2018) Re-claiming rivers from homogenization: meandering and riverspheres. In: Rozzi R, May RH Jr, Chapin FS III, Massardo F, Gavin M, Klaver I, Pauchard A, Nuñez MA, Simberloff D (eds) From biocultural homogenization to biocultural conservation. Ecology and ethics, vol 3. Springer, Dordrecht, pp 49–69
Klaver IJ, Aaron Frith J (2014) A history of Los Angeles's water supply: towards reimagining the Los Angeles River. In: Tvedt T, Oestigaard T (eds) A history of water, series III, vol 1: water and urbanization. I.B. Tauris, London, pp 520–549
Kuhn T (1970) The structure of scientific revolutions. International Encyclopedia of Unified Science, volume 2, number 2, second edition. University of Chicago Press, Chicago
Kull K (1998) Semiotic ecology: different natures in the semiosphere. Sign Syst Stud 26:344–371
Leopold A (1949) A Sand County Almanac and sketches here and there. Oxford University Press, New York, p 1966
Moreno-Terrazas R, Díaz-Gómez Z, González-Galván H, Cariño M, Monteforte M (2022) Starfishes and sky stars: field environmental philosophy education and ecotourism experiences in Baja California Sur, Mexico. In: Rozzi R, Tauro A, Wright T, Avriel-Avni N, Klaver I, Berkowitz A, Brewer C, May RH Jr (eds) Field environmental philosophy: education for biocultural conservation, vol 5. Springer, Dordrecht, pp nnn
Morse D (2005) Dry times for UConn. Hartford Courant, Hartford
Open Letter (2021) Stop attempts to criminalise nonviolent climate protest. Available via: https://www.unpublishedottawa.com/letter/334274/open-letter-stop-attempts-criminalise-nonviolent-climate-protest
Opprezzo, Schwartz (2014) Give your idea some legs: the positive effect of walking on creative thinking. J Exp Psychol Learn Mem Cogn. https://doi.org/10.1037/a0036577
Park MY, Stenstrom M, Pincetl S (2009) Water quality improvement policies: lessons learned from the implementation of proposition O in Los Angeles, California. Environ Manage 43:514–522
Pincetl S (2013) Linking ecology and ethics for a transition to the sustainable city: values, philosophy, and action. In: Rozzi R, Pickett ST, Palmer C, Armesto JJ, Callicott JB (eds) Linking ecology and ethics for a changing world: values, philosophy, and action. Ecology and ethics, vol 1. Springer, Dordrecht, pp 323–331
Rozzi R (2018) Biocultural homogenization: a wicked problem in the Anthropocene. In: Rozzi R, May RH Jr, Chapin FS III, Massardo F, Gavin M, Klaver I, Pauchard A, Nuñez MA, Simberloff

D (eds) From biocultural homogenization to biocultural conservation, Ecology and ethics, vol 3. Springer, Dordrecht, pp 21–47
Rozzi R, Massardo F, Anderson C, Heidinger K, Silander J Jr (2006) Ten principles for biocultural conservation at the southern tip of the Americas: the approach of the Omora Ethnobotanical Park. Ecol Soc 11(1):43 [online] URL: http://www.ecologyandsociety.org/vol11/iss1/art43/
Rozzi R, Arango X, Massardo F, Anderson C, Heidinger K, Moses K (2008) Field environmental philosophy and biocultural conservation: the Omora Ethnobotanical Park educational program. Environ Ethics 30(3):325–336
Rozzi R, Armesto JJ, Gutiérrez J, Massardo F, Likens G, Anderson CB, Poole A, Moses K, Hargrove G, Mansilla A, Kennedy JH, Willson M, Jax K, Jones C, Callicott JB, Kalin MT (2012) Integrating ecology and environmental ethics: earth stewardship in the southern end of the Americas. BioScience 62(3):226–236
Rozzi R, Chapin FS III, Callicott JB, Pickett STA, Power ME, Armesto JJ, May RH Jr (eds) (2015) Earth stewardship: linking ecology and ethics in theory and practice, Ecology and ethics, vol 2. Springer, Dordrecht
Tauro A, Ojeda J, Caviness T, Moses K, Moreno R, Wright T, Zhu D, Poole A, Massardo F, Rozzi R (2021) Field environmental philosophy: a biocultural ethic approach to education and ecotourism for sustainability. Sustainability 13:4526. https://doi.org/10.3390/su13084526
Thoreau HD (1862) Walking. Atlantic Magazine, Washington D.C. Available via: https://www.theatlantic.com/magazine/archive/1862/06/walking/304674/

Chapter 10
Inter-species and Inter-cultural Encounters: The Biocultural Education Program of the Omora Ethnobotanical Park

Ricardo Rozzi

Abstract Science helps us to perceive and understand diverse biological and cultural realities. To enhance these perceptions and understandings, today it is essential to overcome physical, conceptual, and institutional barriers that have come to characterize global society. "Face-to-face" encounters with other living beings and cultures are irreplaceable. Physical, emotional, and sensorial experiences in these encounters transform the notion of biological and cultural diversity from a concept into an experience of biocultural co-inhabitation. This is illustrated with stories of inter-species and inter-cultural encounters at the Omora Ethnobotanical Park in the Cape Horn Biosphere reserve in Chile We concisely discuss the philosophical implications of this biocultural education approach, and how it could help to set priorities to conserve biological and cultural diversity not only in Chile but also in other regions of the planet.

Keywords Biocultural conservation · Biodiversity · Ecology · Ecotourism · Ethics · Philosophy

---

Part of this chapter is based on the article "Desde la ciencia hacia la conservación: el programa de educación y ética ambiental del Parque Etnobotánico Omora," originally published in Spanish by Rozzi et al. (2005).

R. Rozzi (✉)
Sub-Antarctic Biocultural Conservation Program, Department of Philosophy and Religion and Department of Biological Sciences, University of North Texas, Denton, TX, USA

Cape Horn International Center (CHIC), Universidad de Magallanes, Puerto Williams, Chile
e-mail: Ricardo.Rozzi@unt.edu

R. Rozzi et al. (eds.), *Field Environmental Philosophy*, Ecology and Ethics 5,
https://doi.org/10.1007/978-3-031-23368-5_10

10.1 Introduction

Since the mid-twentieth century, the period known as the Great Acceleration has amplified processes of ecosystem degradation and the extinction of a growing number of biological species on a global scale (McNeill and Engelke 2016). Concomitantly, local peoples and cultures have been displaced from their original habitats, replacing both their life habits and relationships of co-inhabitation with local ecosystems and biological species (Rozzi 2012). In less than 100 years, modern civilization characterized by its *anthropocentrism* and *Eurocentrism* transmuted into a global society characterized by its high *anthropogenic* impact. However, it is not the human species as a whole that is equally responsible for its anthropogenic social-environmental footprint (Rozzi 2007; Figueroa 2011; Dunn 2017).

Consequently, to better understand the social-environmental impacts of specific human groups (Rozzi 2015a), we need to more precisely identify the diversity of cultures, their languages, values, and practices in heterogeneous habitats of the planet, so that we can find paths for more sustainable and just futures. To undertake this task and better understand the interrelations between biophysical and cultural heterogeneity, in this chapter we use the perspective of the biocultural ethic that values the vital links between (i) the well-being and identity of the co-inhabitants (humans and other-than-humans),[1] (ii) their life habits, and (iii) the habitats where they take place (Fig. 10.1).

The "3Hs" (co-inhabitants, habits, and habitats) conceptual framework has a heuristic significance because it helps to visualize, understand, and value the interdependencies between the lives of diverse co-inhabitants. Understanding and valuing these interdependencies should guide decision-making and generate greater awareness of the consequences of decided actions, both for one's own individual well-being, as well as for the well-being of the community of human and other-than-human beings. Hence, the "3Hs" conceptual framework has also an ethical significance to foster an intercultural and inter-species solidarity and a socio-environmental justice. Conservation of the habitats and access to them is a matter of justice, because it is a necessary condition for the continuity of life habits linked to the well-being and identity of the broad community of co-inhabitants (Rozzi 2012).

The concept of co-inhabitants alludes to sharing the same habitat. It has a sense analogous to the concept of "companion" that, in its origin, referred to sharing bread (from Latin, *cum* = with; *panis* = bread). The understanding that we share habitats with vertebrates, invertebrates, and a multitude of other living and ecological beings (such as rivers, mountains, rocks, or oceans), has ontological, epistemological, and

[1] The expression "other-than-humans" avoids the dichotomy derived from the more usual expression: "non-humans." It overcomes this dichotomy for two reasons. First, it alludes to the set of biotic and abiotic beings that form different levels of organization and interactions in the ecosystems they co-inhabit. Second, the expression "other-than-humans" allows us to understand that these beings inhabit not only biophysical nature but also the images, symbols, and values of our cultures. Therefore, they are co-inhabitants in our biocultural communities, which encompass biophysical and linguistic domains of reality, and wakeful and oniric phases of our lives (see Rozzi 2018).

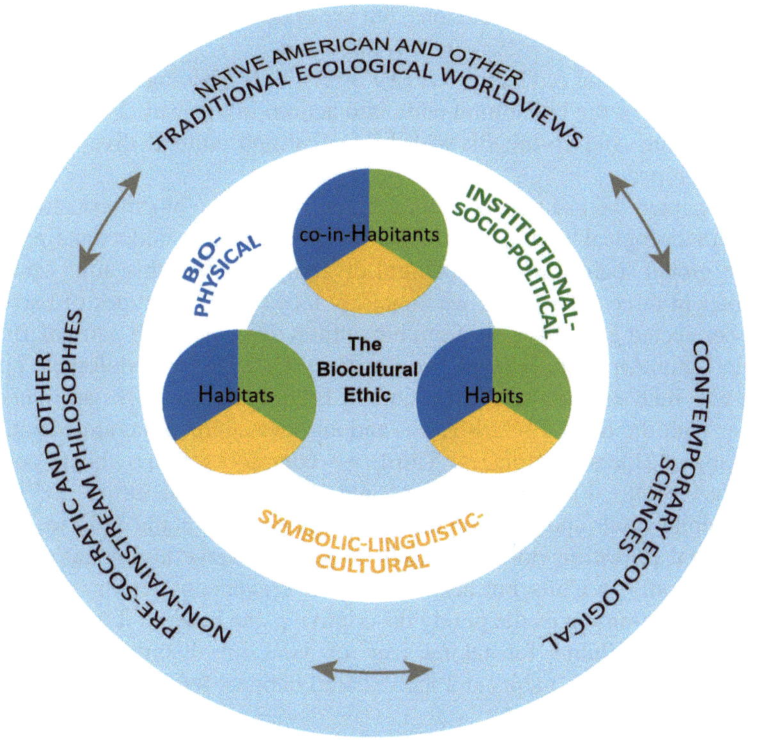

Fig. 10.1 Representation of the "3Hs" model of the biocultural ethic that values the vital links between specific habitats and their co-inhabitants with their habits of life. The colors in each circle illustrate that each of the "3H" includes biophysical dimensions (blue), symbolic-linguistic-cultural dimensions (yellow), and institutional-socio-political, infrastructural-technological dimensions (green). The green color (a mixture of blue and yellow) of this last dimension was chosen to indicate the need to carefully combine the biophysical and symbolic-linguistic-cultural dimensions in the design of policies, decision-making, and infrastructure. The outer circle (in light blue color) indicates that biocultural ethics is informed not only by a diversity of ancient and contemporary forms of knowledge and worldviews of indigenous peoples, philosophies, and sciences, along with a plethora of social, political, and cultural movements. The bidirectional arrows and the circular shape of the figure indicate the multiple active exchanges that occur between worldviews and knowledge in the context of a global society that is, at the same time, dynamic and rapidly changing. Modified figure of Rozzi (2012)

ethical implications. *Ontological*, because human and nonhuman beings do not exist as isolated individuals, but rather exist together in co-inhabitation interrelationships. *Epistemological*, because to understand human beings and other animals it is necessary to consider the co-inhabitation relationships that forge their identities and well-being. *Ethical*, because human beings share a common habitat, the biosphere, which we must take care of for the well-being of all animals by cultivating life habits that recover a sense of being co-inhabitants with myriads of living beings, most of which go unnoticed to citizens, but are critical for the health of humans and ecosystems (Rozzi 2019).

Field environmental philosophy (FEP) offers a methodological approach to put the "3Hs" model of the biocultural ethic into action. In this way, FEP aims to better appreciate, value, and co-inhabit with biological and cultural diversity, and their interrelationships.

In this chapter, we use the conceptual framework of the biocultural ethic and the initial methodological approach of FEP to contribute to understanding how to reconnect global society with biocultural diversity. Toward this aim, we organize this chapter in three parts. First, we concisely introduce three types of barriers that have disconnected global society from biocultural diversity, with particularly severe impact in Latin America. Second, we illustrate the FEP methodological approach through two field experiences at the Omora Ethnobotanical Park: an inter-species encounter with the owls of Cape Horn, and an inter-cultural encounter with Indigenous Yahgan elders and places. Third, we concisely discuss the philosophical implications of the FEP methodology and the biocultural ethic. A biocultural systemic, contextual, and co-evolutionary understanding could help to set social-environmental priorities, policies, and actions to conserve biological and cultural diversity not only in Chile but also in other heterogeneous regions of the planet. FEP's approach transcends the purely descriptive plane because it implies assuming an ethical responsibility for taking care of vital (or disrupted) links between co-inhabitants, their life habits and their shared habitats found in contrasting localities and regions of the world.

10.2 Barriers Hindering Connections Between Global Society and Biocultural Diversity

Why is it that authorities and other members of global society have so many difficulties perceiving, understanding, and appreciating biological and cultural diversity today? We distinguish three main types of barriers that drive the rapidly growing homogenization of life habits and habitats inhabited by people worldwide. Contemporary global society is characterized by an explosive growth of urban population, which leads to a drastic homogenization of the habitats inhabited by most humans. The urban enclosure of contemporary society generates a *physical barrier* that hinders the awareness and understanding of biocultural diversity. Second, *conceptual barriers* due to the loss of diversity of languages and forms of ecological knowledge and the concentration of information in digital media. Third, *institutional and educational barriers* associated with a concentration of economic and political power and homogenous educational programs that have displaced vernacular forms of knowledge These barriers synergistically foster biocultural homogenization (Rozzi 2018).

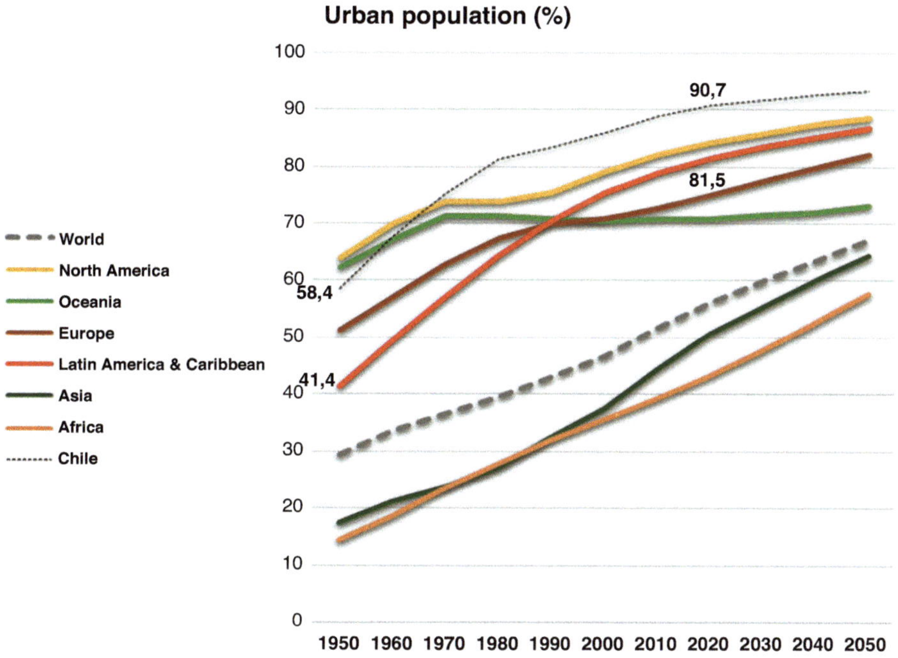

Fig. 10.2 Relative percentages of urban populations since 1950, and estimated percentages until 2050, in Latin America and the Caribbean, Europe, Oceania, Asia, Africa, and the world as a whole (Modified from Rozzi 2012; data from Heilig 2012)

10.2.1 Physical Barriers

At the beginning of the twenty-first century, for the first time in the history of the human species, more than 50% of the world's human population lives in cities (Heilig 2012). This concentration of the population in cities has created physical barriers that distance global society from daily contact with biocultural diversity. The massive rural-to-urban migration is a recent event linked to the Great Acceleration, which has been especially marked in Latin America where the urban population has grown from 41% in 1950 to 79% in 2010 (Fig. 10.2). This rural-urban migration has severe consequences for both the native habitats and the well-being of displaced human communities.

Regarding the native habitats, the rural-urban migration generates a loss of the ancestral human stewards of the land. As indigenous people and other local communities are displaced to cities, native habitats are left open today to accelerated processes of land-use changes, including large-scale mining and expansion of monocultures associated with a concentration of the land property (Ceccon and Miramontes 1999; Tabasura-Acuña 2006; Finer et al. 2008; Borras et al. 2011) During the last five decades, Latin American governments have been subject to an increasingly prevailing neoliberal economic, development model, which, driven by narrow technological and market parameters, has promoted the consolidation of land ownership in service of economies of scale (Escobar 2008, 2010). Both national and international development pressures act over regional habitats displacing and/or eliminating their native human populations. The explosive increment in the concentration of land ownership since the 1970s has had severe negative socio-ecological impacts (Borras et al. 2012). Indigenous and other local communities are keenly aware that the well-being of human communities and other-than-human communities go hand in hand (Rozzi 2015b). In their territories, these populations frequently act as guardians of the land, protecting its biological and cultural diversity. They attempt to resist their territorial displacements caused by development projects such as mining, dams, and monocultures, in order to conserve their autonomy, their traditional habits and native habitats. Consequently, conservation should not be considered merely a luxury of rich people and rich nations. Instead, it is a vital need for the health of the local inhabitants and their culture (Rozzi 2001).

Regarding human habits and well-being, in the cities, displaced indigenous people, peasant and fishermen communities frequently lack access to basic services, such as food, water, shelter, and sanitary conditions; hence, they face extreme poverty conditions that are rapidly expanding in the marginal neighborhoods of metropolis areas in Latin America (Parentelli 1996; Gebara 1999; Rozzi 2001). Additionally, these populations lose their everyday contact with their regional biological diversity. The knowledge that most teachers, authorities, new generations of students, and the majority of citizens have about biological and cultural diversity is acquired in urban contexts, distanced physically, emotionally, and ethically from the regional habitats, their communities of co-inhabitants and diverse life habits (Leopold 2004; Poole 2018).

In summary, regional ecosystems become distant regarding the everyday experience, but they are heavily impacted by new urban lifestyles, with growing levels of energy and material consumption, and production of waste. Hence, losses of regional biological and cultural diversity are coupled with the degradation of environmental and social sustainability. To counteract this trend, participatory conservation of habitats with the inclusion of local communities represents an ethical imperative that must be incorporated into development policies as a matter of eco-social justice (Rozzi 2013).

10.2.2 Conceptual Barriers

At the beginning of the twenty-first century, for the first time in the history of the human species, more than half of the world's population inhabits symbolic worlds that are defined by less than ten languages. According to the data of the Ethnologue (2010), today 52% of the world population speaks one of the seven dominant languages: Mandarin, English, Hindi, Spanish, Russian, Arabic, and Bengali (Fig. 10.3). These seven dominant languages represent a minimal fraction (0.1%) of the 6909 languages that are still spoken around the globe. This linguistic

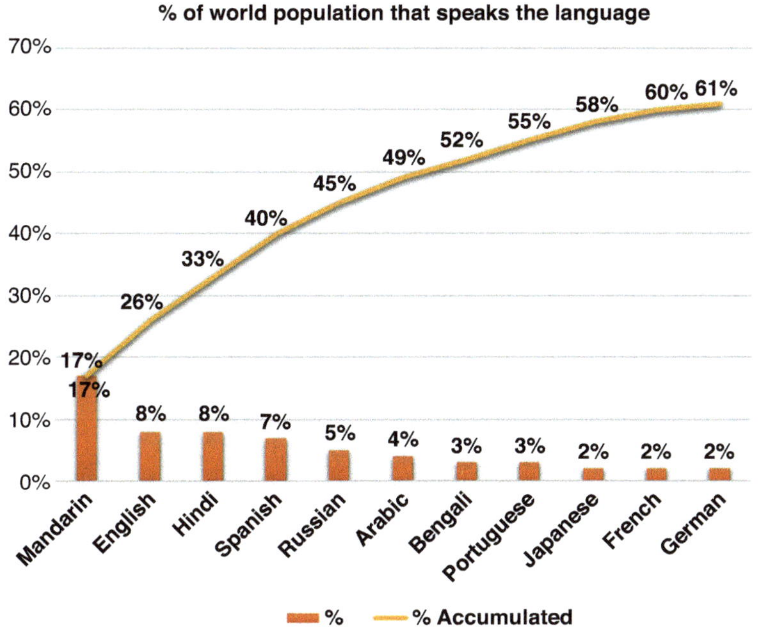

Fig. 10.3 Dominant Languages Spoken in the World in 2000. Bars illustrate the relative percentage of the world population that speaks each of the dominant languages, and the line depicts the accumulated percentage of the world population that speaks one of these languages (Modified from Rozzi 2012; data from Ethnologue 2010)

homogenization drastically reduces the spectrum of concepts and worldviews with which biological and cultural diversity are perceived, understood, and valued. Today the extinction risks for languages exceed those for biological diversity worldwide and particularly in Latin America (Lizarralde 2001). In Chile, for example, over 50% of the native languages are already extinct (Rozzi 2013).

In summary, global society has forged *conceptual barriers* derived from the displacement of linguistic diversity and the widespread mediatization of perceptions about and experiences with biocultural diversity (Rozzi 2012). To counteract this trend, the conservation of vernacular forms of knowledge and life habits represents another pressing need to foster human well-being, autonomy of local communities and their complex and dynamic relationships with biodiversity (Rozzi 2013).

10.2.3 Institutional Educational Barriers

Global society has forged *institutional and educational barriers* derived from a concentration of economic and political power and educational programs that have displaced vernacular forms of knowledge (Escobar 1995). Since the 1970s, several

Latin American countries are making efforts to recover indigenous languages and cultures. However, they are still ignored or only marginally incorporated into formal education, and Spanish continues to be taught as the unifying language of the nation-states (López-Bascuas 2007; Fajardo-Salinas 2011). In this region, and worldwide, monolingualism of the colonial language has been imposed, and formal education represents a major driver of linguistic homogenization (Krauss 1992; Maffi 2001, 2005). Homogenous formal education programs hinder a plural understanding of human natures with cultural habits linked to particular local habitats and ensembles of co-inhabitants. This biocultural homogenization conveys epistemic injustices because the knowledge of indigenous and local communities is systematically and arbitrarily ignored or oppressed.

The biocultural ethic demands overcoming epistemic injustice (*sensu* Fricker 2007) which may derive from the globalization of a hegemonic knowledge linked to Eurocentrism. Globalization of hegemonic knowledge implies exclusion and oppression of forms of knowledge of indigenous and other local communities (Gosfoguel 2007). At the same time, this direct oppression of local communities also implies an indirect oppression of the biodiversity with which they co-inhabit. Epistemological hegemony "makes invisible" ways of knowing, which are intimately linked with forms of co-inhabitation with multiple biological species.

10.3 Inter-species and Inter-cultural Field Encounters

To contribute to reorienting the trends of biocultural homogenization toward biocultural conservation, in this chapter I introduce the biocultural education approach developed at the Omora Ethnobotanical Park at the southernmost archipelagoes of South America, protected by Cape Horn Biosphere Reserve, Chile (Fig. 10.4). A central aim of Omora Park's field environmental philosophy program is the recovery of face-to-face encounters with diverse living beings with whom we co-inhabit, as illustrated by the following field experience.

10.3.1 The Owls of Cape Horn: An Inter-species Encounter at Omora Park

During the first summer outing of the University of Magallanes in Omora Park, we camped in a forest of high-deciduous beeches or Lenga trees (*Nothofagus pumilio*) at the edge of the Róbalo River. Shortly before midnight, we were talking softly around the campfire when we heard a strange vocalization from the canopy. When we heard it again, we were surprised because it was the call of the Rufous-legged Owl (*Strix rufipes*), a rare owl species characteristic of the old-growth forests of southern Chile and Argentina. Although at the time we were still unsure of the call's identification

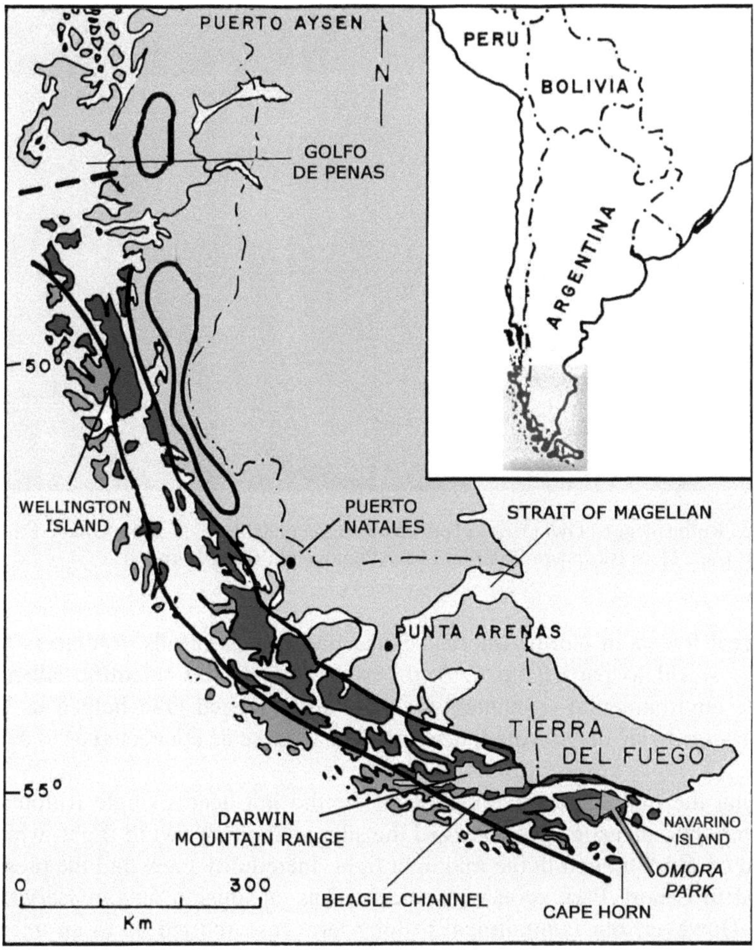

Fig. 10.4 The Omora Ethnobotanical Park (55°S) is located in the Magellanic sub-Antarctic ecoregion dominated by forests (dark strip between the Gulf of Penas and Cape Horn), on Navarino Island, 3 km to the west of Puerto Williams, capital of the Cape Horn County and the Chilean Antarctic Province

because this species was not previously recorded on Navarino Island, its presence pleased us for two reasons: (i) the surprise of encountering this co-inhabitant of which we were not aware was pleasurable, and (ii) it was a gratifying sign that Omora Park's forests have a conservation value for the Rufous-legged Owl and other rare species restricted to old-growth forests. Even more, this bird is a sister species of the Spotted Owl (*Strix occidentalis*) that has been so important for the protection of old-growth forests in the Pacific coast of northwest Canada and the United States. So, the Rufous-legged Owl encouraged us to launch an Earthwatch Watch Program to study "the owls of Cape Horn" and compare them with owls inhabiting temperate

Fig. 10.5 Rufous-legged Owl (*Strix rufipes*) perching in plain daylight at the Omora Ethnobotanical Park, Cape Horn Biosphere Reserve, Chile (Photograph: Omar Barroso)

and boreal forests in North America. Since Earthwatch enrolls volunteers from all over the world to participate in field-based research and scientific education to promote environmental sustainability, the Rufous-legged Owl helped us become part of a worldwide conservation network. The local scale connected us to the global scale.

During the first Earthwatch campaign we did not hear a single Rufous-legged Owl. Only two investigators observed the silhouette of an owl in flight, which later perched on a dry branch in the midnight light. Incredulity grew and the presence of this owl in Omora Park soon was perceived as an illusion or a conservationist's dream. However, our commitment to long-term research led us to an unexpected encounter: at midday, 6 months later, we were surprised by the sighting of a Rufous-legged Owl perched on a Lenga branch. We were able to photograph it in full light, thus proving that it was not an illusion! (Fig. 10.5).

Since then, we have heard the Rufous-legged Owl on several occasions in twilight hours. Moreover, the Rufous-legged Owl has become so accustomed to our presence that it sometimes comes down from the canopy to where we work with mistnets to band forest birds. The owl once settled on a branch and watched us curiously. In our mutual staring, we experienced an unexpected communication with a living being of another species, of whom we only recently had become aware of their presence.

Sleeping under the Lenga trees that summer helped to establish a sympathetic relationship with a living being that was unknown to us shortly before. The repeated sounds of the Rufous-legged Owl's calling at dusk and the "eye-to-eye" contacts have transformed the initial perceptions into a deeper experience of co-inhabitation at Omora Park. The discovery that we share the same habitat with the owl awakened emotions that influence our decision-making regarding habitat. For example, in the layout of interpretive trails, we prevent visitors from passing through areas where the Rufous-legged Owl nests, and other places where the old-growth Lenga trees need protection. In this way, an initial ecological perception and understanding has led to conservation actions and to an ethical relationship of co-inhabitation.

10.3.1.1 Learning from Perceiving and Co-inhabiting Biocultural Diversity

In summary, spending the night under the Lenga trees made us aware of the existence of a living being that had been "invisible" to us, and opened an understanding that would not have occurred in the comfort of the classroom or in a hotel room. Encountering the owl and experiencing downpouring rain and freezing wind, differ radically from watching a nature documentary on the television in our temperature-controlled rooms. In a hotel we can artificially maintain the temperature of our rooms, determine the moment we want to watch the documentary, and the time at which we take a shower. Beneath a tree, on the other hand, our experiences are not predetermined: one never knows if the owl will show up or not, if it will rain or if there will be wind. In the field, we experience integrally the habitat where the Rufous-legged Owl lives (instead of an idealized and simplified conceptual model of its habitat). In this way, upon leaving the classroom or hotel room, we cross physical barriers. We detach ourselves from a conceptual model and an attitude (a mindset) of controlling our world. As a result, we gain access to an experience of co-inhabitation, a form of dialogue based on perceptions and open interactions with the co-inhabitants of the ecosystems or habitats we share.

To solidify this experience of co-inhabitation, the recurrence of these interactions is necessary. Just as the flute player requires practice to master his or her instrument, co-inhabitation with an owl requires encounters that enable the discovery of multiple facets of the owl's life. Through this continuous process of discovery, it is possible to conceive forms of co-inhabitation among humans, owls, and other co-inhabitants in the forests.

In this recurrent experience of co-inhabitation, students gain an understanding of the Rufous-legged Owl and other living beings, that transcend their scientific names and numerical descriptions. This is especially relevant considering that scientific papers, baselines for environmental impact assessments, and textbooks frequently focus only on species lists, population censuses, metabolic measurements, and other quantitative characterizations of biodiversity. The recurrence of direct encounters with the owl enables students to broaden their understanding of biodiversity; by crossing glances with the bird, they progressively discover the similarity of the eyes and glances of humans and birds (Fig. 10.6). The human-bird "eye contacts" have stimulated new questions about "how" is, and "who" is the owl. Through these "direct encounters" in the forest, the owl has become a co-inhabitant, a non-human person, such as when we co-inhabit with our cats or dogs.

Fig. 10.6 The similarity between the eyes of the Rufous-legged Owl and the ornithologist Silvina Ippi reminds us how similar our species *Homo sapiens* is with other living beings. Through "direct encounters," and "eye contacts," birds face us as non-human people, whose existences can be subjected to moral considerations that, to a certain degree, are equivalent to those with which we judge the value of human life (Photographs: Patricio Riquelme [left] and Juan Carlos Torres-Mura [right])

Through this recurrent experience of co-inhabitation, the changes in our perception and understanding of the owl continuously modify our ethical relationship with it. In turn, the ethical experience broadens and transforms the spectrum of our perceptions and comprehension of "who the owl is," its behavior and interactions with its habitat and the biotic community. In this way, ecological sciences and environmental ethics inform each other in a dynamic circle of perceptions and attitudes, which configure our "environmental ethos" or way of co-inhabiting in the midst of biodiversity (Rozzi 1999). Daring to leave the cognitive isolation caused by the logical or mathematical structures that prevail today in written and audiovisual texts and daring to leave the physical isolation caused by the walls of the classrooms or hotel rooms, allows students and visitors to access conceptual and physical spaces for encounters with biodiversity and the reality of Cape Horn, which would otherwise remain "invisible" to us.

The conservation crisis of biological and cultural diversity is largely a cognitive and experiential crisis because we do not perceive the majority of living beings and their life processes; moreover, if we lack the experience of co-inhabiting with them, we do not perceive the completeness of other living beings. Because the experience of co-inhabiting with other-than-human beings is so rare for most citizens today, it is fundamental for us to open the educational experience of Omora Park toward nonformal educational practices, such as ecotourism, that span larger sectors of society. For example, as the experience of the students from the University of Magallanes shows, sleeping under a tree in Cape Horn allows students and visitors to feel the downpour of rain and the gale-force wind of the "sub-Antarctic" climate. Through this experience, they discover the heterogeneity of the microclimatic; i.e., how climate varies from moment to moment, from site to site. This variability is not captured by the data of annual mean temperature or annual rainfall; however, these are frequently the only climate-related data of Cape Horn that are provided by educational texts or tourism brochures. Thus, the educational approach of Omora Park that invites students and visitors to sleep under the Lenga trees provides a richer cognitive experience (that is complementary to understandings gained by reading a textbook or touristic brochure) for perceiving one of the most idiosyncratic aspects of Cape Horn's climate: its unpredictable and variable character. Field experiences lead to broader forms of understanding, which allow members of the twenty-first-century global society to establish closer relationships with the owl and other co-inhabitants of the austral forests.

10.3.1.2 Field Environmental Philosophy

For the integration of ecological sciences and environmental ethics, Omora Park's "field environmental philosophy" program places special emphasis on the living beings that are more difficult to perceive, those that are active at night, live underwater, or are small size. The discovery and observation of these "invisible" beings feed our awareness of how little we know our environment. The type of knowledge that arises through field experiences, such as sleeping under a tree, invites us to review the limitations of language and of the prevailing methodologies of formal education, which have a marked bias toward teaching based on a single alphabet and numeric system. This universalizing approach to teaching has been inherited from modernity, especially the Age of Enlightenment, and is now globalized. This universalizing way of teaching spreads a "physical and conceptual veil" that blocks our ability to "see" and "co-inhabit" with particular beings and unique habitats at specific sites and moments. Removing this veil enriches our learning experiences and awakens awareness about our ignorance of the diversity and uniqueness of the beings with whom we co-inhabit at each site and the planet as a whole. This greater understanding of the limitations of our ways of knowing invites us, in turn, to act with caution. These field experiences stimulate questions about how we want to, and how we should, live and coexist. This transition from understanding toward action and co-inhabiting is the second central aim of the education programs at Omora Park. We illustrate this second aim, and its methodological phase, with the following experience.

10.3.2 *The Yahgan Grandmother Cristina Calderón and the* Upush *Shrub: An Inter-cultural Encounter at Omora Park*

The Yahgan grandmother Cristina Calderón accompanied us during one of the first walks with students from the local school of Puerto Williams through a newly formed interpretive trail in Omora Park. The Yahgan are the original inhabitants of Cape Horn and have their own language. At the beginning of the journey, the grandmother—as elder women are known among the Yahgan—took a moment to show us the abundance of the *zaparrilla* or *upush* shrub. She explained to us that in the Yahgan language Puerto Williams might have been called *Upushwaia* because it is a bay (*waia*) where the *upush* is abundant. The students then investigated how the Spanish conquerors observed that this shrub grows in such abundance in some areas of southern Chile that it acts like a weed or *zarza*, and that its leaves are similar to the leaves of a grapevine or *parra*, that the Spaniards cultivated in Europe. Hence, they called this shrub *zarzaparrilla*. To the Anglican missionaries who came to the area in the nineteenth century, on the other hand, this species looked like a "wild currant," because its fruit looks like small, dried grapes or raisins (= *currant*) that grow wildly (= *wild*). Complementarily, scientists have determined that this type of shrub pertains to the genus *Ribes*, which is distributed globally and includes some 200 species, and since this species is found in the *Magellanic* Region, it is called *Ribes magellanicum*.

The experience of sharing a language with grandmother Cristina and researching the origin of the names given by different European colonizers contributed to the understanding that the names of plants, animals, and places express different cultural traditions and worldviews. Our knowledge about biodiversity possesses an unavoidable cultural dimension: names entail ways of understanding and of ethical relationships with places and living beings with whom we co-inhabit. For example, in the Yahgan toponymy, *upushwaia* refers to the *upush* plant that is abundant in the bay (= *waia*) at the entrance of Omora Park. In the 1950s, this toponym was replaced by the current name of Puerto Williams, referring to the frigate captain Juan Williams who helped take possession of the Strait of Magellan for Chile. The teaching of grandmother Cristina and the research on toponomy led students to understand the values in which languages express different senses of co-inhabiting. The Yahgan denomination maintains the reference to those beings with whom we co-inhabit in a given place; conversely, the current denomination remembers those that helped take possession of the place. In the new name of Puerto Williams, the original inhabitants (humans and other-than-humans) of *upushwaia* are absent.

Linguistic and cultural diversity tend to be as or more "invisible" than biological diversity. The leaves, the fruit, and the Yahgan and European names of the shrub would have remained unperceived if we had not taken the time to observe, investigate, discover, and compose meanings and narratives. The educational approach of Omora requires us to continuously cultivate our dispositions to observe, to listen, and to practice the integration of our perceptions and knowledge to generate our own understandings. For example, the disposition to observe the shape of the leaves of a shrub and associate this shape with the leaves of the grapevine drove us to investigate and compose possible narratives about the origin of the name *zarza-parrilla*. The observation that the fruit resembles small grapes in the spring, which later dry into what look like small raisins in the fall and remain hanging in clusters on the branches during the winter, gave us a clue to understanding the meaning of the English name wild currant. Attentively listening to the word *upush*, and to associate it with the abundance of that shrub in the bay and with our previous knowledge about the Yahgan culture, which often gives names to places in association with abundant plants or animals, allowed us to discover, or co-create, with grandmother Cristina the name: *upushwaia*.

In this educational experience at Omora Park, students are no longer "mere consumers" of biocultural knowledge or enchanting stories; instead, they are actors within the history and webs of biocultural co-inhabitation. An early slogan of the Omora educational programs was "teach children to be citizens and not consumers" (see Hargrove 1997; Rozzi 1997). The sense of amazement in the experiences and stories no longer depends exclusively on the beauty of the landscape, the components of biodiversity, and the names and forms of knowledge about them, but depends on the interactions that we establish with them: on the attention of our observations and on the associations of our narratives. So, students, researchers, and other visitors will not only receive but also give a sense of wonder and responsibility, which nourishes and orients the desire to live and to co-inhabit ethically within the biocultural web.

This attitude of reciprocity is undervalued when we tend to live isolated and blinded within cities of postindustrial society. Global-urban models extend "insensibly" beyond cities into the planetary web of biocultural diversity, installing their languages and lifestyles even in remote places such as Cape Horn. The wonder that arises in field environmental philosophy workshops with discoveries such as the diversity of names, the experiences of discovering or composing metaphors and narratives, and the experiences of co-inhabitation, stimulate educational processes that resist and counterbalance the currently prevailing processes of biocultural homogenization.

Through these learning experiences, the students and researchers become actors and social communicators of biocultural diversity. For example, students, teachers, researchers, and volunteers adopt actions like composing narratives or creating pyrographies and other interpretive signs that invite the Park's visitors to explore the diversity of plant and bird names that can be observed along the interpretive trails (Fig. 10.7). The educational experience at Omora Park culminates with students assuming active roles as receptors and generators of knowledge and narratives, as protagonists—and not mere observers—of these webs of biocultural co-inhabitation. The learning experiences lead students and all participants to assume their responsibility as co-inhabitants, members of the biocultural community, acting in ways that contribute to the continuity of the life of the diverse living beings, cultural traditions, and biocultural processes that participate within it.

Fig. 10.7 The education program of Omora Park consists of two phases: in the first students and ecologists investigate biological and cultural diversity in the field (**a**, **b**), and in the second they undertake biocultural conservation action (**c**, **d**). In the sequence, (**a**) the Yahgan grandmother Cristina Calderón with her sister, niece, and grand-nephews, prepare for a walk with students through Omora Park; (**b**) students of the local school of Puerto Williams and botanists study the leaves of the *zarzaparrila* or *upush* shrub; (**c**) students, ecologists, and art teachers prepare pyrographies for the park's interpretive trails; (**d**) pyrography installed by the studied shrub, indicating its names in Yahgan (*upush*), Spanish (*zarzaparrilla*), English (wild currant), and scientific Latin (*Ribes magellanicum*) (Photos: Oliver Vogel [**b**, **c**] and Ricardo Rozzi [**a**, **d**])

The essential sense of Omora's educational program does not lie in a supposedly objective truthfulness of its biocultural narratives, or in the delight that these stories can produce, but rather lies in assuming the continuous practice of discovering and inventing comprehensions and practices that allow us to re-encounter the beauty of those beings with whom we co-inhabit.

In formal and nonformal education in large cities, students and citizens are too accustomed to being receivers and consumers of information. In the dynamic of the free market, citizens are lulled by a hedonist practice of "consuming and enjoying" stories that are told by the market and the media.

Omora Park invites students and other participants to change from being the passive receivers of attitudes to becoming actors, like a sculptor who in each shrub (or other living beings) discovers-composes forms and narratives that decipher the singularity and the elusive richness of the existences of those living (and non-living) beings with whom they coexist. Such practices allow us to have the experience of re-encountering the beauty and the symbolic-linguistic meaning of living beings, such as the leaves, fruit, and names of the *zarzaparrila* or *upush* shrub. Through these re-encounters, biophysical and symbolic-linguistic meanings fill us with understanding and re-enchantments that inspire participants to cultivate life practices that contribute to coexistence and the conservation of cultural and biological diversity.

The experience of researching the ethical implications of the plants' names clearly illustrates the diverse sources and forms of learning. The grandmother Cristina, the teachers of the Puerto Williams local school, the researchers of Omora Park, the shapes of the shrub's leaves, and the history and biology textbooks offer complementary information and clues to build our understanding. The educational approach of Omora does not negate the value of classroom learning, of mathematics and textbooks. Rather it reaffirms that it is urgently necessary to complement such practices with an education embedded in the everyday environment, which can allow us to recover and appreciate the encounters with so many beings and vital expressions that today are not contained in the names and/or mathematical characterizations of biocultural diversity included in formal education.

10.4 Philosophical Background and Implications

As I argued at the beginning of this chapter, the world changed after World War II. And not only on the political level, but also in the field of science and technology. These changes have generated a cognitive revolution driven above all by information technologies. With the so-called "digital world" the mediatization of knowledge has been installed in formal and non-formal education, in the globalized culture. In the predominantly urban global society, communication is carried out through media communities, to the point of turning face-to-face encounters with biological and cultural diversity into an atavistic memory (Bilbeny 1997).

Communication without face-to-face encounters and sensory experiences, however, leads to a knowledge about biological and cultural diversity that ends up not being felt. For this reason, amidst the Anthropocene, Omora Park's field environmental philosophy programs emphasize experiences of "direct encounter" with mosses, lichens, trees, rocks, water, birds, and other living beings (humans and non-humans). With its emphasis on direct encounters and the immediate and holistic perception of nature, Omora Park's educational experience aims to compensate for the current excess of mediated information (Fig. 10.8). Today, most of our knowledge about nature is mediated by equations and mathematical models, by technology (e.g., television, computers, digital and social media), and by logical frameworks and/ or established scientific narratives (Fig. 10.8).

Direct exposure to natural habitats is an increasingly rare experience. To counter this trend, the educational approach of Omora Park stimulates students to cross cognitive barriers imposed by the prevalence of mediated knowledge. It stimulates them to practice direct encounters with terrestrial, marine, and cultural elements, learning in situ, immersed in the habitat and its web of ecological and cultural interactions. In this scenario, the experiential knowledge of biological and cultural

Fig. 10.8 The Field Environmental Philosophy Program at Omora Park emphasizes "direct encounters" with other living beings. This pedagogical approach aims to compensate for the excess of secondary sources and "mediated information" that prevail today in formal and non-formal education. Currently, the majority of our knowledge of nature is mediated by: (a) scientific narratives established historically (for example, the accounts of the Voyage of the Beagle where Darwin refers to the flight of the hummingbird in the midst of a storm in Magallanes [Darwin 1839]); (b) by mathematical equations (for example, the hummingbird *Sephanoides sephaniodes*, "omora" in the Yahgan language, is studied through an equation of its metabolism (MR = metabolic rate; López-Calleja and Bozinovic 1995); (c) by computer models and audiovisual technology, such as television documentaries (for example, video-recordings of the hummingbird visiting the flowers of the Magellan Copihue or Coicopihue (*Philesia magellanica*); and (d) by cellular phone applications or social media (Figure modified from Rozzi et al. 2006)

diversity surpasses its computational or mathematical representations. In contrast to reductionist universality, students perceive and investigate how each moment, each sector of space, and each living being seems to be different from the other. The heuristic objective of this field environmental philosophy practice is to provide students and other participants with a phenomenological experience and understanding of biodiversity, cultural diversity, and biocultural diversity.

The direct relationship with nature and its diversity is particularly relevant, as it trains students to critically assess the prevailing modern scientific paradigm that studies nature under the mediation of mathematical representations and computer models. According to modern science, it is possible to express the essence of the natural world and its laws through mathematical formulas. Therefore, according to this modern paradigm, it would be not necessary to leave the classroom to know and

Fig. 10.9 The biologist Margaret Sherriffs and the Magellanic architect Fernando Haro observe the diverse shapes of lichens of the genus *Cladonia* that grow on rocks during one of the field environmental philosophy workshops at Omora Park (Photographs: Ricardo Rozzi [left] and Silvina Ippi [right])

understand nature, given that its essence is discovered better through logical and mathematical representations. This modern science approach has led to a "mathematized world." Such distancing from contingent reality has provoked the "disenchantment" of the world. Under these modern assumptions, nature loses its mystery. One can, in principle, control the world and "direct everything through calculus" (Weber 1958). Omora Park's educational program deconstructs these universal mathematical assumptions through philosophical readings and field experiences. It guides students to re-discover the particularities of every living being, and the inapprehensible symbolic-linguistic and biophysical diversity of human and other-than-human beings in places with unique biocultural processes and patterns.

In contrast to the emphasis on information and content that prevails in current formal and non-formal education, Omora Park's pedagogical approach emphasizes the learning and education process. Attitudes and practices of inquiry are more important than knowing the scientific names of biological species and other types of specialized information. Questions are established in simple language to overcome the problem of specialization and technical language, allowing the participation of a broad spectrum of people.

The experience involves the discovery and description of a variety of life forms in a specific site and moment. For example, we can investigate the distribution patterns and learn about the life histories of some barely "visible" but very diverse organisms in Cape Horn. In this way, we have discovered that lichens that grow on the bark of trees often differ in color, shape, and size from those that grow on rocks (Fig. 10.9). With this pedagogical approach, the students are initiated in the basic understanding of distribution patterns of biodiversity. Later, to be familiarized with the biodiversity of the site, they draw images of the lichens observed. An interesting fact usually happens in this activity: the colors of pencils do not coincide with the colors of the lichens. This leads students to experientially learn about the technical difficulties of "capturing" the colors and other dimensions of the living beings with whom we co-inhabit. This experience helps us to reflect on the limitations of our ability to comprehend biodiversity within our conceptual categories and with our technical tools. At the same time, this experience stimulates our awareness of the unlimited singularities of each living being.

10.5 Concluding Remarks

In summary, Omora Park's field environmental philosophy methodological approach emphasizes: (a) field experiences, *in situ*, with biological and cultural diversity in its integral context (in contrast with today's emphasis on mediated knowledge); (b) educational processes (rather than an emphasis on content and information); (c) reflection on learned understanding of biocultural diversity and of our ways of relating with diverse human and other-than-human beings; (d) the adoption of biocultural conservation actions oriented by continuous learning experiences, experiences of co-inhabitation, and inter-cultural dialogue and reflection. This educational process invites us to evaluate our life-ways and eventually to modify them, in the continuous cultivation of a biocultural ethic.

Acknowledgments I thank the valuable comments by Francisca Massardo, Alejandra Tauro, Kelli Moses, and Roy May, and acknowledge the support from the Cape Horn International Center (CHIC) - ANID PIA/BASAL (PFB210018).

References

Bilbeny N (1997) La revolución en la ética: Hábitos y Creencias en la Sociedad Digital. Anagrama, Barcelona
Borras SM, Hall, Scoones J, White B, Wolford W (2011) Towards a better understanding of global land grabbing. J Peasant Stud 38:209–216
Borras SM, Franco JC, Gómez S, Kay C, Spoor M (2012) Land grabbing in Latin America and the Caribbean. J Peasant Stud 39:845–872
Ceccon E, Miramontes O (1999) Mechanisms and social actors in the deforestation of the Brazilian Amazon. Interciencia 24:112–119
Darwin C (1839) The zoology of the voyage of H.M.S. Beagle, under the command of Captain Fitzroy, R.N., during the years 1832 to 1836: published with the approval of the Lords Commissioners of Her Majesty's Treasury. Smith, Elder & Co., London
Dunn CP (2017) Climate change and its consequences for cultural and language endangerment. In: The Oxford handbook of endangered languages. Oxford University
Escobar A (1995) Encountering development. The making and unmaking of the third world. Princeton University Press, Princeton
Escobar A (2008) Territories of difference: place, movements, life, redes. Duke University Press, Durham
Escobar A (2010) Latin America at a crossroads: alternative modernizations, post-liberalism, or post-development? Cult Stud 24(1):1–65
Fajardo-Salinas DM (2011) Educación intercultural bilingüe en Latinoamérica: un breve estado de la cuestión. Liminar 9(2):15–29

Figueroa RM (2011) Indigenous peoples and cultural losses. In: Dryzek JS, Norgaard RB, Schlosberg D (eds) Oxford handbook of climate change and society. Oxford University Press, Oxford, pp 232–246

Finer M, Jenkins CN, Pimm SL, Keane B, Ross C (2008) Oil and gas projects in the western Amazon: threats to wilderness, biodiversity, and indigenous peoples. PLoS One 3(8):e2932

Fricker M (2007) Epistemic injustice, power and the ethics of knowing. New York

Gebara I (1999) Longing for running water: ecofeminism and liberation. Fortress Press, Minneapolis

Gosfoguel R (2007) The epistemic decolonial turn: beyond political-economy paradigms. Cult Stud 21:211–223

Hargrove E (1997) Ética ambiental y educación ambiental. Ambiente y Desarrollo 13:47–51

Heilig GK (2012) World urbanization prospects the 2011 revision. United Nations, Department of Economic and Social Affairs (DESA), Population Division, Population Estimates and Projections Section, New York

Krauss M (1992) The world's languages in crisis. Language 68:4–10

Leopold C (2004) Living with the land ethic. BioScience 54:149–154

Lizarralde M (2001) Biodiversity and loss of indigenous languages and knowledge in South America. In: Maffi L (ed) On biocultural diversity: linking language, knowledge, and the environment. Smithsonian Institution Press, Washington, DC, pp 265–281

López-Bascuas LE (2007) Bilingüismo, educación superior y formación docente. In: Cuenca R, Nucinkis N, Zavala V (eds) Nuevos maestros para América Latina. Morata, Madrid, pp 259–292

López-Calleja MV, Bozinovic F (1995) Maximun metabolic rate, thermal insulation and aerobic scope in a small-sized Chilean hummingbird (Sephanoides sephaniodes). Auk 112:1034–1036

Maffi L (ed) (2001) On biocultural diversity. Linking language, knowledge, and the environment. Smithsonian Institution Press, Washington, DC

Maffi L (2005) Linguistic, cultural, and biological diversity. Annu Rev Anthropol 34:599–617

McNeill JR, Engelke P (2016) The great acceleration: an environmental history of the Anthropocene since 1945. Harvard University Press

Parentelli G (1996) Latin America's poor women. In: Ruether R (ed) Women healing earth: third world women on ecology, feminism, and religion. Augsburg Fortress Press, Minneapolis, pp 29–38

Poole AK (2018) The UN sustainable development goals and the biocultural heritage lacuna: where is goal number 18? In: Rozzi R, May RH Jr, Chapin FS III, Massardo F, Gavin M, Klaver I, Pauchard A, Nuñez MA, Simberloff D (eds) From biocultural homogenization to biocultural conservation. Ecology and ethics, vol 3. Springer, Dordrecht, pp 315–331

Rozzi R (1999) The reciprocal links between evolutionary- ecological sciences and environmental ethics. BioScience 49:911–921

Rozzi R (2001) Éticas ambientales latinoamericanas: raíces y ramas. In: Primack R, Rozzi R, Feinsinger P, Dirzo R, Massardo F y colaboradores, Fundamentos de Conservación Biológica: Perspectivas Latinoamericanas. Fondo de Cultura Económica, México, pp 311–362

Rozzi R (2007) Future environmental philosophies and their biocultural conservation interfaces. Ethics Environ 12(2):142–145

Rozzi R (2012) Biocultural ethics: the vital links between the inhabitants, their habits and regional habitats. Environ Ethics 34:27–50

Rozzi R (2013) Biocultural ethics: from biocultural homogenization toward biocultural conservation. In: Rozzi R, Pickett STA, Palmer C, Armesto JJ, Callicott JB (eds) Linking ecology and ethics for a changing world: values, philosophy, and action, ecology and ethics book series, vol 1. Springer, Dordrecht, pp 9–32

Rozzi R (2015a) Implications of the biocultural ethic for Earth stewardship. In: Rozzi R, Chapin FS, Callicott JB, Pickett STA, Power ME, Armesto JJ, May RH Jr (eds) Earth Stewardship: linking ecology and ethics in theory and practice. Ecology and ethics book series, vol 2. Springer, Dordrecht, pp 113–136

Rozzi R (2015b) Earth stewardship and the biocultural ethic: Latin American perspectives. In: Rozzi R, Chapin FS, Callicott JB, Pickett STA, Power ME, Armesto JJ, May RH Jr (eds) Earth Stewardship: linking ecology and ethics in theory and practice, Ecology and ethics book series, vol 2. Springer, Dordrecht, pp 87–112

Rozzi R (2018) Biocultural homogenization: a wicked problem in the Anthropocene. In: Rozzi R, May RH Jr, Chapin FS III, Massardo F, Gavin M, Klaver I, Pauchard A, Núñez MA, Simberloff D (eds) From biocultural homogenization to biocultural conservation. Ecology and ethics book series, vol 3. Springer, Dordrecht, pp 21–47

Rozzi R (2019) Taxonomic Chauvinism, no more! Antidotes from Hume, Darwin, and biocultural ethics. Environ Ethics 41(3):253–288

Rozzi R, Feinsinger P, Riveros R (1997) La Enseñanza de la Ecología en el Entorno Cotidiano. Ministerio de Educación de Chile, Santiago de Chile

Rozzi R, Draguicevic JM, Arango X, Sherriffs M, Ippi S, Anderson C, Acevedo M, McGehee S, Plana J, Cortés E, Massardo F (2005) Desde la ciencia hacia la conservación: el programa de educación y ética ambiental del Parque Etnobotánico Omora. Ambiente y Desarrollo XXI(2): 20–29

Rozzi R, Massardo F, Anderson C, Heidinger K, Silander J Jr (2006) Ten principles for biocultural conservation at the southern tip of the Americas: the approach of the Omora Ethnobotanical Park. Ecology & Society. http://www.ecologyandsociety.org/vol11/iss1/art43/

Tabasura-Acuña I (2006) Ambientalismos y Ambientalistas: El Ambientalismo Criollo a Finales del Siglo XX, Cuadernos de Investigación, no 21. Universidad de Caldas, Cali

Weber M (1958) In: Gerth HH, Mills CW (eds) Essays in sociology. Oxford University Press, New York

Chapter 11
Beyond Field Environmental Philosophy: Integrating Science Education and Technology

Brant G. Miller and R. Justin Hougham

Abstract To reconnect education with biocultural diversity, Field Environmental Philosophy (FEP) proposes a four-step cycle that integrates (1) transdisciplinary research, (2) composition of metaphors and narratives, (3) field activities guided with an ecological and ethical orientation, and (4) biocultural conservation. In this chapter, we provide an assessment of FEP's effectiveness for reconnecting education with biocultural diversity. At the same time, we introduce novel dimensions that broaden and enhance the effectiveness of FEP; for example, by incorporating digital technologies embedded in field experiences. We explore Adventure Learning and Digital Observation Technology Skills as educational curricular supplements that transcend geographies and connect people to place. We address how to incorporate technologies and digital media without suppressing the concentration and engagement in the intimacy of corporeal-sensorial-emotional experiences of co-inhabitation that are emphasized by FEP.

Keywords Corporeal-sensorial-emotional experiences · Digital media · Education technologies · Innovation · STEM education

11.1 Presentation

In this book, we travel around the world exploring meaningful, place-based experiences using the Field Environmental Philosophy (FEP) methodology. We anchor our exploration in the far south of Chile at the Omora Ethnobotanical Park (Rozzi et al. 2006), but also examine how the FEP methodology built from biocultural ethics (Rozzi 2012) is used in other parts of the world. Uniting the chapters of this book are

B. G. Miller (✉)
Department of Curriculum and Instruction, University of Idaho, Moscow, ID, USA
e-mail: bgmiller@uidaho.edu

R. J. Hougham
Department of Youth Development, University of Wisconsin – Madison, Madison, WI, USA
e-mail: justin.hougham@wisc.edu

R. Rozzi et al. (eds.), *Field Environmental Philosophy*, Ecology and Ethics 5,
https://doi.org/10.1007/978-3-031-23368-5_11

the underlying precepts of the FEP. As a methodology, FEP broadly links education with art, the generation of local economic alternatives (for example, ecotourism), and the consolidation of actions over time through institutional infrastructures.

Here we weave together the chapters of Part I by touching upon their salient themes. This will build toward an offering of ideas for consideration that centers on FEP methodologies and invites a more explicit use of technology to facilitate activities and projects for deeper engagement among participants, in particular students and teachers in formal and informal education environments. We close the chapter with a commentary on possible future directions for the FEP methodology beyond the COVID era (2020–2022) with suggestions on how to leverage opportunities for expansion when presented, such as embracing technologies that support participants in delving deeper into local issues.

11.2 Weaving Chapters

In Part I, we learned of many fascinating and powerful treatments of the FEP methodology from practitioners dedicated to authentic explorations for meaningful learning situated in localized contexts. In the introduction to Part I, Rozzi (2023a) concisely summarized the epistemological, ontological, and ethical dimensions of FEP practices. In Chap. 2, Rozzi (2023a) highlights the need to integrate biophysical and cultural dimensions of a landscape to better understand, support, and mitigate adverse effects associated with the Anthropocene.

In Chap. 3, "The multiple lenses of Ecotourism with a Hand-Lens: Fundamental Concepts and Practices," Rozzi and Tauro (2023) present a set of metaphorical or conceptual lenses, including ethical, aesthetic, economic, and biocultural "hand-lenses" to explore the micro-cosmos. They present educational concepts and practices to apply each of these lenses. These didactic practices enable students and other participants, including tourists, to experience and reflect on reconnections with biological and cultural diversity, which orient ways for respectfully and sustainably coinhabiting biocultural diversity. The activities are guided by the FEP methodology and are built on the biocultural ethic that greatly enhances and makes visible the interconnectedness of human and other-than-human co-inhabitants.

In Chap. 4, "Underwater with a hand-lens in the rivers of Cape Horn: ethical, ecological, and aesthetical values of freshwater ecosystems and their co-inhabitants," Contador et al. (2023) melds Aldo Leopold's land ethic with Ricardo Rozzi's biocultural ethics to explore the freshwater ecosystem and the invertebrates therein of the Cape Horn Biosphere Reserve. Contador and collaborators effectively consider how invertebrates play a role and have tangible value not only within the other-than-human landscapes but also to the human communities in which the invertebrates inhabit and impact. Using FEP, new methodologies have been developed that translate into novel conservation efforts.

In Chap. 5, "Sub-Antarctic High Andean 'Gardeners': Cultivating Caring Relationships," Méndez-Herranz et al. (2023) use the FEP four-step methodology to take a closer look at a high-altitude Andean area of sub-Antarctic Navarino Island. Through the FEP lenses, Mendez et al. reclassify the high-Andean habitats previously viewed as "deserts" to a more accurate characterization that accounts for the abundance and diversity of bryophytes and other little organisms. By effectively using FEP, Méndez-Herranz et al. (2023) bring attention to a landscape that was previously discarded as uninteresting and begin to explore activities that support sustainable conservation for the other-than-human inhabitants of mountainous regions. To invite active consideration and participation with the landscape, Méndez-Herranz et al. (2023) coin the metaphor "High-Altitude Sub-Antarctic Gardens" to foster sensitivity to and responsibility for the other-than-human inhabitants.

In Chap. 6, "Pay attention, dive with eyes wide open: Field Environmental Philosophy activity to foster reciprocity between people and the sea," Jaime Ojeda et al. (2023) take a systematic approach to understanding the seaweeds and mollusks by using the lenses of ecological, ethnoecological, and philosophical theory. By taking this approach, Ojeda et al. are able to associate empirical evidence for the annual rhythms and life histories of both the other-than-human inhabitants of the Cape Horn Biosphere Reserve and the interplay with the Yaghan people's livelihood. In particular, they focus on seaweeds and mollusks as staples of sustenance that are impacted by their seasonal abundance. By developing metaphors, Ojeda's work provides understanding and offers respectful ways for exploring land- and seascapes for biocultural conservation.

In Chap. 7, "The eyes of the Tree: Applying Field Environmental Philosophy to Tackle Conservation Problems at Long Term Socio-ecological Research sites," Crego et al. (2023) identify limitations of the International Long-term Ecological Research (ILTER) network and propose ways forward using FEP for mitigating the identified limitations. Among the limitations identified are (i) little consideration of cultural dimensions alongside the biophysical data; (ii) privileging theory rather than application in study design; and (iii) geographical bias toward the northern hemisphere. Crego et al's. contribution addresses the stated limitations through a carefully crafted study that looked at the impacts of invasive mammal species using an FEP approach to reduce biocultural homogenization.

In Chap. 8, "Starfishes and sky stars: A Field Environmental Philosophy education and ecotourism experiences in Baja California, México," Moreno-Terrazas et al. (2023) bring attention to the abundantly diverse region of Baja California Sur and consider important sociocultural factors that impact these ecosystems. Using the 3Hs framework (habitats, habits, co-inhabitants) of the biocultural ethic, Moreno-Terrazas et al. (2023) use FEP to compose the metaphor "from the starfish to the stars in the sky" and to create activities for ecotourism. This metaphor and activities guide participants to view and move through the landscape to understand the complex factors needed to conserve these sensitive landscapes and seascapes.

In Chap. 9, "Biocitizen's Approach to Biotic Wonder, Citizenship, and Field Environmental Philosophy," Kurt Heidinger et al. (2023) highlight the efforts and impacts of the Biocitizen non-profit educational institution and how using FEP and Aldo Leopold's land ethic, students have been introduced to the wonders of nature. This chapter highlighted an FEP success story and how it can be implemented in different regions of the United States for biocultural conservation.

And lastly, in Chap. 10, Rozzi (2023b) outlined central concepts and some of the first experiences that originated FEP. This chapter lays out how FEP can be utilized within short-term (e.g., class-room science teaching) and long-term (e.g., LTSER study sites) contexts, as well as across levels of education and research entities from schools, higher education institutions, Indigenous communities, and governmental programs. Simply, Rozzi (2023b) makes a compelling case for us to consider our actions and our positionality toward the other-than-human co-inhabitants of this Earth.

Each of these chapters has in common a deep intention to honor and understand a landscape through the biocultural ethic and to invite participants in their respective communities to know and understand their landscape on a more intimate level. The landscapes represented in Part I are as diverse as the authors who crafted each chapter. But each story adheres to the principles in the FEP methodology. As a result, authors have created a new space for considering the carefully orchestrated actions of the other-than-human co-inhabitants of the landscapes and have provided guidance for the human inhabitants responsible for management and conservation. Through FEP and the novel uses of tools both inherent (the five senses) and designed (e.g., hand lenses), we are beginning to understand a new way forward for comprehending, protecting, and conserving critical landscapes locally and globally.

11.3 FEP Pivot to Expansion

As noted, and expanded upon in the previous chapters, FEP integrates ecological research and environmental ethics into biocultural education and conservation through an interrelated four-step cycle: (i) transdisciplinary ecological and philosophical research; (ii) composition of metaphors, and communication through simple narratives; (iii) design of guided field experiences with an ecological and ethical orientation; and (iv) implementation of *in situ* conservation areas (Rozzi et al. 2010). Each of the chapters in this text has highlighted powerful ways in which FEP has been enacted within a context with promising outcomes.

In this final chapter of Part I, we pose some "food for thought" regarding FEP and discuss complementary approaches such as Adventure Learning (Doering 2007), Adventure Learning @ (Hougham et al. 2012), and the Digital Observation Technology Skills (DOTS) kits philosophy (Hougham et al. 2018). This allows us to explore how FEP may benefit from a pivot toward a more technologically mediated framing, which supports participants in FEP style projects to engage more deeply as a result of an enhanced ability to see various aspects of the natural and the biocultural worlds. We begin by laying out the various parts and pieces. Starting with the most rudimentary of technologies, our senses, we will unpack how the most powerful forms of education are already with us because they are built in as part of our being. We expand to consider technologies such as hand lenses and more complex forms such as thermal imagers with the undergirding philosophy of how these tools can support resilience. This resilience of individuals and communities to engage with localized phenomena in meaningful ways furthers our connection to place (Avriel-Avni 2023).

11.3.1 Solving Mysteries with Inherent Tools

It is worth taking a moment to recognize how we humans collect data. We have five senses built in that represent our original operating equipment: sight, smell, touch, hearing, and taste. These powerful data collection instruments open up a world of incredible complexity and wonder, especially for young people. As educators we may lament that somewhere between elementary and high school, often students develop cynicism and apathy for the incredible wonders all around them. Our communities may be impacted by the effects of such disengagement. Of course, many other factors contribute to engagement in schools, formally, and other educational activities informally. The fact remains that we are constantly and inescapably collecting data. We are also conducting science all the time based on that data. Much of it subconsciously (e.g., What should I wear today because it is snowing?), but sometimes explicitly (e.g., How does my diet and sleep impact the way I feel?). When was the last time you slowed down and consciously checked in on the data your senses were collecting?

The FEP methodology is in effect an extension of our senses and a way to frame realities in our communities and world. Thus, let us slow down and tune in to what the other-than-human landscapes are wanting to tell us. A classic environmental education exercise is a series of questions/prompts for participants to engage with in any landscape: I notice, I wonder, It reminds me of (The Regents of the University of California n.d.). It is an open invitation to engage our senses—our inherent data collection instruments. We may tend to diminish the import of our senses, but one could argue they are critical to any inquiry. Thus, building on our senses—enhancing them through analogous and technological tools supports greater depths in which we can consider the issues facing our communities locally, regionally, and globally.

11.3.2 The Hand Lens

In 1962 Richard Headstrom wrote a book called *Adventures with a Hand Lens*. The book is outdated as one can imagine. Lines such as "In many families breakfast is followed by a cigarette" pave the way for a closer exploration of cigarette tobacco for example. But what this book does beautifully is to invite the reader to take a closer look at the world around them. Each chapter invites the inquisitive and adventurous on an exploration accompanied by such chapter titles as "We Examine a Feather," "We Go Botanizing," and "We Marvel at Nature's Ingenuity." The point here is that through a simple shift in framing and perspective we are invited into veritable endless opportunities to see our immediate surroundings, objects, and phenomena anew. To pause and intentionally look, observe, and understand where before we did not see, starting simple and expanding as our curiosities dictate.

Smith (2008) takes a subtly different tact than Headstrom. In her book *How to be an explorer of the world*, Smith similarly invites us into a way of seeing everyday things in new and compelling ways. For example, activities are framed as "explorations" and invite the participant to look closer at their favorite street, water, light, etc. All around us there are powerful and interesting things to learn about, but oftentimes we do not know to ask the right question or even to think about the questions. We often go through our days, weeks, and years wearing blinders. The FEP methodology is an invitation to remove the blinders and engage with the complexity of a place.

Rozzi et al. (2020) explicitly brought the hand lens into FEP when exploring how ecotourism can be a powerful mechanism for such engagement. Rozzi relays a story that led to an epiphany where he was separated from a group of colleagues while in the field in the far south of Chile. A misstep plunged Rozzi into a swamp where he had much difficulty extracting himself. While in the swamp he was struck by the diversity of mosses which got him thinking about things figuratively and literally under his feet that he had not seen before. Hence, the hand lens is a vehicle for seeing the familiar and accessible in heretofore unrealized ways. With this foundation of simple yet profound uses of technologies, we expand into embracing technologies that allow greater insight into our natural and built worlds by using Digital Observation Technology Skills (DOTS) kits.

11.3.3 Digital Observation Technology Skills (DOTS)

Digital Observation Technology Skills (DOTS) kits were invented in 2012 to connect field science to students' STEM (Science, Technology, Engineering, and Math) education. Tools that were selected for these kits met technological specifications as well as retained pedagogical affordances; that is, they have multiple possibilities for how they can be used for learning within an environmental context. In short, the technological specifications continue to evolve as technologies in instrumentation change. Simply, these are tools selected to work in the field, are simple to operate, connect datasets to other tools in the kit, and have a high portability. Pedagogically, these kits remain unchanged over time. The selection and use of the kits and their tools are intended to (1) help users/students ask and answer questions with digital, mostly quantitative results, and (2) help students to tell scientific stories about the processes and phenomena that they experienced in qualitative as well as quantitative ways.

With respect to intention (1) to ask and answer questions, the inspiration for these collections of tools came from a goal to collect climate science data locally and by students themselves. The hope being that by localizing climate data collection and by getting that to be a direct student experience, climate science concepts, data, and variables would be better understood and less abstract (Hougham et al. 2018). In early instances of these applications, educators, students, and researchers applied these kits to arctic, equatorial, coastal, and intercontinental mountainous ecosystems to refine the toolset selections and sampling.

With respect to intention (2) to tell scientific stories, these kits at their core are to support authentic, first-person science communication that affirms cultural perspectives as well as multiple worldviews while engaging in the scientific process and data literacy. To address cultural bias in science education as well as to address wide achievement gaps in science literacy, the research team behind DOTS concluded that approaches to move this notion forward needed to support decolonizing science education and field science data collection for students. As science educators, we need to figure out how all the experiences that we are asking students to have fit into the story of what is going on around them. To do that we need to collect information and process it into a scientific story that students can understand. We also need to provide a context for the data that they collect. In the field, we have found it supportive to start with our senses, such as is encouraged in *How to Be an Explorer of the World* (Smith 2008). From there we move to technology-enhanced observations. Question asking and question forming are best situated as an initial skill set that scaffolds data collection and the use of appropriate technologies. For example, we ask students and teachers to think of a question they would like to answer. Then we ask, what senses would you use to answer your question? How could technology assist you in answering that question? How would your enhanced observation affect your questions?

Technology, and more specifically when technologies are adopted for the classroom, has a long history of being successful and exciting for learners when those tools are hands-on and extending the senses. For example, boundaries of human perception are limited by the extent of the five senses, and also the timeliness of the phenomena in which they are used. For generations, beyond the boundary of human perception is where technology that enhances our understanding and ability to observe comes into the classroom or field experience. Analog examples of this are well known and centuries old, including telescopes and hand lenses. Digital enhancements, more modern examples of instrumentation or data collection are versions of tools that can help learners with some of the same questions as their analog predecessors. This includes digital cameras, acoustic monitoring equipment, X-ray imagery, thermal imagers, and many more. Technology can be about extending our senses to be appropriate adaptations of our scientific processes. For example, a student can see a grain of sand with the naked eye, and the moon or Venus or Jupiter. But we clearly have examples of where tech has enabled us to see and experience well beyond the abilities of our inherent senses. What can be seen is

Fig. 11.1 **a**. Leatherback turtles hatching (right); **b**. Digital thermometer (central); **c**. Temperature ranges (left). Photographs by R. Justin Hougham

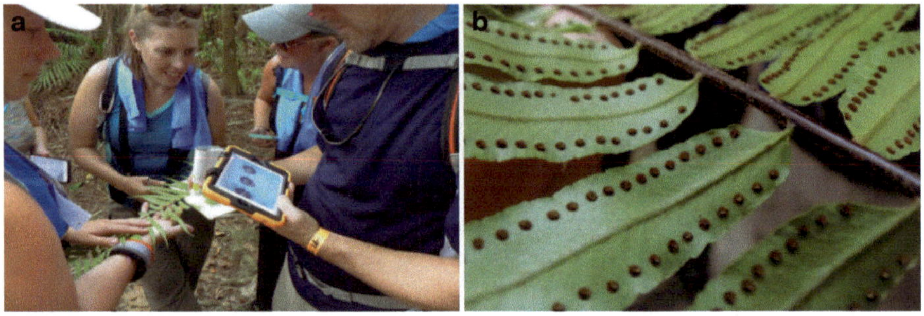

Fig. 11.2 **a**. An educator using a digital microscope in the field (left). **b**. A close-up of a fern viewed through the digital microscope (right). Photographs by R. Justin Hougham

an example of one of our senses that can be pushed wider and wider by advances in technology as well as access to technology. Technology, then, opens to a greater extent of what can be known through observation. The same can be said for our auditory senses through recording equipment as well as sub or ultra-sonic tools.

To highlight how the DOTS philosophy unfolds in practice, we share an example from a successful program where DOTS kits were used and where scientific storytelling was facilitated. The images below (Fig. 11.1) are taken directly from the educator's experience with a teacher professional development program in Tobago, an island of the Republic of Trinidad and Tobago located off the northeast coast of Venezuela. Images in Fig. 11.1 depict digital artifacts that illustrate the enhanced resolution and expanded bandwidth that carefully selected digital tools can offer a field-based science program. In Fig. 11.1a you can see a standard image of leatherback turtles hatching from their nest. Figure 11.1b shows a digital thermometer that gives greater context to the phenomenon. Lastly, Fig. 11.1c shows a thermal imager that illustrates context more fully by capturing multiple ranges of temperature across the site. This progression offers an example of where selected tools set a scaffold and enhances observations as well as inquiry into new surroundings.

The next example is similar in that educators are using a digital microscope in the field to observe ferns (Fig. 11.2). Figure 11.2a depicts the digital microscope, and Fig. 11.2b shows what they observed closely at magnification, a detail that might be lost without the support of technology that allows close-up viewing. Importantly, both examples are cases for selected and limited technology use as well as observation methods that leave sensitive samples or organisms as undisturbed as possible.

As noted, DOTS kits are a philosophical approach to technologies that enhance our senses and support users/students to delve deeper into questions they have about the natural world. In the next section we introduce a technologically supported approach that can be enacted as a vehicle for pursuing questions in formal and informal educational settings. The idea is that DOTS kits, hand lenses, and other appropriate technologies can serve as vehicles for creating stories when a question drives the pursuit. This approach is known as Adventure Learning.

11.3.4 Adventure Learning

Adventure Learning (AL) was originally defined by Doering (2006, p. 197) as "a hybrid distance education approach that provides students with opportunities to explore real-world issues through authentic learning experiences within collaborative learning environments." Veletsianos and Kleanthous (2009, p. 84) explained AL as "an approach for the design of digitally-enhanced teaching and learning environments driven by a framework of guidelines grounded on experiential and inquiry-based learning." AL more tangibly is a series of design principles that when built will facilitate activities in person, in part through an online environment. The AL approach was originally based on seven principles (Doering 2006, 2007) with two additional principles being added to increase accessibility and localized narratives (The Learning Technologies Collaborative 2010). AL projects should:

1. Be adventure-based;
2. Be research-based;
3. Be supported through collaborative and interactive online environment tools;
4. Use the Internet to connect varied audiences;
5. Use media and text within the online environment to educate, engage, and inspire;
6. Have pedagogical guidelines that support teachers in implementation and engagement;
7. Include synchronous and asynchronous learning opportunities that engage students, teachers, and experts in the content being explored;
8. Include an authentic narrative that unifies the expedition, curriculum, student activities, media, and learning experience under a common purpose and theme; and
9. Include the identification of a location and issue to explore (including investigation of the contextual factors surrounding the location and issue)

Early AL projects were both labor and resource intensive, and oftentimes exclusive to a small group of participants in the adventure. Students would "follow along" with the adventure in remote, typically arctic, regions of the world from classrooms using the affordances of the online environment. To democratize AL, Hougham et al. (2012) coined the term AL@. AL@ is the coupling of the AL approach and place-based education principles with the explicit intention of empowering students to cultivate a narrative based on their experiences and not those of a select group of adventurers in a remote part of the world. AL@ projects have gained traction and experienced success through grant-funded opportunities for students (Miller et al. 2015) and teachers (Olsen et al. 2020). But there are more opportunities to explore.

Many critiques can and have been pointed at AL. From resource intensiveness to the exclusivity of who can and cannot participate, AL has been difficult to sustain beyond a one-time project or event. AL@ has taken some important steps in expanding the approach to include greater participation of students and teachers in the crafting of the narrative from the lived experiences predicated by the adventure and curriculum. But still more remains to be done.

11.4 A Timely Opportunity

We see great potential in expanding FEP to include the technological affordances of the DOTS kit's philosophy in conjunction with AL principles to engage students and teachers more explicitly, regardless of level, in the FEP methodology. What might this look like? We will unpack our vision by framing possibilities within the four-step FEP methodology for educational institutions, both formal and informal (Rozzi et al. 2010). As we reintroduce each step, we will offer possibilities and examples of how FEP can be expanded to include technologically mediated approaches as outlined through DOTS kits and AL. We will conclude with an invitation to readers to continue utilizing FEP and to also consider how technological tools can enhance this already powerful methodology.

11.4.1 FEP Step 1: Transdisciplinary Ecological and Philosophical Research

As individuals in our respective roles at the various educational institutions we find ourselves operating within, we tend to become deeply acquainted with the cultural norms of a discipline (e.g., biology or philosophy). A critique of academia may be accurately stated as perpetuating ivory towers of knowledge that prioritize one way of knowing over another. We get so focused on the theories and philosophies of our disciplines that new and different approaches to a problem at first appear to be anathema to the goals and objectives that we feel are best suited to a particular situation or problem. Therein lies the opportunity for expansion regarding this first step in the interrelated 4-step methodology of FEP, to open to the possibilities of bringing interdisciplinary and transdisciplinary teams together to solve the pressing and complex issues facing our communities.

FEP's transdisciplinary research brings together experts associated with the disciplines gathered for the research (e.g., watershed health). These researchers can serve as content experts to communicate in either synchronous or asynchronous modalities to participants from school-age students to community members. This approach coincides with the prescription of the *seventh AL principle* of including synchronous and asynchronous learning opportunities that engage students, teachers, and content experts in the content being explored. The research question can serve as an invitation for adventure. Adventure can be exotic such as dog sledding in the arctic or it can be the engaging pursuit of answering a question.

The *first AL principle*, that projects should be adventure-based, can easily be derived from a driving research question. As a practitioner begins to tune into a local context, a plethora of possibilities emerge for educational opportunities that delve into a particular issue. For example, a local watershed could be impacted by development, agriculture, or recreation. Herein lies an opportunity to do research through the posing of questions and partnerships with content experts and students to bring awareness and possible solutions that adhere to conservation goals. Additionally, the focus on research as noted in this FEP step aligns with the *second AL principle* of projects being research-based.

11.4.2 FEP Step 2: Composition of Metaphors and Communication Through Simple Narratives

Metaphors and narrative represent powerful ways to communicate and engage audiences with research and issues impacting a community, locality, or region through *scientific storytelling*. Combining AL and DOTS with FEP's metaphors and narrative composition through scientific storytelling can be bolstered by the use of technological means (Fig. 11.3). That said, as practitioners we need to be cognizant of how we adopt and use technologies to communicate and engage participants. Technologies, if uninterrogated, will continue to privilege a dominant narrative that is existent in science as we see it today.

Science is a cultural perspective. We need, in science education at least (if not science as a human enterprise), methodologies that affirm the student/learner/teacher's cultural perspective while engaging in these technological enterprises that move us forward on scientific pursuits. By affirming voices while carefully vetting technological tools, we will see greater equity and diverse perspectives in science, and subsequently the outcomes will be more socially just. Scientific storytelling, as imagined here, is an intentional approach that elevates the learner's perspective and in so doing, affirms community and cultural perspectives that are in the context of the learner (Zocher & Hougham 2020).

The inclusion of an authentic narrative that unifies the expedition, curriculum, student activities, media, and learning experience under a common purpose and theme is the *eighth AL principle* and aligns perfectly with this second step of the FEP

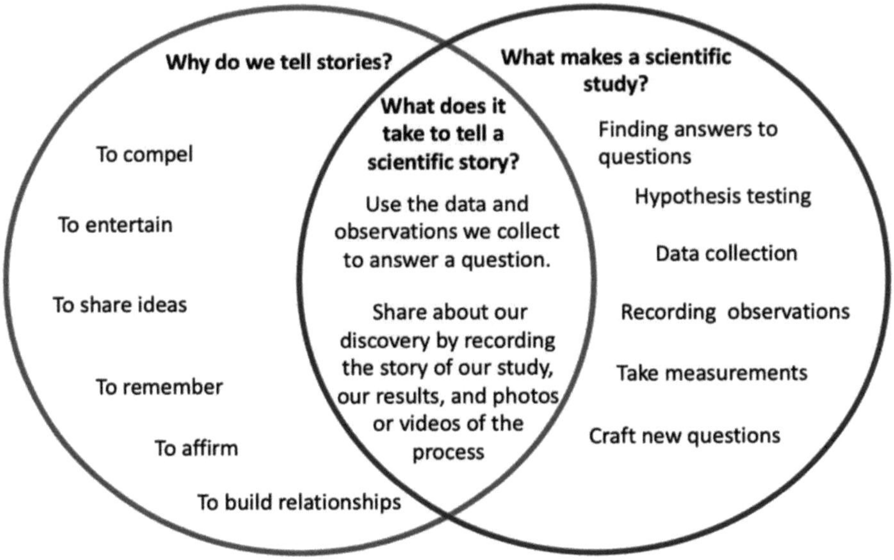

Fig. 11.3 Scientific storytelling is based on overlapping spheres: *Why do we tell stories?* and *What makes a scientific study?* Each has specific content, that when merged, makes a third sphere: *What does it take to tell a scientific story?* The third sphere's content brings into focus the content of the other two spheres to create a scientific story

methodology. The narrative engages the human element and spirit and when coupled with the unknown of adventure it makes for a highly engaging experience that in turn promotes learning. By embracing AL as a mechanism for conducting a project through the FEP methodology the reach can effectively be exponential using the Internet to connect varied audiences (*fourth AL principle*). The narrative in an AL-facilitated project can be portrayed through using media and text within the online environment to educate, engage, and inspire (*fifth AL principle*).

11.4.3 FEP Step 3: Design of Guided Field Experiences with an Ecological and Ethical Orientation

Hands-on experiences situated in place are critical to fully understanding and appreciating the complexities of a given issue. Hands-on experiences that focus on both the scientific issues and the ethical considerations in the spirit of biocultural ethics will support a deeper appreciation for the complexities presented within a field experience. From an AL standpoint, when thinking about how we might bring FEP into either a formal or informal science education context, the *sixth AL principle* of including pedagogical guidelines that support teachers in implementation and engagement will orient the practitioner to effectively deliver an experience to participants.

The pedagogical guidelines can be enacted in person but also just as effectively at a distance. AL projects are facilitated through collaborative and interactive online environment tools (*third AL principle*). These tools open the opportunity to explore field experiences of others that are in a different location and who are exploring a similar question (Olsen et al. 2020). One can think of this AL affordance like the jigsaw cooperative learning method. Participants at each location become experts about their context and associated research. By sharing their experiences through the AL environment, the collective is strengthened and effectively embodies a larger picture of the nuances of a question.

11.4.4 FEP Step 4: Implementation of **in situ** *Conservation Areas*

Rozzi et al. (2010) affirm the necessity of implementing FEP programs *in situ* because it provides the participant an opportunity to coexist in a landscape with the other-than-human co-inhabitants. In so doing, the participant experiences both the habitat and habits of its residents and in turn supports the participants to "acquire a sense of responsibility as citizens who are ecologically educated and ethically active" (Rozzi et al. 2010, p. 23). The *ninth AL principle* states that an AL project should include the identification of a location and issue to explore. This should incorporate an investigation of the contextual factors surrounding the location and issue. To secure a sense of place, a person must intimately engage with a place over time. The place, the context, is important when the intent is to intimately know the factors interplaying around an issue. If everyone can be located physically in the place to experience the place that is great. But how can we expand our understanding of issues that are not isolated to a particular location? Take climate change for example. Every corner of the globe is being impacted in common and unique ways by climate change.

Through AL and the leveraging of technological affordances and online environment support, a question can be explored in place by individuals and communities within the spirit of FEP. However, the question can be shared more broadly through the AL approach than FEP alone. In this way, the conservation efforts and lessons learned in one place can assist and fortify parallel efforts halfway around the world. An outcome of using technology in this way is an infrastructure of resilience where the efforts in one location can enhance not only the local but potentially the global collective. This is particularly relevant to confront challenges and tribulations facing our world such as climate change.

11.5 Conclusion

How do we move forward in this time, post-COVID, as so many norms have been disrupted? Careful adoption of technology was an important consideration for educators and learners before the pandemic. In the post-pandemic educational landscape, we argue, it will be even more important. Our early experimentation with Digital Observation Technology Skills (DOTS) and Adventure Learning (AL), for example, used video chat as well as asynchronous learning content to supplement science education in and out of the classroom. Now, after a long period of accelerated use of online meeting software and downloadable virtual learning, learners as well as educators are fatigued by the multitude of options that now exist for education. We theorize that as a result, in the future we will be more discerning while having higher standards for what content is delivered and used online. Anticipating this future scenario, we expect that tool selection as well as platform selection for digital learning will be highly scrutinized. Meeting that need is the framework laid out here, not just for tool selection, but also for the intentionality of the learning model that it employs. We need to ask questions such as:

Does it elevate the student's voice?
Does it draw from authentic student narratives?
Are the tools used minimally invasive to the environment or to the learners' experience?

Models that move forward in this way will be differentiated from video calls and course downloads because they elevate the students' experience in their environment. Aldo Leopold wrote that "I cannot avoid the conclusion that perhaps the only vicious thing in modern technology is the kind of people who wield its gadgets, and think they are civilized for this reason" (in Meine & Knight 1999, p. 277). While it is certain that technology will continue to evolve, it must also be maintained that educators play a key role in mitigating the potentially damaging or distracting roles technology can afford if not well vetted (Greenwood & Hougham 2015). This commentary on technology is significant. Like the ideas represented here, in short, we cannot conflate the use or adoption of technology with the pedagogical goals that we are working to advance.

The FEP methodology and biocultural ethic have proven to be powerful vehicles for understanding and interrogating issues that directly impact our communities. Taken as a whole, our communities are only as good as the quality of educational opportunities we provide our students and the support our teachers have in facilitating them. FEP, in conjunction with DOTS kits affordances and a technologically mediated learning environment through the AL approach, represents a compelling opportunity for us as thoughtful practitioners to bring meaningful learning into greater focus beyond the COVID era.

References

Avriel-Avni N (2023) Introduction to platforms for integrating the sciences, arts, and humanities into participatory education. In: Rozzi R, Tauro A, Wright T, Avriel-Avni N, May RH Jr (eds) Field environmental philosophy: education for biocultural conservation. Ecology and ethics, vol 5. Springer, Dordrecht, pp 447–456

Contador TA, Rendoll-Cárcamo J, Gañan M, Ojeda-Villarroel J, Kennedy JH, Convey P, Rozzi R (2023) Underwater with a hand lens: ecological sciences and environmental ethics to value freshwater biodiversity. In: Rozzi R, Tauro A, Wright T, Avriel-Avni N, May RH Jr (eds) Field environmental philosophy: education for biocultural conservation. Ecology and ethics, vol 5. Springer, Dordrecht, pp 53–70

Crego RD, Ward N, Rozzi R (2023) The eyes of the tree: applying field environmental philosophy to tackle conservation problems at long term socio-ecological research sites. In: Rozzi R, Tauro A, Wright T, Avriel-Avni N, May RH Jr (eds) Field environmental philosophy: education for biocultural conservation. Ecology and ethics, vol 5. Springer, Dordrecht, pp 101–112

Doering A (2006) Adventure learning: transformative hybrid online education. Distance Educ 27(2):197–215

Doering A (2007) Adventure learning: situating learning in an authentic context. Innovate: J Online Educ 3(6)

Greenwood DA, Hougham RJ (2015) Mitigation and adaptation: critical perspectives toward digital technologies in place-conscious environmental education. Policy Futures Educ 13(1):97–116

Headstrom R (1962) Adventures with a hand lens. Courier Corporation, North Chelmsford

Heidinger K, Carmichael J, Jacobs-Schwartz B, Borucinska-Begg M (2023) Biocitizen's approach to biotic wonder, citizenship, and field environmental philosophy. In: Rozzi R, Tauro A, Wright T, Avriel-Avni N, May RH Jr (eds) Field environmental philosophy: education for biocultural conservation. Ecology and ethics, vol 5. Springer, Dordrecht, pp 131–152

Hougham RJ, Eitel KB, Miller BG (2012) AL@: combining the strengths of adventure learning and place-based education. CLEARING, pp 38–41

Hougham RJ, Nutter M, Graham C (2018) Digital Observation Technology Skills (DOTS): incorporating modern digital technology into outdoor experiential education. J Connect Sci Learn (3). Available via: (PDF) Digital Observation Technology Skills: Incorporating Modern Digital Technology Into Outdoor Experiential Education (researchgate.net)

Meine CD, Knight RL (eds) (1999) The essential Aldo Leopold: quotations and commentaries. University of Wisconsin Press, Madison

Méndez-Herranz M, Marini G, Rozzi R (2023) Sub-Antarctic High Andean "Gardeners:" cultivating caring relationships. In: Rozzi R, Tauro A, Wright T, Avriel-Avni N, May RH Jr (eds) Field environmental philosophy: education for biocultural conservation. Ecology and ethics, vol 5. Springer, Dordrecht, pp 71–86

Miller BG, Cox CJ, Hougham RJ, Walden VP, Eitel KB, Albano AD (2015) Adventure learning as a curricular approach that transcends geographies and connects people to place. Curric J 26(2): 290–312

Moreno-Terrazas R, Díaz-Gómez Z, González-Galván H, Carinño-Olvera M, Monteforte-Sánzhez M (2023) Starfishes and sky stars: field environmental philosophy education and ecotourism experiences in Baja California Sur, Mexico. In: Rozzi R, Tauro A, Wright T, Avriel-Avni N, May RH Jr (eds) Field environmental philosophy: education for biocultural conservation. Ecology and ethics, vol 5. Springer, Dordrecht, pp 113–130

Ojeda J, Tauro A, Contador T, Mansilla A, Rosenfeld S, Ban N, Rozzi R (2023) "Pay attention, dive with your eyes wide-open:" a field environmental philosophy activity to foster reciprocity between people and nature. In: Rozzi R, Tauro A, Wright T, Avriel-Avni N, May RH Jr (eds) Field environmental philosophy: education for biocultural conservation. Ecology and ethics, vol 5. Springer, Dordrecht, pp 87–100

Olsen SK, Miller BG, Eitel KB, Cohn TC (2020) Assessing teachers' environmental citizenship before and after an adventure learning workshop: a case study from a social ecological systems perspective. J Sci Teach Educ 31(8):869–893. https://doi.org/10.1080/1046560X.2020.1771039

Rozzi R (2012) Biocultural ethics: the vital links between the inhabitants, their habits and regional habitats. Environ Ethics 34:27–50

Rozzi R (2023a) Introduction Part I. The field environmental philosophy methodological approach. In: Rozzi R, Tauro A, Wright T, Avriel-Avni N, May RH Jr (eds) Field environmental philosophy: education for biocultural conservation. Ecology and ethics, vol 5. Springer, Dordrecht, pp 17–26

Rozzi R (2023b) Inter-species and inter-cultural encounters: the education and biocultural ethics program of the Omora Ethnobotanical Park. In: Rozzi R, Tauro A, Wright T, Avriel-Avni N, May RH Jr (eds) Field environmental philosophy: education for biocultural conservation. Ecology and ethics, vol 5. Springer, Dordrecht, pp 153–174

Rozzi R, Tauro A (2023) The multiple lenses of ecotourism with a hand-lens: fundamental concepts and practices. In: Rozzi R, Tauro A, Wright T, Avriel-Avni N, May RH Jr (eds) Field environmental philosophy: education for biocultural conservation. Ecology and ethics, vol 5. Springer, Dordrecht, pp 27–52

Rozzi R, Massardo F, Anderson C, Heidinger K, Silander J Jr (2006) Ten principles for biocultural conservation at the southern tip of the Americas: the approach of the Omora Ethnobotanical Park. Ecol Soc 11(1):43. [online] URL: http://www.ecologyandsociety.org/vol11/iss1/art43/

Rozzi R, Anderson CB, Pizarro JC, Massardo F, Medina Y, Mansilla A, Kennedy JH, Ojeda J, Contador T, Morales V, Moses K, Poole A, Armesto JJ, Kalin MT (2010) Field environmental philosophy and biocultural conservation at the Omora Ethnobotanical Park: methodological approaches to broaden the ways of integrating the social component ("S") in long-term socio-ecological research (LTSER) sites. Revista Chilena de Historia Natural 83(1):SM1–SM35. Available via: http://rchn.biologiachile.cl/suppmat/2010/1/MC_Rozzi_et_al_2010.pdf

Rozzi R, La Valle MT, Russell S, Goffinet B, Massardo F (2020) Ecotourism with a hand-lens: a field environmental philosophy experience from the south of the world. In: Brister E, Frodeman R (eds) Philosophy for the real world: an introduction to field philosophy. Taylor & Francis/Routledge, New York, pp 222–239

Smith K (2008) How to be an explorer of the world. Penguin Group, New York

The Learning Technologies Collaborative (2010) "Emerging": a re- conceptualization of contemporary technology design and integration. In: Veletsianos G (ed) Emerging technologies in distance education. Athabasca University Press, Edmonton, pp 91–107

The Regents of the University of California (n.d.) I Notice, I Wonder, It Reminds Me of. BEETLES Science and Teaching for Field Activities. Available via: http://beetlesproject.org/cms/wp-content/uploads/2015/12/I-Notice-I-Wonder-It-Reminds-Me-Of.pdf

Veletsianos G, Kleanthous I (2009) A review of adventure learning. Int Rev Res Open Distributed Learn 10(6):84–105

Zocher JL, Hougham RJ (2020) Implementing ecopedagogy as an experiential approach to decolonizing science education. J Exp Educ. https://doi.org/10.1177/1053825920908615

Part II
Education "On the Razor's Edge"

Chapter 12
Introduction to Part II. Education "on the Razor's Edge"

Roy H. May Jr.

Abstract This chapter introduces Part II. Education is "on the razor's edge" between two options. On the one hand, it is easiest for education to conform to the hegemonic norms established by powerful social and economic interests; this avoids controversy and change. On the other, subordinate social groups, disappearing cultures, and environmental crisis pressure for justice, cultural respect, and a healthy environment; these require change and generate controversy as the interests of dominate groups are challenged. A liberating biocultural pedagogy finds itself maneuvering between these two. It is "on the razor's edge." The chapters in this part are on the liberating edge of the razor. They present historical, theoretical, and philosophical foundations, as well as practical experiences in community-based education, that are both forerunners of and contributors to field environmental philosophy (FEP). These chapters take FEP into account as each develops the theme from diverse perspectives and experiences. They also discuss concepts for furthering FEP methodology. Various approaches complement FEP while others introduce concepts for the theoretical grounding of FEP. These chapters effectively address these problems and complement other pedagogical traditions that value local knowledge, biodiversity, and life habits.

Keywords Aldo Leopold · Biocultural tourism · Collaborative action research · Extinction of experience · Liberation philosophy · Radical renaissance

12.1 Introduction

The major interest of this book is an educational approach that concerns itself with the integral relationship among language, traditional knowledge, environment, and justice in such a way that people are enabled to defend and shape their own local communities. These are threatened by global processes of cultural and economic homogenization "to the detriment of local peoples' sovereignty, and control over

---

R. H. May Jr. (✉)
Departamento Ecuménico de Investigaciones (DEI), San José, Costa Rica

R. Rozzi et al. (eds.), *Field Environmental Philosophy*, Ecology and Ethics 5,
https://doi.org/10.1007/978-3-031-23368-5_12

their ancestral territories and resources, but also to the detriment of their ancestral languages and cultural traditions" (Maffi 2001, p. 5) as well as severe loss of biodiversity. These losses are *biocultural* because culture and environment are integrally combined.

Clearly, contemporary formal and non-formal education, so-often based on standardized, one-size-fits-all curriculums, drive biocultural homogenization, but if differently oriented as active, "learning by doing" and "place-based," education can effectively drive biocultural conservation. Pedagogy is key, for depending on its focus, pedagogy either promotes homogeneity and uniformity or it opens learners to novel and exciting biocultural diversity that enriches their lives and the lives of others. Education is "on the razor's edge" between these two options. On the one hand, it is easiest for education to conform to the hegemonic norms established by imposing social and economic interests; this avoids controversy and change. This is the educational approach increasingly imposed by public authorities in many countries. Emphasis is on quantitative, measurable results. These tend to be highly standardized pedagogies that leave little space for contextual innovation and the needs and abilities of students as individuals. Difference and diversity are shunned (Rial 2016; Sparapani and Callejo-Perez 2015). On the other hand, subordinate social groups, disappearing cultures, and environmental crisis pressure for justice, cultural respect, and a healthy environment; these require change and generate controversy as the interests of dominant groups are challenged. This kind of education cannot be standardized or imposed. Rather, it will embrace diversity and understand difference as stimulating creativity. A liberating, biocultural pedagogy finds itself maneuvering between these two. It is "on the razor's edge."

The previous part presented case studies of Field Environmental Philosophy (FEP). They illustrate place-based educational approaches that actively involve different social groups in novel situations and that commit them to critically appraising environmental reality, as well as their own sense of meaning and relatedness to other inhabitants including other-than-human inhabitants. They illustrate FEP as a liberating, biocultural pedagogy. There are, of course, historical antecedents to place-based, active-learning pedagogies that have influenced directly or indirectly FEP.

12.2 Antecedents to Place-Based, Active-Learning Pedagogies

The great eighteenth-century humanist Jean-Jacques Rousseau (1712–1778) (Aloni et al. 2023 in this volume) advocated for education based not on imposing societal norms, but rather on stimulating the capacity to reason in harmony with nature through active learning processes as opposed to passively receiving information from a teacher. Even before Rousseau, a Western tradition fostering imagination, understanding, and appreciation of diversity emerged in the Radical Renaissance (Gare 2023 in this volume).

The twentieth century witnessed numerous critiques of mass, institutionalized education. Early on Maria Montessori (1870–1952) sought a child-centered approach (Montessori 1912) that emphasized active learning based on "scientific pedagogies." The theory of stages of cognitive development argued by psychologist Jean Piaget (1896–1980) greatly influenced educational theory. His idea that mental development obeyed different stages according to age maturity argued against standardized pedagogies that did not account for such differences (Piaget 1950). In the 1960s and 1970s, Carl Rogers (1902–1987), an existentialist humanist psychologist, proposed "person-centered" learning to foster self-actualization. He based this on what he understood as the innate curiosity and sense of meaning experienced by people (Rogers 1969). Paulo Freire (2005a, b) in the 1960s profoundly influenced education with his "pedagogy of the oppressed" which emphasized learners as active participants in social change for justice (May 2023 in this volume).

There also have been radical critiques of the whole concept of institutionalized education, most notably by Ivan Illich (1926–2002). An anti-institutionalist, Illich questioned many aspects of modern social institutions including the church (he was a Roman Catholic priest) and modern medicine. He valued inter-cultural learning and warned against the imposition of specific cultural values and concepts. In *Deschooling Society* (Illich 1971) Illich criticizes the very idea of mass, institutional-based education. Instead, he advocated life-long self-directed education through multiple informal arrangements.

However, arguably the most important philosopher and educator to advocate for active learning was the US pragmatist John Dewey (1859–1952) (Dewey 1916, 1997). Dewey believed in learning by doing, that is, he believed that learning emerges from and with experience. He rejected top-down pedagogies that assume students to be passive recipients of knowledge. Instead, he argued that teachers and students actively learn together. He also advocated interdisciplinary pedagogies that corresponded to real-world realities. For Dewey, as for the philosophy of pragmatism, the criterion of truth or the verification of knowledge is not correspondence to preconceived notions, doctrines, or authority, but actualization in experience. That is, knowledge is disclosed in doing. Knowledge and action are merged as, later according to Paulo Freire, praxis: action for change (May 2023). This Deweyan educational philosophy has widely influenced contemporary place-based, interdisciplinary, action-oriented pedagogies such as FEP. Although perhaps unrecognized, Dewey's ideas are latent in the ideas and experiences developed in the chapters that make up this part.

These pedagogical approaches and critiques emphasize active, experiential learning in real-life contexts. Importance is placed on qualitative learning. They break away from standardized curriculum by focusing on the individual and their active involvement in learning. Indeed, they are "on the razor's edge" between staid tradition and active creativity, and so not surprisingly these educational approaches have been severely criticized by hegemonic political, religious, and social sectors because they are viewed as undermining authority and social cohesiveness or homogeneity. They also largely focus on individuals and are not directly concerned with cultural or social issues, much less the natural environment. Like these pioneering, active pedagogies, FEP also will be challenging for hegemonic social order, and, while it does not obviate the individual, it is fundamentally a collective pedagogy that takes into account culture and environment, based on the biocultural ethic. It too is "on the razor's edge."

12.3 An Overview of Part II

The chapters in this part are on the liberating edge of the razor. Since FEP aims to foster intercultural dialogue and biocultural conservation at multiple scales, networking from local to global or *glocal* (Aloni et al. 2023 in this volume) is an important dimension. The chapters in this part (as in the other parts) represent perspectives and experiences from diverse geographical regions. They present historical, theoretical, and philosophical foundations, as well as practical experiences in community-based education, that are both forerunners of and contributors to FEP. These chapters take FEP into account as each develops the theme from diverse perspectives and experiences. They also discuss concepts for furthering FEP methodology. Various approaches complement FEP while others introduce concepts for theoretical grounding of FEP. All the chapters seek to effectively address these problems and to complement other pedagogical traditions that value local knowledge, biodiversity, and life habits. In summary, chapters in Part II correspond to biocultural reality and present alternative educational models, experiences, and practices that foster biocultural conservation.

Specifically, FEP is a liberative pedagogy developed in South America and is rooted in Latin American liberation philosophy (Tauro et al. 2021). This philosophy emerged from the underside of power in Latin America to critique the structures of injustice that oppress the poor and subaltern social groups and turn nature into a mere commodity. Although liberation philosophy was developed in the 1960s, it provides a philosophical foundation for biocultural education and FEP that have been developed in the 2000s.

In the opening chapter of this part, "Liberation Philosophy and Biocultural Education. A Latin American Journey" (Chap. 13), **Roy May** develops this theme through a personal narrative of his experiences during nearly 50 years in both formal and non-formal education while living in Latin America. As basic to liberation educational philosophy, May introduces fundamental concepts such as place-based methodology, the "coloniality of power" as developed by Anibal Quijano (2000), and "pedagogy of the oppressed" based on "conscientization" or "awareness," core concepts proposed by Paulo Freire (2005a, b). May joins these educational concepts and practices with decolonial theory for valuing and recovering local places, Indigenous people, and their knowledge. However, the coloniality of power that shapes social structures as well as mindsets and limits life ways is pervasive. This is the major obstacle to liberation. May argues that although liberation philosophy and education as pedagogy of the oppressed emerged before FEP, they are encompassed in FEP as a research, educational, and action methodology that leads to "biocultural liberation."

Indeed, manifold cultural traditions in different regions of the world in the past have contributed, and still contribute today, to biocultural conservation. These cultures hold what is known as Traditional Ecological Knowledge (TEK) or increasingly, also known as Traditional Indigenous Knowledge (TIK). This knowledge is "traditional" because it has deep roots in the cultural history of a group, is unwritten, and is passed from generation to generation as necessary for life. It is "ecological" because it is concerned with the natural environment and how to use it sustainably not only in the present but also for future generations. It is "knowledge" because it means having skills, facts, and information necessary for maintaining and reproducing basic needs for living well. Such knowledge functions as an ongoing circle: (a) Indigenous knowledge is (b) the basis for Indigenous decision making which (c) is operationalized through Indigenous organizations, in turn (d) generating Indigenous experimentation and innovation that produces (a) Indigenous knowledge (Warren 2001, p. 451). Traditional ecological knowledge plays a central role in FEP because of its importance for maintaining the integrity of Indigenous cultures, but also because it is important for "modern" conservation of biodiversity.

In their chapter, "Collaborative Action Research for Biocultural Heritage Conservation" (Chap. 14), **Stanford Zent** and **Egleé Zent** emphasize the importance of TEK. As anthropologists, they share their experience with the Jotï Amerindian group of the Venezuelan Amazon in collaborative action research (CAR). They begin by affirming that biological diversity and cultural diversity are inextricably linked and therefore must be one of the guiding principles of conservation thought and practice. They observe that people perceive and act upon their surroundings through inherited cultural filters. "Such filters are complex and multidimensional," they explain, "consisting of language, ethnoscience, folklore, technology, occupations, food habits, exchange systems, settlement pattern, social organization, customary law, spiritual beliefs, moral values, ethnomedical systems, and more" (2023, in this volume). This TEK joins biological diversity and cultural diversity and shapes the ways that people interact with their environment. In this sense, Zent and Zent argue, biological diversity and cultural diversity are interdependent or co-evolved. TEK occupies the active interface between culture and nature at the local level.

With this theoretical foundation, in the spirit of FEP, collaborative action research (CAR) is conducted jointly by "outside" researchers and "local" people. It is focused on assessing and ameliorating community-defined problems. This chapter, then, describes the authors' experience of CAR with local communities of the Jotï Amerindian group and researchers from the Human Ecology Laboratory at the Venezuelan Institute for Scientific Research. The primary focus of this research was on TEK transmission and biocultural heritage conservation. Although not specifically understood as FEP, CAR resonates with FEP with its emphasis on local communities, collaborative endeavors, and cultural concepts and traditions for responding to local problems and environmental conservation.

However, TEK, local knowledge, and everyday familiarity with nature increasingly are threatened by growing urbanization and digitally mediated experiences. In her chapter, "Education as a Driver of Extinction of Experience or Conservation of Biocultural Heritage" (Chap. 15), **Alexandria Poole** presents "extinction of experience" (EoE), an idea first developed by Robert Pyle, as a driver of biocultural homogenization and environmental crisis. She proposes that "EoE is an indirect driver of biodiversity loss. The lack of knowledge about this loss and the loss of knowledge to co-exist with biodiversity in sustainable ways exacerbate biodiversity losses. Therefore, vectors that control spaces in which knowledge, engagement, and experience of nature are drivers of EoE that should be acknowledged in education" (2023, this volume). Doing this, of course, requires significant curriculum changes that will be "on the razor's edge" between "localized knowledge" and "globalized knowledge."

Any discussion of biocultural conservation and education will have Aldo Leopold as a reference because he was a watershed figure in the history of conservation in the United States and his influence is observed in conservation throughout the world. Leopold's approach to field education and restoration work has been a major inspiration for the development of the FEP methodological approach. As **Baird Callicott** explains in "Aldo Leopold as Educator: His Legacy for Field Environmental Philosophy" (Chap. 16), Leopold had much to say about education, and most of it was gently but sharply critical. Leopold felt that given a limited curriculum and limited schedule of classes, biology students, for example, were better served by more field-oriented courses, more natural history, and less by laboratory techniques.

Leopold also seems to have been ahead of his time in advocating what is now called "interdisciplinary education." Callicott believes that this is a natural consequence of Leopold's ecological habit of mind. Ecology focuses on the relationships among plants and animals and the physical and chemical factors of their environments. "It is natural," Callicott says, "that Leopold, a self-taught ecologist, would be more sensitive to and concerned with relations and connections in general" (Callicott 2023, in this volume). In terms of higher education, this would translate into an emphasis on the relations and connections between and among academic disciplines rather than on their separateness and autonomy. Overall, Aldo Leopold advocated more field and outdoor education and less passive and sedentary indoor study. Again, educational curriculum will teeter "on the razor's edge" between active and passive learning.

Leopold was not unaware of cultural diversity, but, as Callicott says, "he did not seem as concerned about conserving it as he was about conserving what eventually came to be called biodiversity" (Callicott 2023, in this volume). Biodiversity is now understood to consist of variety at every level of biological organization—genes, phenotypes, populations, species, biotic communities, ecosystems, landscapes, and biomes. Leopold seems to consider cultural diversity to be derived from biodiversity at the landscape and biome scales of biological organization. Leopold's affirmation of diversity, connectedness, and interdisciplinary approaches especially has inspired the design of FEP to foster earth stewardship at the southern end of the Americas, specifically through programs at the Omora Ethnobotanical Park in Cape Horn, Chile (Rozzi 2015).

Based on Leopold's idea of "cultural harvest," **Hugh Breakey** and **Noreen Breakey,** in their chapter "Leopold's Cultural Harvest, Biocultural Tourism and Field Environmental Philosophy" (Chap. 17), introduce the "three-stage land ethic," a theory of moral development applicable to individuals and societies, drawn out of the work of Aldo Leopold. For Leopold, emotions, values, and sense of identity are fundamentally important because, as Leopold (1968, p. 48) put it, we cherish only what we "see and fondle." This gives rise to Leopold's idea of the "cultural harvest"—the goods that we can achieve as we engage with the land. Drawing, then, on Leopold's thought, the "three-stage land ethic" provides an account of how moral development occurs, both at an individual and societal level. It incorporates both cognitive and experiential elements that lead to changed values and virtues. It also widens the sphere of moral concern to include animals, plants, and the land itself. As a pedagogy, it combines *reasons* for valuing and respecting the land with *experiences* and *practices* that connect us personally and emotionally to it. As individuals and cultures progress through the three stages, they come to value the land and are drawn into changed attitudes and behavior.

With this three-stage land ethic, the authors introduce *biocultural tourism* as tourism that provides education within natural and/or cultural heritage experiences and prioritizes biocultural conservation through the broadening of ethical care to new ecological and/or cultural phenomena. This innovative understanding of tourism provides meaningful, cultural, and environmental experiences for tourists, but at the same time enriches the places and communities they visit. Although FEP methodology is an educational approach, this three-stage ethic, and its applicability to educational tourism, clearly interests FEP.

The following two chapters, which close Part II, move the discussion in a different direction by proposing philosophical and historical foundations for education for biocultural diversity and FEP.

Enrique Dussel, a founder of the liberation philosophical tradition in Latin America, proposes in his chapter, "A Material Ecological Ethic for Biocultural Education: Relations Between Life on Earth and Humanity" (Chap. 18), a material ecological ethic for biocultural education. An ecological ethic, he argues, is about the absolute condition of the possibility of living beings. Ultimately, it is played out as respect for the universal right of all human beings to survive, especially those most affected and excluded: the poor of present times as well as those of future generations who will inherit a dead Earth if an ecological conscience is not acquired quickly and globally. Dussel (2023, in this volume) explains:

> My standpoint takes material ethics (necessary but not sufficient) and formal morality (necessary but not sufficient either) within a critical process of liberation that subsumes them, by putting them critically, socially, historically, and diachronically into movement. This is highly topical because ecological destruction (as a condition of possibility) and poverty (the effect) are two correlated phenomena, which have the same cause. Both require a material understanding and, simultaneously, the mediation of formal consensuality by the community.

Destruction of biodiversity is articulated concomitantly with misery, poverty, and the oppression of the majority of humanity. This is material reality, Dussel says, must be the basis for dealing with the problem. Hence, in education as well as in biocultural conservation, we need a material understanding and simultaneously the mediation of formal morality, critical thinking, and consensuality.

As shown by the examples of Rousseau, Dewey, and others, Western civilization is not monolithic and hosts multiple and contrasting pedagogical schools for educating citizens that are responsible to society and the natural environment. However, these pedagogical schools have been largely ignored. In the concluding chapter of Part II, "Challenging the Dominant Grand Narrative in Global Education and Culture" (Chap. 19), **Arran Gare** critically examines the historical antecedents of Western modern education. The dominant tradition, originating in the seventeenth-century scientific revolution effected by René Descartes, Thomas Hobbes, Isaac Newton, John Locke, and others, privileges science and technology in the service of economic growth, as the only genuine knowledge. This is the foundation of a globalized, homogenizing culture that reduces everything and everyone to instruments of the globalized economy. By so doing, this kind of education is an indirect—even direct—driver of biocultural homogenization. Notably, Gare also shows that there is an alternative Western tradition fostering imagination, understanding, and appreciation of diversity in the quest for wisdom. Rooted in the Radical Renaissance, this tradition is represented by thinkers such as Giordano Bruno, Giambattista Vico, Gottfried Herder, and Friedrich Wilhelm Schelling. As Gare (2023, in this volume) concludes, a radical redirection of education must be taken "to effectively confront the crises of civilization, most importantly, the ecological crisis." This required redirection has already been indicated by these Radical Renaissance thinkers: "education aimed at wisdom, fostering imagination, and the quest for a comprehensive understanding of the world, ourselves, and our place within it, cultivating humanity, and acknowledging and appreciating diversity and the significance of all life" (2023, in this volume). Wisdom, including appreciation by people of their unique habitats and cultures, is now required more than ever to overcome the global ecological crisis.

12.4 Conclusion

The experiences and ideas presented in these chapters express a common concern for change toward more just relationships. They challenge power structures that foster homogeneity and subaltern status of humans and other-than-humans and emphasize empowering local people—or even tourists—to act for changing the configuration of society toward one that respects biocultural autonomy and diversity. They also show that the current generation is not the first to have these concerns and, therefore providing rich resources for acting today. We are not prisoners of the past nor of homogenized life ways. We may be "on the razor's edge," but these chapters show that we have options. Together the various themes of these chapters provide grounding for FEP.

References

Aloni N, Gan D, Alkaher I, Assaf N, Baryosef-Paz N, Adiv Gal A, Lev N, Margaliot A, Segal T (2023) Nature, humans, and education: ecohumanism as an integrative guiding paradigm for values education and teacher training in Israel. In: Rozzi R, Tauro A, Avriel-Avni N, Wright T, May RH Jr (eds) Field environmental philosophy: education for biocultural conservation. Ecology and ethics, vol 5. Springer, Dordrecht, pp 519–536

Callicott BJ (2023) Aldo Leopold as Educator: his legacy for field environmental philosophy. In: Rozzi R, Tauro A, Avriel-Avni N, Wright T, May RH Jr (eds) Field environmental philosophy: education for biocultural conservation. Ecology and ethics, vol 5. Springer, Dordrecht, pp 263–280

Dewey J (1916) Democracy and education. An introduction to the philosophy of education. MacMillan, New York

Dewey J (1997) Experience and education. A Touchstone Book, New York

Dussel E (2023) A material ecological ethic for biocultural education: Relations between life on earth and humanity. In: Rozzi R, Tauro A, Avriel-Avni N, Wright T, May RH Jr (eds) Field environmental philosophy: education for biocultural conservation. Ecology and ethics, vol 5. Springer, Dordrecht, pp 295–308

Freire P (2005a [1970]) Pedagogy of the oppressed (trans: Ramos MB). Continuum, New York

Freire P (2005b [1974]) Education for critical consciousness (trans: Ramos MB). Contiuum, New York

Gare A (2023) Challenging the dominant grand narrative in global education and culture. In: Rozzi R, Tauro A, Avriel-Avni N, Wright T, May RH Jr (eds) Field environmental philosophy: education for biocultural conservation. Ecology and ethics, vol 5. Springer, Dordrecht, pp 309–326

Illich I (1971) Deschooling society. Harper & Row, New York

Leopold A (1968) A Sand County Almanac: with other essays on conservation from Round River. Oxford University Press, New York

Maffi L (2001) Introduction. On the interdependence of biological and cultural diversity. In: Maffi L (ed) On biocultural diversity, linking language, knowledge and the environment. Smithsonian Institution Press, Washington, DC, pp 1–50

May RH Jr (2023) Liberation philosophy and biocultural education. A Latin American journey. In: Rozzi R, Tauro A, Avriel-Avni N, Wright T, May RH Jr (eds) Field environmental philosophy: education for biocultural conservation. Ecology and ethics, vol 5. Springer, Dordrecht, pp 203–228

Montessori M (1912) The Montessori method. Scientific pedagogy as applied to child education in "the children's houses" with additions and revisions by the author. Frederick A. Stokes, New York

Piaget J (1950) The psychology of intelligence. Routledge and Kegan Paul, London

Poole A (2023) Education as a driver of extinction of experience or conservation of biocultural heritage. In: Rozzi R, Tauro A, Avriel-Avni N, Wright T, May RH Jr (eds) Field environmental philosophy: education for biocultural conservation. Ecology and ethics, vol 5. Springer, Dordrecht, pp 247–262

Quijano A (2000) Colonialidad del poder, eurocentrismo y América Latina. In: Lander E (ed) La colonialidad del saber: eurocentrismo y ciencias sociales. Perspectivas latinoamericanas. CLASCO, Buenos Aires

Rial K (2016) Standarized testing and public education. Washington University Political Review. Available via: Standarized Testing and Public Education – Washington University Political Review. https://www.wupr.org
Rogers C (1969) The freedom to learn. Charles E. Merrill, Columbus, OH
Rozzi R (2015) Implications of the biocultural ethic for earth stewardship. In: Earth stewardship. Linking ecology and ethics in theory and practice. Springer, Cham, Ch 9, pp 113–136
Sparapani EF, Callejo-Perez DM (2015) A perspective on standardized curriculum and its effect on teaching. J Educ Soc Policy 2(5):78–87. Available via: Microsoft Word - 11.docx. https://www.jespnet.com
Tauro A, Ojeda J, Caviness T, Mosses K, Moreno R, Wright T, Zhu D, Poole A, Massardo F, Rozzi R (2021) Field environmental philosophy: a biocultural ethic approach to education and ecotourism for sustainability. Sustainability 13(8):4526. https://doi.org/10.3390/su13084526
Warren DN (2001) The role of global network of indigenous knowledge resource centers in the conservation of cultural and biological diversity. In: Maffi L (ed) On biocultural diversity, linking language, knowledge and the environment. Smithsonian Institution Press, Washington, DC, pp 446–461
Zent S, Zent E (2023) Collaborative action research for biocultural heritage conservation. In: Rozzi R, Tauro A, Avriel-Avni N, Wright T, May RH Jr (eds) Field environmental philosophy: education for biocultural conservation. Ecology and ethics, vol 5. Springer, Dordrecht, pp 229–246

Chapter 13
Liberation Philosophy and Biocultural Education. A Latin American Journey

Roy H. May Jr.

Abstract Liberation philosophy emerged from the underside of power in Latin America to critique structures of injustice that oppress poor and subaltern social groups and turn nature into a mere commodity. It is driven by an action-reflection model in which theory follows practice to produce a new synthesis or praxis. For me, this has been more than just an intellectual pursuit. I moved from the USA to Latin America and for nearly 50 years, I was involved in both formal and non-formal education. But I was as much a learner as a teacher, especially given my original social location. My experiences of repressive military governments in Bolivia and with *campesinos* of the Andean Altiplano, and later in Costa Rica and Guatemala, as well as my involvement in human rights work, brought me into intimate contact with the "coloniality of power" (Quijano). "Conscientization" or "awareness" of the coloniality of power is basic to liberation educational philosophy. In this chapter, I review educational experiences, both formal and non-formal, in which I was involved in South and Central America, primarily with grassroots Indigenous communities, but also with visiting Europeans and US Americans who had little understanding of biocultural injustice and the role of their own countries in fostering it. I also review liberation philosophy and decolonial theory as they urge the "pedagogy of the oppressed" (Freire) for valuing and recovering local places, Indigenous people, and their knowledge. With this background, I assess the methodological approach of field environmental philosophy to address pressing obstacles in education, and more broadly in socio-environmental problems. Finally, I suggest that the "literacy worker" and "extension agent" as proposed by Freire provide pedagogical models for liberative biocultural education. The coloniality of power that shapes social structures as well as mindsets and limits lifeways is pervasive and is the major obstacle to liberation.

Keywords Biocultural liberation · Coloniality of power · Decolonial · Liberation ethics · Paulo Freire

R. H. May Jr. (✉)
Departamento Ecuménico de Investigaciones (DEI), San José, Costa Rica

R. Rozzi et al. (eds.), *Field Environmental Philosophy*, Ecology and Ethics 5,
https://doi.org/10.1007/978-3-031-23368-5_13

13.1 Introduction

Liberation philosophy emerged from the underside of power in Latin America to critique the structures of injustice that oppress the poor and subaltern social groups and turn nature into a mere commodity (Garcia and Vargas 2018; Mendieta 2020). The biocultural liberation ethic joins socio-cultural order and ecological order for a holistic or integral liberation (Rozzi 2009). Environmental and justice issues are merged. Field environmental philosophy (FEP) and biocultural ethics, also emerging from Latin America, provide tools to confront the coloniality of power that turns people and nature into subalterns to be exploited by powerful social, economic, and political interests at local and international scales. The biocultural ethic integrates into a single moral community "habitats," "co-inhabitants," and "habits," that is, where people and other living creatures live, who they are, and their ways of life. It emphasizes heterogeneity and seeks the conservation of biological and cultural diversity (Rozzi 2013). FEP originated in the Omora Ethnobotanical Park located at the southern tip of Chile as an educational program (a) to defend and revitalize local indigenous culture; (b) gain knowledge important for conservation; and (c) establish place-based environmental projects. FEP is concerned with the intentional and unintentional suppression of biophysical, symbolic-linguistic, and socio-political domains and ways of conserving and enhancing them. As Rozzi (2013, p. 11) explains, "Biocultural ethics aims to contribute to ecosocial justice through the conservation of biological and cultural diversity." Ancestral ecological knowledge is affirmed and conserved. The dignity of the other is affirmed because "[b]iocultural ethics demands incorporating the values of the co-inhabitants as subjects into development policies as a matter of socio-environmental justice" (Rozzi 2013, p. 26). In this sense, FEP and biocultural ethics are linked with liberation and decolonialism for biocultural liberation.

13.2 A Latin American Journey in Biocultural Education

Biocultural ethics and FEP are relatively recent intellectual conceptions, having emerged in the 2000s. For me, however, shaped by the ideas of Paulo Freire (2005a, b) and the "pedagogy of the oppressed" and "education for critical consciousness," during many years these biocultural and educational concepts have been more than just intellectual pursuits. They have been part of a long journey. I moved from the USA to Bolivia in the 1970s when liberation thinking was maturing in Latin America, especially in theology (Gutiérrez 1973). For nearly 50 years, I was involved in both formal and non-formal education. But I was as much a learner as a teacher, especially given my original social location and background. My experiences of repressive military governments in Bolivia and with *campesinos* of the Andean Altiplano, and later in Costa Rica and Guatemala, among other places, as well as my involvement in human rights work, brought me into intimate contact with the "coloniality of power" (Quijano 2000).

These experiences forced me to think out of an action-reflection model, definitive for the liberation tradition, in which theory follows practice to produce a new synthesis or praxis (Freire 2005a). From the perspective of liberation philosophy, the point of departure for "philosophical thinking" or "teaching" is the concrete, historical and cultural, situation, that is, "places," and must be "critical reflection" from and on those "places." The "first step," then, is thinking about one's own reality and practice as a subaltern in the broader hegemonic political and cultural economy, so the following "second step" develops a theorical framework that explains subalternity and points to ways of changing it. The "first step" includes analyzing the possibilities and limits of local knowledge and the "second step" consists in finding other "knowledge" that can usefully complement, augment, or even replace "local knowledge" for a new synthesis of knowing that is liberating. This "new knowledge" is produced at the border between subalternity and hegemony. This methodological sequence "reinforces the importance of human action as the point of departure for all reflection" (Gutiérrez 1973, p. 11).[1] In this sense, it is not so much the content that is liberating, but the method that begins in practice leading to theory and then back to action. Theory and practice are united as praxis, as Paulo Freire (2005a) insists.

As I worked through the Methodist Church in Bolivia with Andean *campesinos* in numerous development projects, and, above all, engaged with them in workshops on human rights and political economy during periods of political and economic instability, including repressive military governments and, some years later, through similar engagements in Central America, my own thinking and being were shaped leading me to a new, personal "synthesis" of values and commitments. However, at the same time, I realized how important critical reflection on politics and culture was for the *campesinos* themselves and, even though an outsider and fellow learner, I also was a teacher who could provide useful knowledge. The resulting relationship was mutually enriching.

These workshops or *cursillos* on "political economy," nearly always conducted in isolated, rural areas, were spaces for learning, support, and remembering history. Participants often came from villages with long histories and rich stories of resistance. I especially remember one story. A village on the northern shore of Lake Titicaca dated to before Incan times. It was never conquered by the Spanish nor incorporated into a *hacienda* in later years. When I asked how their forebears kept their freedom, they laughed and explained, "Our *abuelos* [grandparents] tell us that when the Spaniards came to the village, everybody rose up against them and chopped off their heads. We've been free ever since!" Resistance in face of discrimination was often a workshop theme. Many dealt with questions related to national policies, such as currency devaluation and hyperinflation; or social structure and militarism and how different social sectors were differentially affected. We looked at

[1]For an ample discussion of this method for environmental ethics, see May (2015a).

Fig. 13.1 Guatemalan villagers study I Kings 21 during a workshop on land (in-)justice in the early 1990s (photo by Roy H. May Jr.)

the role of international imperial interests in contemporary policies. Since the government often seriously abused human rights, these were always a connecting thread in all the workshops.

Cursillos also were spaces for verbally articulating often repressed feelings and experiences. One workshop in Guatemala during the early 1990s (Fig. 13.1) is memorable. We worked on questions related to land justice: use and distribution. As we reflected together on texts about land (in-)justice found in the Hebrew Bible (i.e., the murder of Naboth so that king Ahab can take his vineyard [I Kings 21]), examining them as "codes" or "metaphors" for contemporary situations, older participants shared their stories of struggle and resistance that sometimes had been at great personal cost to them.

Much of this history of resistance was unknown to younger participants. As we ended the nearly 3 days together, one of the participants told me as we gathered informally with several others, "Now that we have finished, you should know that in Guatemala, there are three words that we are prohibited to use: land, justice, and *campesinos*. But for 3 days, we've been able to say them out loud and without fear!" Everyone laughed and we all took a deep breath. The time together was more than about learning facts. It permitted a kind of catharsis for releasing and sharing pent-up feelings and experiences.

13.2.1 High-Andean Experiences in Bolivia

In the late 1970s in the rural highlands of Bolivia, we put together a development model framed by the ideas of Freire, particularly conscientization, facilitators as fellow learners, and place-based agency (Freire 2005a, b). Our permanent team included two Aymara rural leaders who were especially helpful. One had been trained in village health, the other was a political representative—*dirigente sindical*—who did not especially like outsiders. We also involved leaders from Indigenous, mainly Aymara, political-cultural organizations. They brought a clearly *Indianista* focus. It was evident to all that effective, non-paternalist, and culturally self-affirming "development" would have to be rooted in the villages themselves, not dependent on outsiders for determining local needs and project possibilities. Efforts would be guided by the principles of conscientization, self-reliance or sustainability, and organization for empowerment. We recognized that critical analysis and awareness of community problems and the forces impinging on those problems from both the outside and inside, and the building of trust and solidarity while acting on the world, were crucial for development. We also knew that this would be a constant process of reflection on engagement.

With a Freirian framework, we could not separate development from the overarching political economy that fostered militarism as the power base of hegemonic "white" social groups that maintained structured injustice. Furthermore, self-reliance or sustainability, the optimal use of local and other resources and broad grassroots participation, were vital not only for success but for self-determination. This meant that the locus of decision-making and program support shift from international agencies or central offices to the local communities themselves. Organization for empowerment meant that networking local organizations, such as the *sindicato* or the area's chapter of the National Human Rights Assembly, was required if pressure for change was to be generated. Implicitly, we hoped to connect them to the nationwide *campesino* union. This was the political dimension required for development. Finally, we knew that the meaning of development worker or agent would have to change. Instead of having the role of imparting knowledge, it would be to dialogue with local communities, to push them for deeper questioning about their reality, and for clarification of what they were saying and wanting, and then to facilitate their decisions through contacts and technical expertise (May 1986).[2] To do this, we organized village-level "technical teams" (named by the villagers) to assume responsibility for analyzing community development needs and for overseeing projects that would be established. We brought these teams together for training workshops on how to analyze community needs, aspects of oversight, and, above all, for jointly planning community efforts (Fig. 13.2).

[2]This theoretical framework was inspired by the work of the Commission on the Churches' Participation in Development of the World Council of Churches. See Dickinson (1975) for an overview of the Commission's thinking.

Fig. 13.2 A Technical Team analyzes a community problem during a workshop conducted on the Altiplano of Bolivia in 1980 (photo by Roy H. May Jr.)

We also endeavored to show how local efforts were inevitably caught up in a network of social, economic, and political structures that would affect any local effort. The foreign and local professionals, mostly agronomists, who were working in various development projects, were integrated into these workshops as equals to the village technical teams. As a result, local efforts toward self-determination were enhanced and grassroots proposals emerged as never before. Local communities formed action groups to defend their interests.

Unfortunately, this effort at village-based development was cut short by a violent, right-wing military coup. The program was declared by military authorities as "Marxist oriented" and subversive. Many of the professionals and villagers involved were either imprisoned, expelled from the country, or otherwise incapacitated. I, too, was forced to leave Bolivia. Sadly, the program never recovered.

13.2.2 Rainforests, Rural, and Urban Experiences in Costa Rica

In the 1990s, after leaving Bolivia, as a university professor of ethics and theology at the Latin American Biblical University and research associate at the Ecumenical Research Department (DEI) in Costa Rica, I "applied" this liberation pedagogy. Although in a quite different context, again I had mutually enriching experiences. I taught ethics and theology from a contextual, action-reflection model that discouraged "dogma" as the beginning point. Rather, I encouraged students to engage their own historical and cultural experience to launch their theological and ethical thinking, and then, afterward, to articulate "dogma" (May 2015b). Especially African-decent and Indigenous students often focused their thesis work on aspects of their cultural and historical heritage for re-thinking Christian theology. In addition to classrooms, in order to focus on environmental concerns, we incorporated field experiences at the La Selva Biological Station, administrated by the Organization for Tropical Studies, near San José. Students spent a weekend with local naturalist guides, but mostly just roamed the rainforest trails by themselves. For nearly all, this was their first experience in the rainforest. Frequently, years later when I would meet a former student, I was greeted with, "Professor, remember when we went to the rain forest!" This direct field experience was impactful in a way reading about the rainforest could never have been. This reaffirms the value of *direct encounters* with other-than-human living beings and habitats (Rozzi et al. 2006). Although it is not easy to incorporate "field" into the classroom, sometimes it can be done.

Once in San José, Costa Rica, shortly after beginning an evening class, a Tropical Screech Owl (*Megascops choliba*) began calling just outside the classroom windows. I called on the students to be silent and I cut off the lights. For several minutes we just listened. Afterwards we reflected on that beautiful sound before resuming the discussion of whatever book we were studying. Did they learn something by listening to the owl? I do not know, but it was the first time any of the students had ever listened to a bird, at least in a classroom!

The university also organized "environmental justice field trips" for students (and often for visitors from the USA and Europe as well as exchange students). These consisted of visiting various parts of the country to see "development" in action, especially industrialized, agrichemical intensive, banana and pineapple plantations to observe working and environmental conditions. When companies realized what we were doing, our groups were prohibited from entering their plantations and packing plants on pain of trespassing; we stayed on public roads and people observed and took photographs from tour bus windows, or sometimes everyone lined-up along the public right-of-way and viewed what was going on among the banana "trees" or packing sheds. During these field trips, we also visited organic coffee and banana plantations, never with opposition from their owners! Seeing and then reflecting on the political economy and policy connections that supported these industries and development models did much to raise awareness of environmental justice.

During this period, the Emmaus Forum, a grassroots organization supported by the Roman Catholic Church's social ministry office in Costa Rica's Caribbean region, along with other non-governmental organizations and labor unions, actively advocated for the land rights of *campesinos* and Indigenous people, the labor rights of banana company workers, and the region's environmental health. The organization brought together workers, *campesinos*, and Indigenous people to pressure for labor and environmental reforms, especially regarding agrichemicals and pollution of groundwater. We took politicians and leading citizens on "environmental justice

Fig. 13.3 *Campesinos* exchange ideas about environmental questions in the Talamanca region of Costa Rica in the early 2000s (photo by Roy H. May Jr.)

field trips" to see firsthand some of these problems and we interviewed top company officials about working and environmental conditions. We conducted training workshops with *campesinos*, labor organizers, and Indigenous people on land use and tenure, labor, environmental, and human rights issues, and brought pressure on both company and governmental authorities for changes (Fig. 13.3).

A perceptive workshop participant summed up the organization's purpose when he said, "Land doesn't fall from heaven. You have to fight for it!" (May 1996)

These educational experiences were inspired by Paulo Freire's action-reflection model, especially as "conscientization" or critical awareness about the structural realities of injustice as well as sensitivity to nature that is often deadened in urban contexts or intentionally overlooked in the name of development. The underlying motive was to move people to act against injustice in its many manifestations. Intuitively, we were working out of what later would be called biocultural ethics and field environmental philosophy.

13.3 Education for Biocultural Liberation

Education for liberation is founded on action-reflection. It actualizes processes of critical analysis or "conscientization" based on one's own, local situation, through which the oppressed are enabled to understand their subordinate status and are empowered to act for changing their situation. The oppressed discover what limits their life possibilities and ponder how to change them. This "pedagogy of the oppressed," as termed by its founder, the Brazilian educator Paulo Freire in the 1960s, is a dialogical process in which questions are posed rather than information given, and teachers are facilitators and fellow learners (Freire 2005a).

"Local knowledge" is respected, encouraged, and connected to or augmented by "other" knowledge. Coherence as well as contradictions between these kinds of knowledge is revealed, enhancing the co-creation of knowledge that is both useful and liberating. This educative approach, or critical pedagogy is participatory and intercultural, emphasizing diversity. Theory and practice are united as praxis or new and creative ways of being that undermine subaltern status. It moves "learners" from being the objects or "the clients" of education, to being active "subjects" or agents empowered to make their own decisions.

Similarly, Field Environmental Philosophy (FEP), developed in the mid-2000s by Ricardo Rozzi et al. (2010) for implementation at the Omora Ethnobotanical Park in Cape Horn, Chile, begins in local reality. It proposes a four-step cycle including (1) transdisciplinary ecology and philosophy research; (2) composition of metaphors and communication through simple narratives; (3) field activities guided with an ecological and ethical orientation; (4) identification and implementation of *in situ* conservation projects. Each step is carried out by teams composed of local people, students, and professionals. Together they research, compose the metaphors, and actions to be taken. This methodological approach permits the consideration of local knowledge as well as life histories of underperceived organisms that most often remain outside of formal education (Moreno-Terrazas et al. 2023; Crego et al. 2023; Heidinger et al. 2023). At the same time, working with local people empowers them as agents for change and political recognition. This biological and cultural diversity and their interrelationship are essential for the well-being of human communities and ecosystems (Méndez-Herranz et al. 2023; Contador et al. 2023; Ojeda et al. 2023).

13.3.1 Liberation Philosophy

Liberation philosophy, which underlies the "pedagogy of the oppressed" and "field environmental philosophy," emerged in Latin America and elsewhere to critique structures of injustice that oppress the poor by turning them into subalterns in their own countries in the service of powerful social, economic, political, and international interests. These structures, in turn, tend to be internalized by the oppressed themselves, thus continuing their oppression in the interests of others. As the Peruvian sociologist Aníbal Quijano (2000) proposes, this "coloniality of power" establishes discriminatory social order as natural or inevitable, and that which is non-indigenous as superior to native. Thus "thinking" is alienated from local or national reality and is embedded in what is imposed from outside. People and their thinking are colonialized, much to the benefit of the hegemonic cultural, socioeconomic, and political sectors (Freire 2005a). Official education, whether formal or informal, tends to reproduce these structures of alienation, thus re-enforcing the condition of subalternity and oppression. In so doing, people are colonialized by superordinate interests embedded in socioeconomic, political, and even cultural structures. The practical consequence is the loss of Indigenous languages and cultures, traditional ecological knowledge, destruction of biodiversity, and continuing social injustice.

However, not only people are colonized; nature is also oppressed. The "coloniality of nature," that is, turning the resources of the natural world into mere commodities for the enrichment of the few by ripping them from the earth and the people nearest to them, is the consequence of the same structures of injustice that oppress people. Nature is reduced to "subaltern space" to be exploited at will by powerful interests (Alimonda 2011, p. 22). Indeed, the "*founding* phenomenon" (Alimonda 2011, p. 23; italic in the original) of the colonization of Latin America, Africa, and Asia was the extraction of natural resources and the exploitation of native peoples as the crucially needed labor force. At the same time, the exploitation, often genocide, of the original populations meant the loss of local knowledge and traditional ways of interacting with nature (Alimonda 2011, p. 49). The "coloniality of nature" is the major driver of the extinction of biodiversity and important ecological knowledge. This biocultural colonization continues today, notably through national and international extractive enterprises that pollute the environment and take the land of Indigenous people (LAM 2021; Parrilla 2021).

Liberation philosophy responds to these issues, yet it hardly is a unified, coherent body of thought. Rather it is "the collective name for a philosophical movement and method of doing philosophy" (Mendieta 2016), which is rooted in historical situations of coloniality and deep, structural social injustice. It takes on diverse dimensions including theology. Some emphasize political aspects, others socioeconomic, cultural, and increasingly, environmental or biocultural dimensions (Rozzi 2015). Because of this diversity, "the philosophy of liberation was and remains from the outset an internally heterogeneous movement" (Mendieta 2016). In many ways it is, as Eduardo Mendieta (2016) argues, a "counter-philosophical discourse" because it is intentionally "a critique of colonialism, imperialism, globalization, racism, and sexism." He concludes:

> Collectively, philosophers of liberation affirm that . . . human freedom cannot be understood in the abstract, but only against . . . very specific historical conditions that are material because they take the form of socio-political institutions. For philosophers [of] liberation, human liberty must be embodied and material precisely because it is part of a dynamic historical reality. (Mendieta 2016)

An articulating concern of this philosophy is to reveal cultural dependency in its many manifestations and then to recover autochthonous intellectual and other cultural resources for authentic ways of thinking appropriate to specific sociohistorical realities. In a small but impactful book, Peruvian philosopher Augusto Salazar-Bondy (1968) asked, "Dose [Latin] America have its own philosophy?" He launched an important debate about whether Latin America thought for itself or if it was dependent on Europe. Every culture must ask the same question about itself. Does it have its own "thought" or philosophy, or is it dependent on others? The task of liberation philosophy is to recover what is distinctly one's own. Field environmental philosophy builds on this tradition of liberation philosophy (Tauro et al. 2021).

13.3.2 Decolonial Theory

More recently, "decolonial" theory has merged with liberation philosophy. What "liberation" and "decolonial" have in common is that each, from its own perspective, criticizes structures that limit the life possibilities of the poor—and subaltern racial, cultural, and social groups—and seek their empowerment to overcome these structures. These limiting structures emerged with capitalism and European colonial imposition that created a hegemonic Eurocentric vision of the world. In this vision, the only legitimate ways of doing and knowing are universalized, and local places, cultures, and languages are marginalized as either impediments to true knowledge or quaint objects for study. This pervasive vision penetrates the whole of reality, even being internalized by the colonialized themselves. Thus the "coloniality of power" implies the imposition of a hegemonic worldview (Quijano 2000). In this sense, Western epistemology cannot be separated from the history of capitalism and colonial expansion (Mignolo 2000, p. 55).

Decolonial education, then, not only challenges Eurocentrism, but de-links from Western hegemonic ways of thinking and doing by reclaiming Indigenous thought and knowledge systems through place-based pedagogies (Zavala 2016). Crucial to this is "the need, indeed the strong need, for building macronarratives from the perspective of coloniality" (Mignolo 2002, p. 22). This means reconstructing history from the experience of subalternity that is "geared toward the search for a different logic" (Mignolo 2002, p. 22). Indeed,

> [w]ithout such macronarratives told from the historical experiences of multiple local histories (the histories of modernity/coloniality), it would be impossible to break the dead end against which modern epistemology and the reconfiguration of the social sciences and the humanities since the eighteenth century have framed hegemonic forms of knowledge (Mignolo 2002, p. 22).

Constructing such narratives is a goal of decolonial education for liberation. Miguel Zavala (2016) proposes three broad goals for decolonial education. The first, following Walter Mignolo, is to "deconstruct Modernity or Eurocentrism" and its logic that legitimates colonialism and the deculturization of Indigenous peoples. The second goal is to encourage "border thinking," that is, thinking and knowledge creation that is done from the margins, from outside the centers of power. Mignolo (2000, p. 13) also emphasizes "border thinking" or "border gnosis" in order "to bring to the foreground the force and creativity of knowledge subalternized during a long process of colonialization." It is "border" because it is done "between" the history of subalternization and hegemonic social order, or between Indigenous tradition and colonial history, which are presently active in the lives of Indigenous and other "colonialized" peoples. Tied to transformative praxis, creative "border thinking" produces knowledge that is created and owned by subalterns themselves. The third is "epistemological diversity" which "attempts to recover repressed and latent knowledges while at the same time generating new ways of seeing and being in the world."

All decolonial or liberation pedagogies—critical Indigenous pedagogies—will be participatory, based on dialogue and reciprocity, drawing from and oriented to specific places and situations. These critical Indigenous pedagogies resist injustices caused by colonization and oppression and are committed to Indigenous self-determination and sovereignty in the educational context (Garcia and Shirley 2012). As Zavala (2016) concludes, "Education is a site of struggle and rupture: It comes into being as people engage in dialogue and in response to the coloniality of power." This is what decolonial education and education for liberation are tasked to do.

13.4 Problems of Education for Liberation

However, impediments to education for liberation are multiple. The overarching social structures themselves leave little space for critical analysis of how they function to maintain oppression. Indeed, they *suppress* critical thinking. The legacy of capitalism and European colonialism is a hegemonic Eurocentric vision of the world in which local places, cultures, and languages are marginalized. Only knowledge that furthers the interests of Eurocentric hegemony is viewed as valid. This pervasive grand narrative organizes educational endeavors and pulls colonized subalterns into the orbit of the powerful by shaping the way they live and think. This "coloniality of power" (Quijano 2000)—or "Totality" according to Enrique Dussel (1980)—ontologically impedes education for liberation by imposing, often consciously unseen and unfelt, parameters of thinking. "Coloniality" becomes a mindset, inevitable because it is "natural." Materially, it is achieved through the suppression of native language in favor of imposed languages; references to places outside local realities; employment of teaching illustrations that rely on non-native flora and fauna; reliance on foreign scholarly works and ideas; as well as promotion of obedience to established authorities, whether they be political, religious, or intellectual. It also means not respecting or belittling local knowledge, as well as affective ties to specific places and traditions. Diversity is shunned in favor of biocultural homogenization. Many of the problems that hinder biocultural education, especially formal education, are reviewed by Rozzi (2018) and chapters in this volume, e.g., O'Donoghue and Sandoval-Rivera (2023); Gare (2023); Zent and Zent (2023).

However, as indicated, "coloniality" often is internalized by subalterns themselves; the oppressed participate in their own oppression (Weenie 2000, p. 67). So, the other side of these problems is resistance by the oppressed to liberative pedagogies. For example, even though loss of language is closely tied to loss of culture, in some places, it is not uncommon for Indigenous parents to reject the use of their native language for their children's education and likewise are hesitant to incorporate cultural traditions, such as religion and ceremonies, into the curriculum (Bachelder 2000). Nor is there agreement as to who should be allowed to teach, some demanding only Indigenous persons from their specific community, or that they just be Indigenous, or anyone so long as they are certified (Bachelder 2000). At the same time, however, many parents demand that native language and traditions be taught. Still others completely avoid native cultural traditions because they do not want to be set apart from hegemonic society. They want "to get ahead" and tradition holds them back. The diversity of attitudes, and frequent resistance, undermines the success of place-based or liberative education. These too are problems that liberative pedagogies must deal with.[3]

In short, these historical processes establish an exclusionary moral community that marginalizes certain groups of people and the whole of nature, and as such, is the major impediment to education for liberation. Moral community refers to those to whom we have moral obligations. However, there is no single moral community that is universal, rather there are many moral communities based on religion, race, sex and gender, social and economic class, even culture, nationality, and language, that "justify" discrimination and subalternity. Furthermore, historically Western moral communities include only humans, not other living creatures. Consequently, moral obligations to nature have not been part of the Western mindset.

Unfortunately, moral communities exclude more than they include. This affects education and the production of knowledge because moral community determines who, how, and what is "educated" and "taught" and what is considered legitimate ways of thinking and knowing. The hegemonic moral community imposes itself on the educative process so that outcomes favorable to its continued domination are assured. For this reason, constructing non-exclusionary moral community is a major liberative pedagogical task.

None of this means that outside sources are not to be used, rather, that their use should always be in the context of specific historical and cultural realities. In all cases it means that local knowledge and realities are to be considered as basic for the educative process. Education always is place-based, so "places"—understood as geographical location, cultural settings, social location—are fundamental for liberative education.[4]

13.5 Biocultural Liberation Philosophy

FEP and biocultural ethics can be re-dimensioned as biocultural liberation philosophy because they endeavor to fulfill the tasks of decolonial education and education for liberation. Even though the concepts are posterior to the idea of education for

[3] These studies were conducted mostly among the Diné (Navajo) in the US Southwest.
[4] On the importance of "sense of place," see Semken (2011) and Avriel-Avni et al. (2023).

liberation, FEP and biocultural ethics methodologically are categorized as "liberation" because of their affinity with the action-reflection model of "Step One, Step Two," for their linking local scale with international scale, and because they are intentionally oriented toward socio-environmental justice.

The strength of FEP is that it is place-based and corresponds to local cultural contexts (cfr. Avriel-Avni et al. 2023). It draws on local cultural systems and beliefs, rather than imposing alien ideas and images (cfr. Costa-Carvalho 2023). It intentionally engages local people in all aspects of fieldwork and understands the researcher or professional as learner as well as teacher. Indeed, requiring the professional to work alongside of, and to learn from, local people and students is an important attribute of FEP. The process is dialogical, thereby breaking down the social hierarchy of "professional expert" who teaches "ignorant locals" and "blank slate" students (Rozzi et al. 2008). FEP takes local knowledge not only as worthy of respect, but as contributing to knowledge in general and to solving local environmental needs. In so doing, it affirms and conserves local cultures and contributes to resolving socio-environmental problems. This fits Freire's model of the literacy worker and extension agent very well. Neither are bringers of knowledge, but "problematizers" and fellow learners embedded in a community. Together they do "border thinking" that produces knowledge.

This process, as Freire emphasizes, is never neutral. It always has implications for structures of power. However, by focusing primarily on culture and environment, the deeply political nature of these concerns might be overlooked in environmental education and conservation programs. "Culture" too often hides the political or "historical" by emphasizing customs and beliefs, rather than political frameworks—the complex of interests and power relations—that shape polities, cultures, and their liberating possibilities.

Indeed, focusing exclusively on culture and environment can be a way of maintaining the coloniality of power. For example, during the 1990s in Bolivia, a nation largely of Indigenous people, a strong neo-liberal government implemented far-reaching cultural reforms by reframing the legal system to recognize ancient landholding systems based on the *ayllus* or traditional villages and their common lands. This allowed highland "rural areas to rebuild past governing structures, rescue traditions and customs from obscurity, and gain new rights recognized by the government" (Dangl 2019, p. 146). However, this "neo-liberal multiculturalism," while granting cultural rights to indigenous people, intentionally undermined the rural unions that had long been the source of considerable political power by prohibiting communities recognized as *ayllus* to have a union or *sindicato*. This, in turn, greatly debilitated the nation-wide *campesino* union and rural radicalism that opposed privatization and other neo-liberal policies. The purpose of cultural affirmation was to de-mobilize the peasantry (Dangl 2019, p. 156).

Deep cultural values, however, must be called upon if power is to be decolonialized and authentic thinking made possible. In Bolivia, no longer dominated by neo-liberal politicians, Aymara thinkers and Indigenous politicians are reinterpreting ancient beliefs and using them as "metaphors" or "codes" as Freire (2005b) proposed, not only to affirm their ancient cultures, but to put forth a political vision of the country that emphasizes unity in diversity (Appendix). For example, *chacha-warmi*, literally a male-female married couple, becomes a metaphor for understanding opposing political viewpoints and cultural traditions not as antagonists but as complementary and *Qhaoak Ñan*, the ancient roadways that connected the Inca Empire, the metaphor for unity amidst the nations' geographical and cultural differences. This work corresponds to the methodology of FEP that insists on the importance of composing metaphors drawn from local cultures because "analogical thought . . .leads to a conceptual synthesis of facts, values and action" (Rozzi et al. 2010, p. SM19). Doing so also corresponds to Mignolo's insistence on the importance of reconstructing history from the experience of subalternity and creating a liberating macronarrative.

Biocultural ethics and education examine the overarching sociopolitical and economic framework of structural injustice. Its focus on the conservation of both the integrity of biocultural habitats and the access of biocultural communities to them is an indispensable condition for social-environmental justice (Rozzi 2013). For the biocultural ethic and liberation philosophy, this is essential due to the deeply political nature of culture and the environment. If this is not addressed, there will be gaps in efforts to confront socio-environmental questions; the coloniality of power will remain unaddressed in its political dimension. From the perspective of liberation philosophy, this must be addressed. Eroded culture, loss of local knowledge, and environmental degradation are consequences of political interests and social structures that assure the hegemony of certain sectors of society. Unless this structural framework is undermined, socio-environmental injustice will continue. Environmental education is inevitably political education.

Finally, importantly biocultural ethics and FEP conceive moral community as widely embracing. They recognize alterity, so important to some versions of liberation philosophy (Dussel 1980, 1988), by promoting "face-to-face" encounters with the "other," whether the "other" be human cultures or native animals and plants. In this way,

> biocultural diversity becomes an experience of cohabitating with diverse living beings and their life histories . . . [E]xperiences of "direct encounters" in the terrestrial and marine [and cultural] habitats that transform not only the knowledge about biodiversity [and culture], but also, the ethics of living and coexisting in regional ecosystems with its diverse inhabitants (Rozzi et al. 2010, p. SM 23).

The Argentine/Mexican philosopher Enrique Dussel argues that the proximity or nearness of the "other" is the key to ethics. "The experience of the nearness of persons as persons," he explains, "is what constitutes *the other* as one's 'neighbor' (someone 'neighboring,' our 'near one,' a 'someone'), rather than as merely a thing, an instrument, a mediation" (Dussel 1988, pp. 9–10). Biocultural ethics broadens the "other" to include other-than-human forms of life and even whole landscapes. These become "neighbors" in a new moral community. As "community" it is inherently political, so this new kind of community in which moral obligations are broadly understood and in which diversity or heterogeneity is encouraged and affirmed implies new configurations of power that respect the variety of co-inhabitants. This will be biocultural liberation.

13.6 Conclusion

This has been a long journey. I learned much and hopefully contributed something. I am grateful that no one chopped off my head! Sadly, many have given their lives in struggles for biocultural liberation (May 2018). Fortunately, years of struggle by subalterns have forced significant changes in educational programs in numerous places in Latin America and around the globe. Indigenous peoples are demanding that their languages, cultures, and lands not only be respected, but be taken seriously in policy debates. Fissures are opening in Eurocentric hegemony and place-based, participative, and dialogical pedagogies are being implemented.[5] The ideas of Freire and like-minded educators have been widely disseminated and influential. Moral community has been opened more widely and inclusively. But the problems that liberative education faces still abound. Structures of injustice oppress people and nature. Moral community too often remains exclusionary. Biocultural education continues to confront the all too pervasive "coloniality of power." There still is much to be done to achieve biocultural liberation. We journey onward.

Appendix. Ancient Metaphors and Contemporary Politics in Bolivia

To illustrate how ancient cultural metaphors continue to inspire contemporary thinking in Bolivia,[6] in this Appendix, we reproduce the address delivered by David Choquehuanca on the occasion of his inauguration as vice president of the Plurinational State of Bolivia, La Paz, November 8, 2020:[7]

---

[5]For examples, see Semken et al. (2017) and the many case studies presented in this volume. The DEI (Departamento Ecuménico de Investigaciones http://www.deicr.org) in San José, Costa Rica is a leading promoter of this kind of education in Latin America.

[6]For interested readers we recommend the essays by Vicenta Mamani-Bernebé (2015), May and Roy (2017), and Xavier Albó (2018).

[7]This annotated English translation of the address by Vice-President David Choquehuanca has been done by Roy H. May Jr. based on the verbal recording of the speech [in Spanish], complemented with revisions, corrections, and annotations. Source: Government of the Plurinational State of Bolivia. Spanish text published on line 8 November 2020 by La Razón (La Paz, Bolivia): https://www.la-razon.com/nacional/2020/11/08/lea-el-discurso-completo-de-la-posesion-del-vicepresidente-david-choquehuanca/ and verbal recording of the address. Other English translations are available via: Debates Indígenas: Speech of David Choquehuanca—Debates Indígenas (https://www.debatesindigenas.org); TML Weekly: New President and Vice President Address People of Bolivia (https://cpcml.ca); Proyecto Esperanza: https://www.esperanzaproject.com/2020/native-

With the permission of our gods, of our elder brothers and sisters, and of our *Pachamama*,[8] of our ancestors, of our *Achachilas*,[9] with the permission of our *Patujú*,[10] of our rainbow,[11] of our sacred coca leaf;[12] with the permission of our peoples, with the permission of everyone present and not present in this chamber:

Today in a few minutes, I want to share some thoughts because we are obligated to communicate, obligated to dialog, for this is a principle of living well.[13]

The peoples of millennial cultures, the cultures of life, have maintained their origins since the dawn of remote times. We children have inherited an ancient culture that understands that everything is interrelated, that nothing is divided, and that nothing is neglected and left out.

Let Us Go Together

That is why they tell us that we must all go together, that no one is to be left behind, that everyone is to have all that is needed, and that no one is to lack anything. And the well-being of all is the well-being of oneself, that helping motivates growing and happiness, that giving up for the benefit of others makes us feel strengthened, that uniting and recognizing ourselves as part of everything is the pathway of yesterday, today, tomorrow, and always from which we have never strayed.

---

american-culture/bringing-the-pachamama-to-the-presidency/;. Chicago Alba Solidarity: Speech by David Choquehuanca on Assuming the Vice-Presidency of the Plurinational State of Bolivia, November 8, 2020 | Chicago ALBA Solidarity (https://wordpress.com).

[8] *Pachamama* is understood as a great maternal womb, whose life-giving power is materialized in cultivable soil and home space; through her fecundity, the continuation of the cosmic process of regeneration and fundamental relatedness and cosmic order are assured. She is understood as the universal feminine principle. Pachamama often is translated as Mother Earth.

[9] Aymara (*apus* in Quechua), high mountains, are understood as the homes of masculine, spirit "grandparent" protectors. These "grandparents" must be cared for and venerated.

[10] *Heliconia rostrata*, a flowering plant characteristic of Amazonia; it is one of the two national flowers of Bolivia.

[11] The rainbow is rooted in traditional Andean culture; each color has symbolic significance. In recent years, the rainbow is identified with the Wiphala or Indigenous peoples' flag, which is composed of multiple blocks of rainbow colors and symbolizes the resurgence of the Aymara and Quechua people in Bolivia and throughout the Andean region. The Wiphala is an official flag of Bolivia.

[12] Coca, one of four native Andean plants of the Erythroxyladeae family, is believed to have divine powers for healing and for foretelling the future.

[13] "Living well" or *buen vivir* in Spanish and *suma kamaña* in Aymara, is an Andean philosophical concept understood in contradistinction to the Western idea of living better or the "good life." In recent years it has emerged as the framework for social and environmental ethics throughout the Andean region.

The *ayni*, the *minka*, the *tumpa*, our *colka*, and other codes of ancient cultures are the essence of our life, of our *ayllu*.[14]

Ayllu is not only an organization of society of human beings, *ayllu* is a system that organizes the life of all beings, of everything that exists, of everything that flows in balance on our planet or Mother Earth.

For centuries, the civilizing canons of *Abya Yala*[15] were taken apart, de-structured, their original meanings altered, and many of them exterminated. Original thought was systematically subjugated to colonial thought. But they could not stamp us out. We are alive, we are from *Tiwanaku*;[16] we are strong, we are like stone, we are *cholke*;[17] we are *sinchi*, we are *Rumy*;[18] we are *Jenecherú*,[19] the fire that never went out. We are from *Samaipata*;[20] we are jaguar;[21] we are *Katari*.[22] We are Ainus, we are Maori, we are Comanches; we are Mayas; we are Guaraní; we are Mapuches and Mojeños; we are Aymara; we are Quechua; we are Hopis;[23] and we

---

[14] *Ayni* means reciprocity and is the form of all social relationships and is the logic that connects material and spiritual reality; without *ayni*, nothing holds together. This concept is fundamental to the traditional Andean worldview. *Minka* is a form of reciprocal labor relations and materializes *ayni*. It is popularly defined as "Today for me, tomorrow for you." *Tumpa* is to seek out someone who is "lost" or to remember those who are not present. *Colka* expresses the importance of life or living. *Ayllu* is the traditional village and its agricultural and pastoral lands. It is the political structure of *ayni*. It also is a concept applied to personal or collective areas of life, work, or thought as sector domains, such as an "ayllu of writers."

[15] Derived from the Guna (Panama) word meaning land full of life, or full of goodness, or full of maturity. The term is widely used as the non-colonial name for the Americas.

[16] An ancient city on the shores of Lake Titicaca, Bolivia. Thriving between, 500 CE and 1000 CE, it is the cultural cradle of various pre-Hispanic Andean civilizations.

[17] *Cholke* means hard as stone and refers to the endurance of Indigenous people.

[18] In Quechua *sinchi* refers to strength, force, or to a knowledgeable and valiant leader; *romy* is stone or rock.

[19] Literally a smoldering log maintained to re-ignite cooking fires and fires for warmth. Now a Bolivian colloquialism, especially associated with the Santa Cruz region of the eastern part of the country, it is used to express optimism, happiness, or cheerfulness in face of diversity or irritation. Its root is a Guaraní word signifying "the fire never goes out."

[20] An ancient ceremonial site in central Bolivia active between 300 CE and 1500 CE.

[21] A powerful feline is ubiquitous in ancient Andean culture. Lake Titicaca is perceived as being shaped like a wild cat (*titi* in Aymara is wild cat or "tiger"); this mythological feline denotes power and is represented in stela and ceramics found at ancient sites such as Tiwanaku.

[22] Tupac Katari (Julián Apasa), his wife Bartolina Sisa, and sister Gergoria Apasa, led an Aymara insurrection against the Spanish in 1781 that included a 6-month siege of the City of La Paz. In the mid-1500s in Peru, rebellions against the Spanish were led by leaders named Katari and Tupac. Julián Apasa took their names to inspire the insurrection he led over two centuries later in Bolivia. He is considered a national hero by the Aymara and Quechua people.

[23] The Ainus are an ethnic group indigenous to Japan; the Maori are the original people of New Zealand; the Comanches are a Native American tribe in the USA; the Mayas are the ancient inhabitants of Guatemala and parts of Mexico and Central America; the Guaraní people dominate Paraguay and southeastern Bolivia; the Mapuches are located in central Chile; the Mojeños live in the savannahs of northern Bolivia; the Aymara and Quechua are the principal indigenous people of Bolivia; the Hopi, are an ancient people who still live in Arizona, USA.

together are all the peoples of the culture of life who awaken *larama*,[24] equals, rebellious, and wise.

A Transition Every 2000 Years

Today Bolivia and the world are living through a transition that is repeated every 2000 years, within the framework of the cyclicity of time. We go from timelessness to time, thus beginning a new dawn, a new *Pachakuti*[25] in our history, a new dawning sun and a new expression in the language of life in which empathy for others, or the collective good, replaces selfish individualism.

This is the time when all Bolivians look at each other as equals and we know that when we are united, we are worth even more. We are returning to being *Jiwasa*[26] again, not me but us. *Jiwasa* is the death of egocentrism. *Jiwasa* is the death of anthropocentrism and it is the death of theocentrism.

We are in the time of returning to being *Iyambae*,[27] a code that our Guaraní brothers and sisters have protected. *Iyambae* is a person who has no owner. Nobody in this world should feel like they are the owner of anyone or anything.

In 2006[28] in Bolivia we began the hard work of connecting our individual lives and our collective roots, of returning to being ourselves, of returning to our center, to *taypi*,[29] to *pacha*,[30] to the balance from which emerges the wisdom of the most important civilizations our planet.

We are in the process of fully recovering our knowledge, the codes of the culture of life, the civilizing standards of a society that lived in intimate connection with the cosmos, with the world, with nature, and with individual and collective lives. We are building our *suma kamaña*, our *suma akalle*,[31] to guarantee the individual good and the collective or community good.

---

[24] In Aymara the color blue; in common usage *larama* refers to someone who is rebellious, even disrespectful toward others.

[25] *Pachakuti* is the end and beginning of cosmic epochs; the end of time and the beginning of time.

[26] In Aymara "us" or "we."

[27] A Guaraní term meaning "freedom."

[28] The year that Evo Morales was elected president of Bolivia. He is the first Indigenous person to hold that office.

[29] Aymara for middle or center; refers to the convergence of opposing forces or as a place for social gathering such as a plaza.

[30] *Pacha* is the Aymara and Quechua concept of time and space that encompasses the universe and existence itself; it is time, place, nature, cosmos, beginning and end of existence. In *Pacha* everything is related, and existence is coherent and harmonious.

[31] Reiteration of *suma kamaña.*

Chacha-warmi[32]

We are in the time of recovering our identity, our cultural roots, our *sake*.[33] We have cultural roots, we have philosophy, history, we have everything. We are people, and we have rights.

One of the un-moveable canons of our civilization is the wisdom we have inherited that is centered around the *Pacha*, wisdom guaranteeing balance in all time and space by knowing how to manage complementary energies: the cosmic one that comes from Heaven which joins with the one that emerges from deep within the Earth. These two telluric cosmic forces interact, creating what we call life as a visible (*Pachamama*) and spiritual (*Pachakama*)[34] totality.

By understanding life in terms of energy, we have the possibility of modifying our history, converging matter and life through the strength of *chacha-warmi*, the complementarity of opposites.

The new time that we are beginning will be sustained by the energy of the *ayllu*: community, consensus, horizontality, balanced equilibriums, and the common good.

Historically, revolution is understood as a political act to change the social structure, thereby transforming the life of the individual. No revolution has managed to modify the power structure that keeps control over people.

Our Revolution Is the Revolution of Ideas

Changing the nature of power has not been achieved, but power has managed to twist the minds of politicians. Power can corrupt, and the strength of power and its institutions are very difficult to modify, but it is a challenge that we will assume by drawing on the wisdom of our own peoples. Our revolution is the revolution of ideas, it is the revolution of balances, because we are convinced that in order to transform society, the government, the bureaucracy and laws, and the political system, we must change as individuals.

We are going to promote the coincidences of opposites to seek solutions between the right and the left, between the rebellion of the young and the wisdom of the old, between the limits of science and unyielding nature, between the creative minorities and the traditional majorities, between the sick and the healthy, between the rulers and the governed, between the cultured leadership and the gift of serving others.

---

[32] *Chacha-warmi*, literally a heterosexual couple, conceptually refers the joining of complementary opposites in perfect symmetry; one part without the other is broken and dis-functional. Only when complementary symmetry or mutuality is respected is well-being assured.

[33] Literal meaning not clear but refers to cultural identity.

[34] A universal masculine principle. *Pachakamak* is the creator god in Andean mythology.

Our truth is quite simple. The condor[35] takes flight only when its right wing is in perfect balance with its left wing. The task of shaping us into being balanced individuals was brutally interrupted centuries ago, yet we have not concluded this task and the time, the era of the *ayllu*, the community, is still with us.

We must be free and balanced individuals so that we will be able to build harmonious relationships with others and with our surroundings. It is urgent that we be capable of sustaining balance for ourselves and for the community.

We are in the times of the brothers and sisters of *Apanaka Pachakuti*, brothers and sisters of change, where our struggle was not only for ourselves, but also for them and not against them. We seek delegated authority, we do not seek confrontation, we seek peace. We are not part of the culture of war or domination. Our struggle is against all kinds of subjugation and against the single-mindedness of colonial, patriarchal thought, from wherever these come.

The idea of the encounter between spirit and matter, Heaven and Earth, *Pachamama* and *Pachakama*, assures us that as renewed women and men we will be able to heal humanity, the planet, and the beautiful life that is on it, returning beauty to our Mother Earth.

We will defend the sacred treasures of our culture from all meddling interference, we will defend our peoples, our natural resources, our freedoms, and our rights.

We Will Return to Qhapak Ñan

We will return to *Qhapak Ñan*,[36] the noble path of integration, the path of truth, the path of brotherhood, the path of unity, the path of respect for our traditional authorities and our sisters, the path of respect for fire, the path of respect for rain, the path of respect for our mountains, the path of respect for our rivers, the path of respect for our Mother Earth, the path of respect for the sovereignty of our peoples.

Brothers and sisters, in conclusion, Bolivians must overcome division, hatred, racism, and discrimination among compatriots. No more persecution of the freedom of expression, no more judicialization of politics. No more abuse of power. Power must be exercised in function of helping and power must be circulated among all. Power, as well as the economy, must be redistributed. It must circulate, it must flow, just as blood flows within our body. No more impunity—justice, brothers and sisters. But justice must be truly independent. Let us put an end to intolerance, to the humiliation of human rights and of our Mother Earth.

The new time means listening to the message of our peoples that comes from the bottom of their hearts. It means healing wounds, seeing ourselves with respect,

[35] The Andean Condor (*Vultur gryphus*), once ubiquitous in the Andes Mountains, is a powerful being in traditional Andean thought; it is the *mallku* or leader and is the king of birds.

[36] The network of roads that connected all parts of the Inca Empire. Many of these ancient roads continue to be used today.

recovering the homeland, dreaming together, building brotherhood, harmony, integration, and hope in order to guarantee the peace and happiness of the new generations.

Only then will we be able to live well and govern ourselves.

Jallalla Bolivia![37]

References

Albó X (2018) Suma Qamaña or living well together: a contribution to biocultural conservation. In: Rozzi R et al (eds) From biocultural homogenization to biocultural conservation. Springer, Dordrecht

Alimonda H (2011) La colonialidad de la naturaleza. Una aproximación a la ecología política latinoamericana. [English translation available via: http://www.alternautas.net/blog/2019/6/10/the-coloniality-of-nature-an-approachto-latin-american-political-ecology]. In Alimonda H (coord.) La naturaleza colonizada. Ecología política y minería en América Latina. CLACSO y Ciccus, Buenos Aires

Avriel-Avni N, Lavi-Neeman M, Holzer JM (2023) Reframing sense of place: a goal for environmental education and an indicator supporting social-ecological resilience. In: Rozzi R, Tauro A, Wright T, Avriel-Avni N, May RH Jr (eds) Field environmental philosophy: education for biocultural conservation. Ecology and ethics, vol 5. Springer, Dordrecht, pp 447–456

Bachelder A (2000) Teaching Diné Language and Culture in Navajo Schools: voices from the community. In: Reyner J, Martin J, Lockard L, Sakiestewa Gilbert W (eds) Learn in beauty. Indigenous education for a new century. Northern Arizona University, Flagstaff

Contador T, Rendoll-Carcamo J, Gañan M, Ojeda J, Kennedy J, Convey P, Rozzi R (2023) Underwater with a hand lens: ecological sciences and environmental ethics to value freshwater biodiversity. In Rozzi R, Tauro A, Wright T, Avriel-Avni N, May RH Jr (eds) Field environmental philosophy: education for biocultural conservation. Ecology and ethics, vol 5. Springer, Dordrecht, pp 53–70

Costa-Carvalho ME (2023) Small, silent and (in)significant: childhood as an experience of education. In: Rozzi R, Tauro A, Wright T, Avriel-Avni N, May RH Jr (eds) Field environmental philosophy: education for biocultural conservation. Ecology and ethics, vol 5. Springer, Dordrecht, pp nnn

Crego RD, Ward N, Rozzi R (2023) The eyes of the tree: applying field environmental philosophy to tackle conservation problems at long term socio-ecological research sites. In: Rozzi R, Tauro A, Wright T, Avriel-Avni N, May RH Jr (eds) Field environmental philosophy: education for biocultural conservation. Ecology and ethics, vol 5. Springer, Dordrecht, pp 101–112

Dangl B (2019) The five hundred year rebellion. Indigenous movements and the decolonization of history in Bolivia. AK Press, Chico

Dickinson RDN (1975) To set a liberty the oppressed. Towards an understanding of christian responsibilities for development/liberation. Commission of the Churches' Participation in Development, World Council of Churches, Geneva

Dussel E (1980) Filosofía de la liberación. Universidad de Santo Tomás, Centro de Enseñanza Desescolarizada, Bogotá

Dussel E (1988) Ethics and community (trans: Barr R). Orbis Books, Maryknoll

Freire P (2005a [1970]) Pedagogy of the oppressed (trans: Ramos MB). Continuum, New York

[37] *Jallalla* is similar to "long live!" and is used as an affirmation of Bolivia and the Aymara and Quechua cultures.

Freire P (2005b [1974]) Education for critical consciousness. (trans: Ramos MB). Contiuum, New York

Garcia J, Shirley V (2012) Performing decolonization: lessons learned from indigenous youth, teachers' and leaders' engagement with critical indigenous pedagogy. J Curriculum Theorizing 28(2):76–91

Garcia J, Vargas M (2018) Latin American philosophy. In: Zalta EN (ed) The Stanford encyclopedia of philosophy. Available via: https://plato.stanford.edu/archives/sum2018/entries/latin-american-philosophy/

Gare A (2023) Challenging the dominance of the grand narrative of economic growth in global education and culture. In: Rozzi R, Tauro A, Wright T, Avriel-Avni N, May RH Jr (eds) Field environmental philosophy: education for biocultural conservation. Ecology and ethics, vol 5. Springer, Dordrecht, pp 309–326

Gutiérrez G (1973 [1971]) A theology of liberation (trans: Inda C, Eagleson J). Orbis Books, Maryknoll

Heidinger K, Camichael J, Jacob-Schwartz B, Borucinska-Begg M (2023) Biocitizen's approach to biotic wonder, citizenship, and field environmental philosophy. In: Rozzi R, Tauro A, Wright T, Avriel-Avni N, May RH Jr (eds) Field environmental philosophy: education for biocultural conservation. Ecology and ethics, vol 5. Springer, Dordrecht, pp 131–152

LAM (2021) Latin American mining. Available via: Overview – LAM http://latinamericanmining.com

Mamani-Bernebé V (2015) Spirituality and the Pachamama in the Andean Aymara worldview. In: Rozzi R et al (eds) Earth stewardship. Linking ecology and ethics in theory and practice. Springer, Dordrecht

May RH Jr (1986) The church and Latin American community development. In: LeFevre P, Schroeder WW (eds) Pastoral care and liberation praxis. Studies in personal and social transformation. Exploration Press, Chicago

May RH (1996) La tierra no cae del cielo, hay que luchar por ella. La pastoral de la tierra en la Diócesis de Limón. ASEPROLA, Pastoral Social de la Diócesis de Limón, San José

May RH Jr (2015a) Dorothy Stang: monkeys cry and the poor die, earth stewardship as liberation ecology. In: Rozzi R, Chapin FS III, Callicot JB, Picket STA, Power ME, Armesto JJ, May RH Jr (eds) Earth stewardship. Linking ecology and ethics in theory and practice. Springer, Dordrecht

May RH Jr (2015b) Ethics without principles. Another possible ethics—perspectives from Latin America. Pickwick, Eugene

May RH Jr (2017) Pachasophy: landscape ethics in the Central Andes Mountains of South America. Environ Ethics 30(3):301–319

May RH Jr (2018) Land grabbing and violence against environmentalists. In: Rozzi R, May RH Jr, Chapin III FS, Massardo F, Gavin MC, Klavder IJ, Pauchard A, Nuñez MA, Simberloff D (eds) From biocultural homogenization to biocultural conservation, ecology and ethics 3. Springer, Cham, pp 109–123

Méndez-Herranz M, Marini G, Rozzi R (2023) Sub-Antarctic high Andean "gardeners:" cultivating caring relationships. In: Rozzi R, Tauro A, Wright T, Avriel-Avni N, May RH Jr (eds) Field environmental philosophy: education for biocultural conservation. Ecology and ethics, vol 5. Springer, Dordrecht, pp 71–86

Mendieta E (2016) Philosophy of liberation. In: Zalta EN (ed) The Stanford encyclopedia of philosophy (winter 2016 edition). Available via: https://plato.stanford.edu/archives/win2016/entries/liberation/

Mendieta E (2020) Philosophy of liberation. In: Zalta EN (ed) The Stanford encyclopedia of philosophy. Available via: https://plato.stanford.edu/archives/win2020/entries/liberation/

Mignolo WD (2000) Local histories/global designs. Coloniality, subaltern knowledges, and border thinking. Princeton University Press, Princeton

Mignolo WD (2002) The geopolitics of knowledge and the colonial difference. S Atl Q 101(1): 59–96

Moreno-Terrazas R, Cariño MM, Díaz Z, González H, Monteforte-Sánchez M (2023) Starfishes and sky stars: a field environmental philosophy education and ecotourism experiences in Baja California, México. In: Rozzi R, Tauro A, Wright T, Avriel-Avni N, May RH Jr (eds) Field environmental philosophy: education for biocultural conservation. Ecology and ethics, vol 5. Springer, Dordrecht, pp 113–130
O'Donoghue R, Sandoval-Rivera JC (2023) Hand-print CARE: intergenerational and plural knowledge in schools. In: Rozzi R, Tauro A, Wright T, Avriel-Avni N, May RH Jr (eds) Field environmental philosophy: education for biocultural conservation. Ecology and ethics, vol 5. Springer, Dordrecht, pp 415–426
Ojeda J, Tauro A, Contador T, Mansilla A, Rosenfeld S, Ban N, Rozzi R (2023) "Pay attention, dive with your eyes wide-open:" a field Environmental Philosophy activity to foster reciprocity between people and nature. In: Rozzi R, Tauro A, Wright T, Avriel-Avni N, May RH Jr (eds) Field environmental philosophy: education for biocultural conservation. Ecology and ethics, vol 5. Springer, Dordrecht, pp 87–100
Parrilla J (2021) Social conflict blights mining projects in Latin America. Diálogo Chino. Available via: https://dialogochino.net/en/extractive-industries/43744-social-conflict-blights-mining-projects-across-latin-america/
Quijano A (2000) Colonialidad del poder, eurocentrismo y América Latina. In: Lander E (ed) La colonialidad del saber: eurocentrismo y ciencias sociales. Perspectivas latinoamericanas. CLASCO, Buenos Aires
Rozzi R (2009) La filosofía ambiental. In: Dussel E, Mendieta E, Bohorquez C (eds) El pensamiento filosófico latinoamericano, del caribe y "latino," historia, corrientes, temas y filosofías. Siglo XXI editores, Mexico, pp 434–445
Rozzi R (2013) Biocultural ethics: from biocultural homogenization toward biocultural conservation. In: Rozzi R, Pickett STA, Palmer C, Armesto JJ, Callicott JB (eds) Linking ecology and ethics for a changing world: values, philosophy, and action, Ecology and ethics 1. Springer, Dordrecht
Rozzi R (2015) Implications of the biocultural ethic for earth stewardship. In: Rozzi R, Chapin FS III, Callicott JB, Pickett STA, Power ME, Armesto JJ, May RH Jr (eds) Earth stewardship: linking ecology and ethics in theory and practice, Ecology and ethics, vol 2. Springer, Dordrecht, pp 113–136
Rozzi R (2018) Biocultural conservation and biocultural ethics. In: Rozzi R, May RH Jr, Chapin FS III, Massardo F, Gavin MC, Klaver IJ, Pauchard A, Nuñez MA, Simberloff D (eds) From biocultural homogenization to biocultural conservation. Ecology and ethics. Springer, Cham
Rozzi R, Massardo F, Anderson C, Heidinger K, Silander J Jr (2006) Ten principles for biocultural conservation at the southern tip of the Americas: the approach of the Omora Ethnobotanical Park. Ecol Soc 11(1):43. [online] url: http://www.ecologyandsociety.org/vol11/iss1/art43/
Rozzi R, Arango X, Massardo F, Anderson C, Heidinger K, Moses K (2008) Field environmental philosophy and biocultural conservation: the Omora Ethnobotanical Park educational program. Environ Ethics 30(3):325–336
Rozzi R, Anderson CB, Pizarro JC, Massardo F, Medina Y, Mansilla AO, Kennedy JH, Ojeda J, Contador T, Morales V, Moses K, Poole A, Armesto JJ, Kalin MT (2010) Field environmental philosophy and biocultural conservation at the Omora Ethnobotanical Park: Methodological approaches to broaden the ways of integrating the social component ("S") in Long-Term Socio-Ecological Research (LTSER) Sites. Revista Chilena de Historia Natural 83:27–68
Salazar-Bondy A (1968) ¿Existe una filosofía de nuestra América? Siglo XXI, Mexico City
Semken S (2011) A sense of place in the American Southwest. Place-based earth systems science for diverse students. In the Trenches 1(3):1–4
Semken S, Ward EG, Moosavi S, Chinn PWU (2017) Place-based education in geoscience: Theory, research, practice, and assessment. J Geosci Educ 65(4):542–562. https://doi.org/10.5408/17-276.1

Tauro A, Ojeda J, Caviness T, Mosses K, Moreno R, Wright T, Zhu D, Poole A, Massardo F, Rozzi R (2021) Field environmental philosophy: a biocultural ethic approach to education and ecotourism for sustainability. Sustainability 13(8):4526. https://doi.org/10.3390/su13084526

Weenie A (2000) Post-colonial recovery and healing. In: Reyner J, Martin J, Lockard L, Sakiestewa Gilbert W (eds) Learn in beauty. Indigenous education for a new century. Northern Arizona University, Flagstaff

Zavala M (2016) Decolonial methodologies in education. In: Peters MA (ed) Encyclopedia of educational philosophy and theory. Springer, Singapore

Zent S, Zent E (2023) Collaborative action research for biocultural heritage conservation. In: Rozzi R, Tauro A, Wright T, Avriel-Avni N, May RH Jr (eds) Field environmental philosophy: education for biocultural conservation. Ecology and ethics, vol 5. Springer, Dordrecht, pp 229–246

Chapter 14
Collaborative Action Research for Biocultural Heritage Conservation

Stanford Zent and Egleé Zent

Abstract Given the rapid pace of environmental and cultural change all over the world, including Indigenous hotspots like the greater Amazon-Orinoco basin, innovative strategies are needed to meet the challenge of biocultural conservation. Collaborative Action Research (CAR) focused on the preservation and revitalization of traditional and local environmental knowledge (TEK/LEK) represents a promising approach in this regard. CAR entails cooperation, exchange, and complementary labor between different stakeholders—local communities, scientific researchers, and potentially others—for the purpose of studying and understanding issues of mutual concern and then taking informed measures to address those issues. In contrast to basic research, which is guided by academic questions and geared toward information extraction, CAR is focused on assessing and ameliorating community-defined problems. This chapter argues for the particular importance of CAR for the challenge of intergenerational TEK/LEK enculturation and maintenance in the face of rapid cultural and environmental change. The customary processes of such knowledge transmission and the contextual factors that alter or affect such processes are very site-specific. CAR offers a participatory and bioculturally appropriate strategy for bolstering traditional mechanisms of transmission and for developing new ones that actually incorporate ongoing changes. Examples of long-term CAR between local communities of the Jotï Amerindian group and researchers from the Venezuelan Institute for Scientific Research (IVIC) are described. The specific application of this approach in support of TEK/LEK transmission in this biocultural context is described in a companion chapter.

Keywords Biocultural conservation · Collaborative action research · Indigenous people · Jotï · Traditional ecological knowledge

---

S. Zent (✉) · E. Zent
Laboratorio de Ecología Humana, Centro de Antropología, Instituto Venezolano de Investigaciones Científicas (IVIC), Altos de Pipe, Miranda, Bolivarian Republic of Venezuela

R. Rozzi et al. (eds.), *Field Environmental Philosophy*, Ecology and Ethics 5,
https://doi.org/10.1007/978-3-031-23368-5_14

14.1 Introduction

The notion that biological diversity (BD) and cultural diversity (CD) are inextricably linked has become one of the guiding principles of conservation thought and practice in the twenty-first century (Redford and Brosius 2006; Rozzi et al. 2006; Maffi and Woodley 2010; Gavin et al. 2015; Rotherham 2015; Bridgewater and Rotherham 2019). This in turn is based on the growing recognition that human sociocultural systems cannot be separated ontologically from their biophysical milieus (Descola and Palsson 1996; Ingold and Palsson 2013; Rozzi 2013, 2018). Given the importance of education for conservation, the fundamental insight of CD and BD connectivity has important implications for formal educational design and execution. Integrated biocultural curriculums rooted in local ecological realities and cultural knowledges may be important for meeting the challenges of rapid social and environmental change. This is especially relevant for the surviving Indigenous populations and territories. Yet in much mainstream formal education, Indigenous youth are often taught subjects and facts that are foreign to their traditional physical and intellectual contexts.

The vital connection between biosphere and ethnosphere can be understood as a dialectical process with reciprocal effects. On the one hand, human populations adapt to environmental limitations and affordances by cultural means. Thus cultural differences are to some extent the product of adaptive radiation in a heterogeneous biogeographical space. On the other hand, humans perceive and act upon their surroundings through inherited cultural filters. Such filters are complex and multidimensional, consisting of language, ethnoscience, folklore, technology, occupations, food habits, exchange systems, settlement pattern, social organization, customary law, spiritual beliefs, moral values, ethnomedical systems, and more. All of these variables, alone or in combination, go into shaping the ways that people interact with environmental components. Such interactions leave unique landscape footprints and the cumulative effect is to augment environmental variability (Pretty et al. 2009). From a processual perspective, it seems evident that BD and CD are interdependent or co-evolved, at least in part. The concept of *biocultural diversity* (BCD) was developed to theorize this web of interdependency (Baer 1989; Maffi 2005; Pretty et al. 2007).

Traditional ecological knowledge (TEK) occupies the active interface between culture and nature at the local level. It represents the complex matrix of understanding and doing where the semiotic realm of ethnoecological categories and meanings converges with the material realm of living organisms, biotic communities, and abiotic entities. Such knowledge encompasses diverse cognitive and behavioral operations: folk biological classification, the natural history traits attributed to each ethnotaxon, recognition of ecological interactions and associations among organisms, the symbolic or ritual significance attached to environmental entities, the tools and techniques for producing, procuring, and processing resources for human use, the recipes or scripts for carrying out coordinated actions, the embodied skills needed to perform these operations, and the mental models which explain how the world works in terms of the organization of its parts and the relationships among them. Besides empirical knowledge, complementary facets of TEK include aspects of language, spiritual belief, ethics, cosmology, social institutions, and resource practices (Posey 1999; Toledo 2002; Berkes 2008; Zent 2009a). The centrality of TEK within the BCD nexus, as both producer and product of diversity, is expressed by the concept of *biocultural heritage*, which has been defined as:

> Knowledge, innovations and practices of indigenous and local communities that are collectively held and are inextricably linked to: traditional resources and territories, local economies, the diversity of genes, species and ecosystems, cultural and spiritual values, and customary laws shaped within the socio-ecological context of communities (IIED 2021).

Besides highlighting the holistic, hereditary, collective, and custodial qualities of TEK (Bridgewater and Rotherham 2019), this concept also provides a more definite target for conservation efforts.

While TEK constitutes an inherently dynamic body of information that is updated over time as environmental parameters and signals change, historical longevity is another one of its dynamic properties (Zent 2013b). Intergenerational continuity depends on effective means of knowledge transmission, cultural reinforcement, and sustained utility. In many traditional knowledge settings, the learning process is characteristically informal, oral, observational, experiential (i.e., discovered by experiencing it for oneself), active (i.e., "learning by doing"), additive, sequential, and learner-directed (Zent 2009b; Lancey et al. 2010). Beyond these common characteristics, the exact mechanisms and rhythms of TEK transmission can be quite variable from place to place (Ruddle 1993). Such variation reflects the holistic nature of TEK transmission and its deep embeddedness in different spheres of a sociocultural system. In addition to peripheral participation in subsistence and domestic tasks, ecological knowledge may be passed down through a variety of channels and situations: storytelling, songs, proverbs, conversations, ceremonies, taboos, play, walkabouts, meals, arts and crafts, hygiene, and basically any common occurrence of daily life (Lave and Wenger 1991; Si 2020).

Despite its often deep historical roots, TEK can be very vulnerable to erosion or extinction. Well-documented reports of the rapid loss or simplification of local ethnoecological wisdom have become notoriously common and come from all over the globe. A variety of factors associated with cultural modernization, economic development, environmental change, and other large-scale processes have been identified as drivers of this degenerative trend (Zent 2013b). Less well known is how external factors of change disrupt internal processes of knowledge transmission (but see Cristancho and Vining 2009; Kodirekkala 2017; Turvey et al. 2018; Si 2020). Judging from the speed, severity, and ubiquity of TEK erosion in recent decades, we may surmise that this trend is very distinct from the local processes of change that are typically gradual and incremental in nature. Numerous analysts lament this catastrophic loss of intellectual capital because it is regarded as having much value for contemporary policy agendas, such as grassroots development, appropriate technology, food security, primary health coverage, climate change mitigation and adaptation, sustainable land management, and community-based conservation (cf. CBD 1992; Hassan et al. 2005; Vinyeta and Lynn 2013; WHO 2019). For the affected parties, TEK erosion often represents a loss of cultural distinctiveness and economic self-sufficiency. Another significant outcome is diminished use and value attached to local biodiversity, and hence lower incentive to preserve it. Therefore, a good place to begin biocultural conservation is by targeting the maintenance or revitalization of TEK.

Indigenous peoples are seen as crucial caretakers of global BCD because they are responsible for 2/3 of the languages and 80% of the biodiversity in the world today (FAO 2017; Eberhard et al. 2020). In particular, those who follow a traditional lifestyle and depend directly on their local environment for their livelihood and subsistence usually maintain a more extensive and intensive set of relationships with biodiversity—biocultural richness, if you will—than do westernized groups. Such place-based groups typically display an encyclopedic ethnoecological knowledge of their habitat as well as an ecocentric environmental ethic that emphasizes the moral rights and responsibilities of human and nonhuman creatures alike (Rozzi 2013). However, these populations also tend to be highly vulnerable to rapid cultural change and TEK erosion due to their small population size, economic marginality, and subaltern political status.

Without any external support system, Indigenous people may be unable to resist the corrosive cultural repercussions of exogenous change forces ranging from land invasions and forced resettlement, to market integration and urbanization, to religious colonization and modern consumerism, to formal education and language shift. In dynamic social contexts, where hybrid cultural realities have already begun to emerge, the question of how to achieve biocultural heritage conservation may require innovative approaches that seek to incorporate rather than shut out some of the transformations going on. It is probably not feasible to block or alter the drivers of change stemming from the extra-local context. One possibility for intervention at the local level would involve bolstering community processes of TEK transmission across generations (Zent 2009a; Marrie 2019). At least two general fields of action can be identified in this regard: (1) assessing the vitality of traditional means of transmission and (2) developing new mechanisms of transmission. Another strategy would be to foster collaboration between native and nonnative actors and integration of their respective knowledge systems to find co-managed conservation solutions (cf. Berkes and Turner 2006; Bussey et al. 2016).

14.2 Collaborative Action Research

Collaborative action research (CAR) focused on TEK transmission offers a promising approach for tackling the challenge of biocultural conservation. CAR is the terminology used here to refer to a broad body of work in which different stakeholders cooperate and work together to study and understand issues of mutual concern and then take informed measures to address those issues. This type of research approach goes by different names according to the discipline in which it is applied. Some of the more common ones are participatory action research, collaborative research, action anthropology, citizen science, and scholarly activism. One can detect differences between them, but the description given here focuses on the common aspects. The investigative process is frequently iterative, involving an ongoing cycle of observation, action, experience, and reflection (Bryant 1995; Fluehr-Lobban 2008; Stapp 2014; Rubinstein 2018).

Applied to an intercultural, community-based program setting, the notion of collaboration refers minimally to a partnership between local and scientific actors based on the principles of equality and complementarity. Collaborative projects often emerge naturally out of the prior experiences and relationships that have formed between the researcher(s) and study community. Partners are equal in the sense that they have vested interests in the program aims and participate actively in different phases from formulating research questions to collecting and analyzing data to generating products. They also share in decision-making through a democratic or consensus-building process. Participants are complementary in that each one brings different capabilities and fulfills different roles in the program execution. Local participants are the intermediaries between their communities and others, acting as gatekeepers that control and facilitate access to the local knowledge system. They are intimately familiar with community dynamics and norms, and are able to convey their constituents' wishes, concerns, and protocols. They also contribute labor to different tasks and coordinate whatever action plans are implemented at the local level. Scientific team members lend their expertise in conducting research, analyzing data, and communicating the results to relevant outsiders. They are the primary providers of research methods, training, resources, and information about the world around them.

The action component of CAR refers to the general goal of making social change to improve the lives of the study community(ies). In contrast to basic research, which is guided by academic questions and geared toward information extraction, action research is focused on assessing and ameliorating community-defined problems. The investigative process is largely concerned with elucidating the possible cause(s) and corrective(s) of those problems. The insights gained through systematic study serve as input for decision-making. The research design is also fundamentally experimental, open to considering additional matters based on the outcomes of actions taken. This type of research program is typically very long-term, often with no definite beginning or end. Below we illustrate the CAR approach by briefly describing some collaborative projects involving sustained cooperation between local communities of the Jotï Amerindian group of the Venezuelan Amazon and researchers from the Human Ecology Laboratory at the *Instituto Venezolano de Investigaciones Científicas* (Venezuelan Institute for Scientific Research) (HE-IVIC). A more detailed description of CAR projects carried out in recent years involving the same principal actors is also provided in Zent and Zent (2023) of this volume.

14.3 Jotï Biocultural Context

The Jotï are one of 52 ethnic groups of Indigenous origin presently found living in Venezuela (INE 2013). Their population stands at approximately 1800, spread among 25–30 local groups in northeastern Amazonas and southwestern Bolívar States (Fig. 14.1). Until 1969, they had no direct contact with non-Indigenous

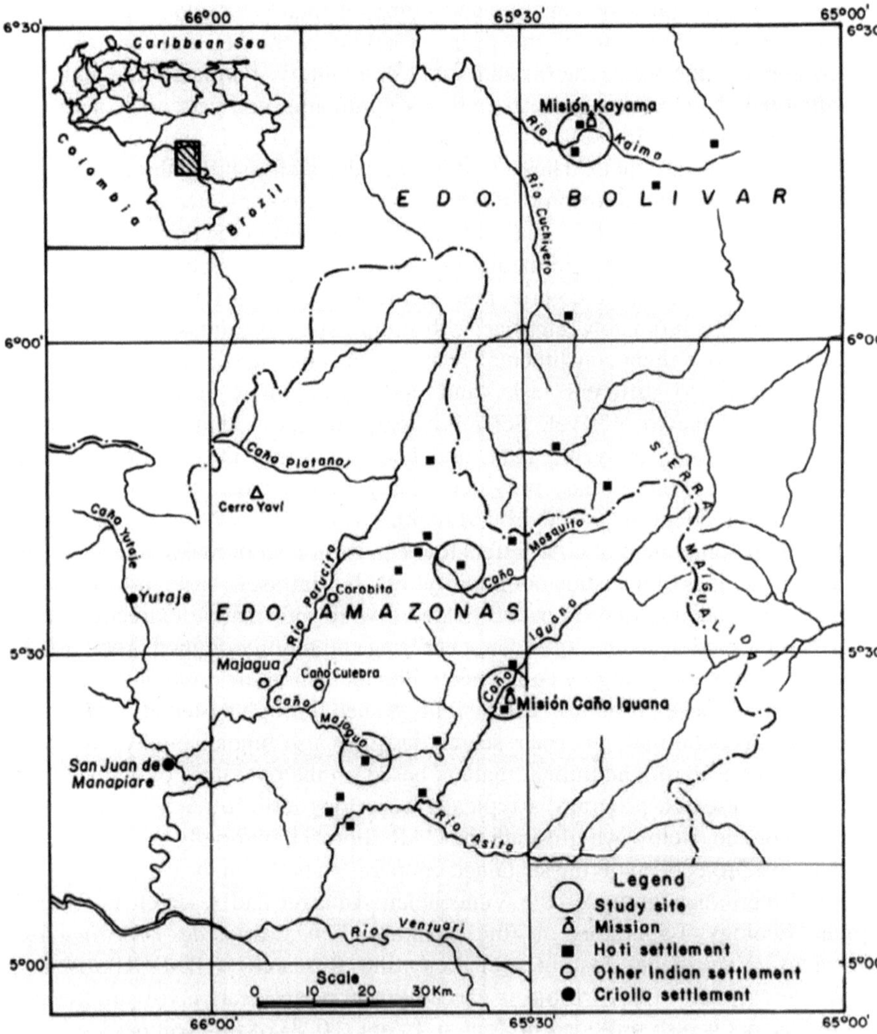

Fig. 14.1 Map of Venezuela (inset upper left), and the states ("estados," EDO) of Bolivar and Amazonas showing the location of Jotï settlements, other Indian settlements, the former missions of Caño Iguana and Kayama, and Creole settlements. Elaborated by Carlos Quintero and Stanford Zent, Anthropology Center, IVIC

people. But then foreign missionaries arrived and set up mission stations within the Jotï homeland, thus ending centuries of voluntary isolation. The first of these was established in 1970 at Caño Iguana in Amazonas state by US missionaries affiliated with the evangelical Christian organization New Tribes Mission (NTM). A second mission was founded in 1983 at the confluence of the Kayamá and Moyá rivers by Catholic nuns belonging to the Lauritas religious order based in Colombia. The Kayamá mission was originally built to serve isolated subgroups of the Eñepa (Panare) ethnic population but ended up attracting both Eñepa and Jotï followers. Over time, the mission settlements grew, drawing formerly dispersed, nomadic bands into their orbits of material and ideological influence.

Besides disseminating the Christian religion, the missionaries initiated a series of changes that would transform the Jotï cultural landscape forever: nucleated settlements, western trade goods, biomedicines, clothing, exotic crops, paper money, organized games, new rituals, literacy, schools, and the Spanish language, among others. Other outsider groups also began to show up on the scene in the wake of the missionaries (e.g., military personnel, government officials, pilots, foreign tourists, other Amerindians, adventurers, fortune-seekers, scientists), introducing their own peculiar sociocultural influences. Such changes were absorbed selectively and unevenly by the Jotï, such that some groups became more heavily acculturated and other groups much less so. The missionized Jotï were obviously exposed to greater contact and transcultural diffusion, and some individuals have spent considerable time in distant cities of the region or country. Meanwhile there are still nomadic bands in the uplands that have absolutely no contact with any outsiders except for sporadic visits to neighboring Amerindian groups. After two generations of intercultural contact, the Jotï still remain one of the most culturally- and linguistically-independent groups in lowland South America. The local language continues to be the universal mother tongue and three-quarters of the population are entirely monolingual. At the same time, the outside world has made some serious cultural inroads in certain areas.

The missionaries no longer maintain a physical presence in the immediate area. The missionaries at Caño Iguana (Amazonas) were expelled by the federal government in 2006 and those at Kayamá (Bolívar) pulled out of their own accord in 2016. Nevertheless the biocultural integrity and survival of the Jotï are currently threatened as never before. Land insecurity is a deep concern frequently mentioned by the tribal leaders. Although the national government issued a document explicitly recognizing the rights of the Jotï to occupy their traditional lands, such recognition refers to a single community in Amazonas state and covers only a minor fraction of the total land area currently occupied by them (see Zent et al. 2011). In the last decade, wildcat goldminers and armed guerrilla or paramilitary groups have invaded some parts of the Jotï territory. By contrast, the Venezuelan state military or security forces are nowhere to be seen. Illegal mining is responsible for deforestation, mercury contamination of water and fish, and expansion of the breeding grounds for disease-transmitting mosquitos. Health conditions have worsened in recent years due to a combination of the proliferation of environmental pathogens and the deterioration of the state-sponsored health care network. One indicator of this is the exponential growth of malaria morbidity and mortality (Zent and Zent 2019). After decades of concentrated settlement around the mission stations, wild resources are now depleted in those areas, thus creating nutritional stresses and pressure to intensify agricultural production to make up for the deficiencies. Access to western money and industrial products, present yet in short supply, has led to incipient socioeconomic stratification in some places. Another legacy of missionization is religious pluralism. There were two distinct religious orders that proselytized the Jotï, Latin American Catholics in the north and North American Protestants in the south, and each had their own pedagogical style and emphasis. The ideological differences they implanted have given rise to social clashes between and within certain communities. One of the primary effects of contact with outsiders has been erosion of traditional knowledge and customs. In addition to a dependence on imported technology, contact with outsiders introduced bilingualism and formal education. This acculturative process is especially evident in the ex-mission communities, where the knowledge gap between younger and older generations is most obvious (Zent 2009a).

14.4 From Conventional to Collaborative Action Research

Like most researchers working with relatively unacculturated peoples in remote areas, the authors' engagement with the Jotï began with a conventional research agenda. The initial research project entailed a field study of their ethnobiology and behavioral ecology in four communities that was carried out in 1996–99 (Zent 1999a). An explicit objective was to contribute to biocultural conservation by documenting the inventory, amounts, use values, densities, and distributions of subsistence resources. This information would then be used as input for a proposal to create a biosphere reserve or protected land area which would be large enough to sustain the population and their culture moving forward (Zent et al. 2001). While this research had a combined academic-applied purpose, it followed the conventional top-down model of research in the sense that the research design and intended products were defined exclusively by the researchers. A process of previous informed consent (PIC) to the best of our abilities was realized in all communities at the outset but, admittedly, these were limited by our rudimentary competence in the local language. Spanish speakers were present in only one of the four communities. In this type of context, PIC was really an extended process constructed day-by-day through living together, sharing meals and other things, social interaction, observation and participation in our activities, and improving language fluency. The project concluded with a report to the National Parks Institute and Ministry of Environment of the Venezuelan national government giving a detailed accounting of the natural resources used by the Jotï and a map of the proposed area that we recommended be set aside as a reserve for them. We never heard back whether this proposal ever had any impact on state environmental or *indigenista* policy or protected area programs.

This plan, as well as our whole approach to research, then changed because the Jotï had other ideas. Sometime in 2001, persons from the Kayamá community contacted us with a research proposal of their own: they wanted researchers from IVIC-HE to help them self-demarcate their land pursuant to claiming their territorial rights. That call represented a direct response to the new national constitution of 1999, which recognized the land rights of Venezuela's Indigenous nations. A short time later, laws were passed (the Indigenous Peoples' Lands and Habitat Guarantee and Demarcation Law and the Decree creating the National Commission on Demarcation of Indigenous Lands and Habitats, both in 2001) to create an institutional framework for implementing the constitutional mandate and to define the terms and conditions for filing a land claim. In view of this opportunity and considering their historical track record of coexisting with high levels of biodiversity, it made sense to place our conservation bets with direct Jotï ownership of the land rather than administrative protection by a distant governmental authority.

To make a long story short, we came to an agreement to work together toward achieving a collective land title or similar type of legal recognition by the Venezuelan state. Consequently, from 2001 to 2006, IVIC-HE researchers and members of several Jotï communities in both Amazonas and Bolívar states carried out a collaborative project which entailed doing all the groundwork and documentation needed to support the Jotï application (Fig. 14.2). This work included defining the territorial boundaries, mapping all sites and objects of cultural significance within those boundaries, and producing georeferenced maps. Ancillary activities involved the transfer of data recording and processing technology such as computers, GPS devices, and photographic cameras to the communities, along with the building of local capacities to use such instruments through on-site training. Additionally, we collected data about their origins and residential histories, patterns of land use and occupation, toponyms and place-based semantic constructs, land ethic and ecogonic philosophy. Ecogony refers to the origins or root causes of particular interactions between human groups and their local environments. As conceived from an Amerindian viewpoint, such causes are frequently holistic, which is to say simultaneously material, ideological, spiritual, and functional (Zent 2013a).

Work on the project eventually encompassed the active participation of hundreds of individuals of the Jotï population in one capacity or another. All of this material was put together to form the land claim application file along with other supporting documents and requirements specified under the law (e.g., full census, proof of identity documents, affidavits, life plan, basic cultural information). Following the established procedure, separate claims were filed in Amazonas and Bolívar states. Thus far, only the land claim in Amazonas has been legally certified, although the land area granted was less than 60% of the self-demarcated map area and refers only to the community of Caño Iguana. In Bolívar, even though 15 years have passed since the first filing date, the petitioners are still waiting for official resolution (Zent et al. 2004, 2011, 2016). A similar result (i.e., general lack of progress and unfulfilled expectations) describes the situation of Indigenous land rights vindication in the country in general (IACHR 2019: 51–52).

Fig. 14.2 (**a**) Jotï children lounging on a tree branch growing over the Jkutile River at a canoe embarking spot, Amazonas, 2012. (**b**) Jotï young boy climbing a tree to harvest edible fruit. (**c**) Family group collecting honey, Amazonas, 1998. (**d**) Neliwa caring for her niece, Amazon 2022. Photographs by E. and S. Zent

The self-demarcation project was just the beginning of our collaborative engagement with the Jotï. Other CAR projects that have involved us include the following:

(A) Elaboration of a consensus-based alphabet and other orthographic conventions (e.g., punctuation, spelling) for writing the Jotï language through a series of community-based workshops. The alphabet was then employed to compile a bilingual Jotï-Spanish dictionary (Quatra 2008) and other written pedagogical materials for use in the schools. Today it is widely used by a majority of the younger population possessing literacy skills and is a key resource for language documentation and the independent production of written texts in the local language.

(B) A descriptive account of Jotï understandings of health-related topics: wellness and illness, human anatomy and physiology, disease categories and healing

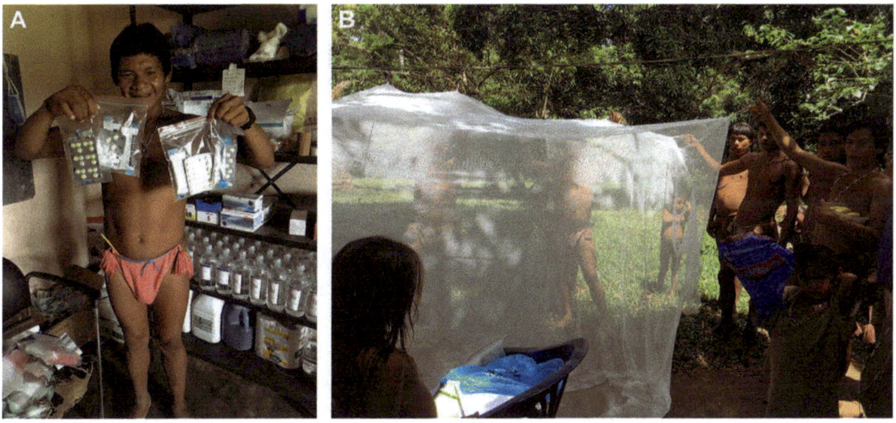

Fig. 14.3 Collaborative Project in Action: Mitigating malaria infection with insecticide-impregnated mosquito nets. (**a**) Bekato holding the malaria kit that he prepared (RDTs rapid diagnotic tests), treatments, indications, records. (**b**) Installing mosquito nets in the field. Amazonas, 2022. Photographs by Egleé Zent

(Zent and Zent 2007). This study was originally solicited by the Indigenous People's Health Coordinator of the Ministry of Health of Venezuela in order to create a bridge between ethnomedical and biomedical health care systems. The results were published in a set of volumes describing the ethnomedical beliefs of numerous different native groups within the country (Freire and Tillett 2007). It was intended for use by health care professionals working with Indigenous populations in order to facilitate better communication between providers and patients, and more culturally sensitive health care services.

(C) Facilitation of networking activities for leaders and delegates of the Jotï civil association *Jodena U̲*. Such networking is instrumental for establishing strategic connections and alliances with national and international agents. The range of contacts included government officials, scientists, educators, possible funding sources, and other Indigenous leaders and organizations. In 2010, one of the authors (Egleé) accompanied Jotï and Eñepa leaders to attend a meeting on participatory mapping and territorial management attended by 100 Amazonian Indigenous representatives, held in Rio Branco do Sul, Brazil (Goulart and Barretto Filho 2012). This type of activity forms part of IVIC-HE's ongoing commitment to advocate on behalf of the Jotï with respect to their human rights, land claims, health care, education, community development, self-determination, and other matters.

(D) Attention to ongoing medical emergencies. In recent years, these include the dramatic increase in malaria incidence among the Jotï, Huottöja, and Eñepa in the Venezuelan Amazon (Fig. 14.3). This work has involved fund-raising and delivering medicines and other critical medical supplies to different communities. Among the assistance provided were insecticide-treated mosquito nets delivered in 2019 to every man, woman, and child in both southern (Amazonas) and northern (Bolívar) population blocs (McCandless 2019). This initiative included setting up a monitoring system to measure the impact of the nets on infection rates.

The collaborative projects and results recounted above were made possible only through sustained dialogue, idea exchange, and cooperative labor between the local and scientific actors. The research process included critical assessments of the acculturation process and its effects on the Jotï population, environment, culture, health, resilience, and sovereignty moving forward. The activities undertaken were aimed toward enhancing the Jotï capacity to resist or adapt to ongoing threats in line with their own wishes and goals, which also include embracing aspects of change that they perceive as being positive. Effective CAR is ultimately based on mutual trust that is established through long-term relationship between the cooperating parties. Another crucial factor is the prioritizing of local perspectives on how to evaluate and address diagnosed problems. This type of intercultural scientific engagement contrasts significantly with the more conventional research programs that treat local actors as merely passive objects or sources of information and are focused more on obtaining data for academy-centered research agendas. The concept of biocultural conservation was a common denominator throughout this extended and multi-chapter history. The position taken here is that the multidimensional and complex nature of biocultural diversity and heritage demands a more interactive and dialogical participatory perspective, as embodied in the CAR approach.

14.5 Discussion

The emerging paradigm of biocultural heritage conservation, predicated on the mutualism of memes and genes, presents the challenge of developing more holistic and pluralistic solutions to the problem of sustainability and social-environmental justice (Rozzi 2013, 2018). As more and more researchers, practitioners, and policymakers embrace this philosophy, there has been a virtual explosion in recent years of different plans and proposals seeking to integrate cultural and biological variables into conservation actions (Gavin et al. 2015; Wali et al. 2017; McCarter et al. 2018; Hanspach et al. 2020). One discernible trend in this regard is the emphasis on decentralized, context-specific, community-based, participatory approaches (Maffi and Woodley 2010). A common feature in many of these is the importance assigned to safeguarding traditional and local ecological knowledges (TEK and LEK). TEK/LEK is directly relevant for maintaining ecological stability and resilience inasmuch as it is the basis for decision-making at the local level in regards to livelihood, food production, primary health care, dwelling construction, education, social organization, rights allocation, and natural resource management in general (Warren 1991). It also has a strong impact on identity, social solidarity, values, feelings of well-being, self-determination, and other intangibles that exert an influence on people's disposition or capacity to regulate resource exploitation patterns (UNPFII 2019; Marrie 2019).

Although the preservation and promotion of TEK occupy a prominent place in recent academic and policy discourses on biocultural conservation, the road from theoretical recognition to empirical practice is still not very straight and smooth. In socio-ecological contexts where TEK is subject to rapid change or endangerment, the issue of *in situ* transmission looms large as a focal point for action. For example, the United Nations Permanent Forum on Indigenous Issues (UNPFII) selected traditional knowledge generation, transmission, and protection as the main theme of discussion during its Eighteenth Session (UNPFII 2019). Yet effective action depends on a good understanding of the dynamics and drivers of TEK change at the local level. While a spate of recent research has focused on this very issue (Zent 1999b; Ross 2002; Reyes-García et al. 2015), if there is any generalization that can be made from this body of work, it is that no two situations look alike.

In view of the incredible variability of local contexts, cultures, and concerns, a site-specific approach to adaptive management is pressingly needed (Wali et al. 2017). It has been suggested that co-management regimes involving multi-stakeholder, multi-perspective, cross-scale cooperation offer a promising path forward for sustainability science and practice (Bussey et al. 2016; Garavito-Bermúdez 2020). In order to turn this vision into action, CAR will be a necessary ingredient. CAR based on dialogue, information-sharing, and the respectful collaboration among local knowledge holders, academic researchers, and potentially other institutional actors, represents a useful tool for adapting conventional conservation measures to particular local conditions. Participation in CAR programs is also a source of empowerment for community groups in that it gives them experience in dealing with outsiders and can open the door for new political alliances and collaborations to promote their rights, resources, and resilience (Berkes 2004). The use of TEK indicators can play a useful role in this type of cooperative undertaking by sharpening understanding of the complex and dynamic local processes of TEK transmission and facilitating communication between partners whose backgrounds and worldviews are very different (Sterling et al. 2017).

The application of a collaborative action approach to biocultural conservation has an important educational dimension that should be highlighted. CAR resonates with Field Environmental Philosophy (FEP), considering the latter's emphasis on transdisciplinary research, active participation by different stakeholders, including local communities, and support of educational designs rooted in local historical, sociocultural, and biophysical realities (Tauro et al. 2021). It is also consistent with emerging perspectives in education science that promote more holistic, locally appropriate learning processes in contexts where "people have nature as a workplace" (Garavito-Bermúdez 2020) or those which seek to overcome the systemic disconnect between formal educational curriculums and Local Ecological Knowledge (LEK)/Traditional Ecological Knowledge (TEK)/Indigenous Ecological Knowledge (IEK) systems in heavily colonized regions (Ezeanya-Esiobu 2019). Both CAR and FEP approach the challenge of sustainability as a problem of maintaining or reinforcing biocultural connections in the face of rapid-fire change in which accurate and appropriate knowledge is crucial for taking effective actions. A biocultural perspective on education puts the focus on the multidimensionality of learning landscapes, both in space, incorporating the knowledges from diverse cultural sources, and in time, looking to integrate traditional and modern understandings.

References

Baer A (1989) Maintaining biocultural diversity. Conservat Biol 3(1):97–98

Berkes F (2004) Rethinking community-based conservation. Conservat Biol 18(3):621–630

Berkes F (2008) Sacred ecology: traditional ecological knowledge and resource management. Taylor & Francis, Philadelphia

Berkes F, Turner NJ (2006) Knowledge, learning and the evolution of conservation practice for social-ecological system resilience. Hum Ecol 34(4):479–494

Bridgewater O, Rotherham ID (2019) A critical perspective on the concept of biocultural diversity and its emerging role in nature and heritage conservation. People Nat 1:291–304

Bryant P (1995) Collaborative action research "on the cutting edge". Master in education thesis. The University of Lethbridge, Lethbridge, AB

Bussey J, Davenport MA, Emery MR, Carroll C (2016) "A lot of it comes from the heart": The nature and integration of ecological knowledge in tribal and nontribal forest management. J Forest 114(2):97–107

Convention on Biological Diversity (CBD) (1992) Convention Text. CBD Secretariat, United Nations Environment Programme. Available at: http://www.cbd.int/convention/articles/

Cristancho S, Vining J (2009) Perceived intergenerational differences in the transmission of traditional ecological knowledge (TEK) in two indigenous groups from Colombia and Guatemala. Culture Psychol 15(2):229–254

Descola P, Palsson G (1996) Nature and society: anthropological perspectives. Routledge, London, UK

Eberhard DM, Simons DF, Fenning CD (eds) (2020) Ethnologue: languages of the world, 23rd edn. SIL International, Dallas, TX. Online version: http://ww.ethnologue.com

Ezeanya-Esiobu C (2019) Indigenous knowledge and education in Africa. Springer Open. eBook: https://doi.org/10.1007/978-981-13-6635-2

Fluehr-Lobban C (2008) Collaborative anthropology as twenty-first-century ethical anthropology. Collab Anthropol 1:175–182

Food and Agriculture Organization (FAO) (2017) 6 ways indigenous peoples are helping the world achieve #ZeroHunger. Available at: http://www.fao.org/zhc/detail-events/en/c/1028010/

Freire G, Tillett A (eds) (2007) El estado de la salud indígena en Venezuela, vols. I–II. Coordinación Intercultural de Salud de los Pueblos Indígenas (CISPI), Ministerio de Salud y Desarrollo Social, Caracas, VE

Garavito-Bermúdez D (2020) Biocultural learning—beyond ecological knowledge transfer. J Environ Plan Manag 63(10):1791–1810

Gavin MC, McCarter J, Mead A, Berkes F, Stepp JR, Peterson D, Tang R (2015) Defining biocultural approaches to conservation. Trends Ecol Evol 30(3):140–145

Goulart A, Barretto Filho HT (2012) Mapeamentos Participativos e Gestaô de Territórios Indígenas na Amazônia. America Indígena, Brasília, BR. https://www.academia.edu/5694281/Anais_do_Semin%C3%A1rio_Internacional_Mapeamentos_Participativos_e_Gest%C3%A3o_de_Territ%C3%B3rios_Ind%C3%ADgenas_na_Amaz%C3%B4nia_Legal_organizador

Hanspach J, Haider LJ, Oteros-Rozas E, Stahl Olaffson A, Gulsrud NM, Raymond CM, Torralba M, Martín-López B, Bieling C, García-Martín M, Albert C, Beery TH, Fagerholm N, Díaz-Reviriego I, Drews-Shambroom A, Plieninger T (2020) Biocultural

approaches to sustainability: A systematic review of the scientific literature. People Nat 2:643–659
Hassan R, Scholes R, Ash N (eds) (2005) Ecosystems and human well-being: current state and trends, Volume 1. Findings of the condition and trends working group of the millennium ecosystem assessment. Island Press, London, UK
Ingold T, Palsson G (2013) Biosocial becomings: integrating social and biological anthropology. Cambridge University Press, Cambridge, UK
Instituto Nacional de Estadística (INE) (2013) La población indígena de Venezuela. Boletín 1(1): 1–15. XIV Censo de población y vivienda 2011. Gerencia general de estadísticas demográficas
Inter American Commission of Human Rights (IACHR) (2019) Situation of human rights of the indigenous and tribal peoples of the Pan-Amazon region. ISBN 978-0-8270-6931-2. Available at: http://www.oas.org/en/iachr/reports/pdfs/Panamazonia2019-en.pdf
International Institute for Environment and Development (IIED) (2021) Biocultural heritage: an evolving concept. Available at: https://biocultural.iied.org/evolving-concept
Kodirekkala KR (2017) Internal and external factors affecting loss of traditional knowledge: evidence from a horticultural society in South India. J Anthropol Res 73(1):22–42
Lancey DF, Bock J, Gaskins S (eds) (2010) The anthropology of learning in childhood. AltaMira, Walnut Creek, CA
Lave J, Wenger E (1991) Situated learning: situated peripheral participation. Cambridge University Press, Cambridge, UK
Maffi L (2005) Linguistic, cultural, and biological diversity. Annu Rev Anthropol 34:599–617
Maffi L, Woodley E (2010) Biocultural diversity conservation: a global sourcebook. Earthscan, London, UK
Marrie H (2019) Emerging trends in the generation, transmission and protection of Traditional Knowledge. Paper prepared for the Eighteenth Session of the United Nations Permanent Forum on Indigenous Issues, April-May 2019, New York. https://www.un.org/development/desa/indigenouspeoples/wp-content/uploads/sites/19/2019/04/TK-Emerging-trends-in-the-generation-transmission-and-protection-of-TK-final-paper.pdf
McCandless S (2019) A local project with global support helps reduce malaria-related deaths. Available via https://www.global-diversity.org/a-local-project-with-global-support-helps-reduce-malaria-related-deaths/. Accessed 25 Apr 2021
McCarter J, Sterling EJ, Jupiter SD, Cullman GD, Albert S, Basi M, Betley E, Boseto D, Bulehite ES, Harron R, Holland PS, Horning N, Hughes A, Jino N, Malone C, Mauli S, Pae B, Papae R, Rence F, Revo O, Taqala E, Taqu M, Woltz H, Filardi CE (2018) Biocultural approaches to developing well-being indicators in Solomon Islands. Ecol Soc 23(1):32. https://doi.org/10.5751/ES-09867-230132
Posey DA (ed) (1999) Cultural and spiritual values of biodiversity. United Nations Environment Program. http://www.unep.org/Biodiversity/unep.pdf
Pretty J, Ball A, Benton T, Guivant J, Lee DR, Orr D, Pfeffer M, Ward H (2007) The SAGE handbook of environment and society. Sage, Thousand Oaks, CA
Pretty J, Adams B, Berkes F, Ferreira de Athayde S, Dudley N, Hunn E, Maffi L, Milton K, Rapport D, Robbins P, Sterling E, Stolton S, Tsing A, Vintinner E, Pilgrim S (2009) The intersections of biological diversity and cultural diversity: towards integration. Conservat Soc 7(2):100–112
Quatra M (2008) Bajkewa Jkwĩkĩdẽwa-Jya. Diccionario Básico Castellano-Jodï. Ediciones IVIC, Caracas, VE
Redford KH, Brosius PJ (2006) Diversity and homogenization in the endgame. Global Environ Change 16:317–319
Reyes-García V, Broesch J, TAPS Bolivian Study Team (2015) The transmission of ethnobotanical knowledge and skills among the Tsimane' in the Bolivian Amazon. In: Ellen RK, Lycett SJ, Johns SE (eds) Understanding cultural transmission in anthropology: a critical synthesis. Methodology and history in anthropology series, vol 26. Berghahn Books, Oxford, UK, pp 191–212

Ross N (2002) Lacandon-Maya intergenerational change and the erosion of folkbiological knowledge. In: Stepp J, Wyndham FS, Zargar RK (eds) Ethnobiology and biocultural diversity. International Society of Ethnobiology, Athens, GA, pp 585–592

Rotherham ID (2015) Bio-cultural heritage and biodiversity: emerging paradigms in conservation and planning. Biodivers Conservat 24:3405–3429

Rozzi R (2013) Biocultural ethics: from biocultural homogenization toward biocultural conservation. In: Rozzi R, Pickett STA, Palmer C, Armesto JJ, Callicott JB (eds) Linking ecology and ethics for a changing world, Ecology and ethics, vol 1. Springer, Dordrecht, pp 9–32

Rozzi R (2018) Biocultural homogenization: a wicked problem in the Anthropocene. In: Rozzi R, May RH Jr, Chapin FS III, Massardo F, Gavin M, Klaver I, Pauchard A, Nuñez MA, Simberloff D (eds) From biocultural homogenization to biocultural conservation, Ecology and ethics, vol 3. Springer, Dordrecht, pp 21–47

Rozzi R, Massardo F, Anderson C, Heidinger K, Silander JA Jr (2006) Ten principles for biocultural conservation at the southern tip of the Americas: The approach of the Omora Ethnobotanical Park. Ecol Soc 11(1):43. https://doi.org/10.5751/ES-01709-110143

Rubinstein R (2018) Action anthropology. In: Callan H (ed) The international encyclopedia of anthropology. Wiley, Oxford, UK. https://doi.org/10.1002/9781118924396.wbiea2230

Ruddle K (1993) The transmission of traditional ecological knowledge. In: Inglis JT (ed) Traditional ecological knowledge: concepts and cases. International Program on Traditional Ecological Knowledge, International Development Research Centre, Ottowa, ON, pp 17–31

Si A (2020) Patterns in the transmission of traditional ecological knowledge: a case study from Arnhem Land, Australia. J Ethnobiol Ethnomed 16:52. Available at: https://doi.org/10.1186/s13002-020-00403-2

Stapp DC (2014) Action anthropology. In: Coghlan D, Brydon-Miller M (eds) The SAGE encyclopedia of action research. Sage, Los Angeles, CA, pp 4–5

Sterling E, Ticktin T, Morgan TKK, Cullman G, Alvira D, Andrade P, Bergamini N, Betley E, Burrows K, Caillon S, Claudet J, Dacks R, Eyzaguirre P, Filardi C, Gazit N, Giardina C, Jupiter S, Kinney K, McCarter J, Mejia M, Morishige K, Newell J, Noori L, Parks J, Pascua P, Ravikumar A, Tanguay J, Sigouin A, Stege T, Stege M, Wali A (2017) Culturally grounded indicators of resilience in social-ecological systems. Environ Soc Adv Res 8:63–95

Tauro A, Ojeda J, Caviness T, Moses K, Moreno R, Wright T, Zhu D, Poole A, Massardo F, Rozzi R (2021) Field environmental philosophy: a biocultural ethic approach to education and ecotourism for sustainability. Sustainability 13(8):4526. https://doi.org/10.3390/su13084526

Toledo VM (2002) Ethnoecology: a conceptual framework for the study of indigenous knowledge of nature. In Stepp J, Wyndham F, Zarger R (eds) Ethnobiology and biocultural diversity: proceedings of the 7th international congress of ethnobiology, Athens, Georgia, USA, October 2000. International Society of Ethnobiology, pp 511–522

Turvey ST, Bryant JV, McClune KA (2018) Differential loss of components of traditional ecological knowledge following a primate extinction event. Roy Soc Open Sci 5:172352. Available at: https://doi.org/10.1098/rsos.172352

UNPFII (2019) Traditional knowledge: generation, transmission and protection. Discussion paper for the Eighteenth Session of the United Nations Permanent Forum on Indigenous Issues. New York, 22 April–3 May 2019. Available via https://www.un.org/development/desa/indigenouspeoples/unpfii-sessions-2/18-2.html. Accessed 1 May 2021

Vinyeta K, Lynn K (2013) Exploring the role of traditional ecological knowledge in climate change initiatives. General Technical Report PNW-GTR-879. U.S. Department of Agriculture, Forest Service, Pacific Northwest Research Station, Portland, OR, 37 p. Available at: https://www.fs.fed.us/pnw/pubs/pnw_gtr879.pdf

Wali A, Alvira D, Tallman P, Ravikumar A, Macedo MO (2017) A new approach to conservation: using community empowerment for sustainable well-being. Ecol Soc 22(4):6. https://doi.org/10.5751/ES-09598-220406

Warren D M (1991) Using indigenous knowledge in agricultural development. World Bank Discussion Paper 127. World Bank, Washington, DC
World Health Organization (WHO) (2019) WHO global report on traditional and complementary medicine 2019. World Health Organization, Geneva. Available at: https://www.who.int/traditional-complementary-integrative-medicine/WhoGlobalReportOnTraditionalAndComplementaryMedicine2019.pdf?ua=1
Zent E (1999a) Etnobotánica Hotï: Explorando las interacciones entre la flora y el ser humano del Amazonas Venezolano. PhD dissertation. University of Georgia, Athens, GA
Zent S (1999b) The quandary of conserving ethnoecological knowledge: a Piaroa example. In: Blount B, Gragson T (eds) Ethnoecology: knowledge, resources, and rights. University of Georgia Press, Athens, GA, pp 90–124
Zent S (2009a) Traditional ecological knowledge (TEK) and biocultural diversity: a close-up look at linkages, delearning trends, and changing patterns of transmission. In: Bates P, Chiba M, Kube S, Nakashima D (eds) Learning and knowing in indigenous societies today. UNESCO, Paris, FR, pp 39–58
Zent S (2009b) A genealogy of scientific representations of indigenous knowledge. In: Heckler S (ed) Landscape, process and power: a new environmental knowledge synthesis, Studies in environmental anthropology and ethnobiology, vol 10. Berghahn, Oxford, UK, pp 19–67
Zent E (2013a) Jodï Ecogony, Venezuelan Amazon. Environ Res Lett 8:015008. http://stacks.iop.org/1748-9326/8/015008
Zent S (2013b) Processual perspectives on traditional environmental knowledge: continuity, erosion, transformation, innovation. In: Ellen RK, Lycett SJ, Johns SE (eds) Understanding cultural transmission in anthropology: a critical synthesis, Methodology & history in anthropology series, vol 26. Berghahn, Oxford, UK, pp 213–265
Zent E, Zent S (2007) Los Jodï. In: Freire G, Tillett A (eds) El estado de la salud indígena en Venezuela. Coordinación Intercultural de Salud de los Pueblos Indígenas (CISPI), Ministerio de Salud y Desarrollo Social, Caracas, VE, pp 77–130
Zent E, Zent S (2019) "Malaria among the indigenous people of Venezuela: the Jotï of Amazonas and Bolívar States." Symposium: malaria resurgence in Venezuela and its regional implications, American Society for Topical Medicine & Hygiene 68th annual meeting, Gaylord National Resort and Convention Center, National Harbor, MD, USA, 23 Nov 2019
Zent S, Zent E (2023) Collaborative action research with the Jotï in Venezuela: experiences in autoethnography and TEK vitality assessment. In: Rozzi R, Tauro A, Wright T, Avriel-Avni N, May Jr. RH (eds) Field environmental philosophy: education for biocultural conservation. Ecology and ethics, vol 5. Springer, Dordrecht, pp 229–246
Zent S, Zent E, Marius L (2001) "Informe Final del Proyecto 'Etnobotánica Cuantitativa de los Indígenas Hotï de la Región Circum-Maigualida, Estados Amazonas y Bolívar, Venezuela'". Research project report prepared for Consejo Nacional de Investigaciones Científicas y Tecnológicas (CONICIT), Ministerio de Ciencia y Tecnología, Republica Bolivariana de Venezuela. 275 pp
Zent E, Zent S, Marius L (2004) Autodemarcando la Tierra: Explorando las Ideas, los Árboles y Caminos Hotï. Boletín Antropológica 58(2):313–338
Zent S, Zent E, Juae Molo L (2011) Un largo y sinuoso camino: breve historia de la autodemarcación territorial Jodï en su décimo aniversario. In: Bello LJ (ed) El Estado ante la Sociedad Multiétnica y Pluricultural: Políticas Públicas y Derechos de los Pueblos Indígenas en Venezuela (1999-2010). Copenhagen, DK, IWGIA, pp 97–117
Zent S, Zent E, Juae Mölö L, Chonokó P (2016) Reflexiones sobre el Proyecto Auto-Demarcación y EtnoCartografía de las Tierras y Hábitats Jodï y Eñepa. Revue d'ethnoécologie [En ligne] 9. Available at: http://ethnoecologie.revues.org/2670

Chapter 15
Education as a Driver of Extinction of Experience or Conservation of Biocultural Heritage

Alexandria K. Poole

Abstract The concept of extinction of experience has increasingly garnered attention in environmental education literature. "Extinction of experience" (EoE) is a neologism articulated by nature writer and lepidopterist Robert Michael Pyle to capture the somewhat intangible loss that occurs when biodiversity is removed from key experiences in our daily lifeworld, and it refers to the cultural and experiential loss that ultimately occurs following the abstention of nature experience. In this paper, I introduce Pyle's landmark concept and propose that it has significant implications as an additional indirect driver within formal education. With the increasing loss of local species, the rapid extinction crises, and the impacts of climate change shifting ecological systems, there is significant loss and disruption of ecological communities. I argue that EoE is an indirect driver of biodiversity losses. Within formal school settings, knowledge of biodiversity losses and knowledge to co-exist with biodiversity in sustainable ways are not adequately addressed. Therefore, formal education contributes to losses of local ecological knowledge and nature experiences and undermining biocultural heritage. To reverse this trend, it is necessary to identify key mechanisms within formal education that can serve as drivers to protect, promote, and engage biocultural heritage. This approach can also be applied to consider ways to remediate processes that would otherwise drive EoE within dominate practices in our society for biocultural conservation.

Keywords Formal education · Informal education · Environmental education · Conservation · Indirect drivers · Urban

---

A. K. Poole (✉)
Faculty of Behavioural, Management and Social Sciences (Philosophy), University of Twente, Enschede, The Netherlands
e-mail: a.k.poole@utwente.nl

R. Rozzi et al. (eds.), *Field Environmental Philosophy*, Ecology and Ethics 5,
https://doi.org/10.1007/978-3-031-23368-5_15

15.1 Introduction: The Need to Address the Extinction of Experience Within Formal Learning Experiences

The concept of extinction of experience has increasingly garnered attention within environmental education and related fields concerned with conservation of biodiversity, culture, or language, and their interdynamics such as conservation biology, ethnoecology, and biocultural conservation. A neologism articulated by nature writer and lepidopterist Robert Michael Pyle (1978) to capture the somewhat intangible loss that occurs when biodiversity is removed from key experiences in our daily lifeworld, extinction of experience (EoE) refers to the cultural, psychological, and personal lacking that ultimately occurs following the abstention of nature experience. The complex relationships that humans share with nature in everyday life in relation to the form of engagement, of course, depends upon the technologies, access, and habits that those communities have with the non-human world. However, the concept has been adopted by practitioners ranging from urban environmental educators to indigenous movements.

There are many vectors which influence the ways in which we engage the world, including agriculture, economy, landscapes, information technologies, media, and education. When these systems drive us away from engagements with the natural world and the ecological systems that produce food, air, water, and the rich biodiversity of life, human cultures can become unaware of biodiversity losses, and ecological disruption follows. This is perhaps most striking in the climate change denial movements seen in the USA. This denial of climate change and biodiversity losses has profound consequences if it becomes permanent because species and nature experiences will be impossible as they disappear entirely.

In this chapter, I discuss the concept of "extinction of experience" which refers to the corresponding experiential loss occurring as a result of this significant disruption of ecological systems (Randall 2009; Lawler et al. 2006; McKinney 2006; MA 2003; Poole 2015) and the experience of the absence the remains behind (Pyle 1993, 2001, 2014). I propose that EoE is an indirect driver of biodiversity loss. The lack of knowledge about this loss and the loss of knowledge to co-exist with biodiversity in sustainable ways exacerbate biodiversity losses. Classically expressed in terms of local ecological (LEK) or traditional ecological knowledge (TEK), the significance of the loss of the deep ecological knowledge for sustainability and environmental ethics needs to be understood for maintaining sustainable societies (Berkes et al. 2000). Therefore, vectors that control spaces in which knowledge, engagement, and experience of nature are drivers of EoE that should be acknowledged in education. This is particularly important as we engage new innovations to address climate change and sustainability innovations such as artificial intelligence, synthetic biology, geo-engineering, and other socially disruptive technologies that have the potential to greatly shift cultural and ecological systems.

The loss of nature experience impacts cultural, psychological, and ethical aspects of ecological knowledge and will further exacerbate the loss and disregard for biodiversity and ecological deterioration within culture and values (Kimmerer

2002; Berkes et al. 2000; Miller 2005; Orr 1996, 2004). Emerging literature on the extinction of experience has focused on urban environments and citizen science, often overlooking the importance of agrarian communities and indigenous interrelations with nature. These communities also experience EoE in unique ways that should be understood and recognized. In this chapter, I articulate ways in which formal education and spaces of learning and knowledge transmission can remediate EoE across the urban-rural gradient by recognizing the role that biocultural heritage and local ecological knowledge play in understanding a sense of place. To address this anthropogenic gradient, I discuss the EoE concept in terms of biocultural heritage; that is, in the context of its interrelation to the culture, language, and biodiversity of the complex landscape in which human-nature relationships take place and methods to remediate these trends within education (Gavin et al. 2015; Poole 2018).

The book chapter introduces the section on education for the fourth volume of the Springer *Ecology and Ethics* series "Field Environmental Philosophy: Education for Biocultural Conservation." By discussing the implications for education, I hope to focus on education policy as a driver for change and ways we may protect our communities in systematic ways for environmental values, our relationships to nature, and capacity to better cultivate our role in sustainability cultures.

15.2 The Extinction of Experience Cycle as an Indirect Driver of Change: Learning from Where We Live

Misleadingly, urban spaces and the built environment are often presented as vacant for nature experience, largely because of the dominance of technology, and the prevalence of weedy or unwanted vegetation (Poole 2019; Colléony et al. 2017). On the one hand, when nature is present, a common argument is that it is highly managed, often considered to be a simulacrum of true complex "wild nature," an impossible version of the complexity of biodiversity, and therefore, a false image of the much larger natural world beyond the city. On the other hand, the nature that is present is often exotic, or ornamental, but rarely the native or endemic species that originally occurred in the region. How then, does the urban dweller gain experience of the natural world, or an understanding of its workings? And as more and more of the human population moves to urbanized settings, what is the overall impact of these settings on the environmental consciousness of human cultures and values?

The first use of the term "extinction of experience" appeared in Pyle's short essay originally printed in the popular magazine *Horticulture* (Pyle 1978). Pyle's inaugural introduction of the concept argued that the loss of *opportunities* of experience can have irreversible consequences for knowledge and care of nature as a personal experience, and even as a cultural phenomenon. Further, this lacking exacerbates trends to reduce local instances of biodiversity through management practices, continuing a cycle of biodiversity loss as the losses becomes unnoticed, because community members lack the knowledge to note the absence.

> The majority of extinctions involves not the eradication of species on a global scale, but the disappearance of portions of them wherever we look. A colony goes extinct here, a subspecies drops out there—two varieties of butterflies vanish from the High-line Cana. They add up, and the consequence is a drastically undermined flora and fauna. It seems to me that the impact of these partial extinctions upon our natural base and collective psyche, as well as our ability to withstand future assaults on the environment, are very grand indeed. In the long run they may affect more of that than will the disappearance of entire species. For what these local extinctions represent is the loss of opportunities—the extinction of experience (Pyle 1978, p. 56).

> So it goes, on and on, the extinction of experience sucking the life from the land, the intimacy from our connections. This is how the passing of otherwise common species from our immediate vicinities can be as significant as the total loss of rarities. People who care conserve; people who don't know don't care. What is the extinction of the condor to a child who has never known a wren? (Pyle 1993, p. 146–147).

In the second iteration, *The Thunder Tree: Lessons from an Urban Wildland,* published almost 20 years later, Pyle (1993, 146–147) tells the story of the moment he discovers that a beloved woodland has been cut down for a parking lot. Here he describes this personal and intimate moment when his sense of place is also harmed, as it is not only species that are displaced, but an entire biotic community. Pyle's emphasis is that an extinction of experience can occur when opportunities for engagement are lost, whether it be a physical or conceptual barrier, or the physical disappearance of an endemic species, that is, the absence of its *local* presence (as opposed to an extinction wide event). Pyle argues that EoE is caused by loss of individual experience of biodiversity in one's own neighborhood. The young, elderly, poor, or disabled are particularly vulnerable to local biodiversity losses as they do not have the capacity to move beyond their home range to seek out nature experiences. While the loss of local biodiversity has evolutionary and ecological consequences, it has experiential consequences for the human community, too: "a different type of depletion," because "the loss of neighborhood species endangers our experience by nature" (Pyle, 1993, p. 260).

Pyle's initial definition regarding the extinction of experience clarifies the difference between (1) an actual extinction and disappearance of the thing itself, so that no experience of it is possible, and (2) the loss of certain lifestyle practices or habits or types of engagements which allow for the presence or absence of an experience with that thing. In the case of biodiversity extinction, our range of experiences with a species might change depending on its physical presence in a particular region, or a change in our habits to such an extent that we do not engage that being. It is important to explicitly add the justice dimension that biodiversity is being exterminated and landscapes decimated. Further, this becomes a self-perpetuating cycle as experiences with nature diminish and understanding and emphasis on nature decrease, exacerbating the tendency to create conditions in which nature experiences are unlikely. This estrangement has become so systematic that after lasting beyond two generations Peter Kahn described the cultural loss as a kind of "environmental generational amnesia" in which "children who know about pollution in general and live in a polluted city [are] unaware of their own city's pollution" (Kahn and Kellert 2002; Kahn 2002).

What Pyle's discussion shows is that urban and built environments are not biodiversity deserts, but complex spaces where engagement with nature and natural wonder is indeed quite possible. However, ecological literacy and understanding of the environment along urban-rural gradients and associated complex ecological knowledge systems, will be undermined without proper educational support. Additionally, these experiences are tied to ethical and emotional associations with nature, as learning is not an ethically or emotionally neutral process (Pyle 1993; Leopold 2004; Kahn and Kellert 2002; Noddings 2013; Callicott 1987).

Environmental education requires an understanding of landscape and place as educational spaces, in which a human community learns about the non-human world. Michael J. Samways (2007) citing Miller (2005), drew attention to this and proposed that urban society lives on the edge of the "real and virtual world." Samways and Miller popularized Pyle's EoE concept within conservation biology and environmental education. There is a surprisingly diverse use of this term across conservation, ethnoecology, environmental education, and even within popular culture blogs and magazines (Maffi 2001; Miller 2005; Stokes 2006; Soga and Gaston 2016). Considering the term's deictic usage, it can generally be understood to refer to the abstention of experience with complex nature, the reduction of green spaces, biodiversity, or non-human life. Pyle (2003, p. 206) described EoE as an "inexorable cycle of disconnection, apathy and progressive depletion" (Fig. 15.1). —Pyle's cycle of EoE is spiraling as opportunities for positive experiences with an intact nature are lost, and its presence in urban or consumer life discounted. Because of this, Pyle urged us to remember that any natural experience can be life changing, and that even the smallest and most "humble" habitats can fulfill the role of awakening a child's care for the environment.

Pyle (1993, p. 262) emphasizes that "[t]he extinction of experience is not just about losing the personal benefits of the natural high. It also implies a cycle of disaffection that can have disastrous consequences." Hence, EoE self-perpetuates; this is a point that is vital for environmental education, and education more generally. Pyle points out that nature conservation is not only about beneficial experiences a child might have with nature, but is also key for the conservation of biodiversity itself. Consequently, we should ask two pedagogical questions: How can we conserve these experiences? How can we protect the body of knowledge that reflects these fundamental experiences?

While never presented by Pyle as a philosophically developed theory, the extinction of experience (EoE) has remained a device used by environmental educators and conservationists to describe this social phenomenon of experiential loss. This concept has gained popularity in use by the general public as well, appearing in public lectures and inspiring Richard Louv's "No Child Left Insides" social movement in the USA and expressed in House Bill "No Child Left Inside" which has been proposed numerous occasions, though an iteration has yet (2022) been made into law (Sarbanes 2022; ACA 2015). This concept is a particularly significant because it

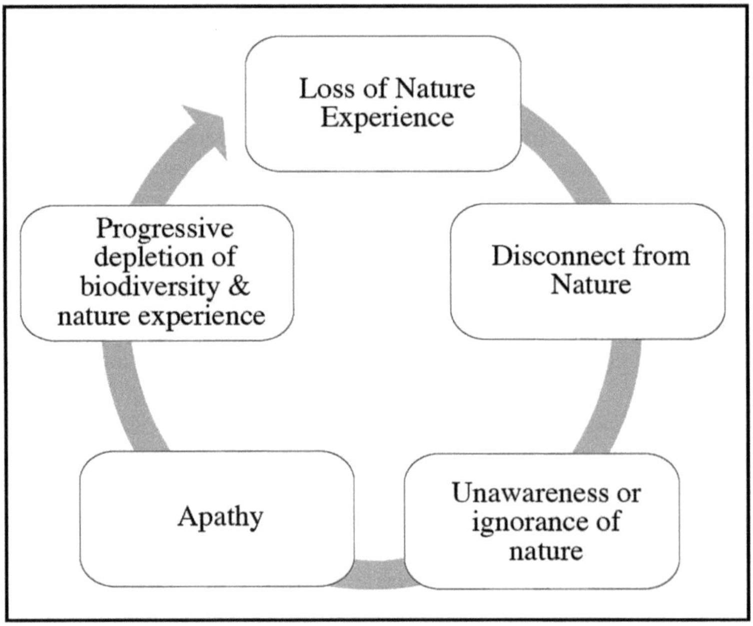

Fig. 15.1 Robert Pyle's cycle of extinction of experience. Modified from Pyle (2003)

illuminates a phenomenon occurring systematically within urban and educational communities. Consequently, I propose that if left unaddressed, the EoE cycle may continue to self-perpetuate as a hidden phenomenon impacting societal values, cultures and underlying the loss of biodiversity (Poole 2015).

EoE involves at least two factors: (1) the degree to which a community cares for the well-being of non-human entities and (2) the ecological knowledge the community has to ensure the integrity of that ecosystem (Poole 2015). EoE certainly seems to impact both aspects of engaging with ecological systems. Each of these senses raises important questions about ethical and ecological literacies within environmental ethics and biocultural conservation. Not only should we be concerned about the well-being of the ecological systems, but we must maintain the ecological knowledge that is needed to be good environmental stewards. This of course raises further questions as to what ethical obligations we have within our educational systems to teach about our local ecosystems—how much knowledge should we have about our local, regional, and global environments, and about our local habitats? Who should be teaching this knowledge? Should it be taught in the local languages of the people? Whose intellectual property is this knowledge?

Consequently, EoE is a potential driver within education and other institutional processes that erodes experiential, cultural, and linguistic knowledge and understanding of the natural world, and therefore the human embeddedness with the non-human world. This is particularly pressing for cultures and identities that conceive of themselves as part of nature or inseparable from the natural world, because their ways of being in the world are disrupted, undermined, or alienated. Their memory, culture, and knowledge of those ways are being lost as well. This will be discussed next in the context of the extinction of experience cycle as a driver within formal education and environmental education.

15.3 Formal Education as a Driver of Extinction of Experience and Biocultural Homogenization

There are many drivers of EoE, and a central one is formal education (Poole 2015; Pyle 1978, 2001, 2002; Miller 2005; Nabhan and St. Antoine 1993; Soga and Gaston 2016). With its standardized methodologies, languages, and content, formal education has two outcomes that drive EoE and are problematic for biocultural conservation. First, it creates the conditions for EoE, as outlined in the previous section: (a) that is the loss of knowledge of local species; (b) experience with them; and (c) awareness of these losses. Second, what philosopher and conservationist Ricardo Rozzi (2013) has called *biocultural homogenization*. These two dynamics will be elaborated in this section.

The EoE cycle within education disconnects the transmission of intergenerational knowledge and linguistic knowledge within cultural groups, damaging their linguistic and cultural knowledge, further detaching them from their sense of place. This detachment is becoming a standardized process that is recreated as a systematic and monocultural process (Poole 2015; Rozzi 2013; Pretty et al. 2009). The loss of cultural heritage—and respectively the cultural sovereignty of distinct groups—is a major contributor to environmental degradation and incidentally the creation of poverty more generally further exacerbating these dynamics (Escobar 1995; Hunn 2007; IPCCA 2009; Maffi 2005; Posey and Dutfield 1996).

Jules Pretty et al. (2009) identified significant threats to both cultural and biological diversity, including: (1) resource use by new commercial sectors such as biofuels or timber industries; (2) extended commodification of natural resources; (3) immigration of new economic actors into long-standing community structures. The list also includes important drivers that influence values and spaces in which communities live, such as (4) the aspirations of consumer lifestyles worldwide; (5) the continuing globalization of food systems; (6) urbanization and rural to urban migration trends (though there are notable reversals of these patterns); (7) modernization of healthcare; (8) language erosion and loss; (9) assimilation, and (10) homogenization of formal education and the expansion of dominant belief systems.

The role of formal education driving erosion of local languages and local ecological knowledge is clearly documented. Stanford Zent, a US-Venezuelan ethnobiologist, explores the social dimension of biocultural experience and transmission of knowledge. In a UNESCO report on "delearning trends and changing patterns of [knowledge] transmission," Zent (2009) argues the importance of local participation and engagement within learning patterns. Synthesizing many studies, Zent focuses his report on the knowledge transmission of learning (or delearning patterns) within indigenous communities, as influenced by Western pedagogical practices, by reviewing case studies in multiple countries. From this synthesis, he concludes that traditional patterns of knowledge transmission are usually "informal, context-dependent, activity-situated and participatory in nature" (Zent 2009, p. 51). For instance, in Venezuela, when a mission institutionalized formalized education structures, such as indoor classroom environments or structured recess, children were less likely to participate in traditional, local-based activities with the environment (Ishizawa 2006). These students demonstrated less local ecological knowledge and less interest in their traditional community practices. Zent (2009, p. 51) reports that:

> According to our experience of Jotï daily life, most talk about plants occurs in contexts in which there is direct contact and interaction with them. These include: walking through the forest, harvesting plant products, processing or eating the catch, painting the body with vegetable dyes, performing rites with magical plants to enhance hunting success, curing the sick, etc. We did not witness formal or consciously planned teacher-led instruction about plants, with the exception of one of the schoolteachers at Kayamá who decided to teach his pupils about plants after we completed our study and then informed the community of the gap between children's and adult's knowledge on the subject. *Only when children asked questions first* to parents, older siblings or other caregivers, usually in the course of subsistence activities, was specific and directed verbal instruction provided. In that sense, we would characterize ethnobotanical knowledge transmission among Jotï as *learner initiated or motivated*: information is verbally transmitted from expert to apprentice upon the latter's request.

The importance of knowledge transmission being "learner initiated or motivated" and taking place during day-to-day activities within the community emphasizes two aspects often missing within formal educational structures. First the local contact with biodiversity and emotional engagement is missing, and further, the absence of connectivity with the community's day-to-day priorities creates a gap between the student and the world in which they find themselves as subsistence ways of life are replaced by neoliberal economies. These two dimensions are not only missed in formal education, but also within informal education and the prevalent of cell phones and digital media that mediate experiences today (Zent 2009, p. 51).

In an earlier study exploring the loss of cultural ecological knowledge from one generation to the next, Gary Nabhan and Sara St. Antoine (1993) compared the intergenerational knowledge of O'odham and Yaqui tribal elders and schoolchildren of the US/Mexico desert borderlands. The elders and schoolchildren were two generational groups who had considerably distinct environmental experiences. This resulted from changing technological, environmental, and educational conditions, including habitat degradation, loss of oral traditions, and loss of visceral contact with local flora and fauna, meaning that the younger generations suffered a significant "extinction of experience" (Nabhan and St. Antoine 1993, p. 235). The tribal elders, "who have engaged in considerable hunting and gathering activities during their lifetimes" had different ethnobiological knowledge than ". . .their grandchildren who have grown up fully exposed to television, prepackaged foods, and other trappings of modern life" (Nabhan and St. Antoine 1993, p. 231).

Nabhan and St. Antoine (1993, p. 241) further observed, "[n]ow that global electronic media dominate knowledge of nature, these children are losing the kind of local awareness that television documentaries cannot supply." Their analysis demonstrated that children were losing local knowledge of their desert surroundings. They found that children did not know that birds sang more frequently in the morning than at dusk, that it was possible to eat the fruit from the prickly pear cactus, or that the creosote bush was the fragrance that carried in the wind so strongly. Further, children were unaware of a major food source, the prickly pear cactus, despite its popularity in the region for over 8000 years and though it continues to have a strong presence in markets in the area. Their analysis also emphasizes that one of the major distinctions in the types of learning that children are undergoing is that television and similar passive learning does not require the child to engage their surroundings; they just " sit and absorb," ... "[b]oth television and certain formal education approaches run counter to the ways of indigenous education—for example, apprenticeships with elders—that have been more successful in previous generations" (Nabhan and St. Antoine 1993, p. 241). Nabhan and St. Antoine call this trend the "demise of the oral tradition." It is important to note in the discussion of EoE that this dimension is often overlooked. Not all knowledge of the environment is equal. Personal, cultural, and psychological connections provide different connections to biodiversity, ecosystems, and what might generally to be understood as nature.

The loss of local ecological heritage and local ecological knowledge occurs in myriad ways, many of which are intentionally directed by particular agents (certain traditional practices cease and are replaced by a new ethical view, for instance). Since these losses are often obscured and made invisible, they are discounted even as their physical experience is threatened, perpetuating the slow erosion of the capacity to recognize this loss (Nabhan et al. 1996; Orr 1992). The loss of such sensitivity unravels thousands of years of deep co-evolutionary adaptation of humans to their surrounding world, in its living, thriving culture. Examples of acculturation practices include the reduction of the number of languages spoken and the standardization of textbooks and content to represent one cultural worldview.

This tension has also been identified by conservationists as the difference between monoculture and polyculture systems (Shiva 1993; Ishizawa 2006). These tensions between these systems have far-reaching implications for policy and institutional practices. These cultural and institutional structures also influence cultural memory, sense of place, imagination, and appreciation. Therefore, we need to develop an understanding of the underlying causes of such experiential disconnect from knowledge and to comprehend this loss within the complex interrelation that culture shares with biological diversity and the environment as a co-evolutionary aspect of this dynamic. In this case, in order to redress EoE, it is necessary to revise the internal mechanisms within formal education.

By comparison, schoolchildren in the USA are unable to recognize basic plants of their region, though they have a high incidence of product recognition (Louv 2005,

2021). In rural communities, rural migrations have been taking place as younger generations move to cities. For many living in urban environments, local extinction of species or environmental degradation remains unknown or unobserved (Wandersee and Schussler 1999; Miller 2005; Samways 2007). Additionally, especially among indigenous communities, oral traditions become lost (Maffi 2005; Rozzi 2012; Berkes et al. 2000). Informal educational praxis of learning from family, participating in community events and home-economic activities are discounted. In this sense, EoE should be recognized as reiterative process that can become embedded in institutionalized systems if it is not intentionally addressed.

I call attention to the fact that these trends in formal and informal education are exacerbated by standardized institutions, such as industrial agricultural systems and market economy practices. Not only does daily engagement with biodiversity decrease, but the knowledge about these systems is disregarded in practice as well. This standardized urban experience implies a deprecation of engagement with local biodiversity and local knowledge. Nabhan (2001, p. 145) criticized this trend by stating that a prevailing approach in the West is the attempt "to gain insights about the natural world from indigenous peoples, treating them as 'native ecologists' whose traditional ecological knowledge is worthy of respect." However, dominant educational systems still disregard local ecological knowledge, even though in theory this knowledge is often considered important in conservation and sustainable management.

The importance of knowledge transmission being "learner initiated or motivated" and taking place during day-to-day activities within the community emphasizes two aspects often missing within formal educational structures: (1) the local contact with biodiversity and emotional engagement, and further, (2) the absence of connectivity with the community's day-to-day priorities. The latter, particularly affects indigenous and other local communities because it creates a gap between the student and the world in which they find themselves as subsistence ways of life are replaced by neoliberal economies. These two dimensions are not only missed in formal education; they are missed also within informal education driven by a growing prevalence of mediated experiences (Rozzi 1999, 2013).

For example, an insightful analysis of Chilean textbooks and recreational books by Juan Luis Celis Díez et al. (2016) found that only 7.6% (83 out of 1095) of the children's books depicted native fauna. Additionally, books depicted mostly exotic plants and foreign wild landscapes or natural scenarios. Of animals represented, 72.7% were exotic to Chile. Educational or formal institutions that discount the importance of local ecological knowledge undermine cultural integrity and biocultural heritage. Therefore, formal education can be considered an indirect driver of environmental degradation and a contributor to the loss of biocultural heritage (Poole 2015; Poole et al. 2013; Bridgewater and Rotherham 2019).

Ricardo Rozzi (2013) has described the confluent loss of the rich interrelation of biological, cultural, and linguistic diversity as *biocultural homogenization*. Ignoring the erosion of cultural heritage and ecological knowledge as an indirect driver of biocultural homogenization can lead to further degradation of biodiversity and traditional ways of life. Ultimately, this could result in further collective loss and depreciation of biocultural heritage, thus hindering efforts within sustainable development.

Biocultural heritage as a concept reflects the diverse ways of being between human communities and their local environments, including their rich history of language, heritage, cultural memory, ecological knowledge, and values and therefore should be explicitly articulated as a key component to any sustainability agenda (Poole 2018). Consequently, I and others continue to propose that to accomplish any inclusive policy or institutional space for environmental learning and sustainability agenda, it is indispensable to include capacities to address biocultural heritage. Therefore we must acknowledge biocultural heritage in our formal institutions and policies (Poole 2018; Rozzi 2013; Gavin et al. 2015; Cocks 2006). Regarding a biocultural approach for addressing EoE, it is essential to acknowledge multiple dimensions about the complexity of losses of biological and cultural diversity (Poole 2015). This requires multiple viewpoints from both the dominant and marginalized groups (Agrawal 1995; Maffi 2005).

15.4 Theoretical Implications: Limitations and Problems of Formal Education

Since the identification of the concept of EoE by Pyle as an ongoing process, the term first communicated a sense of disconnection and alienation from nature (Pyle 1978). This loss results from the absence of not only the knowledge about the natural world, but additionally, of fundamental opportunities for experience. Further, this alienation can be expressed in forms of discomfort, or disgust, as illustrated by urban schoolchildren who have disdain for rural ways of life or consider packaged foods more palatable than unprocessed ones (Poole 2015). These experiences provide the foundation for that knowledge to be useful for everyday life and to take hold as a fundamental aspect of the individual's life experience.

For these reasons, by addressing EoE, I highlight both the experience that is lost and the drivers or conditions that create this extinction (Poole 2015). As global society is coping with the environmental crisis, it is necessary to address it pluralistically, including distinct perspectives across the rural-urban gradient. Future policies must incorporate a biocultural viewpoint in education, as well as technologies, policies, and societal structure to prevent the EoE cycle. I also argue the biocultural framing has greater sensitivity to the nuanced relationship between culture and place, its multi-generational heritage, and the great time that is required to develop understanding of ecosystem management by local communities (Poole 2015; Maffi 2005; Nabhan and St. Antoine 1993; Miller 2005).

By marking the loss of this biocultural understanding with the concept of EoE, we can explicitly create a conceptual placeholder for this knowledge and where it should be present within education, policy, and practice. By identifying its importance, even when it is lacking in practice, we simultaneously address its invisibility while creating the space to work on the recovery of these more sustainable ways of thinking and practices in the instances they have been lost (Poole 2018). To summarize the theoretical implications of these findings then, we can make the following observations:

The extinction of experience cycle can be understood as occurring when there is loss of:

(a) Knowledge of local species
(b) Experience with local species
(c) Or, awareness of these losses

That the EoE cycle self-perpetuates by varying degrees when tied to educational institutions that formalize practices which:

(a) Decouple cultural knowledge, local languages, and inter-generational knowledge from each other
(b) Embed elements of biocultural homogenization

Conceptualizing EoE as a critical placeholder for biocultural losses reminds us to look at the interrelated dynamics of the structured world around us and the ways formal education, industrial agriculture, and the global economy influence our imagination and capacity to engage the world according to our ecological conscience. Nonetheless more research is needed, and the importance of this concept is increasingly being recognized (Gaston and Soga 2020).

Indeed, the capacity to understand the complexity of ecological systems is also necessary to avoid an idealization of the moral complexity of nature in that all of nature is not ideal or beautiful. The emphasis upon "recreational care" or "utilitarian care" of nature are not the only values that are lost through EoE, nor are they the only tools that can be used to draw public attention to the ecological importance and innate value of nature. Indeed, these are values of nature and should be acknowledged, but other ways of relating, valuing, and engaging persist as well (Kellert 1997). The emphasis on the need to develop an ethic of care in response to the loss of nature experience can undermine the significance of biodiversity loss and its causes to the concerns of the individual's quality of life obscures our attention to the causes that produce systematic loss of biodiversity and alternative cultural ways of life.

We must therefore integrate the discussion of EoE within sustainability discourse throughout the urban fabric—reminding ourselves that every "vacant" lot is really an urban wildland, and in turn, every structure built, and the experiences these spaces create has the potential to drive our global society towards or away from the flourishing of biocultural diversity and engaging biocultural heritage(s). In particular, focusing our discussion on revisions of education, specifically environmental education, will require careful reflection on the elements that engage meaningful experiences with our communities, environments, learning spaces, and the content that is taught at schools. To reverse EoE, education must become a space in which we can engage biocultural heritage to reinforce and support sustainable interconnections between diverse human societies and their distinct environments.

15.5 Mitigating the EoE Cycle: Envisioning Formal Education as a Driver of Biocultural Heritage

The chapter has introduced the extinction of experience cycle and the potential that education has to serve either as an indirect driver of EoE or as driver of biocultural conservation. The biocultural critique is often expressed in terms of proposed alternative management practices and institutionalized policies that alter the human–nature relationship and the capacity for local communities to express and utilize their local and traditional ecological knowledge. As we engage formal education to address consumer and urban communities, these educational contexts must also engage biocultural framings that situate for this new sense of place and consider context and complex colonial and linguistic histories.

The deep connection of culture with ecological place is often obscured by the dominance of cultural views that do not recognize this co-constitutive connection. I propose that extinction of experience can also be understood as the absence of the knowledge of interconnection between culture and nature, and that in the absence of a cultural heritage that is able to articulate such losses, sensitivity to this interrelation can become lost, and never known by generations. That is, for communities that lack a strong intellectual, linguistic, or ecological tradition related to the land, it can be difficult to conceive of what this knowledge might be to a people who have such a tradition. Further, this knowledge is often treated within formal education in global society as less valuable than techno-scientific knowledge and placed lower on the "hierarchy of knowledges" about the world. A biocultural framing of EoE will aid in understanding the ongoing drivers and impacts of heritage, language, and biodiversity losses in the long term by affirming a view of reality that is invisible to many within formal education.

A biocultural framing of EoE aids in understanding the ongoing drivers and impacts of heritage, language, and biodiversity losses in the long term by affirming a view of reality that is invisible to many within formal education. Just as we need to consider biocultural heritage within our policies and sustainable development goals, we must find spaces to acknowledge this type of learning and way of thinking, and ethical engagement within our formal educational models as well. This can be transformative as we revisit our educational policies, but also as we engage learning in our informal learning pathways as well, our media, online and virtual environments as well.

References

Agrawal A (1995) Dismantling the divide between indigenous and scientific knowledge. Dev Change 26(3):413–439

American Camps Association (ACA) (2015) "No child left inside." Available via: No Child Left Inside Act - Reintroduced | American Camp Association. https://www.acacamps.org

Berkes F, Folke C, Colding J (eds) (2000) Linking social and ecological systems: management practices and social mechanisms for building resilience. Cambridge University Press, Cambridge
Bridgewater P, Rotherham ID (2019) A critical perspective on the concept of biocultural diversity and its emerging role in nature and heritage conservation. People Nat 1(3):291–304
Callicott JB (1987) The conceptual foundations of the land ethic. In: Baird Callicott J (ed) Companion to a sand county Almanac: interpretative and critical essays. University of Wisconsin Press, Madison, pp 186–217
Celis Díez JL, Forestier JD, Garcia MM, Lazzarino S, Rozzi R, Armesto JJ (2016) Biodiversity knowledge loss in children's books and textbooks. Front Ecol Environ 14(8)
Cocks M (2006) Biocultural diversity: moving beyond the realm of 'indigenous' and 'local' people. Hum Ecol 34(2):185–200
Colléony A, Prévot AC, Saint Jalme M, Clayton S (2017) What kind of landscape management can counteract the extinction of experience? Landsc Urban Plan 159:23–31
Escobar A (1995) Encountering development: the making and unmaking of the third world. Princeton University Press, Princeton
Gaston KJ, Soga M (2020) Extinction of experience: The need to be more specific. People Nat 2(3): 575–581
Gavin MC, McCarter J, Mead A, Berkes F, Stepp JR, Peterson D, Tang R (2015) Defining biocultural approaches to conservation. Trends Ecol Evol 30(3):140–145
Hunn E (2007) Ethnobiology in four phases. J Ethnobiol 27(1):1–10
IPCCA (2009) The Indigenous Peoples' Biocultural Climate Change Assessment Initiative. http://ipcca.info/. Accessed 5 July 2015
Ishizawa J (2006) Cosmovisions and environmental governance: the case of in situ conservation of native cultivated plants and their wild relatives in Peru. In: Reid W (ed) Bridging scales and knowledge systems: concepts and applications in ecosystem assessment. Island Press, Washington, DC
Kahn PH Jr (2002) Children's affiliations with nature: structure, development, and the problem of environmental generational amnesia. In: Kahn P Jr, Kellert SR (eds) Children and nature: psychological, sociocultural, and evolutionary investigations. MIT Press, Cambridge, MA
Kahn PH, Kellert SR (2002) Children and nature: psychological, sociocultural, and evolutionary investigations. MIT Press, Cambridge, MA
Kellert SR (1997) The value of life: biological diversity and human society. Island Press, Washington, DC
Kimmerer RW (2002) Weaving traditional ecological knowledge into biological education: a call to action. BioScience 52(5):432–438
Lawler JJ, Aukema JE, Grant JB, Halpern BS, Kareiva P, Nelson CR et al (2006) Conservation science: A 20-year report card. Front Ecol Environ 4(9):473–480
Leopold AC (2004) Living with the land ethic. BioScience 54(2):149–154
Louv R (2005) Nature deficit. Orion 70–71
Louv R (2021) No child left indoors. Orion. Available via: Orion Magazine - Leave No Child Inside
Maffi, L (2001) On biocultural diversity: linking language, knowledge, and the environment. Smithsonian Institution Press, Washington, DC
Maffi L (2005) Linguistic, cultural, and biological diversity. Annu Rev Anthropol 34:599–617
McKinney ML (2006) Urbanization as a major cause of biotic homogenization. Biol Conservat 127(3):247–260
Millennium Ecosystem Assessment (MA) (2003) Ecosystems and human well-being: a framework for assessment. Island Press, Washington, DC
Miller JR (2005) Biodiversity conservation and the extinction of experience. Trends Ecol Evol 20(8):430–434
Nabhan P (2001) Cultural perceptions of ecological interactions: an 'endangered people's' contribution to the conservation of biological and linguistic diversity. In: Maffi L (ed) On biocultural

diversity. Linking language, knowledge and the environment. Smithsonian Institution Press, Washington, DC, pp 145–156
Nabhan GP, Mirocha P, Buchmann S (1996) Forgotten pollinators. Island Press, Washington, DC
Nabhan G, St. Antoine S (1993) The loss of floral and faunal story: the extinction of experience. In: Wilson EO (ed) The biophilia hypothesis. Harvard University Press, Cambridge, MA, pp 229–250
Noddings N (2013) Caring: A relational approach to ethics and moral education. Univ of California Press, Berkeley
Orr DW (1992) Ecological literacy: education and the transition to a postmodern world. SUNY Press, Syracuse, NY
Orr DW (1996) Virtual nature. Conservat Biol 10(1):8–9
Orr DW (2004) Earth in mind: On education, environment, and the human prospect. Island Press, Washington, DC
Poole AK (2015) Urban sustainability and the extinction of experience: acknowledging drivers of biocultural loss for socio-ecological well-being. Doctoral dissertation, University of North Texas
Poole AK (2018) The UN sustainable development goals and the biocultural heritage lacuna: where is goal number 18? In: Rozzi R et al (eds) From biocultural homogenization to biocultural conservation. Springer, Cham, pp 315–331
Poole AK (2019) Environmental philosophy in the city: confronting the Antiurban bias to overcome the human-nature divide. In: The Routledge handbook of philosophy of the city. Routledge, pp 335–352
Poole AK, Hargrove EC, Day P, Forbes W, Berkowitz AR, Feinsinger P, Rozzi R (2013) A call for ethics literacy in environmental education. In: Rozzi R et al (eds) Linking ecology and ethics for a changing world. Springer, Dordrecht, pp 349–371
Posey DA, Dutfield G (1996) Beyond intellectual property rights: toward traditional resource rights for indigenous and local communities. IDRC, Ottawa
Pretty J, Adams B, Berkes F, Ferreira De Athayde S, Dudley N, Hunn E, Maffi L et al (2009) The intersections of biological diversity and cultural diversity: towards integration. Conservat Soc 7(2):100–112
Pyle RM (1978) The extinction of experience. Horticulture 56:64–67
Pyle RM (1993) The thunder tree: lessons from an urban wildland. Houghton Mifflin, Boston
Pyle RM (2001) The rise and fall of natural history. Orion People Nat 20(4):16–23
Pyle RM (2002) Eden in a vacant lot: special places, species, and kids in the neighborhood of life. In: Kahn PH Jr, Kellert SR (eds) Children and nature: psychological, sociocultural, and evolutionary investigations. MIT Press, Cambridge, MA, pp 305–327
Pyle RM (2003) Nature matrix: reconnecting people and nature. Oryx 37(02):206–214
Pyle RM (2014) No child left inside: nature study as a raid act. In: Place-based education in the global age. Routledge, New York, pp 179–196
Randall R (2009) Loss and climate change: The cost of parallel narratives. Ecopsychology 1(3): 118–129
Rozzi R (1999) The reciprocal links between evolutionary-ecological sciences and environmental ethics. BioScience 49(11):911–921
Rozzi R (2012) Biocultural ethics. Environ Ethics 34(1):27–50
Rozzi R (2013) Biocultural ethics: from biocultural homogenization toward biocultural conservation. In: Rozzi R et al (eds) Linking ecology and ethics for a changing world. Springer, Dordrecht, pp 9–32
Samways (2007) Rescuing the extinction of experience. Biodivers Conservat 16(7):1995–1997
Sarbanes J (2022) Sarbanes, reed, Collins reintroduce no child left inside act. 7 Apr 2022. Press release. Retrieved from: https://sarbanes.house.gov/media-center/press-releases/sarbanes-reed-collins-reintroduce-no-child-left-inside-act
Shiva V (1993) Monocultures of the mind: Perspectives on biodiversity and biotechnology. Palgrave Macmillan, London

Soga M, Gaston KJ (2016) Extinction of experience: the loss of human–nature interactions. Front Ecol Environ 14(2):94–101
Stokes L (2006) Conservators of experience. BioScience 56(1):6–7
Wandersee JH, Schussler EE (1999) Preventing plant blindness. Am Biol Teach 61(2):82–86
Zent S (2009) Traditional ecological knowledge (TEK) and biocultural diversity: a close-up look at linkages, delearning trends, and changing patterns of transmission. In: Learning and knowing in indigenous societies today. UNESCO, Paris, pp 39–58

Chapter 16
Aldo Leopold as Educator: His Legacy for Field Environmental Philosophy

J. Baird Callicott

Abstract Aldo Leopold advocated more field and outdoor education and less passive and sedentary indoor study. He stressed the importance of the study of plants and animals and their living relationships, beyond a prevailing pigeonhole approach that relies on the abstract conceptual scheme of taxonomical categories. Leopold suggested that the ability to read the land was as important as the ability to read books and that too much of the latter could dull awareness. Leopold understood that the sciences and the humanities are not antagonistic orientations of the mind. He knew that the human mind and spirit are one and that latent in the emerging science of ecology were vast implications for philosophy and religion. Leopold insisted that higher education must recognize this wholeness and courageously explore the connections across disciplinary boundaries. As an educator, Leopold eschewed theatrics and lectured in a conversational tone and manner. His vocabulary was rich, his syntax clear, his diction precise, and he assiduously avoided pedantic speech. Leopold encouraged his students to think creatively and make their own discoveries by asking questions requiring inferences from observable facts. He was an attentive and interested listener. The land ethic is built on Darwin's account of the origin and social evolution of ethics, to which Leopold added the community concept in ecology. He thought that cultural diversity was linked to biological diversity. He thus provides a foundation upon which to build the theory and practice of biocultural conservation and of field environmental philosophy.

Keywords Ecology · Evolution · Diversity · Taxonomy · Wholeness

This chapter began its odyssey as the keynote address for a conference of the Midwest Environmental Education Association, held at Upham Woods Outdoor Learning Center near Wisconsin Dells, Wisconsin, September 24, 1981. It was published in the *Journal of Environmental Education* 14 (1982): 34–41. That essay was edited, augmented, and updated for this iteration in 2021.

J. B. Callicott (✉)
University of North Texas, Denton, TX, USA

R. Rozzi et al. (eds.), *Field Environmental Philosophy*, Ecology and Ethics 5,
https://doi.org/10.1007/978-3-031-23368-5_16

Aldo Leopold was truly a watershed figure in the history of American conservation. His slim volume of elegant essays, *A Sand County Almanac,* belongs on the same shelf with the works of George Perkins Marsh, Henry David Thoreau, and John Muir. It represents the most recent development of fundamental ideas in that great tradition of eco-philosophical literature. Roderick Nash (1967) credits Leopold with powerfully advocating, for the first time in Western intellectual history, broad human ethical responsibility to the non-human natural world. Aaran Gare (2023) indicates that Leopold actually carries forward an anti-reductive, holistic school of Western thought that Gare traces from the Renaissance through the work of Giordano Bruno (in the sixteenth century), Giambattista Vico (in the seventeenth century), Gottfried Herder (in the eighteenth century), Friedrich Schelling and Wilhelm von Humboldt (in the nineteenth century) and José Ortega y Gassett (Leopold's contemporary in the twentieth century). An environmental ethic may also be found latently present in some of the writings of Muir (1916), but Muir neither fully articulated, nor fully grounded it, as Leopold did, in a supporting theoretical matrix of humanistic (especially moral philosophy) and scientific (especially evolutionary biology and ecology) concepts, the latter of which were not available to most of his intellectual forebears.

Aldo Leopold's contribution to environmental thought and value, moreover, is recognized and acknowledged not only in this country, but internationally as well. Over two million copies of *A Sand County Almanac* have been sold and the book has been translated into a dozen languages.

Although an ethic may be articulated and persuasively advocated by a single creative individual, it will remain ineffectual unless it becomes generally distributed in the population—unless, indeed, it becomes a firmly entrenched cultural institution. Aldo Leopold's own literary mode of expression, a style at once simple but profound, concrete yet reaching the most abstract and subtle domains of thought and feeling, has shepherded the land ethic in its first steps toward broad cultural currency. But a wider effort is required if Leopold's dream of a national, even eventually global, ecological conscience is to be realized. The task is primarily one of environmental education.

This chapter is divided into four parts. The first concerns Leopold's own reflections on education, as revealed in *A Sand County Almanac*, *Round River*, and *The River of the Mother of God*. Leopold was himself an educator, a professor of game management, an academic discipline which he pioneered, at the University of Wisconsin. He made a deep impression on his students. The second part concerns Leopold's own practical approach to education. To learn what his style and methods were as a teacher, I have consulted his daughters, Nina and Estella, his youngest son, Carl, and several of his former students, men, and women all of whom have themselves gone on to distinguished careers in environmental conservation. The third part concerns a general educational strategy to impart the land ethic and develop an ecological conscience in contemporary environmental education. The approach that I recommend is implied by the conceptual foundations of the land ethic itself. The fourth part concerns the implementation of this strategy for the conservation of biocultural diversity in the Field Environmental Philosophy program of the Omora Ethnobiological Park at the southern tip of the Americas.

16.1 Aldo Leopold on Education

Aldo Leopold had a lot to say about education; and most of it was gently but sharply critical. Indeed, one complete essay, "Natural History-The Forgotten Science," in *Round River* (1953) is a short dissertation on the state of the art, its past, and its future. Leopold is concerned that education in biology, set in a high school or university laboratory, has usurped the place of natural history. Laboratory biology, in Leopold's day, involved such things as "memorizing the names of the bumps on the bones of a cat" (Leopold 1953, p. 60). It was pursued indoors and involved dead animals. Natural history, on the other hand, involves living plants and animals and more importantly their living context. Leopold felt that given a limited curriculum and limited schedule of classes, biology students are better served by more field-oriented courses, more natural history, and less laboratory technique.

Leopold's friend, renowned ecologist Charles Elton (1927, p. 1), defined "ecology" as "scientific natural history." The influence of the connection Elton draws between ecology and natural history reveals another more deeply theoretical tension at which Leopold hints in "Natural History" lurking beneath these manifest indoor/outdoor dead/alive dichotomies. Natural history, he remarks, suffered some loss of ground earlier in the teaching of biology because it then consisted largely of "labeling species and amassing facts about food habits without interpreting them" (Leopold 1953, p. 61). However,

> modern natural history deals only incidentally with the identity of plants and animals, and only incidentally with their habits and behaviors. It deals principally with their relations to each other, their relation to the soil and water in which they grew, and their relations to the human beings who sing about 'my country' but see little or nothing of its inner workings (Leopold 1953, p. 63).

There was, in other words, within natural history itself, a conflict between taxonomy and ecology. Ecology is built upon taxonomy as upon a foundation, but ecology (scientific natural history) is a more advanced development in understanding the world of Nature.

Taxonomy, the art of labeling and classifying plants and animals according to an elaborate conceptual schematism consisting of species, genus, family, order, class, phylum, kingdom, and (recently) domains (the number and the natures of the latter two remain a matter of controversy). Taxonomy originated as a formal study with Aristotle, the founder of Western biology, in the fourth century BCE; it was greatly refined by Carl Linnaeus in the eighteenth century, and it has remained the backbone of the biological sciences (until, perhaps, eclipsed by molecular biology in the late twentieth century). Ecology itself is, relatively speaking, a newcomer, emerging as a distinct science around the turn of the twentieth century. Taxonomy manages well enough with dead specimens, but that is not its most serious shortcoming as a science that is supposed to deal with living nature. Rather, because the taxonomical identity of an animal or plant locates it within a highly formal and largely abstract system of organization, preoccupation with this system tends to obscure the actual and concrete system of living ecological relations among animals and plants. A plant's or animal's taxonomical niche, in other words, conceptually connects it to other phylogenetically similar species, while its ecological niche may place it in an altogether different pattern of relations. As Leopold (1949, p. 215) remarks, for purposes of ecological understanding, "The species of a [trophic] layer are alike not in where they came from or in what they look like, but rather in what they eat." A clear example is afforded by the few species of plankton-feeding shark. They are, of course, phylogenetically and taxonomically related to other sharks, but from an ecological point of view, they more closely resemble baleen whales. Filter-feeding sharks, in other words, pursue the same "profession" in the "economy of nature" as the Mysticeti.

The exclusive attention paid to taxonomical classification for many centuries may have been a necessary stage in the development of biology and certainly of the theory of evolution, as Leopold (1953, p. 60) notes. Ironically, however, it tended to obscure the complex relations among species and resulted in a picture of the living world as so many logically ordered species arrayed on the landscape like so much furniture in a room. One may rearrange the furniture in a room without untoward consequences, and one may discard pieces of furniture in a room without adversely affecting the remaining pieces. But one rearranges the species on a landscape or extirpates any of them at the risk of cascading effects throughout the whole biotic community.

This observation has direct and concrete implications for contemporary environmental education. Take the naturalist-led nature walk in a state park or biodiversity reserve as a case in point. Practically all of those I have ever been guided along consisted of a catalogue of species. It usually goes something like this. "There's an indigo bunting; here a red oak, yonder a white oak; to the left spiderwort, to the right butterfly weed; on the branch above your head, a vireo; and high in the sky there's a hawk! Is it a red-tailed or a rough-legged?" And so on and on. But how all these species are connected to the sandy loam on which I stand and its proximity to the water table; the mean annual temperature and rainfall; the velocity and direction of prevailing winds; frost-depth, and snow cover in winter; and hundreds of other affective features of the landscape are things I am left to wonder about. Knowing the names and pedigrees of our non-human neighbors is a foundational accomplishment which I do not intend at all to demean, but that should be the beginning, not the end, of a naturalist's lore. How to create a really ecological nature walk is a question I cannot answer, but it is a problem which should be high on the priority list in the field of outdoor environmental education.

Another recurrent theme in Leopold's philosophy of education is summed up in the following remark: "Education, I fear, is learning to see one thing by going blind to another" (Leopold 1949, p. 18). For example, he writes,

> I once knew a lady, banded by Phi Beta Kappa, who told me that she had never heard or seen the geese that twice a year proclaim the revolving seasons to her well-insulated roof. Is education possibly a process of trading awareness for things of lesser worth? The goose who trades his is soon a pile of feathers. (Leopold 1949, p. 18)

Here again is the indoor/outdoor dichotomy, but with a different twist. Achievement in modern American education, as certified by Phi Beta Kappa, is measured almost exclusively in terms of literacy. Learning in general is equated, in other words, with book learning. Therefore, it follows, the more learned we are, the less aware we are of things other than the pages of a book, since time spent reading is time spent with our eyes darting across a line of type and our other senses idling. Aldo Leopold had the daring to ask pointedly if the benefits of this practically universal educational assumption are worth the price. We can be reminded of the peculiarity of modern Western education, largely via the written word, and its cost in the currency of awareness when we compare it to the experiential forms of education in indigenous cultures (Reyhner and Singh 2010).

In order to counter this assumption or, perhaps better, in order to deflect it in a way useful to environmental education, Leopold often employs the concept of reading as a metaphor. Just as reading a book is a process of interpreting signs and recreating a world in imagination, so nature is an open and ever-present book for those who can read her signs and who possess an active intelligence. Here are a few illustrations. "He who owns a veteran bur oak owns more than a tree. He owns a historical library, and a reserved seat in the theatre of evolution. To the discerning eye, his farm is labeled with the badge and symbol of the prairie war [between grassfires and trees]" (Leopold 1949, p. 30). Following the destruction of prairie flora by the mindless mowing of a cemetery, Leopold (1949, p. 46) remarks, "If I were to tell a preacher of the adjoining church that the road crew has been burning history books in his cemetery, under the guise of mowing weeds, he would be amazed and uncomprehending. How could a weed be a book?" But to those who are literate in this broader sense, to those who can read the land, silphium, dock, and bluestem are intelligible runes and teach a lesson in history as well as botany. Again, he writes, "Every farm is a textbook on animal ecology; woodsmanship is the translation of the book" (Leopold 1949, p. 81). Why should education not consist as much in learning to read the land as in learning to read books?

Of course, book learning versus a broader kind of awareness should not be construed simply as a matter of either one or the other, but not both. Leopold (1953, p. 127) does in fact say of "reading sign" that "this skill is rare, and too often seems to be inverse to book learning," but literacy in the ordinary sense is crucially important: indeed, book learning may be foundational to learning to read the land deeply. Daniel Boone was, for his time, a peerless woodsman, but "Boone saw only the surface of things. The incredible intricacies of the plant and animal community—the intrinsic beauty of the organism called America then in the full bloom of her maidenhood—were . . . invisible and incomprehensible to Daniel Boone" (Leopold 1949, p. 291). That's because "Daniel Boone's reaction depended not only on the quality of what he saw, but on the quality of the mental eye with which he saw it. Ecological science has wrought a change in the mental eye"

(Leopold 1949, pp. 173–174). But science is historically a product of literacy and depends upon letters for its transmission and development.

Leopold also seems to have been ahead of his time in advocating what is now called "interdisciplinary education." I believe that this is a natural consequence of his ecological habit of mind. Ecology focuses on the relationships among plants and animals and the physical and chemical factors of their environments. It is natural, thus, that Leopold, a self-taught ecologist, would be more sensitive to and concerned with relations and connections in general. In terms of higher education, this would translate into an emphasis on the relations and connections between and among academic disciplines rather than on their separateness and autonomy. In "The Role of Wildlife in a Liberal Education," Leopold (1991, p. 302) remarks that "All the sciences and arts are taught as if they were separate. They are separate only in the classroom. Step out on the campus and they are immediately fused." This fusion is what Leopold (1991, p. 303) called the study of "land ecology" which is "putting the sciences and arts together for the purpose of understanding our environment." It is, in other words, an ecological approach to education (although some of my ecologist colleagues might chastise me for putting it that way, charging me with vulgarizing their good name).

Leopold gently satirized, with his characteristic wry wit, what I shall call (at the further risk of antagonizing my colleagues in ethology) the "academic territorial imperative." In "Song of the Gavilan" Leopold expanded and made explicit the familiar metaphor, the "harmony" of Nature: "its score inscribed on a thousand hills, its notes the lives and deaths of plants and animals, its rhythms spanning the seconds and the centuries" (Leopold 1949, p. 149). He then goes on to make the following comment:

> There are men [and women] charged with the duty of examining the construction of the plants, animals, and soils which are the instruments of the great orchestra. These men [and women] are called professors. Each selects one instrument and spends his [or her] life taking it apart and describing its strings and sounding boards. The process of dismemberment is called research. The place for dismemberment is called a university.
>
> A professor may pluck the strings of his [or her] own instrument, but never that of another, and if he [or she] listens for music, he [or she] must never admit it to his [or her] fellows or to his [or her] students. For all are restrained by an ironbound taboo which decrees that the construction of instruments is the domain of science, while the detection of harmony is the domain of poets (Leopold 1949, p. 153)

Here clearly Leopold chides his colleagues for their failure to make scientific education something whole and constructive rather than something endlessly analytic, a discontinuous series of fragments. In perfect accord with the anti-Cartesian tradition discussed by Gare (2023), Leopold further suggests that science in isolation from the humanities—from poetry, music, literature, art, and philosophy—is something barren and even destructive. Good scientists must understand the whole picture of Nature into which their specialty fits and, more expansively still, good scientists must be sensitive to the wider emotional, evaluative, and philosophical implications of scientific investigation and discovery. It has always been especially significant to me that "The Land Ethic" begins with an allusion to Homer's *Odyssey* and "The

Geese Return" contains a fragment—"One swallow does not make a summer"—from Aristotle's *Nichomachean Ethics* 1098a18 in McKeon 1941, p. 943, Leopold 1949 p. 18). Clearly Leopold himself was a broadly cultured man. The Greeks generally conceived of ethics not in terms of rules or in terms of consequences but in terms of virtue, ingrained qualities of character—chief among them justice, courage, temperance, piety, generosity, prudence—that guided right behavior. Aristotle was the first Western philosopher to systematically theorize virtue ethics and remains the touchstone for the modern theories of virtue ethics emerging in the mid-twentieth century (Anscombe 1958). To the extent that "The Land Ethic" is cast in the virtue-ethics paradigm—as Breakey and Breakey (2023) indicate and as I (Callicott 2013) trace its source to the moral philosophy of David Hume, who locates the wellspring of virtues in the moral sentiments—the legacy of Aristotle exerted a significant philosophical influence on Leopold.

With respect first to tearing down artificial walls between academic disciplines, he comments, in "The Land Ethic," that

> One of the requisites for ecological comprehension is an understanding of ecology, and this is by no means co-extensive with 'education'; in fact, much higher education seems deliberately to avoid ecological concepts. An understanding of ecology does not necessarily originate in courses bearing ecological labels; it is quite as likely to be labeled geography, botany, agronomy, history, or economics. (Leopold 1949, p. 224)

Scondly, with respect to the integration of ecological understanding with the humanities, he comments that

> No important change in ethics was ever accomplished without an internal change in our intellectual emphasis, loyalties, affections, and convictions. The proof that conservation has not yet touched these foundations of conduct lies in the fact that philosophy and religion have not yet heard of it. (Leopold 1949, pp. 209–210)

At least this circumstance has now been remedied. Although thinkers with a religious bent, such as Ralph Waldo Emerson (1836), had long grappled with the moral issues raised by conservation or the lack thereof, it seems not to have gained wide traction among rank and file congregants, or so it seemed to Leopold in the late 1940s. Religion heard loudly and rudely about conservation in 1967 with the publication in *Science* of Lynn White Jr.'s controversial classic, "The Historical Roots of Our Ecologic Crisis," in which he traced human arrogance, self-importance, and the consequent environmental crisis to the basic tenets of the Judeo-Christian religious tradition. White (1967) thus inadvertently provoked a vigorous debate in religion and theology, which crystalized the formulation of a Judeo-Christian Stewardship environmental ethic (Bauckman 2011). In 2009, the Ecological Society of America endorsed an Earth Stewardship environmental ethic (Rozzi et al. 2015). White also suggested that religions other than Christianity might foster a conservation ethic, a project that I pursued in a comprehensive survey (Callicott 1994). And over the course of the last quarter of the twentieth century, environmental ethics (in many secular forms) became a recognized and well-established subdiscipline in academic philosophy (Callicott and Frodeman 2009).

To sum up: Aldo Leopold advocated more field and outdoor education and less passive and sedentary indoor study. He stressed the importance of the study of *live* plants and animals in their living environments as well as the study of dead ones in sterile and artificial laboratory environments. Above all, he believed that biological education should be ecological; that is, it should strive to impart to students the actual living relationships of plants and animals rather than only their pigeonholes in the abstract conceptual scheme of taxonomical categories. He suggested that the ability to read the land was every bit as important as the ability to read books and that too much of the latter could dull awareness. Finally, he understood that science and poetry are not antagonistic orientations of the mind. As any reader of *A Sand County Almanac* knows, Leopold was accomplished in both. He knew that the human mind and spirit are one and that latent in the emerging science of ecology were vast implications for philosophy and religion. Higher education, he insisted, must recognize this wholeness and courageously explore the connections across disciplinary boundaries, even though some sacred cows may be sacrificed in the process.

16.2 Aldo Leopold as Educator

In *Strictly for the Chickens,* Frances Hamerstrom (1980, pp. 26, 28) writes, “Aldo Leopold was known by many of his students simply as ‘The Professor’ as though there were no other professors on the University of Wisconsin campus.... In those days not everybody knew that Aldo Leopold was a great man, but his students did.” Leopold was not only a great man, as measured by his contribution to the republic of letters and the environmental consciousness (and conscience) that he forged, now international in scope, he was also a great teacher, as measured by the respect, veneration, affection, and, not least, accomplishment of his erstwhile students.

What were the qualities, exactly, that made The Professor a person of such profound impact upon his students? In addition to the information found in several memoirs, I interviewed some of his former students and one of his own children to inquire. I expected, I must admit, to hear from them the adjectives commonly used to characterize a memorable teacher: dynamic, enthusiastic, spellbinding, and so on. Robert McCabe—who joined him in 1939, continued as his “assistant” until his death in 1948—expressly denied that he was an especially dynamic lecturer. That is not to say that he was dull or boring. He was always interesting, according to McCabe, but eschewed the theatrics which lecturers of less substance sometimes use to purchase the interest of their students. According to McCabe, Leopold never used a lectern; rather he sat on the desk, smoked his pipe, and made his point in an unpretentious anecdotal style.

He treated each student as an individual person. He called on students in class to answer questions. These classroom questions and his test questions never emphasized rote memorization but did require thoughtful integration of information. By all accounts, although he was always probing and questioning his students, he was unfailingly gentle and always kind. Nevertheless, according to McCabe, he was not an easy touch for a good grade. His tests were never true/false or multiple choice. They consisted of essay questions, and he fully justified his evaluation with penciled notes in the margins.

His graduate students were personally selected by The Professor. He seems to have looked for students who were strongly field-oriented and showed promise in field techniques. He did not hold his graduate students at arm's length. Indeed, he developed warm personal and enduring relationships, at least with some of them. His daughter Nina told me that he practically considered his graduate students as part of the family. Nevertheless, as a graduate student you did not go to The Professor's office to shoot the breeze. He regarded himself as busy, as indeed he was. Frederick Hamerstrom remembers that when you paid him an office visit, you had his full and complete attention, but McCabe adds that neither did you waste his time. He was warm, but serious, just as he was gentle, but demanding.

Nina Leopold Bradley described for me a typical workday in her father's life. He was an early riser and reached his office by six A.M. or even earlier. He said that he did his best work before eight. His day was filled with appointments, lecture preparations, classes, and the usual paperwork and routine business of university professors. He would arrive home at five for supper. He did not bring his work home. He engaged with his family in intelligent and stimulating supper conversation. He asked the children about their day and listened attentively to their thoughts and accounts. He and his wife, Estella, often read aloud or listened to classical music in the evenings. They retired early.

On Monday nights, seminars were held, after which apples were served for refreshment. Frances Hamerstrom remembers the seminar discussions as being sharply critical. Professor Leopold's questions were always probing, but again, he was never harsh, unkind, or overbearing. Neither did he dominate or overtly direct discussion. His aim was to draw things out of the seminar presenters and stretch their powers of thought and analysis as far as possible and to encourage a critical attitude on the part of all participants.

His conversation apparently had the same mastery of the language that is so amply evident in his writing. His vocabulary was rich, his syntax clear, his diction precise, and he assiduously avoided pedantic speech. According to Frances Hamerstrom, Leopold asked that his students take similar pains to express themselves clearly and directly and to always care deeply for the language. He encouraged them to give radio talks, garden club lectures, and so on, to promote public awareness of wildlife and conservation. Field work was only part of the campaign; communication was also vitally important to him.

Frederick Hamerstrom recalls that Leopold had contempt for what he called "College English"—inflated, muddy, and fatuous prose—and that he often criticized the narrowness of many professionals to whom he disparagingly referred as "first generation scientists"—Ortega's "learned ignoramuses" as noted by Gare (2023). These were scientists who were so overly specialized that they lacked breadth of interest and overall cultural sophistication. Leopold himself displayed an extraordinary breadth of interest and he was a charming, interesting, and interested conversationalist.

It is difficult for even the most articulate and reflective of students to say exactly what qualities make a great teacher like Leopold. I suppose, as in many things human, there is an ineffable something, a bit of magic, an unusual and fortuitous blend of qualities, something hard to pin down. Nonetheless, several characteristics were reiterated by all my informants.

The first requirement of a good teacher is to know something worth imparting to others. Leopold's knowledge and talents in his field were nothing short of astonishing, and they have now become legendary. McCabe once said that The Professor could pick up a handful of forest soil, filter it through his fingers, smell it, and without having previously taken a census of the area, give a remarkably accurate prediction of the flora and fauna that could be found there. In her book, Frances Hamerstrom tells an equally impressive vignette. While helping the Hamerstroms dig an outhouse repository (to ordinary folk a grimy, boring, mindless task), The Professor took the occasion, as each layer of soil was exhumed, to speculate extemporaneously on its origin and subsequent history, from relatively recent timber fires all the way back to the glacial incursion more than 10,000 years ago. These are just two illustrations of his uncanny ability to "read the land" (Hamerstrom 1980, p. 38).

A second characteristic was his genuine interest in his students. Students can immediately sense indifference and an inability to relate personally and become genuinely involved with students will always seriously inhibit the transmission of skill and information.

According to Nina Bradley, as a teacher her father exercised restraint. He was keenly aware of the thrill of personal discovery and did not rob those around him of the pleasure of learning. His method of teaching was, in other words, essentially Socratic. Socrates, if we may trust Plato's portrait of him, though always a paragon of virtue and usually good humored, could also be a truculent, abrasive, and occasionally an unscrupulous quibbler. I certainly do not mean to suggest that Aldo Leopold was like Socrates in these particulars. But Socrates realized a fundamental educational truth, that the best teachers provide their students with a standard of personal and professional excellence and with a stimulating and encouraging environment in which to learn. A good teacher is rather like a catalyst in a chemical reaction. There is the body of information, technique, and theory, and there is the student. The good teacher brings the two together and enables the reaction to take place.

Charles Bradley also compares Leopold with Socrates. As he writes in a brief memoir,

> then I would remember the gentle pressure of the questions coming at me and realize the extent to which my mind had been pushed: pushed for clarity, pushed for accuracy, pushed for depth of understanding, pushed to the point where ignorance emerged as the limiting factor to that line of conversation. (Bradley 1979, p. 7).

While Socrates might have gone further and embarrassed his interlocutor, Leopold, mercifully, "just changed the subject when he felt any trace of defensiveness" (Bradley 1979, p. 7).

This Socratic core of The Professor's teaching technique seems to be confirmed by McCabe and the Hamerstroms. McCabe implied that Leopold was in fact notorious for asking his graduate students unexpected but penetrating questions about their field research during their regular reports in his office. Indeed, according to McCabe, one student who was not identified to me by name, lived in chronic fear (which became acute as the hour arrived) of those periodic interviews.

Leopold left a record of hypothetical questions that this "modern Socrates" (in Bradley's insightful characterization) might ask in "Natural History-The Forgotten Science:"

> To visualize more clearly the lopsidedness and sterility of biological education as a means of building citizens, let's go afield with some typical bright student and ask him some questions. We can safely assume he knows how plants grow and cats are put together but let us test his comprehension of how the land is put together.
>
> We are driving down a country road in northern Missouri. Here is a farmstead. Look at the trees in the yard and the soil in the field and tell us whether the original settler carved his farm out of prairie or woods? Did he eat prairie chicken or wild turkey for Thanksgiving? What plants grew here originally which do not grow here now? Why did they disappear? What did the prairie plants have to do with creating the corn-yielding capacity of this soil? Why does this soil erode now, but not then?
>
> Again, suppose we are touring the Ozarks. Here is an abandoned field in which ragweed is sparse and short. Does this tell us anything about why the mortgage was foreclosed? About how long ago? Would this field be a good place to look for quail? Does short ragweed have any connection with the human story behind yonder graveyard? If all the ragweed in the watershed were short, would that tell us anything about the future of floods in the stream? About the future prospects for bass and trout?
>
> I fear that our Phi Beta Kappa biologist would consider these questions insane, but they are not. Any amateur naturalist with a seeing eye should be able to speculate intelligently on all of them and have a lot of fun doing it. (Leopold 1953, pp. 62–63).

This method of teaching was not just for his students, it was also for his kids. "One instance I particularly remember," writes Leopold's younger daughter, Estella.

> We were standing by the [Wisconsin] river, when he pointed and said, 'Baby, look at that island. Why do you suppose the big cottonwoods are on the upstream end and the little ones are growing at the bottom end?' I struggled with the issue until I remembered that the river was moving sand all the time to the downstream end of the island. That meant that the upstream end was older and the lower end was younger. Ha! That's the answer. In fact, theoretically one could date the upper and lower ends of the island using cottonwoods. Dad was a great storyteller. That made him a good teacher. (Leopold 2016)

It is easy to see how a safe and sure student lulled into complacency and intellectual torpor by rote learning might have dreaded going afield with The Professor. On the other hand, imagine the excitement, the sheer exhilaration the same experience would afford an intellectually active and genuinely curious student. These questions and others like them engage the imagination. What can be seen is interesting not only in its own right but as an indication of what cannot be seen. They conjure the past and the future and that which is present but hidden from view. An apparently static and plain landscape by these questions comes alive with drama, heritage, and portent. Perhaps in just this direction lies the solution to our previous problem: How, in outdoor education, can we transform the passive taxonomical nature walk into the imaginatively active, participatory, ecology walk? Here may be an alternative not only to the Point and Name school but to the blindfolds, ropes, and other gimmicks of the Climatization school. The latter may stimulate the senses and entertain, but it does little or nothing to stimulate the mind and promote land literacy—skill in reading the land.

16.3 The Land Ethic in Contemporary Environmental Education

Aldo Leopold's youngest son, Carl, recounts his childhood memories planting trees and doing other work with his father at what Aldo called their family's "weekend refuge from too much modernity: the 'shack'" (Leopold 1949, p. vii–viii). In Carl's words:

> For his children, the family efforts at ecosystem restoration resulted in a deep enthusiasm for environmental issues. . . . The story of the shack is not just a tale of a happy family project. It is the story of the origin of two concepts in environmentalism: ecological restoration and ethical responsibility toward nature. . . . The story is a metaphor of the contagious value of restoration in bringing people back to the natural world in a mutually beneficial relationship. At the same time, it fortifies a personal sense of ethical responsibility to the natural world. (Leopold 2004, p. 149)

Aldo Leopold's teaching and research field was game management or wildlife management or wildlife ecology. (As the field matured, its name changed.) He was its founder. He wrote the book, both figuratively and literally (Leopold 1933). Yet, if this were his single or greatest achievement, he would be remembered fondly only by his family and students and venerated as Founding Father by succeeding generations of fellow professionals in this special and somewhat arcane field. He is also venerated as the Founding Father of ecological restoration, as his son Carl testifies, and of conservation biology as Michael Soulé (1985) acknowledges. But what has propelled Leopold to posthumous international celebrity is his philosophical achievement—the exposition and inculcation of an evolutionary-ecological worldview and its axiological and normative implications. Those implications are sketched in *A Sand County Almanac*'s capstone essay, "The Land Ethic" (Callicott 2013).

The land ethic is sorely needed in our time; indeed, in the long run, the viability of our civilization, even the future tenure of our species on the planet, may well depend on the capacity of our now global culture to absorb it. Leopold himself, even in the late 1940s, was well aware of this necessity. "There is no other way," he writes, "for land to survive the impact of mechanized man" (Leopold 1949, p. xix).

How can we in environmental education contribute to the dissemination, the promulgation of the land ethic? It is a vitally important task, and it rests more squarely and directly with educators than with any other group or profession.

Everyone's experience will amply indicate, I think, that ethical education, generally speaking, must be indirect to be successful and lasting. One does not, in other words, get very far by climbing on a soap box and announcing, as if a message from Heaven, addenda to the Ten Commandments or an emendation of the Golden Rule. Indeed, the evangelistic approach is more likely to produce the opposite result. People, more commonly than not, react defiantly to self-righteous preaching. Aldo Leopold was as keenly aware of this feature of human psychology as anyone else and thus presented the land ethic—although in urgent (but not strident) tones—as a natural development in cultural evolution: "The extension of ethics to this third element [the land] in [the] human environment is, if I read the evidence correctly, an evolutionary possibility and an ecological necessity" (Leopold 1949, p. 239).

In what sense and to what degree the advent of the land ethic is a cultural evolutionary possibility can be inferred from the theoretical foundations of the land ethic. And, upon drawing out this inference, we shall also discover the most effective educational approach for its promulgation. The land ethic is biological in scope (that is, it includes plants and animals, soils and waters within its purview) and it is biological in its theoretical foundations. This gives it, apart from its other merits, a certain formal beauty, a sort of cognitive harmony.

Where do ethics, generally speaking, come from? They are, after all, personally troublesome. Individually, it would seem, we would be better off without them, for an ethic is in Leopold's words, "a limitation on freedom of action in the struggle for existence" (Leopold 1949, p. 202).

Darwin was the first to tackle the problem from a natural history point of view in his second great work, *The Descent of Man.* At first glance, it would seem that ethical behavior would be weeded out of the human gene pool by the inexorable operation of the principle of natural selection, for Darwin the fundamental mechanism of evolutionary biology. Any limitation on freedom of action in the struggle for existence is apparently advantageous to everyone but the agent him- or herself. Evolutionary theory would seem to require, indeed, that as time goes on, human beings should become more and more aggressive, competitive, mutually hostile, and narrowly selfish, because it is the victors in life's *struggle* who live to pass on their traits to succeeding generations. Yet what we actually find is that, through time, moral behavior has become more complex, more widely cast, and more refined. We are faced, thus, with a paradox, an anomaly. How can evolutionary *theory* be squared with the manifest *fact* of human moral behavior, the general acceptance of "limitation[s] on freedom of action in the struggle for existence"?

Darwin's solution is in principle both direct and simple. Many species of animals, *Homo sapiens* outstanding among them, survive and flourish better in cooperative social organizations than as solitaries. Social existence, however, is not possible unless individuals relinquish certain liberties, unless individuals are to some degree mutually deferential, cooperative, and considerate of one another's welfare. In short, accepting ethical limitations on behavior is the dues one pays in order to join society. These limitations in human societies are articulated in a code or body of custom or usage which we call, in sum, ethics or morality. As Leopold points out, we might alternatively define an ethic as "a differentiation of social from antisocial conduct" (Leopold 1949, p. 202).

This natural-history explanation of the origin of ethics, to which Leopold alludes toward the beginning of "The Land Ethic," clearly issues in a fundamental principle: ethical relations and social organization are correlative. The details of our ethic reflect the structure of our society and, more importantly for the land ethic, the perceived boundaries of our society are also the perceived boundaries of the extent of our moral obligations. History amply confirms this relationship between social boundaries and the extent of moral obligation. When societies were small, merely clan or tribal organizations, an individual's ethical obligations extended only to members of his or her own clan or tribe; anyone else was dealt with according to the dictates of expediency, not conscience. As society grew in extent, as tribes merged into nations, the extent of ethical obligations grew apace. We now live in what Marshall McLuhan (1962) has aptly styled a "global village." Those who acknowledge their membership in this single global human community also acknowledge moral obligations to all members of such a community—that is, to all human beings, nationality, race, or other socially sorting characteristics notwithstanding.

The land ethic in Aldo Leopold's great vision is the next step in this process of social-ethical evolution. He writes, "all ethics so far evolved rest upon a single premise: that the individual is a member of a community of interdependent parts." Ecology "simply enlarges the boundaries of the community to include soils, waters, plants, and animals, or collectively: the land. . . . In short, a land ethic changes the role of *Homo sapiens* from conqueror of the land community to plain member and citizen of it" (Leopold 1949, pp. 203–204).

But how can we help to bring this next step in the ethical sequence to pass? The answer to this question should now be very direct and clear: It is simply through promoting extensive and intensive *ecological understanding*. As Leopold has pointed out, "that land is a community is the basic concept of ecology ... " (Leopold 1949, p. xix). Educating people in ecology is, therefore, teaching them to understand and to perceive the natural world as "one humming community" (Leopold 1953, p. 148). It is, moreover, fostering the perception of ourselves and the human economy as but one part of the larger society and economy of Nature. Ecology is not just one science among many; it is a habit of mind and a way of experiencing. The end result of genuine ecological education is a complete reorientation of a person to his or her surroundings.

When ecological understanding and awareness become generally distributed in our now global culture, then the land ethic will follow as a natural and psychologically necessary consequence. Evolution has endowed us with what we might call, following Hume, a set of "moral sentiments" which are, so to speak, actuated by the recognition of a fellow member of our society or community. The basic concept of ecology is that the myriads of non-human natural beings—soils and waters, plants and animals—are functioning members of a single natural community to which we also belong and upon which we utterly depend for the means to life. When this basic concept of ecology is taught at all levels of education, from story and song in early childhood education to mathematics, the sciences, and the humanities in higher education, the land ethic may be transformed from one human's dream to all humankind's reality.

16.4 Field Environmental Philosophy: Building on Leopold's Legacy

Aldo Leopold was not unaware of cultural diversity, but he did not seem as concerned about conserving it as he was about conserving what eventually came to be called "biodiversity." The term was coined in 1985 by Walter Rosen (Sarkar 2002). Biodiversity is now understood to consist of variety at every level of biological organization—genes, phenotypes, populations, species, biotic communities, ecosystems, landscapes, and biomes. Leopold seems to consider cultural diversity to be derived from biodiversity at the landscape and biome scales of biological organization:

> Wilderness is the raw material out of which man has hammered the artifact called civilization. Wilderness was never a homogeneous raw material. It was very diverse, and the resulting artifacts are very diverse. These differences in the end-product are known as cultures. The rich diversity of the world's cultures reflects a corresponding diversity in the wilds that gave them birth. (Leopold 1949, p. 188).

Leopold goes on to note the impending "exhaustion of wilderness in the more habitable parts of the globe" and "the worldwide hybridization of cultures through modern transport and industrialization" and electronic communication, I might add. He goes on to "plea for the preservation of a few tag-ends of wilderness as museum pieces" (Leopold 1949, p. 188). But he does not also plea for the preservation of threatened and endangered cultures.

Sharing the connection Leopold draws between biodiversity and cultural diversity, Ricardo Rozzi (2013) augmented Leopold's emphasis on biodiversity conservation with equal attention and concern for cultural conservation and has fused the two into one with the concept of biocultural conservation. He has fostered biocultural conservation in southern Chile at the Omora Ethnobotanical Park (OEP) (Rozzi et al. 2006). To enhance understanding of biocultural diversity, at OEP (55°S), Rozzi and colleagues developed the methodological approach called field environmental philosophy (FEP). FEP emphasizes ecologically and philosophically guided field experiences in local habitats, sociocultural communities, and regional institutions and is designed to stimulate the perception and value of biological and cultural diversity in specific places and moments.

Today, fairly pristine high-latitude regions offer humanity a unique opportunity to make an ethical shift (Rozzi et al. 2012). The FEP methodology provides an orientation for graduate students and other participants to research and respect the "otherness" in such remote wildernesses—the expression of ancient cultures, life forms, and habitats not yet immersed in global society. This can help re-contextualize the global economy, politics, and culture. This research could stimulate an ethical–ecological shift from the current tendency to overlook vital bonds between humans and nature toward a new understanding of humans as co-inhabitants of ecosystems (Rozzi 2013). In the twenty-first century, the biocultural ethic values a culturally and biologically diverse array of human and other-than-human life forms that sustain ecosystem processes, as envisioned by Aldo Leopold.

References

Anscombe GEM (1958) Modern moral philosophy. Philosophy 33:1–19

Bauckman R (2011) Living with other creatures: green exegesis and theology. Baylor University Press, Waco, TX

Bradley CC (1979) A short story of a man hunt. Wisconsin Acad Rev 7–9

Breakey H, Breakey N (2023) Leopold's cultural harvest, biocultural tourism, and field environmental philosophy. In: Rozzi R, Tauro A, Wright T, Avriel-Avni N, Klaver I, Berkowitz A, Brewer C, May RH Jr (eds) Field environmental philosophy: education for biocultural conservation. Ecology and ethics, vol 5. Springer, Dordrecht, pp 281–294

Callicott JB (1994) Earth's insights: a multicultural survey of ecological ethics from the Mediterranean Basin to the Australian outback. University of California Press, Berkeley

Callicott JB (2013) Thinking like a planet: the land ethic and the earth ethic. Oxford University Press, New York

Callicott JB, Frodeman R (2009) Encyclopedia of environmental ethics and philosophy. Macmillan Reference USA, Detroit

Elton C (1927) Animal ecology. William Clowes, London

Emerson RW (1836) Nature. James Munroe, Boston

Gare A (2023) Challenging the dominant grand narrative in global education and culture. In: Rozzi R, Tauro A, Wright T, Avriel-Avni N, Klaver I, Berkowitz A, Brewer C, May RH Jr (eds) Field environmental philosophy: education for biocultural conservation. Ecology and ethics, vol 5. Springer, Dordrecht, pp 309–326

Hamerstrom F (1980) Strictly for the chickens. Iowa State University Press, Ames

Leopold A (1933) Game management. Charles Scribner's Sons, New York

Leopold A (1949) A Sand County Almanac and sketches here and there. Oxford University Press, New York, p 1966

Leopold A (1953) Round river: from the journals of Aldo Leopold. Leopold LB (ed). Oxford University Press, New York

Leopold A (1991) The role of wildlife in a liberal education. In: Flader SL, Callicott JB (eds) The river of the mother of god and other essays by Aldo Leopold. University of Wisconsin Press, Madison

Leopold AC (2004) Living with the land ethic. BioScience 54:149–154

Leopold EB (2016) Stories from the shack: Sand County revisited. Oxford University Press, New York

McKeon R (ed) (1941) The basic works of Aristotle, trans. of EN: Ross WD. Random House, New York

McLuhan M (1962) The Gutenberg Galaxy: the making of typographic man. The University of Toronto Press, Toronto

Muir J (1916) A thousand-mile walk to the gulf. Badè WF (ed). Houghton Mifflin, Boston

Nash R (1967) Wilderness and the American mind. Yale University Press, New Haven, CT. Ch. 11

Reyhner J, Singh NK (2010) Cultural genocide in Australia, Canada, New Zealand and the United States. Indigenous Policy J 21:1–26

Rozzi R (2013) Biocultural ethics: from biocultural homogenization toward biocultural conservation. In: Rozzi R, Pickett STA, Palmer C, Armesto JJ, Callicott JB (eds) Linking ecology and ethics for a changing world: values, philosophy, and action, Ecology and ethics, vol 1. Springer, Dordrecht, pp 9–32

Rozzi R, Massardo F, Anderson C, Heidinger K, Silander J Jr (2006) Ten principles for biocultural conservation at the southern tip of the Americas: the approach of the Omora Ethnobotanical Park. Ecol Soc 11(1):43. [online] URL: http://www.ecologyandsociety.org/vol11/iss1/art43/

Rozzi R, Armesto JJ, Gutiérrez J, Massardo F, Likens G, Anderson CB, Poole A, Moses K, Hargrove G, Mansilla A, Kennedy JH, Willson M, Jax K, Jones C, Callicott JB, Kalin MT (2012) Integrating ecology and environmental ethics: earth stewardship in the southern end of the Americas. BioScience 62(3):226–236

Rozzi R, Chapin FS III, Callicott JB, Pickett STA, Power ME, Armesto JJ, May RH Jr (eds) (2015) Earth stewardship: linking ecology and ethics in theory and practice, Ecology and ethics, vol 2. Springer, Dordrecht

Sarkar S (2002) Defining "biodiversity"; assessing biodiversity. Monist 85(1):131–155

Soulé M (1985) What is conservation biology? BioScience 35:727–734

White L Jr (1967) The historical roots of our ecologic crisis. Science 155:1203–1207

Chapter 17
Leopold's Cultural Harvest, Biocultural Tourism, and Field Environmental Philosophy

Hugh Breakey and Noreen Breakey

Abstract This chapter considers the opportunities for tourism to provide biocultural conservation experiences and outcomes. As the basis for this discussion, we explain the Three-Stage Land Ethic—a theory of moral development applicable to individuals and societies, drawn out of the seminal work of Aldo Leopold. The Three-Stage Land Ethic incorporates both cognitive and experiential elements, leading to changed values and virtues, and a widened sphere of moral concern (including animals, plants, and the land itself). Then, while we acknowledge the relevance of ecotourism to biocultural ethics, we introduce and justify the related concept of *biocultural tourism* as tourism which provides education within natural and/or cultural heritage experiences and prioritizes biocultural conservation through the broadening of ethical care to new ecological and/or cultural phenomena. We then illustrate how the ideals of biocultural conservation can be achieved through biocultural tourism by activating the elements of the Three-Stage Land Ethic and demonstrate the power of one element for progressing such moral development. We conclude by addressing some seeming tensions between the Leopoldian Land Ethic and Field Environmental Philosophy, showing they are complementary approaches.

Keywords Aldo Leopold · Ecotourism · Education · Ethics · Moral development

17.1 Introduction

Sustainable tourism aims to achieve a balance between the needs of tourists, the environment, local people, and businesses. Ethics plays a key role in achieving this balance. For this reason, we have previously explored the land ethic of Aldo

H. Breakey (✉)
Institute for Ethics, Governance and Law, Griffith University, Nathan, QLD, Australia
e-mail: h.breakey@griffith.edu.au

N. Breakey
School of Business, The University of Queensland, Brisbane, QLD, Australia
e-mail: n.breakey@uq.edu.au

R. Rozzi et al. (eds.), *Field Environmental Philosophy*, Ecology and Ethics 5,
https://doi.org/10.1007/978-3-031-23368-5_17

Leopold. In his writings, Leopold described a theory of moral development, explaining how individuals and societies come to value and respect other peoples, cultures, animals, and plants. Crucially for tourism, Leopold argued that the process of moral development required experiences and interactions with cultural and natural phenomena, in order to value and cherish them (Leopold 1968), as also affirmed by his son Carl (Leopold 2004). Understanding Leopold's theory could therefore empower those designing, delivering, supporting, and selling tourism experiences to protect and nurture the most precious sustainability resource: human beings that value and respect the environment (Breakey and Breakey 2015).

This chapter builds upon this prior work by drawing on biocultural ethics and Field Environmental Philosophy. It begins with an overview of the Three-Stage Land Ethic (3SLE), a theory of moral development drawn from Leopold's work. With this analysis in tow, the next section defines "biocultural tourism," showing its extension beyond existing categories like ecotourism, and illustrating how it can partake of key 3SLE elements to drive improved and broadened ethical values and virtues. We then show that both the psychological theory of Leopold's Three-Stage Land Ethic and the practice of Field Environmental Philosophy can support and enrich practices of biocultural tourism.

17.2 The Three-Stage Land Ethic: An Overview

The Three-Stage Land Ethic (3SLE) is a model of moral development (Breakey and Breakey 2015), drawn from the seminal work of Aldo Leopold (1968), *The Sand County Almanac*. The 3SLE provides an account of how moral development occurs, both at an individual and societal level. Specifically, the 3SLE focuses on achieving expansions in the realm of moral considerability or subjecthood—that is, how the circle of ethical concern can be expanded to include new entities (human beings from diverse cultures, animals, plants, and more broadly the land).

The 3SLE combines *reasons* for valuing and respecting the land with *experiences* and *practices* that connect us personally and emotionally to it. (Like Leopold, we use the term "land" to refer to all elements of ecosystems, including animals [human and non-human], flora, and the processes supporting life.) As individuals and cultures progress through the three stages, they come to value the land and are drawn into changed attitudes and behavior.

The first stage consists of learned realizations of human community with the land. We come to think of non-human entities as our *collaborators* (each with our own roles in ecosystem functioning), our *kin* (sharing an evolutionary history and journey), and our *benefactors* (sharing our fate with all life).

The second stage reflects the fact that *mere understanding is insufficient* to create a sense of ethical concern and drive ethical behavior. Our emotions, values, and sense of identity must be engaged—as Leopold (1968, p. 48) put it, we cherish only what we "see and fondle." This gives rise to Leopold's idea of the "cultural harvest"—the goods that we can achieve as we engage with the land. These include

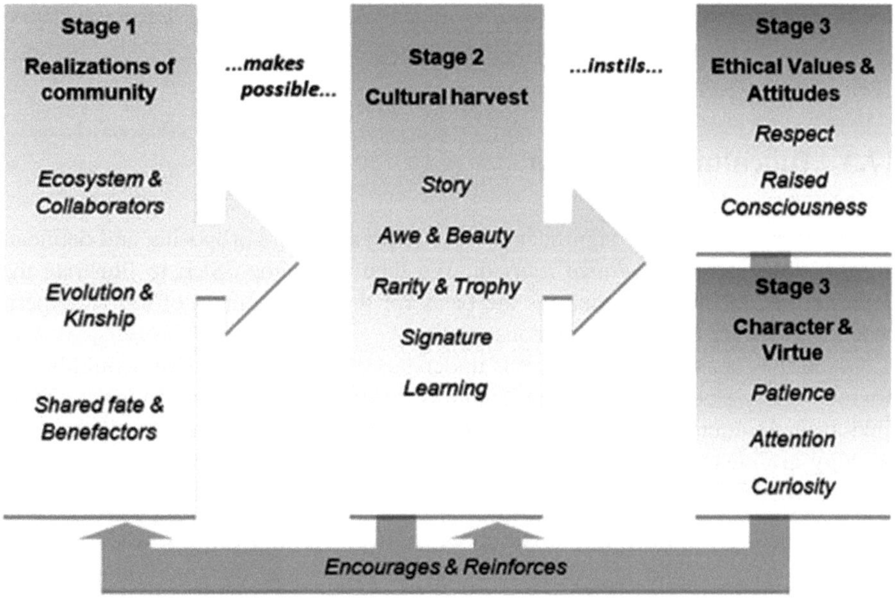

Fig. 17.1 Three-stage land ethic: process of moral development. Adapted from Breakey and Breakey (2015)

story (creating narratives that involve us with the land), *awe* and *beauty* (perceiving the land's majesty and wonder), *rarity* and *trophy* (overcoming challenges and having unique experiences), *signature* (seeing our personal impact on the land), and *learning* (coming to a deepened understanding of the land).

Having progressed through Stage 2, we enter Stage 3, where we have a raised consciousness of the land's value, and we respond respectfully and sustainably to it. Figure 17.1 illustrates the 3-stage process.

In Fig. 17.1 describing the 3SLE, we incorporate two caveats motivated by the Field Environmental Philosophy (FEP) methodological approach (Rozzi et al. 2010, 2012, Rozzi 2013). First, rather than being a strictly staged process, there are feedback and feedforward links between the stages. For example, we can apprehend new realizations of community (Stage 1) from stories and learning (Stage 2). Or we might observe how local people respect their habitat (Stage 3) and then begin looking for the (Stage 1) reasons and (Stage 2) cultural goods they possess for doing so.

Second, the final stage of the 3SLE model includes not only changes in values, attitudes, and actions, but also in character traits (ongoing dispositions such as patience, mindfulness, and thoughtfulness). As Rozzi et al. (2010) shows, FEP activities can aim at changing character *not* merely in the sense of creating new values or attitudes, but also in changing *mentality*, promoting slowness, patience, carefulness, and curiosity. This is incorporated into Stage 3, such that Stage 2 activities can instil character traits and virtues like patience, mindfulness, attention to detail, carefulness, and thoughtfulness in action—all of which Leopold (1968) exemplified in his person and his "shack stories."

17.3 Biocultural Tourism

This section discusses the significance of ecotourism before proposing and delineating the concept of *biocultural tourism.* We then apply the 3SLE to illustrate the potential of biocultural tourism as the basis for the development of tourist experiences that deliver on biocultural conservation.

Given the focus on education, it is understandable that *ecotourism* would be the obvious starting point in any examination of biocultural ethics and tourism. Direct links from ecotourism to education for biocultural conservation and the FEP methodology are clear in the definition of ecotourism, which according to the United Nations World Tourism Organization (UNWTO), has the following characteristics:

1. All nature-based forms of tourism in which the main motivation of the tourists is the observation and appreciation of nature as well as the traditional cultures prevailing in natural areas.
2. It contains educational and interpretation features.
3. It is generally, but not exclusively organized by specialized tour operators for small groups. Service provider partners at the destinations tend to be small, locally owned businesses.
4. It minimizes negative impacts upon the natural and socio-cultural environment.
5. It supports the maintenance of natural areas which are used as ecotourism attractions by:

    (a) Generating economic benefits for host communities, organizations and authorities managing natural areas with conservation purposes.
    (b) Providing alternative employment and income opportunities for local communities.
    (c) Increasing awareness towards the conservation of natural and cultural assets, both among locals and tourists (UNWTO 2001, p. 19).

Furthermore, ecotourism has drawn on education theories since its beginnings around three decades ago. For example, Forestell (1993) developed an interpretive model for marine ecotourism, applying Piaget and cognitive psychology's approach to learning. Formal education approaches applied in various ecotourism settings were reviewed by Walter (2013), who also noted that visitor education can develop in *ad hoc* and informal ways. This aligns with ecotourism drawing on various sources of expert knowledge, including both scientific sources, such as marine biologists for whale-watching, and indigenous local people, who often provide a more holistic, blended interpretation of the close interaction between the people and their environment (Walter 2013).

Additional direct links between ecotourism and education for biocultural conservation are the requirements of both *immersive* experiences and *opportunities to actively support conservation*, in contrast to passive natural encounters that are also included in the broader *nature-based tourism* (TEQ 2021) and are common in mass tourism. Finally, building on the initial ecotourism tenet of generating conservation awareness, tourism experiences have been designed to generate pro-environmental attitudes and behavior. Research has shown that such experiences contribute to knowledge, awareness of issues, and post-experience behavior, but intentions may not result in long-term behavioral changes (Ballantyne and Packer 2005).

Beyond ecotourism, there are also other types of tourism relevant to this discussion. As a result, we propose a focus on what we term *biocultural tourism* as a way of encompassing more fully the relevant types of tourism that a biocultural ethic would support and exemplify. Ecotourism focuses on the environment and then includes the cultural elements within that environment. Coming from the other direction *community-based tourism* focuses first on immersion in authentic cultural experiences within traditional small communities and only then embraces the local interaction within nature. Such tourism experiences commonly do not align with ecotourism, even though they may incorporate many of its tenets. Instead, the key motivation of the tourists is the cultural experience, and a priority outcome is ownership and management of tourism by the local community.

Furthermore, other forms of tourism can be the basis for biocultural conservation experiences, including *cultural tourism*, *historical tourism*, *Indigenous tourism*, *geotourism*, *dark tourism*, *wildlife tourism*, *slow tourism*, and *voluntourism*. Such experiences are only "ecotourism" when they occur in a natural setting, and yet they commonly prioritize education, immersion, and/or moral development. Designing tourism experiences for biocultural conservation can build on knowledge accrued beyond ecotourism, such as:

1. *Free-choice learning*, as applied in visitor studies. While based in education, this form of learning is informal and self-selected, directed, and controlled (Falk 2005). Free-choice learning is not limited to natural settings, such as its application in museums and aquariums. In addition to the education element, there is often a strong focus on generating pro-environmental attitudes and behavior in visitors.
2. *Co-creation*. For immersion, co-creation in tourism is relevant. In their review of co-creation of tourist experiences, Ana Claudia Campos et al. (2018) outline how this approach shifts to the tourist, prioritizing their active role in creating and giving personalized meaning and value to their own tourism experiences.
3. *Voluntourism*. Self-discovery motivated voluntourism can result in moral development. Such shifts are also attributed to the opportunity for voluntourists to immerse themselves in the local life.

Thus, while tourism opportunities for biocultural conservation link directly with ecotourism, there are other relevant forms of tourism that have successfully incorporated educational, immersive, and/or personal development approaches to the design of the tourists' experiences in both natural and/or cultural settings. With this in mind we define biocultural tourism as *tourism which provides education within natural and/or cultural heritage experiences, and prioritizes biocultural conservation through the broadening of ethical care to new ecological and/or cultural phenomena.* Biocultural tourism thereby encompasses:

(i) Pro-environmental behavior experience design (which is commonly included in ecotourism and visitor experiences)
(ii) Cultural awareness and appreciation aspects (which are often offered in cultural / Indigenous tourism)
(iii) Personal development opportunities (which can occur through voluntourism)

Biocultural tourism thus requires and prioritizes these distinct types of ethical development experiences and outcomes. This specific aim of moral development also differentiates biocultural tourism from other forms of tourism. For example, while education is integrated into tourism experiences, the education elements are part of the offering and tourists may choose to only engage in limited ways. By focusing on the opportunities for tourism to drive moral development, and designing the tourist experiences accordingly, biocultural tourism can have greater impact on values, attitudes, and behavioral change. Furthermore, this experience design extends beyond education. The 3SLE demonstrates the variety of activities, which can be included in tourism experiences, that can support such moral development.

Biocultural tourism can therefore employ the 3SLE to progress its aim of driving ethical development for biocultural conservation. This aligns with the FEP methodology, by having people developing philosophically through their activities in the field. Tourism, and in particular biocultural tourism, can engage all elements within Stages 1 and 2 to progress to the Stage 3 moral development elements. Figure 17.2 illustrates the opportunity to utilize tourism experiences for biocultural conservation.

We now turn to an in-depth consideration of one element of the cultural harvest to demonstrate the opportunities of biocultural tourism through the 3SLE to advance moral development. The opportunity to experience *awe* and *beauty* is an exceedingly common driver for participation in leisure tourism. Tourism marketing is primarily based on showcasing beauty—natural phenomena, iconic wildlife (Zhu 2023 in this volume), cultural expressions, and artistically presented food and wine. The opportunity is to then integrate the power of this experience with the meaning behind the awe and beauty. For example, the "time path" in the Copper Coast Geopark, Ireland, provides a "walk through time" using 28 slabs in the ground that take visitors through the major geological stages of Earth (Dowling 2010), which are ultimately responsible for the spectacular coastal vistas.

However, to illustrate the true power of awe and beauty, we will focus on an insect, albeit a beautiful one. Research shows that beauty of the butterfly can be a specific route through the cultural harvest to reach appreciation and value. People prefer certain animals over others, with preference often given to those with human-like traits (Lemelin 2007; Woods 2000). This preference aligns with several parts of the 3SLE Stage 1, particularly "kinship"—but it fails to deliver an encompassing ecocultural ethic working across species and taxa (Rozzi 2019). There are

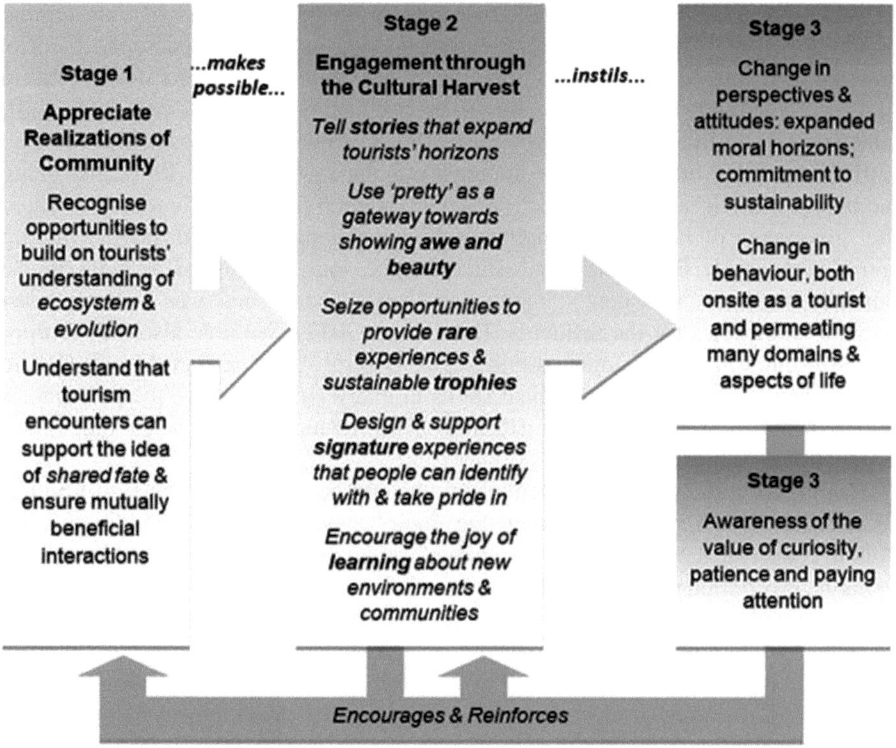

Fig. 17.2 Tourism and the three-stage land ethic: process of moral development. Adapted from the initial model of the 3SLE to tourism (Breakey and Breakey 2015)

exceptions, such as the "inverse appeal" of some dangerous, "un-human" animals like crocodiles (Woods 2000), and variation in perceptions, such as some respondents disliking snails, while others liking them because they are "beautiful" (Barua et al. 2012). While invertebrates, and insects in particular, are virtually unanimously disliked, Shipley and Bixler (2016, p. 65) propose that "insects are widely loathed because a few species are highly irritating." In contrast, butterflies are highly admired, charismatic, even flagship, invertebrates (New 1997; Barua et al. 2012). A national survey in the USA determined that invertebrates were the least preferred taxonomic group; however, the butterfly and ladybug were favorites, with the finding attributed to their aesthetic appeal (Kellert 1996). In their study of the attitudes of nongovernmental organizations (NGOs) towards species for conservation, Edo Knegtering et al. (2002, p. 390) considered an inconsistent finding may have been due to the "charismatic appearance" of the butterflies in the "insect" taxon. Specifically relevant for the 3SLE, research has shown that endangered butterflies receive "substantial public support for their protection," demonstrating that beauty can lead to "admiration and appreciation, though rarely affection" (Kellert 1996, pp. 125–126).

While not focused on butterflies, research by Woods (2000, p. 33) into animal preferences found "beautiful" to be the most popular word used to describe favorite animals, and this "aesthetic attraction can be based on shape (e.g., sleek), texture (e.g., fluffy, cuddly), color (e.g., bright or contrasting), or movement (e.g., fast, athletic)." Butterflies fulfil three of these four considerations. Maan Barua et al. (2012) found that butterflies were the most popular invertebrate, with beauty being the most common reason for selection. Their research into butterfly traits identified the importance of large-wings and/or striking color patterns, with people finding some of the butterflies "dazzling." Interestingly, tourists selected as a favorite a butterfly with an "unusual," "interesting shape," while this was one of "least popular" butterflies for the residents (Barua et al. 2012). We note also the incorporation of butterflies in the ancient cultures of Central America (Lemelin 2007) and Egypt (New 1997), and the mythical status in many cultures of the metamorphosis into a beautiful flying butterfly (Kellert 1996), which all reinforce biocultural connections.

Highlighting the untapped potential of awe and beauty, in the context of butterflies, Kellert (1996, p. 129) acknowledges their "aesthetic appeal" and how this has "often endeared these creatures to people" and that a "more respectful attitude towards invertebrates could perhaps be encouraged by making butterflies... the moral equivalent of charismatic vertebrates like the wolf or whale." More recently, Lemelin and Jaramillo-Lopez (2020, p. 299) consider that the Monarch Butterfly Biosphere Reserve in Mexico has "overlooked the importance of education, the creation and fostering of environmental ethics, and the development of sustainable low-impact facilities."

Furthermore, activities within the 3SLE can work together. For butterfly tourism experiences, there is often an element of rarity (New 1997). This may be due to a species being endangered or threatened, challenging to see due to migration, or being limited to a particular region. This understanding of rarity may occur through education about the endangered or threatened status of butterflies. As a result, a combination of Stage 2 cultural harvest elements may be required to reach Stage 3; i.e., beauty combined with learning and rareness.

Finally, there are two issues to be cognizant of when designing biocultural tourism experiences. First, it is important to acknowledge that biocultural tourism requires dealing with a variety of people at different stages of moral consciousness. The challenge is to attract, and then inform and develop all individuals at their different positions and move them forward while also delivering a positive tourism experience. Second, there can be a temptation to overly focus on the unusual. Tourism commonly "sells" the unusual as the drawcard to a particular destination. This can have advantages, such as highlighting diversity, and drawing on the cultural harvest elements of *rarity* and *wonder*. However, this can potentially also be misleading and exclude the *mundane* and *everyday* (Bramwell and Lane 1993) that make up an important part of the biocultural world.

17.4 3SLE and FEP as Complementary Approaches

The 3SLE complements the core insights of biocultural ethics. *Biocultural ethics* presents an ecological ethic that pushes against biocultural homogenization by stressing the links between ethos as habit and habitat for both human and non-human animals, leading to an understanding and lived practice of seeing other life as our co-inhabitants (Rozzi 2013). The *field environmental philosophy* methodology is a process for researching and respecting the "otherness" in remote wildernesses, as well as in rural and urban settings, in a way that exemplifies biocultural ethics (Tauro et al. 2021). It incorporates "field" because field research is necessary to see all relevant components and processes, allowing full perception of biocultural diversity in habits and habitats, and in engaging experience as well as concepts. It incorporates "environment" because ethics is not a purely human affair, and "philosophy" by integrating epistemological and ethical dimensions. There is a parallel here with Frodeman and Briggle's (2016, p. 66) "field philosophy," where the application of philosophy to concrete situations becomes itself a philosophic question, and where philosophic principles can as much flow *from* embeddedness in a field situation, as be applied *to* that situation.

The four-step field environmental philosophy (4SFEP) is a specific instantiation of the broader field environmental philosophy methodology (Rozzi et al. 2010, 2012). A practical pedagogical model for teaching the teachers, the 4SFEP methodology includes (1) transdisciplinary biocultural research; (2) composition of narratives, metaphors, and communication that synthesizes results from step "1"; (3) design of ethically/ecologically guided field activities that overcome physical and conceptual barriers to perceiving and valuing biocultural diversity; and (4) implementation of conservation sites where these activities and ethical actions can continue sustainably.

How does all this relate to Leopold's 3SLE? The 3SLE is an empirical theory about how ethical development occurs—how the "circle of concern" can be widened—in individual humans and larger cultures. As such, the 3SLE provides an explanation of field environmental philosophy's efficacy and importance. Within the 4SFEP, we can see all three 3SLE stages—and indeed the operation of distinct elements within each stage. Learning (Stage 2) and empirical insights (Stage 1) arise in the initial research. Next, Stage 2 cultural harvest elements of story, awe and beauty, and rarity and trophy are present in the 4SFEP synthesis into narratives and metaphors and the design of ethically-ecologically guided field activities. Finally, Stage 3 ethical action arises both in the design of the field activities, and in the final implementation of the conservation sites. The 3SLE thus explains *how* the 4SFEP works—by providing key realizations of ecological community (working on a cognitive level) intertwined with cultural harvest elements (working on axiological and emotional levels) that lead to ecological ethical values, attitudes, and virtues.

Yet it might be thought there remain important disagreements between these two approaches. From a biocultural point of view, two interrelated concerns with a Leopoldian ethic arise (see Rozzi 2013, pp. 10, 19). First, Leopold focuses on the universal need to move from the view of *Homo sapiens* as a conqueror of the land to a plain member and citizen of it. Second, Leopold tells a story of the historical widening of the circle of ethical concern throughout history, suggestive of linear progress. It might reasonably be objected from a biocultural ethics standpoint that both these points appear peculiarly Western and are inattentive to other cultures valuing the land (and so not perceiving themselves as conquerors), and possessing different (and non-linear) historical trajectories. Both of these concerns warrant careful consideration.

To consider whether they raise significant disagreements between Leopoldian ethics (as understood through the 3SLE) and biocultural ethics, it is worth teasing out the parts of Leopold's thinking that are universal, and those that are more fluid, pluralist, and inclusive. Doing so will reveal that the systems are complementary, with little substantive disagreement.

First, Leopold is committed to the *universal human need* for an ecological ethic. To survive and flourish on Earth, human beings (and ultimately all human cultures) must recognize their fate is intertwined with ecological well-being, and therefore they must come to value the land (if they do not do so already). This is a universal prescriptive claim—but it is a common theme in many environmental ethics, including biocultural ethics (see, e.g., Rozzi 2019, pp. 249–250).

Second, Leopold wants to show that such an ethic is *possible*—that people can come to believe it and value it. This is the pivotal reason for his account of the historical widening of the circle of ethical concern—not because it tells a story of linear progress, but because it shows progress *is* possible. Through employing both principled reasons and emotional connections, human ethical norms can become better and broader.

Third, the 3SLE requires that it be coherent and rational to believe in such an ethic: that there are strong, accessible reasons for valuing and respecting the land. Leopold appeals to three types of reasons (the "realizations of community" in 3SLE Stage 1). It may be that there are other, different reasons available, learned long ago or newly uncovered by other cultures, and capable of being shared through intercultural dialogue. Such reasons would be welcome inclusions into the 3SLE.

Fourth, Leopold located five distinct (Stage 2) cultural harvest elements that could fire the imagination and engage the emotions to encourage valuing of the land. Rozzi (2012, p. 233) makes a similar point, but perceptively ties it more closely to local habitats: "Inhabiting a particular habitat generates recurrent forms of *habitation* over time—that is, the habits that configure the identity of humans and other animals." As such, many different cultures would have their own versions of Leopold's cultural harvest elements—their own unique types of story, beauty, and art, for example, that attach them emotionally to their local habitat. Perhaps entirely new cultural harvest elements, that largely escaped Leopold's attention, could be added here. For example, Leopold's journey, as related in his shack stories, is in many respects a solitary one (alongside his faithful dog). Indeed, Leopold (1968, pp. 171, 176) eulogizes the precious isolation nature can offer us. Yet it might be rightly pointed out that many of the practices valuing the land around the world are not solitary, and that *community*, in the sense of shared practices, rituals, traditions, myths, and collective identification and solidarity, play a central role. At present, we have opted to cleave to the five original cultural harvest elements, but if these collective practices cannot be appropriately captured within these confines, then the addition of another cultural harvest element of *community* would be apt.

Finally, the same may be said for ethical action (Stage 3). While the 3SLE's ethical principles and virtues are broadly fixed (in terms of respect for, and raised consciousness about, people, animals, and the land), they are nevertheless capable of—and might clearly benefit from—different instantiation contextualized to different cultures' habits and habitats. Cultural difference is to be expected and welcomed in such cases, and different cultures may learn much from one another. An analogy might be drawn with psychologist Jon Haidt's "social intuitionism," where ethical intuitions are seen as basic (biologically given), but are shaped and specified by social practices to deliver idiosyncratic pro-social ethical behaviors (Haidt and Bjorklund 2006). In a similar manner, the broad principles and pro-ecological attitudes arising at Stage 3 of the 3SLE can be shaped by the local bio-cultures into unique ethical practices (albeit with overlapping concerns of sustainability and conservation).

Summing up, in developing his ethical position, Leopold drew upon the history and habitat he knew best. Understandably this narrowed his exposition. But the core insights of the 3SLE (the need for an enlarged ethic, and the modes by which both reason- and experiential-based elements drive that ethic) still ring true and can be better harnessed through the contributions of the biocultural ethic on the importance of specific cultural creations in response to local habit and habitat.

17.5 Conclusion

We have defined *biocultural tourism* as tourism which provides education within natural and/or cultural heritage experiences and prioritizes biocultural conservation through the broadening of ethical care to new ecological and/or cultural phenomena. The 4SFEP process is a perfect example of biocultural tourism, and other examples can be found in ecotourism, slow tourism, voluntourism, and other types of tourism. Coming from a different basis, but developing in a common direction, Leopold's 3SLE puts forward a theory about social and psychological ethical development that works to activate and widen the scope of ethical values and virtues. We have shown the strong overlap between the 3SLE and biocultural tourism, such that the 3SLE supports and informs the way biocultural tourist experiences can occur—at their best, developing ethical concern for species like butterflies, far removed from our own.

References

Ballantyne R, Packer J (2005) Promoting environmentally sustainable attitudes and behaviour through free-choice learning experiences: What is the state of the game? Environ Educ Res 11(3):281–295. https://doi.org/10.1080/13504620500081145
Barua M, Gurdak DJ, Ahmed RA, Tamuly J (2012) Selecting flagships for invertebrate conservation. Biodivers 21:1457–1476. https://doi.org/10.1007/s10531-012-0257-7
Bramwell B, Lane B (1993) Interpretation and sustainable tourism: the potential and the pitfalls. J Sustain Tour 1(2):71–80. https://doi.org/10.1080/09669589309450706
Breakey N, Breakey H (2015) Tourism and Aldo Leopold's 'cultural harvest': creating virtuous tourists as agents of sustainability. J Sustain Tour 23(1):85–103. https://doi.org/10.1080/09669582.2014.924954
Campos AC, Mendes J, Oom do Valle P, Scott N (2018) Co-creation of tourist experiences: a literature review. Curr Issues Tour 21(4):369–400. https://doi.org/10.1080/13683500.2015.1081158
Dowling R (2010) Geotourism's global growth. Geoheritage 3:1–13. https://doi.org/10.1007/s12371-010-0024-7
Falk JH (2005) Free-choice environmental learning: framing the discussion. Environ Educ Res 11(3):265–280. https://doi.org/10.1080/13504620500081129
Forestell P (1993) If Leviathan has a face, does Gaia have a soul? Incorporating environmental education in marine eco-tourism programs. Ocean Coast Manag 20:267–282
Frodeman R, Briggle A (2016) Socrates tenured: the institutions of 21st century philosophy. Rowman & Littlefield, New York
Haidt J, Bjorklund F (2006) Social intuitionists answer six questions about moral psychology. In: Sinnot-Armstrong W (ed) Moral psychology, vol 2. The cognitive science of morality: intuition and diversity. MIT Press, Cambridge, MA, pp 181–217
Kellert SR (1996) The value of life. Shearwater Books, Washington
Knegtering E et al (2002) Effects of species' characteristics on nongovernmental organizations' attitudes toward species conservation policy. Environ Behav 34(3):378–400
Lemelin RH (2007) Finding beauty in the dragon: the role of dragonflies in recreation and tourism. J Ecotourism 6(2):139–145. https://doi.org/10.2167/joe161.0
Lemelin R, Jaramillo-Lopez PF (2020) Orange, black, and a little bit of white is the new shade of conservation: the role of tourism in Monarch Butterfly Conservation in Mexico. J Ecotourism 19(4):291–303. https://doi.org/10.1080/14724049.2019.1656726
Leopold A (1968) A Sand County Almanac: With other essays on conservation from Round River. Oxford University Press, New York
Leopold C (2004) Living with the land ethic. BioScience 54:149–154
New TR (1997) Butterfly conservation, 2nd edn. Oxford University Press, New York
Rozzi R (2013) Biocultural ethics: from biocultural homogenization toward biocultural conservation. In: Ricardo R, Pickett STA, Palmer C, Armesto JJ, Callicott JB (eds) Linking ecology and ethics for a changing world: values, philosophy & action. Springer, Dordrecht p, pp 9–32
Rozzi R (2019) Taxonomic Chauvinism, no more! Antidotes from Hume, Darwin, and biocultural ethics. Environ Ethics 41:249–282
Rozzi R et al (2010) Field environmental philosophy and biocultural conservation at the Omora Ethnobotanical Park: Methodological approaches to broaden the ways of integrating the social component. Rev Chil de Hist Nat 83(supplementary material):1–35
Rozzi R et al (2012) Integrating ecology and environmental ethics: earth stewardship in the Southern End of the Americas. BioScience 62(3):226–236
Shipley N, Bixler RD (2016) On the need to interpret insects: an always small but Gargantuan opportunity. J Interpret Res 21(2):65–72. https://doi.org/10.1177/109258721602100205

Tauro A, Ojeda J, Caviness T, Moses K, Moreno R, Wright T, Zhu D, Poole A, Massardo F, Rozzi R (2021) Field environmental philosophy: a biocultural ethic approach to education and ecotourism for sustainability. Sustainability 13:4526. https://doi.org/10.3390/su13084526

TEQ (2021) Nature-based tourism strategy 2021–2024. Tourism and Events Queensland, Brisbane

UNWTO (2001) The British ecotourism market. United Nations World Tourism Organization, Madrid

Walter P (2013) Theorising visitor learning in ecotourism. J Ecotourism 12(1):15–32. https://doi.org/10.1080/14724049.2012.742093

Woods B (2000) Beauty and the beast: preferences for animals in Australia. J Tour Stud 11(2): 25–35

Zhu D (2023) Bridge the channel, enhance the inclusivity: a comparison between flagship species-centered and moss-centered conservation in Chile and China. In: Rozzi R, Tauro A, Wright T, Klaver I, May R, Avriel-Avni N, Brewer C, Berkowitz A (eds) Field environmental philosophy: education for biocultural conservation. Ecology and ethics, vol 5. Springer, Dordrecht, pp 457–482

Chapter 18
A Material Ecological Ethic for Biocultural Education: Relations Between Life on Earth and Humanity

Enrique Dussel

Abstract An ecological ethic is about the absolute condition of possibility of living beings. It is ultimately played out as respect for the universal right of all human beings to survive, especially those most affected and excluded: the poor of present times as well as those of future generations who will inherit a dead Earth if an ecological conscience is not acquired quickly and globally. Life is the absolute condition for human existence, and therefore life on Earth is this extended condition. This chapter presents some principles of an ecological ethics in a cultural moment in which universal consensual formal morals seem to have refuted the possibility of founding ecology in material ethics; that is, an ethics of life. Material ethics and formal morality are both necessary but not sufficient. Having as our horizon the ecological destruction of biodiversity that is articulated concomitantly with misery, poverty, and the oppression of the majority of humanity, we must recover material references, since these "facts" can be discovered critically via contrast (contradiction or non-compliance) to a positive material standard previously stated. Hence, in education as well as in biocultural conservation, we need a material understanding and simultaneously the mediation of formal morality, critical thinking, and consensuality.

Keywords Christian responsibility · Formal morality · Liberation ethics · Marx · Material ethics

---

Edited translation by Roy May (Ecumenical Department of Investigations, DEI, Costa Rica) and Ricardo Rozzi (University of North Texas, USA and University of Magallanes, Chile) of "Algunos principios para una ética ecológica material de liberación (relaciones entre la vida en la tierra y la humanidad)" by Enrique Dussel, and originally published in Pixley, J (ed.), *Por un mundo otro. Alternativas al mercado global*. Ecuador, Consejo Latinoamericano de Iglesias (CLAI) (2003), pp. 29–44. Version approved by Enrique Dussel and published with permission of CLAI.

E. Dussel (✉)
Facultad de Filosofía, Universidad Nacional Autónoma de México, Ciudad Universitaria, México, DF, México
e-mail: dussamb@unam.mx

R. Rozzi et al. (eds.), *Field Environmental Philosophy*, Ecology and Ethics 5,
https://doi.org/10.1007/978-3-031-23368-5_18

Formal education often overlooks the richness and diversity of forms of local knowledge embedded in biological and cultural diversity that is specific to locations in different regions of the world. The spread of the dominant culture is linked to a prevalence of formal over material philosophy (Gare 2023) and ethics (May 2023), and involves the exclusion of the majority of co-inhabitants from having a moral status and from being considered in global formal education (Rozzi 2012, 2019; Poole 2018). If an ecological ethics is to be taught it needs, on the one hand, to take into account the actual biocultural diversity on which life depends and, on the other hand, the social-environmental injustices that lead to oppression and death.

An ecological ethic is about the absolute condition of possibility of living beings. It is ultimately played out as respect for the universal right of all human beings to survive, especially those most affected and excluded: the poor of present time as well as those of future generations, who will inherit a dead Earth if an ecological conscience is not acquired quickly and globally.

Life is the absolute condition for human existence, and therefore life on Earth is this extended condition. In fact, the Earth cannot be destroyed, nor can nature—in its mere physical or purely material constitution. However, the conditions for the existence of life can be destroyed. Life on Earth can be destroyed.

Here I refer only to some principles of an ecological ethics in a cultural moment in which universal consensual formal morals[1] seem to have refuted the possibility of founding ecology in material ethics; that is, an ethics of life. My standpoint takes material ethics (necessary but not sufficient) and formal morality (necessary but not sufficient either) within a critical process of liberation that subsumes them, by putting them critically, socially, historically, and diachronically into movement. This is highly topical because ecological destruction (as a condition of possibility) and poverty (the effect) are two correlated phenomena, which have the same cause. Both require a material understanding and, simultaneously, the mediation of formal consensuality by the community.

Rightly Kant opposed the irrationality of the ethics of the English empiricists who tried to establish the good only in pleasure or happiness as mere sensations, feelings, or emotions, without the intervention of reason.[2] But, erroneously, Kant rejected the ethical meaning of the struggle for life (in its current ecological sense and justice):

[1]I will use the word "moral" to indicate the formal aspect and "ethics" for its material sense. Discourse ethics, for example, would then be called "moral discourse," while our "ethics of liberation" is "ethics" (for being material), and also for being liberation (by critical historical process).

[2]English empiricist John Locke (1975, Book I, chapter 28, § 5) wrote: "Good and bad [...] are nothing but pleasure or pain, or the occasions that seeks pleasure or pain." In another passage, he stated: "Things are good or bad only in reference to pleasure and pain" (1975, Book. II, chapter 20. §2). Later, the founder of utilitarianism Jeremy Bentham (1948, p. 3) expressed: "[...] the fundamental axiom [is] the greatest happiness for the greatest number is the measure of everything right and wrong."

> To preserve one's life is a duty, and besides everyone has an immediate inclination to do so. But on this account the anxious care that most people take of their life still has no inner worth, and their maxim has no moral content (Kant 1968, 23).

Ecological ethics, which essentially has to do with the defense of life, is situated here by Kant in such a way that it cannot be satisfactorily justified. In this dualistic tradition of resolute negation of the material determination of ethics, the "anxious care" to "preserve" life has no material ethical meaning and is relegated to mere egoism or to a pathological or capricious motivation. We must oppose this Kantian view so that we can avoid the cost of leaving ecological ethics hanging in the air.

18.1 An Attempt to Articulate Material Ethics and Formal Morality in Ecology

Jesus' words, "I was hungry, and you gave me food" (Matthew 25: 35), require material ethics: hunger is a type of pain, and food is the fruit of labor and land. In contrast, current formal morality (such as positivist or analytical thought like that of George Moore,[3] or the philosophies of John Rawls[4] and Jurgen Habermas[5]) discard material ethics (such as utilitarianism, the *Sittlichkeit* of Hegel,[6] or a Charles Taylor[7]).

Having as our horizon the ecological destruction of the earth that is articulated concomitantly with misery, poverty, and the oppression of the majority of humanity (taking into account phenomena such as central and peripheral capitalism, racism, and sexism), we must recover material references, since these "facts" can only be discovered critically via contrast (contradiction or non-compliance) to a positive material standard previously stated. For this reason, we need to reconstruct the truth of a material ethics—where ecological destruction and poverty are identified as ethical problems in themselves—and articulate it adequately to a formal morality—from which we can proceed consensually.

Currently in neurobiology consensus defines the human brain as an organ of survival. For example, Gerald Edelman writes:

> The frontal cortex is a prime example of a conceptual center of the brain [...] Given its connections to the basal ganglia and the limbic system, including the hippocampus, the frontal cortex also establishes relations subserving the categorization of values and sensory experiences themselves. In this way, conceptual memories are affected by values—an important characteristic in enhancing survival (Edelman 1992, 109–110).

---

[3] See Moore (1903).

[4] See Rawls (1971) and Rawls (1993).

[5] For example, see Habermas (1983) and Habermas (1981).

[6] See Benhabib (1986).

[7] See especially Taylor (1989) and Taylor (1992).

"To value" or "to categorize according to values" means to assess, taking into consideration life and its reproduction and growth. Indeed, this is the fundamental criterion of all material ethics that satisfies basic needs, as well as the development (which I call "survival"[8]) of human life. We can formulate this criterion as the fundamental-universal ethical principle:

> Whoever acts seriously and honestly already has presupposed, always *a priori*, the ethical demands of the struggle for reproduction and growth of the life of human subjects, concretely in terms of a good life (happiness, survival, and other dimensions), considering all members of a real cultural and historical community,[9] and including the aspiration to share the good life with the whole humanity.

The person who works (praxis in the strict sense) humanly, seriously, and honestly, does so from a universal rational motive: reproduction and growth of human life (and concomitantly, of life on planet Earth). This project is inevitably carried out on the horizon of a "good life" in one's own cultural community. That "good life" includes all the "sources" available for rationality. These obviously are different among the Aztecs, the Spanish conquistadors of the sixteenth century, the English of the eighteenth century, or the US citizens today.[10] Additionally, because such persons accept seriously and honestly this "good life" as intrinsically superior, ideal, and universal, they should accept dialogue or discussion with other conceptions of "the good life," since being serious and honest is a "universal pretension:" ideally or potentially valid for all humanity. Whoever "does not enter into" the discussion would retreat into a dogmatic or fundamentalist position and would stop acting "seriously and honestly." Moreover, the "good life" is not primarily the goal (*Zweck*, *telos*) of instrumental or strategic reason. Rather it is a way of "community" life (ideal or real), which should be the fulfillment of the original ethical intersubjective recognition of the *Other* as other. This opens the possibility for communication and for the exercise of discursive reason. Against formal moralities, which today are fashionable, it is possible to prove that the preservation and growth of the life of the human subject is the absolute condition, whose differentiated mediation are the "good lives" seriously and honestly attempted. These "good lives" are always

---

[8]The prefix "sur-" in the term "survival" indicates: (a) the higher mental functions of life (conceptualization, valuation assessment, language, self-awareness, freedom, autonomy and responsibility); and (b) one's own historical developments of culture, history, religious spirituality, full aesthetic and ethical growth of humanity.

[9]Even the dissident "presupposes" the "good life" that he or she opposes eventually and without which it could not be dissident. Radical nihilism is letting die (letting yourself die would presuppose the motivation for suicide): the performative contradiction here is even more radical than in the case of a skeptic. The skeptic could "not go" to the discussion; but even here the "not go" is still an action: that is, it must have a reason (the reason is theoretical, but the motive is practical) not to enter.

[10]To be rational the "good life" must have the best possible resources. One cannot require the Aztecs to have resources other than their own. For this reason, Bartolomé de las Casas showed that from their cultural horizon, from their rational resources, offering human sacrifice to the sun god was perfectly rational and good: they did not have other resources available and no one would be able to object for not using them properly.

presupposed and claim universality, even if they are (and they can never be) particular due to their historical grounding—even in the case of the current post-conventional morality, which often is Eurocentric and accomplice, because it is not critical of neoliberal capitalism.

The moment of formal morality, whose criterion of validity is effective intersubjectivity or real argumentative consensuality, has a basic moral principle that can be stated as:

> Whoever argues has already testified *in actu*, and thereby recognized [...] a communication community of people who recognize each other as equals (Apel 1980, 161).
>
> Only those norms accepted by all those affected as virtual participants in practical discourse, can be accepted as valid (Habermas 1981, 12).

The ecological dimension thus would be defined materially (as an absolute condition for survival) and formally (as it has been decided intersubjectively in private and in public, nationally and internationally). For a sustainable ecological ethics, the formal moral principle (consensual democratic discussion) "applies" the concrete case (particular maxim to act) in the context (social and historical "totality") based on the material ethical principle of survival. In this manner, we have inverted what is presented today among practitioners of the basic norm of formal morality. It is not formal morality that is "applied" in the concrete case—without interest in the content of the discussion, which is always human life and whose agreement will be valid just because of procedure. Rather it is the principle of material ethics of content (the ecological condition of the survival of the community of human-cultural life), mediated by the principle of formal moral procedure (the community of communication as consensuality).

I want to make one last material reflection. Marx is considered by many to be an "anthropocentric" economist without ecological sensitivity.[11] However, we must remember a text of 1875, the *Critique of the Gotha Program*. In its first paragraph, the Gotha Program says: "Labor is the source (*Quelle*) of all wealth and all culture." Marx replies to this statement:

> Labor is not the source of all wealth. Nature is just as much the source of use values (and it is surely of such that material wealth consists!) as labor, which itself is only the manifestation of a force of nature, human labor power (Marx 1875, 13).

For Marx only two levels of reality have no exchange value (economic), and they are nature and the human person:

> The price of labor or the price of land or soil (or natural forces in general) are the only irrational expressions (*irrationellen*) of this type. The price of land is irrational because an adequate price is the monetary expression of value, but there cannot be value [of exchange] when there is no labor materialized in the thing [...] In the same way the expression *price of labor* is irrational (Marx 1863, 2190).

I should point out, honestly, that this doctrine of Marx is exactly in line with the key principle of all material ethics, Kantian, or in general, of all ethics of the West, in that

[11] I remember a discussion about this with John B. Cobb in California in 1988.

Table 18.1 Ecological value (EV), use value (UV), and exchange value (ExV)

Dignity	Usefulness	Formally economic
Ecological value: Natural, physical, and real	Use value: (a) Natural (b) Produced	Exchange value: Economic
Level 1 (EV)	Level 2 (UV)	Level 3 (ExV)
Nature	Product of human labor	

"the person is an end and not a means"; i.e., the dignity of the person is the criterion of ethics (and the ethics of liberation). But at the same time, "salary" is often spoken of as the price of labor, and it is intended to be "fair", which is a *contradictio terminorum,* as if it could not be inconsistent with the dignity of the person; or talking about private property as a natural right, without indicating that most, in their poverty, are excluded from having property. And because the poor do not have any property, they must sell their own corporeality in the system of employment. Marx, however, in a position consistent with the ethical criterion of Christian ethics, shows that land and labor have no "exchange value," but rather infinite, priceless "dignity" (actually they find themselves in "another order:" in the order of causes, while the "exchange value" or the "price" are located in the order of effects). Who would have said that Marx was more consistent with the very essence of Christian ethics in the West (or the principle of absolute priority of the human person) than many members of the churches!

Paradoxically, and contrary to the opinion of many, for Marx, land and humanity have "dignity." They cannot have exchange value, because they do not contain as such objectified labor, since labor is the "source (*Quelle*)" of all exchange value—and "creative source (*schopferische*)" from nothing (*ex nihilo*) of surplus value. From this, we could propose the following scheme (Table 18.1):

Natural things have "dignity" (the medieval *bonum*); that is, "ecological value" (EV) that can produce "use value" (UV) (natural[12] or produced, material as well as cultural and aesthetic). Only human products have "exchange value" (ExV) or "economic value." Material ethics considers goods with use value to be wealth as such (objective goods of happiness, which is the subjective good). The political management of these public goods is the formal practical moment, which is public and consensual (democracy, for example). Hence, ecology and political economy speak first of the material level of ethics but managed at the formal level of democracy or public morals. This is the moment of sacramental bread and its communitarian distribution from the meaning of hermeneutics of the text revealed to faith.

[12] "The earth [...], which in the original state provides humanity food and means of livelihood ready to consume, exists without human intervention" (Marx 1875 I, chpt. 5).

18.2 The Ethical-Ecological Critique of the Current System: Totality

What human beings honestly want is an established cultural system, a good that is valid and ecologically sustainable (which must have material ethical content and formal, consensual, rational mediation). In times of balance, peace, consensus, and classical stability:

> Nation shall not lift up sword against nation, neither shall they learn war anymore. O house of Jacob, come, let us walk in the light of the Lord! (Isaiah 2:4c-5 NRSV 1990).

This situation of an existing system effective through consensus, that is legitimate, is not preserved indefinitely and diachronically, but rather occurs in certain privileged moments. The social order that keeps legitimacy (Habermas) or hegemony (Gramsci) is one that balances two dimensions: the good life and a lifestyle that is valid as a way of preserving the historical and cultural-ecological life for all. Conversely, when, under the cloak of a "good life," life becomes impossible, empirically becoming oppressive, domineering, repressive, or is irresponsible about the effects future generations will suffer, then it ceases to have validity or legitimacy for the oppressed present or future. In the eyes of the excluded (or their advocates), it becomes only a supposed "good life," because, in reality, it denies life. Consequently, out of the suffering corporeality of those who cannot reproduce their lives emerges the critique that now invalidates the old system that has been in effect. They are the "critics," the prophets.

Critics (or prophets) are those who uncover the contradiction between the "good life" as an ideal or utopian proposal (for example, "liberty, equality, and property for all" in the utopia of nascent capitalism) from the impossibility for most members of reproducing this in real life (due to the lack of freedom suffered as illegitimate coercion, inequality, poverty, and lack of property). Here the system appears as a "Totality" (in the negative sense of Levinas), as "the sin of the flesh" in the New Testament. By not including a material reference, formal moralities do not have the possibility of discovering this contradiction between the universal principle of life—namely "good life" proposed as ideal, and its intended "validity" (the splendor of Pharaonic Egypt)—and what really and empirically happens—the misery and death of slaves.

The critical process can emerge only from the impossibility of living, that is, life experienced in the denied corporeality of the oppressed (or of future generations) or from living in the actual ecological destruction in which one cannot "survive" (nor remain, because of hunger, and even less develop or grow by education, by greater spiritual, artistic, ethical possibilities). The first exercise of an "originating-ethical reason" is recognizing the Other. The Other is the face of the un-happy (the ambiguous intention of utilitarianism), the abused, the poor (theology of liberation), the ones dominated by their libido (Freud), the ones denied as ethical subjects in the micro-structures of power (Foucault), the ones upon whom are imposed inverted values (Nietzsche), and those who live in the wastelands. This recognition of the

Other is mediated by rational critical resources that are available (from intuition in ancient times to today's social sciences or critical hermeneutics[13]), from the utopia of the Kingdom of God , as already begun, and eschatologically, future.

Formal moralities (such as those of twentieth-century German philosophers Jürgen Habermas or Otto-Karl Apel) cannot have a sufficiently critical criterion (neither on the material level nor in their "application"), because the critical criterion has a material component and a formal component: the impossibility of the moral validity of a system that prevents the reproduction of life or the material and ethical destruction of ecology—in front of the private or public conscience of the Aztec or the conqueror:

> I have seen the misery of my people who are in Egypt; I have heard their cry on account of their slave masters. Indeed, I known their sufferings (Exodus 3:7 NRSV 1990).

The critical principle could be stated then, at least in one of its dimensions, as:

> Act critically from a practical point of view—presupposing the ethical demands of the reproduction and growth of the life of the human subject, that is, growth in the "good life" of effective community cultural life (Egypt)—to discover non-compliance with these ethical demands or their negation by members of society (the slaves), in order to invalidate said order and its ethical demands.

Now the critic (the prophet) is on the "exterior of the system," (that is, *bamidbar*, the Hebrew word for desert or wilderness), given that he or she has lost the "accomplice consciousness" or the naive acceptance of its legitimacy (Habermas) or hegemony (Gramsci). The "being outside" of the system that is criticized by the prophet, becomes articulated to the original and first "being outside" of ecological reality[14] in the suffering corporeality of the oppressed as Other[15]: as an ethical subject recognized in their dignity as Other, different from the system, a dignity that has been in their person by the denial of accomplishing the ideal "good life" of the system.

Albrecht Wellmer (1986, pp. 20ff) has shown that maxims that cannot be generalized are universally banned. But, in order to affirm that it cannot be generalized, the maxim must be confronted with the ethical requirements that are presupposed in a certain community of "good life." Otherwise, it cannot be concluded that the maxim "cannot be generalized." That it "cannot be generalized" presupposes an actually existing material "good life." The prohibition confronts the possibility of the "good life"—that is proposed as a utopian ideal—with the

[13]The criteria for distinguishing the "critical social sciences" from the "non-critical" are the same as for material ethics and formal morality. The question is: which sciences take as their basic criteria the survival and intersubjective consensuality to validate their findings? Sciences that naively accept the status quo are not critical.

[14]The poor suffer the ecological death of nature before anyone else (because "the rich" have mediations that postpone suffering).

[15]The ecological dignity of nature also must be recognized, because it a creature of God and not our own, and because it is the condition for the life of the community.

impossibility of generalizing the "unjust life" of some of its members or of the ecological reality as their material condition.

Critical consciousness has as its absolute limit the ultimate prohibition of maxims that cannot be generalized: "Do not destroy life on earth nor that of humanity present or future!" And if this should happen, whatever were the procedural conditions for reaching that conclusion, the act would be invalid (*ungültig*).

Contemporary capitalism, according to Marx's judgment (and in full force even today), with its fundamental criterion of "increasing the rate of profit" (as opposed to the criterion of life experience), implicitly proposes the following principle:

> Whoever acts according to the criterion of "increasing the rate of profit" presupposes, always *a priori*, that neither the material ethical principle of survival nor the formal moral principle of democratic consensuality, will be obstacles or boundaries to obtaining this purpose.

Here we have a fetishized instrumental reason, much more destructive than what Max Weber and Jürgen Habermas have thought. The anti-ecological danger of technology is an effect, and not the cause of the problem. The destructive technology of life (of the Earth and of humanity) is chosen and used according to the instrumental criterion of "increasing the rate of profit," and not according to the material criterion of the "permanence and development of life" of the land (ecology) and the survival of humanity.

Technology is a means of labor, a condition of production, mediation aimed at increasing the productivity of the workforce to decrease the value of the product or to lower the cost of the capacity of work, i.e., increase the relative surplus value (and its rate). But, and here is the central point, the criterion to subsume a new technology (the combustible gasoline automobile motor and not the electric car) must always be "increasing the rate of profit." If a new technology or discovery gives a higher rate of capital gain otherwise, that technology will be chosen. So the technology as we know it, is but one of the possible technologies. The criterion of reality or existence is determined to be the highest rate of gain that occurred in the short term.

Technology that is ecological and ethical is possible in the abstract. Today technology is a moment, a "determination" of capital, subsumed in its fetishized and destructive process. Overall environmentalists turn against technology, just as the workers of the nineteenth century tried to destroy the machines, which appeared to them as the cause of their unemployment. These environmentalists as well as those workers, are fighting against the mere "mask" of capital (fixed capital). They naively ignore the cause. By not knowing the cause, it remains as not guilty, hidden in the dark, out of sight: capital as the process of valuing through the subsumption of human life (workforce) and by means of a technique (means of production) that is destructive because it operates only from the criterion of increasing the rate of profit. The perversity of technique--against what Heidegger and many others thought about "technique"—is capital as the basis for making value-judgements. Former US president George Bush in Rio de Janeiro (1992) could not sign the ecological protocols: an economic crisis (decreased rate of profits) in the USA was for him (for capitalism) in the short term more important than the ecological destruction of the Earth and humanity in the long-term. For the same reason, in 2017 former president Donald Trump withdrew the USA from the Paris Agreement on climate change mitigation.

The "Earth"—from the Greek *physis* to the creature of Yahweh of the Hebrews—is now a "means of production," secularized, exploitable, "at hand." Its "dignity," which has no exchange value (being the source of "use value," for Marx), becomes now a non-renewable object, destructible by instrumental reason, through the mediation of increasing profits, consumed and hoarded as waste, residue, garbage. Marx's vision is clear: capital is Moloch, the fetish, the Antichrist, not the Dragon but the Beast of the Apocalypse.

18.3 Christian Responsibility

Wealth, as useful (use value), is ecologically and humanly created and is a blessing of God (as Calvinism understood the origin of capitalism, according to Weber's version). The consensual use of that wealth in a community of justice is a foretaste of the Kingdom of God:

> They devoted themselves to the apostles' teaching and fellowship, to the breaking of bread [...] (Acts 2:42 NRSV 1990).

We have seen, then, that the criticism of the prophets begins when the utopia of the current system (which now becomes the "system of oppression" or the Egypt of Hebrew slavery) is compared with the reality of the suffering of "slaves." It is the starting point of the messianic community. It is the anti-hegemonic, formal consensuality of the poor who have become aware of their reality. The people set out across the desert (the praxis of liberation) on their way to the Promised Land.

Indeed, the local and global Christian community must realize the need for "conversion." A new consensus about the importance of saving life on Earth must be achieved. This is a responsibility of all religious communities (not only Christian, but also of the world religions: Buddhism, Hinduism, Confucianism, Islam, and others). As in the days of the young Christians who emulated the Manichaean monks in the deserts of Egypt and founded communities away from the decaying cities (such as Alexandria) because Constantinian Christianity had been born, it will be necessary to originate a style of life contrary to the excessive comfort of contemporary consumer society.

Austerity, non-consumerism, non-destruction of required non-renewable resources (oil and plastics, for example), implies a new civilization that is satisfied with what has been already achieved technically, and that develops only in solidarity with what is necessary for science, art, justice, and mysticism. This should be the motivation of deep "Christian spirituality." If this can be required of "the rich," the citizens of the States of Law of late-stage capitalism, what more can be asked of the "poor" of the nations of the South, of peripheral capitalism, of those fighting poverty and injustice? Can these too fight to avoid dire ecological destruction? Indeed, the poor, in order to cook and eat, annihilate tropical rain forests; dry up still virgin rivers in order to drink; occupy the best agricultural land in order to have a place to live. We face dilemmas which demand responsibility, ethical awareness, and creativity. In many cases, as among the Amazon peoples in Brazil or the Maya of Mexico, they are the ones who have best-preserved ecology, as *Terra Mater* with which they have lived for millennia.

A project for the ecological liberation of Earth should integrate the material principles of ethics and the formal consensus of mutual awareness by the community, into the diachronic process of co-solidarity with all humanity. Churches are one of those ultimate ethical "resources" of humanity because they are communities of moral communication that create intersubjective responsibility by reading and rereading of the revealed Text. These communities are irreplaceable in the current time. The Kingdom of God includes as a condition of possibility life on Earth, the life of humanity. Thus, new requirements are necessary, a "New Commandment" for the new journey:

> You have heard that it was said in those ancient times. . .But I say to you. . . (Matthew 5:21a–22a NRSV 1990).

Like the way in the desert, having left Egypt behind, it is time to become aware of new ecological and political-economic demands related materially or sacramentally to the preservation and development of life: survival of humanity present and future.

18.4 Concluding Remark

The former analysis shows the necessity of both material ethics and formal morality. Both need to be socially, historically, and environmentally contextualized in order to overcome exclusion and oppression of the majority of living beings, including humans. This contextualization as well as the integration of material and formal dimensions of ethics are essential to field environmental philosophy (Rozzi 2013). Regarding the cultivation of material ethics, face-to-face encounters with persons from different cultures and with organisms of different biological species in their native habitats open an experiential understanding of material ethics of co-inhabitation (Rozzi et al. 2006, 2008). Regarding the learning experience of formal morality, teaching of philosophy in FEP is also essential to cultivate dialogical practices and hermeneutical skills to critically build dynamic formal epistemological and ethical frameworks considering both biological and cultural diversity, and their interrelationships (Aguirre 2015). Moreover, beyond FEP, liberation philosophy and ecological ethics offer a practical and theoretical foundation for educational programs with indigenous and other local communities around the world (see Mamani-Bernabé 2015; Montoya-Greenheck 2018; Toyoda 2018; Kono 2018).

References

Aguirre JF (2015) Hermenuetics and field environmental philosophy: integrating ecological sciences and ethics into Earth stewardship. In: Rozzi R, Chapin FS III, Callicott JB, Power ME, Armesto JJ, May RH Jr (eds) Earth stewardship. Linking ecology and ethics in theory and practice. Springer, Dordrecht

Apel K-O (1980) Notwendigkeit, Schwierigkeit und Möglichkeit philosophischen Begründung der Ethik im Zeitalter der Wissenschaft. In: Kanellopoulos P (ed) Festschrift für K. Tsatsos. Nomikai Ekdoseis Ant., Athens

Benhabib S (1986) Critique, Norm, and Utopia. A study of the foundations of critical theory. Columbia University Press, New York

Bentham J (1948) A fragment on government. Basil Blackwell, Oxford

Edelman G (1992) Bright air, brilliant fire: on the matter of the mind. Basic Books, New York

Gare A (2023) Challenging the dominant grand narrative in global education and culture. In: Rozzi R, Tauro A, Wright T, Klaver I, Avriel-Avni N, Brewer C, Berkowitz A, May Jr. R (eds) Field environmental philosophy: education for biocultural conservation. Ecology and ethics, vol 5. Springer, Dordrecht, pp 309–326

Habermas J (1983) Moralbewuβtsein und kommunikatives Handeln. Suhrkamp, Frankfurt

Habermas J (1981) Erlduterungen zur Diskursethik. Suhrkamp, Frankfurk

Kant I (1968) Grundlegung zur Metaphysik der Sitten, I, BA 10. Wissenschaftliche Buchgesellschaft, Darmstadt

Kono T (2018) The garden as a representation of nature: a space to overcome biocultural homogenization? In: Rozzi R, May RH Jr, Chapin FS III, Massardo F, Gavin M, Klaver I, Pauchard A, Nuñez MA, Simberloff D (eds) From biocultural homogenization to biocultural conservation, Ecology and ethics, vol 3. Springer, Dordrecht, pp 459–474

Locke J (1975) An essay concerning human understanding. Clarendon, Oxford

Mamani-Bernabé V (2015) Spirituality and the Pachamama in the Andean Aymara worldview. In: Rozzi R, Chapin FS, Callicott JB, Pickett STA, Power ME, Armesto JJ, May RH Jr (eds) Earth stewardship: linking ecology and ethics in theory and practice, Ecology & ethics book series. Springer, Dordrecht, pp 65–76

Marx K (1863) Zur Kritik der politischen Ökonomie (Manuskript 1861-1863). Printed in 1976-1982. MEGA Dietz Verlag, Berlin

Marx K (1875) Critique of the Gotha Programme. Translated and reprinted in: Marx/Engels selected works, volume three (1970). Progress Publishers, Moscow, pp 13–30

May Jr R (2023) Liberation philosophy and biocultural education. A Latin American Journey. In: Rozzi R, Tauro A, Wright T, Klaver I, Avriel-Avni N, Brewer C, Berkowitz A, May Jr. R (eds) Field environmental philosophy: education for biocultural conservation. Ecology and ethics, vol 5. Springer, Dordrecht, pp 203–228

Montoya-Greenheck F (2018) Biocultural diversity and Ngöbe people in the South Pacific of Costa Rica. In: Rozzi R, May RH Jr, Chapin FS III, Massardo F, Gavin M, Klaver I, Pauchard A, Nuñez MA, Simberloff D (eds) From biocultural homogenization to biocultural conservation, Ecology and ethics, vol 3. Springer, Dordrecht, pp 361–378

Moore G (1903) Principia ethica. Cambridge University Press, Cambridge

NRSV (1990) The Holy Bible, new revised standard version. Thomas Nelson Publishers, Nashville

Poole AK (2018) The UN sustainable development goals and the biocultural heritage lacuna: where is goal number 18? In: Rozzi R, May RH Jr, Chapin FS III, Massardo F, Gavin M, Klaver I, Pauchard A, Nuñez MA, Simberloff D (eds) From biocultural homogenization to biocultural conservation, Ecology and ethics, vol 3. Springer, Dordrecht, pp 315–331

Rawls J (1993) Political liberalism. Columbia University Press, New York

Rawls J (1971) A theory of justice. Harvard University Press, Cambridge

Rozzi R (2012) Biocultural ethics: the vital links between the inhabitants, their habits and regional habitats. Environ Ethics 34:27–50

Rozzi R (2013) Biocultural ethics: from biocultural homogenization toward biocultural conservation. In: Rozzi R, Pickett STA, Palmer C, Armesto JJ, Callicott JB (eds) Linking ecology and ethics for a changing world, Ecology and ethics, vol 1. Springer, Dordrecht, pp 9–32

Rozzi R (2019) Taxonomic Chauvinism, no more! Antidotes from Hume, Darwin, and biocultural ethics. Environ Ethics 41:253–288

Rozzi R, Massardo F, Anderson C, Heidinger K, Silander J Jr (2006) Ten principles for biocultural conservation at the southern tip of the Americas: the approach of the Omora Ethnobotanical Park. Ecol Soc 11(1):43 [online]. http://www.ecologyandsociety.org/vol11/iss1/art43/

Rozzi R, Arango X, Massardo F, Anderson C, Heidinger K, Moses K (2008) Field environmental philosophy and biocultural conservation: the Omora Ethnobotanical Park educational program. Environ Ethics 30(3):325–336

Taylor C (1992) The ethics of authenticity. Harvard University Press, Cambridge

Taylor C (1989) Sources of the self. The making of the modern identity. Cambridge University Press, Cambridge

Toyoda M (2018) Revitalizing local commons: a democratic approach to collective management. In: Rozzi R, May RH Jr, Chapin FS III, Massardo F, Gavin M, Klaver I, Pauchard A, Nuñez MA, Simberloff D (eds) From biocultural homogenization to biocultural conservation, Ecology and ethics, vol 3. Springer, Dordrecht, pp 443–457

Wellmer A (1986) Ethik und Dialog. Suhrkamp, Frankfurt

Chapter 19
Challenging the Dominant Grand Narrative in Global Education and Culture

Arran Gare

Abstract This chapter critically examines the dominant tradition in formal education as an indirect driver of biocultural homogenization while revealing that there is an alternative tradition that fosters biocultural conservation. The dominant tradition, originating in the Seventeenth Century scientific revolution effected by René Descartes, Thomas Hobbes, Isaac Newton, John Locke and allied thinkers, privileges science, seen as facilitating the technological domination of the world in the service of economic growth, as the only genuine knowledge. This is at the foundation of a globalized, homogenizing culture that reduces everything and everyone to instruments of the globalized economy. The alternative, now recognized as the Radical Enlightenment, has its roots in the Renaissance. Represented by thinkers such as Giordano Bruno, Giambattista Vico, Gottfried Herder, and Friedrich Wilhelm Schelling, it challenges such dogmatic scientism and upholds the value of diverse cultures, past, and present. It offers an alternative tradition and model of education fostering imagination, understanding, and appreciation of diversity in the quest for wisdom. It is a model of education which engenders respect for and appreciation of the value of different cultures, including indigenous wisdom with very different attitudes toward nature, thereby developing the capacity to reflect on, question, criticize, and overcome the homogenizing imperialism of mainstream modernist culture.

Keywords Education · Herder · Humanities · Scientism · Vico

19.1 Introduction

As Elio Gianturgo, the translator of Giambattista Vico's *On the Study Methods of Our Time,* wrote in his introduction to this work:

A. Gare (✉)
Swinburne University of Technology, Hawthorn, VIC, Australia
e-mail: agare@swin.edu.au

R. Rozzi et al. (eds.), *Field Environmental Philosophy*, Ecology and Ethics 5,
https://doi.org/10.1007/978-3-031-23368-5_19

> We live in a Cartesian world, a world of scientific research, technology, and gadgets, which invade and condition our lives; of new disciplines, like kybernetics, sociometry, biometry; of new machines, like electronic computers, videos, satellites…In our milieu so intensely penetrated on the one hand by mathematical intellectualism, by science-worship, and, on the other, by an exacting pragmatic utilitarianism… with consequences that are plainly visible in the educational curriculum and the criteria governing our young men's choice of profession (Vico 1990, p. xxi)

Vico's address was published in 1709, 72 years after Descartes published in 1637 his immensely influential *Discourse on the Method of rightly conducting one's reason and seeking the truth in the sciences* and Newton, strongly influenced by Descartes, published in 1657 his *Mathematical Principles of Natural Philosophy.* The tone of Descartes' philosophy was conveyed in *Meditations on First Philosophy* published in 1641. In this work, Descartes began by claiming to make a clean sweep of all past influences, including what he had learned as a student, and to begin again from the very foundations, only accepting what can be known with absolute certainty, doubting everything, and only accepting claims to knowledge justified through a rigorous method of reasoning. The wisdom of the ancients, along with those who studied them, was dismissed as little worthy of credence.

Giambattista Vico (1668–1744), perhaps lesser known among philosophers, identified and diagnosed the totalitarian tendencies in Descartes' philosophy (Gare 2006). These tendencies were manifest in the expansion of European civilization and are manifest today in pressures for biocultural homogenization, destroying, or dissolving indigenous cultures throughout the world of people adapted to their unique ecosystems (Rozzi et al. 2018). One of the most pernicious ways this takes place is through our education systems. Biocultural homogenization's drivers include a lack of everyday contact with biodiversity due to the rapid growth of urban population, the omission in formal and non-formal education of native languages that contain a broad spectrum of traditional ecological knowledge and values, as well as a rapid elimination or reduction of the teaching of philosophy in formal education during the twentieth century (Rozzi 2012). In this chapter, I emphasize the relevance of reintroducing philosophy, history, and the humanities into education.

It is clear, however, from Vico's writings, including his writing on education, that Cartesian thought was challenged by preserving and then developing older modes of thought. The nature of this challenge and how it developed and how it was and continues to be resisted is still not well understood. In this chapter, I will examine the historical background to current debates and contemporary struggles. The education system is a major site for the imposition of cultural homogenization and for resistance to this homogenization (May 2023). At present, the Cartesian tradition is associated with a commitment to economic growth through expansion of the market through techno-science (Gare 1998), while resistance to it is now focused on efforts at ecological and cultural conservation (Rozzi 2020). By telling the story of the reactions against the Cartesian tradition of thought, I will attempt to convey the story's strength and how it has provided the foundation for opposition to this Cartesian tradition, but also how this challenge has been, and still is being, neutralized, particularly in the field of education. In this way, I will attempt to identify the educational principles that could be built upon by those promoting biocultural conservation.

19.1.1 The Problematic Cartesian Tradition

While Descartes did not write explicitly on education, his claims for how knowledge is obtained had as much relevance to pedagogy as to science, and were duly recognized for this. Vico's philosophy, directed against Descartes and all he stood for, began with his work on education. Along with the work of Bacon, Galileo, Hobbes, Newton, and Locke, Descartes laid the philosophical foundations of modern civilization by elevating experimental science and its claims to certainty based on a supposedly rigorous scientific method that supersedes all other discourses. The differences among these founding figures are minimal. Descartes argued that knowledge should be acquired through the application of a rigorous method, and that the purpose of knowledge is to control nature. Following Galileo, Hobbes argued for the application of a "resolutive-compositive" method by which anything can be understood by breaking it down into its components and seeing how they can be put back together. Hobbes argued that all reasoning is adding and subtracting in the service of gaining control over things to satisfy appetites and avoid aversions. Subsequently, in the seventeenth and eighteenth centuries, the empiricists, beginning with Locke, gave a greater place to sense impressions in the foundations of knowledge, and instead of appetites and aversions, referred to pleasure and pain as motives. Later, in the nineteenth and twentieth centuries, the positivists and then the logical positivists and other logical empiricists, combined Descartes' rationalism with empiricism, using advances in symbolic logic to achieve this synthesis. The hidden agenda of all this is the basis of Modernity. As Stephen Toulmin (1990) argued in his study of Descartes, it was to counter the Renaissance quest for liberty and its associated civic humanism and to establish efficient control not only over nature but over people and society.

In contrast to the Cartesian method, civic humanism was produced by the humanities. It was associated with the revival of rhetoric, history, and the arts. Initially the civic humanists ignored natural philosophy but later it was associated with the rise of Nature Enthusiasm. The foremost exponent was Giordano Bruno (1548–1600), who radicalized Renaissance thought and along with republicanism, called for greater egalitarianism (Jacob 2003). The political opposition to this was developed by Hobbes in a radical way. Hobbes defended the concentration of power in the hands of an enlightened despot, with the rest of the population left to focus on commerce (Skinner 1998). In chapter 24 of his book *Leviathan*, Hobbes (1651) virtually invented modern economics, taking the economy as an entity which distributes money as nutrients to its component organs. Newton and Locke diluted Hobbes' political philosophy by promoting rule by an oligarchy of the wealthy. The Renaissance's goal to free people from slavery to achieve liberty and to develop their full humanity was altered by Descartes and these British philosophers to be the maximization of the production of commodities to satisfy people's appetites.

The call by Descartes for people who could ignore the weight of tradition and think and view the world afresh, relying only on their own reason and powers of observation, obliterated the values of the civic humanists. It generated what José Ortega y Gasset (1961) characterized as a new technicism and "mass-man," a conformist, selfish, and spoiled individual whose main concern is himself. "Modern technicism," Ortega y Gasset wrote, "springs from the union between capitalism and experimental science" (p. 82). Technicism in turn has produced mass man. "The actual scientific man," he claimed, "is the prototype of the mass-man. Not by chance, not through the individual failings of each particular man of science, but because science itself—the root of our civilization—automatically converts him into mass-man, makes of him a primitive, a modern barbarian" (p.83). As Ortega y Gasset observed:

> [i]n order to progress, science demanded specialisation, not in herself, but in men of science... [E]xperimental science has progressed thanks in great part to the work of men astoundingly mediocre, and even less than mediocre... The specialist "knows" very well his own tiny corner of the universe; he is radically ignorant of all the rest... We shall have to say that he is a learned ignoramus, which is a very serious matter (pp. 83–86).

Ortega y Gasset observed that because scientists have genuine knowledge of their little corner of the universe, and do genuinely contribute to advancing this knowledge, they have the self-assurance characteristic of the genuinely learned of the past and impose their ignorant views on others in a way that ignoramuses of the past would not have dared.

After having achieved major results in the natural sciences, the "scientific method" was brought to bear on the study of humans. This method marginalized the humanities, including history, and carried through modernity's hidden agenda by fitting human nature and society into exact rational categories to achieve full control of people. First economics, then psychology, and later sociology became positive sciences deploying what they took to be the scientific method. Apart from the language of power and law, making known our will to others, as Hobbes (1651, p. 102) put it, discourses that were "unscientific" were deemed to be mere speculation, expressions of personal opinions, of emotions, or just forms of entertainment and therefore not to be taken seriously.

Modern education has been transformed accordingly. The humanities have been on the defensive and, except for a revival in late eighteenth and early nineteenth century Germany, have been losing ground ever since. Since the last quarter of the twentieth century, public universities have been pressured to transform themselves into transnational business corporations run by managers, competing for students on the international market (Readings 1996). Consequently, in some universities, the humanities and the sciences aligned with them are on the verge of being eliminated, as also are anthropology and humanistic forms of the human sciences.

The strongest defense of the humanities has been provided by historians of science who have shown the self-image of science to be fallacious. The great achievements of science, including the work of Descartes and Newton, were built on ideas going back to the Ancient Greeks and beyond, and were achievements of great imagination. Positivistic forms of the human sciences based on this fallacious self-image of science, exemplified by neo-classical economics and behaviorist psychology, have been shown to be pseudo-sciences (Mirowski 1989). However, with academic specialization, even the capacity to understand this is lost. Formal education is dominated by a debased form of science and by "scientism" that has no place for revolutionary science, equating such science with the accumulation of true knowledge. This is the triumph of what Karl Popper (1902–1994) criticized as the bucket view of knowledge. Scientism reproduces this learned ignorance with fragmented knowledge, understood against background assumptions built upon an inconsistent mixture of positivism, atomism, and mechanistic materialism. Combined with an ignorance of history, this is taken to be the scientific worldview.

Today fragmented scientific knowledge is widely sold as a commodity. It is valued because it facilitates the development of profitable technology, and so people are willing to pay for it. Scientific experts are employable because they can develop military technology or make profits for companies, and education in science, technology, engineering, and mathematics can be sold because supposedly, it produces employable scientific and technical experts. Business studies teach students how to exploit these experts and other workers efficiently, while economics legitimates the commodification of every aspect of life, thus legitimating the reduction of scientific expertise and knowledge to commodities. The success of such scientism is manifest in the exponential growth of the economies, now equated with progress, and measured as growth of Gross Domestic Product (GDP). Supported by government policies, scientism now dominates the curricula of students around the world, marginalizing all alternative claims to knowledge. The arts and humanities survive for the most part only as components of the entertainment industries. As noted in Harvard report on the decline in humanities majors and enrolments in the USA (Harvard Working Group Arts on the and Humanities 2013, p.5):

> Research has demonstrated that university disciplines must do at least one of three things to draw the support of university administrators. To be successful, the discipline must either (i) be devoted to the study of money; or (ii) be capable of attracting serious research money; or (iii) demonstrably promise that its graduates will make significant amounts of money.

Along with the humanities, local histories, and local traditions of thought are disappearing. This makes it almost impossible for people to understand themselves and their place in history or in nature. With the world facing a global social and ecological crisis due to the economic growth that this scientism is designed to serve, this one-dimensional culture could prove more disastrous than the cultural decay described by Ortega y Gasset in the 1930s.

There has been a long history of resistance to this modern transformation of education, mostly, although not always, associated with defense of the humanities. "Humanities" as a form of education originating with Petrarch was the product of the Florentine Renaissance quest to inspire people to defend their liberty, and to develop the virtues required of citizens to govern themselves. While Renaissance thinking had been subordinated by the seventeenth-century Scientific Revolution, it had not been extinguished. It survived as the Radical Enlightenment, opposing the mainstream Enlightenment inspired by Newton and Locke (Jacob 2003). The Radical Enlightenment came to fruition in Germany in what has appropriately been referred to as the German Renaissance (Watson 2010). Consequently, the most concerted opposition to Cartesian thought came from Germany, where the Humboldtian model of the university developed in reaction to the atomistic, mechanistic, and utilitarian thought of France and Britain.

Wilhelm von Humboldt (1767–1835), the architect of this model, was committed to reconciling the sciences, the humanities, and the arts to provide students with an integral understanding of the world as central to their self-cultivation or *Bildung*. German education subsequently was recognized as the best in world (Ben-David 1971, ch. 7). However, here also, specialization destroyed this ideal, even within the humanities. It was this that ignited Nietzsche's furious reaction conveyed in his 1872 lectures on education, *Anti-Education: On the Future of our Educational Institutions*. However, while influential, Nietzsche did not really provide a solution to the problem. In heralding the "overman," he simply promoted an extreme elitism and contempt for "ordinary" people.

To get to the root of the debasement of education, and for culture to find a way out, the most promising path is to look at where things went wrong. Von Humboldt and those who influenced him clearly saw the problems facing modernity and the role that education should play in dealing with these problems. It would be worthwhile to investigate these thinkers and their views on education. However, before doing so it is necessary to go back in history to Vico, who confronted Cartesian thought at an earlier stage of its development when its assumptions could still be clearly seen, and alternatives more easily formulated.

19.2 Giambattista Vico's Concept of Education: Defending the Humanities

Giambattista Vico (1668–1744) was a professor of rhetoric in the law faculty at the University of Naples. He was not only defending but also further developing Renaissance thought, the most important part of which was the revival of rhetoric and history. His views on education and his criticisms of Descartes were developed from this perspective. However, Vico did not merely counterpose a humanist legal education to a scientific education. He fully recognized the achievements of seventeenth-century thinkers, including developments in mathematics and science. He followed Aristotle in arguing that different domains required different approaches, but he was against the fragmentation of knowledge. He argued that the ultimate end of education is self-knowledge. He proclaimed in his first inaugural oration, "[a]s a sphere rotates on its axis, so does my argument hinge on this: knowledge of oneself is for everyone the greatest incentive to acquire the university of learning" (Vico 1993, p. 37f.) To this end, Vico argued that education should aim at developing a comprehensive knowledge of the world and that "young men should be taught the totality of sciences and arts" (Vico 1990, p. 19). Vico firstly criticized Descartes' interpretation of his own achievements. Secondly, but more fundamentally, he criticized Descartes' failure to understand the preconditions for these achievements and what was required to go beyond them. These preconditions, historical and in each individual, are imagination and memory, and the capacity to deploy metaphors.

In his tract on education Vico first targeted Descartes, and in doing so developed his most important ideas. He criticized Descartes' claim to have a method of invention, and that there is only one method of inquiry for all disciplines. He rejected Descartes' claim that his analytic approach to mathematics, which was associated with his analytic geometry, was superior to the synthetic method, which was exemplified by Euclidean geometry. More fundamentally, he rejected Descartes' characterization of human intellect in abstraction from human fantasy, passion, emotion, and imagination as well as its social and historical context. It was in defending these claims that Vico developed his most important idea—that we can only really know what we ourselves create.

Descartes believed that he had achieved certainty through only accepting clear and distinct ideas as in mathematics and mathematical reasoning, and extended this claim to certainty from mathematics to physics by claiming that the physical world is mathematical. Vico accounted for this certainty in mathematics by arguing that it is certain because we created it. However, the claimed identity of mathematics and the physical was belied by the failures in Descartes' physics, as pointed out by Leibniz. Descartes had no justification for assuming that nature would conform to his mathematics. Only God who created nature could truly know its principles. What humans could really know is the social world produced by humanity. So in opposition to Descartes who evinced nothing but contempt for history, Vico argued that it is through history, studying the creation by humans of their institutions and themselves, that knowledge can be most fully obtained. To properly understand mathematics and science, we are required to understand their history. The new science referred to in his *Principles of New Science,* was a science of history.

In this history, Vico was concerned above all with the "birth" or genesis of "nations," that is, the genesis of a people with a common language and other institutions, and then with how these evolved, not due to external causes but to internal stresses. He attempted to show how language itself originated and then how the different forms language took with this evolution. The ancients, he argued, were very different from the people of his own time, although "modern" people were ultimately products of the world the ancients had created. While Vico was particularly concerned to characterize the genesis and evolution of legal institutions and the development of language associated with these, his interests extended far beyond. He was also concerned to explain the genesis of philosophy, mathematics, and modern science. It was through the study of Ancient Greece and Rome and the aboriginal people of America that he sought to show the preconditions of these, beginning with the development of language by people originally under the impulse of violent passions for whom "fables" were true narrations of things (Vico 1982, p. 145). Vico argued:

> [P]oetic wisdom, the first wisdom of the gentile world, must have begun in a metaphysics which was not rational and abstract, like that of the learned of today, but sensed and imagined, as that of these first men, devoid of reason and wholly composed of powerful senses and vigorous imaginations ... must have been" (p. 209of).

The most necessary and frequent trope of this poetic logic was metaphor, with the first metaphors drawn from the human body and its parts (p. 223). The next tropes were synecdoche and metonymy, with synecdoche passing into metaphor, by the raising of particulars into universals.

The method for comprehending such genesis and to reconstructing the immanent dynamics of nations involves above all *fantasia* or imagination. Such comprehension reveals that not only does the abstract thinking of the modern world originate in the passions and imagination of early people, but is still dependent upon these, "sharing the poetic nature of their mother" (Vico 1984, p. 123). This is evident in synthetic mathematics and is only disguised by the analytic mathematics of Descartes and its derivatives. In the twentieth century, Vico's argument in this regard has been supported by the constructivists or intuitionists in the philosophy of mathematics. Among these, Mark Johnson, George Lakoff, and Rafael Núñez have shown how mathematics is based on the imaginative use of metaphors, including very basic metaphors associated with body schema (Johnson 1987; Lakoff and Núñez, 2000). Vico argued that "abstraction is in itself but a dull and inert thing" (Vico 1990, p. 39). Those who have inherited modern physics, the Cartesians, are like people who have inherited a "gorgeous mansion" leaving them only able to move the furniture around and add ornaments (Vico 1990, p. 21). They are incapable of creative thinking. In practice, technology designed entirely on abstract principles associated with analytic thinking tends to result in "constant failure" (Vico 1990, p. 29).

Following these observations, Vico argued that the most important precept for education is the injunction "know thyself," which was equated by Vico to "know your own spirit" (Vico 1993, p. 38f.). This is not achieved through introspection but through wisdom, requiring the entire curriculum of thought. Education of the young should be based on psychology, recognizing stages in psychological development that parallel the cultural stages traversed by humanity in its evolution. It should not introduce abstract reasoning too early into a child's education, but rather should foster their imaginations and memory, thereby kindling their wills and sparking their enthusiasm. Imagination should be preserved even when students develop more abstract reasoning, preserving the ability to deploy metaphors and cultivating "the capacity to see analogies existing between matters lying far apart and, apparently, most dissimilar" (Vico 1990, p. 24). The aim should be to grasp the whole of any situation and to put this into words. While arguing that students should be "well versed on all fields of knowledge" (p. 78), Vico complained that education paid too much attention to science and not enough attention to ethics, most importantly, "that part of ethics which treats of human character, of its dispositions, its passions, and the manner of adjusting these factors to public life" (p. 33). History is central to this.

Vico acknowledged that different nations originated at different places and in different times in different circumstances having very different experiences; consequently, their customs and thinking and very natures differed from each other (Vico 1984, p. 148). This is manifest in the differences between languages. While the study of history is required to comprehend human character as the foundation for ethics, this requires imagination that enables students to appreciate people different from themselves. It is not difficult on this basis to use Vico's ideas on education to argue that each region of the world, having different customs and ways of thinking, requires a different education, one that acknowledges their unique histories, circumstances, and associated practices, and customs of where they were born. However, Vico did not take this step.

19.2.1 Herder and the Concept of Cultures

It was Johann Gottfried Herder (1744–1803) in Germany who acknowledged the diversity of cultures and defended this diversity, with important implications for education. Along with Johann Gottlieb Fichte, Herder was one of the most eminent students of Kant and immensely influential toward the end of the eighteenth century and a progenitor of the ideas which were embraced by Wilhelm von Humboldt. This influence has continued to the present but has been marked by controversies and misinterpretations of his work by confusing it with later chauvinistic forms of German nationalism (Arnold et al. 2009). Recent scholarship has eliminated most of this confusion, and Herder is now recognized along with Vico as one of the greatest proponents of education in the humanities and its importance for life and society (Wiborg 2000; Sikka 2011, 98ff.).

Herder also had a major influence on the development of post-mechanistic science, including geography, and through this, ecology (Tang 2008, Chap. 3). He also had an enormous influence on the development of cultural anthropology, leading to appreciation of the cultures of indigenous populations. As H.B. Nisbet (1970) has shown, he did not ignore developments in the natural sciences but embraced and further developed challenges to Cartesian and Newtonian cosmology. While this involved an alliance with Leibniz and those influenced by him, as part of the Radical Enlightenment Herder was reviving the cosmology of the late Renaissance natural philosopher Giordano Bruno (Singer 1950, 195). Herder was central to what has been called "Europe's Third Renaissance" and "The Second Scientific Revolution" (Watson 2010).

When he formulated his own ideas, Herder had not read Vico's work, although he would have heard of him. Clearly, their concerns and their philosophies had much in common. Both were reacting against the Cartesian tradition of thought and, privileging history over science, both emphasized the differences among peoples. Like Vico, Herder attempted to explain the emergence of language and then the development of humanity through history (von Herder 2002). Both argued for the primordial role of metaphors in language and thought. When it came to history, both focused on the life of communities rather than the exploits of individuals. In opposition to the atomic individualism of Britain and France, Herder argued for the supreme value of belonging to a group. It is only through membership of groups, most importantly, a "nation," that people develop their individuality. A "nation" as Herder understood it, is an ethnic community and can include indigenous people. The development by individuals of their individuality is part of the development of the individuality of their nation.

While Vico focused on institutions and their development, Herder developed the notion of "culture," using this word in the plural for the first time to acknowledge that different nations had different cultures. For the most part, this difference was terminological, since "culture" for Herder included the whole way of life of a people and so also included their language, their institutions, their traditions, and collective memories. Both Vico and Herder regarded language as the core of human communities. However, Herder gave an even greater place to the differences among peoples, which included contemporaries as well as peoples of past eras. While Herder's notion of culture was a major contribution to history and to the humanities, Herder understood people in their geographical contexts as part of nature. Differences among them were seen by him to be related to "climate," including geographical and physical environments as well as biological needs that contributed to the formation of nations.

While being one of the original philosophers of nationalism, as Isaiah Berlin (1976, p. 153) argued, Herder's nationalism was cultural rather than political. While ancient philosophers such as Plato, Aristotle, and Cicero had argued that people should be striving to realize their potential as human beings to become fully human, Herder argued that each nation and each individual person has its own center of gravity and its own unique potential to be realized. The challenge for nations through individuals is to discover this center of gravity as part of the process of their self-realization. In this process of self-realization people express themselves in what they produce. Self-expression is the essence of being a human being, and in the quest for self-realization the challenge is to express oneself fully in one's work, being committed to it rather than divided and uncommitted. In this way, each nation, through the quest for self-realization by its individual members, makes its own contribution to humanity.

Herder is generally known as an originator of the notions of historicism, nationalism, and the *Volkgeist* or spirit of the people, and for his call for respect for all cultures by acknowledging the incommensurability among the values of different cultures and by appreciating these differences. However, he also argued that there is a general tendency through history toward greater "humanity," that is, the realization of the human potential for reason and justice. Progress was seen as the joint product of creative legislators, poets, artists, philosophers, inventors, and educators through the ages, leading to more benevolent government (von Herder 1968, Chap. II; Sikka 2011). Growing respect for these differences should be recognized as part of the development of humanity.

Consequently, Herder argued against imperialism, not only of the present but also of the past. He was unsympathetic to Roman imperialism. He wrote of Rome:

> The walls that separated nation from nation were broken down, the first step taken to destroy the national character of them all, to throw everyone into one mold called "the Roman people." ... Foreign peoples were judged according to customs with which they were unfamiliar, presented with vices and punishments that they had never even heard of. And was not the ultimate ... outcome of this entire legislation, which was really only appropriate to the constitution of Rome, such a diminishment and debasement of the conquered peoples' characters, after a thousand oppressions, that instead of their original features, nothing remained in the end but the Roman eagle who, after pecking out their eyes and devouring their entrails, covered the sad corpses of the provinces with its feeble wings? (von Herder 2004, p. 22).

The advance of European civilization parallels what happened in Rome. Herder asks,

> Can you name a land where Europeans have entered without defiling themselves forever before defenceless, trusting mankind, by the unjust word, greedy deceit, crushing oppression, diseases, fatal gifts they have brought? Our part of the earth should be called not the wisest, but the most arrogant, aggressive, money-minded: what it has given these peoples is not civilization but the destruction of the rudiments of their own cultures wherever they could achieve this. (quoted in Berlin 1976, p. 160f.)

All this had implications for education. Herder was a superintendent of schools and strove to reform education (Müller-Michaels 2010). He was an originator of the notion of education as *Bildung*—formation of character as a process of self-creation. Teachers should not be seen as imparting knowledge, but as engaging with their students to stimulate their curiosity, and to encourage them to realize their potential, thereby achieving self-knowledge by understanding their unique place in history and in nature. While this should involve a quest for comprehensive knowledge that includes science as well as history, this should not obliterate the experience of each student, which should be recognized as the starting point for this quest.

As with Vico, the development of language was accorded major significance in education. However, for Herder, each language is different. This difference is extremely important because language is the unique repository of the wisdom of cultures, embodying the images, metaphors, and values of a people. "Has a nation anything more precious than the language of its fathers?" Herder wrote. "In it dwells its entire world of tradition, history, religion, principles of existence; its whole heart and soul" (Berlin 1980, p. 165). While the quest for *Bildung* involves a quest to advance the whole of humanity, this quest originates in the specific culture and specific circumstances of each individual person, with their own unique conditions and potentialities. This does not imply complete relativism. Herder argued that the quest for wisdom, including the development of science, can never be completed. Moreover, this quest can only advance through diverse perspectives being developed and by challenging each other while learning from each other. Herder was defending what, since the end of the twentieth century, has been defended by Mikhail Epstein (1995, p. 299ff.) as "transculturalism" rather than "multiculturalism."

19.3 Von Humboldt, Schelling, and the Philosophy of Nature

Both directly, and indirectly, through German theologian Friedrich Schleiermacher (1768–1834) and philosopher Friedrich Schelling (1775–1854), Herder was a major influence on Wilhelm von Humboldt's views on education when he established the University of Berlin in 1810, privileging the humanities while supporting and aligning it with developments in the sciences. Von Humboldt's ideas on education made explicit and further developed Herder's notion of *Bildung*. To this end, von Humboldt argued that university education must combine teaching and research, with students participating in research, working with their lecturers rather than simply learning from them. The starting point for all enquiry, von Humboldt argued, is the quest for self-understanding, which ultimately leads to the quest to understand the cosmos. Education had to be an active, dialectical, process; it should involve and respect the perspectives of students. As von Humboldt argued:

> ... the inward organization of these institutions must produce and maintain an uninterrupted cooperative spirit, one which again and again inspires its members, but inspires without forcing them and without specific intent to inspire. ... It is a further characteristic of higher institutions of learning that they treat all knowledge as a not yet wholly solved problem and are therefore never done with investigation and research. This ... totally changes the relationship between teacher and student from what is was when the student still attended school. In the higher institutions, the teacher no longer exists for the sake of the student; both exist for the sake of learning. (von Humboldt 1963, p. 132f.)

While von Humboldt was more directly influenced by Schleiermacher, Schelling (1803) did have some influence through his work *On University Studies,* published at the beginning of the nineteenth century. However, Schelling's significance for advancing the humanist ideal of education derives from the success of his quest to produce a philosophy which could reconcile science with history and art. Building on Kant and Fichte, Schelling developed Herder's philosophy of nature and ideas on culture far more rigorously than had Herder, and in doing so, inspired developments in science that challenged and eventually replaced Cartesian dualism and Newtonian cosmology.

Kant had accepted Vico's argument that mathematics is a human construction, arguing that it requires productive imagination. Imagination was then given a central place in all cognition, including the development of science. Kant developed the notion of concepts to characterize the forms of intuition and categories of understanding. He claimed that he could show through transcendental deductions which concepts must be accepted to make our experience of the world intelligible. Imagination, however, is also required to organize the sensory manifold, synthesizing appearances into a unity as the condition for subsuming these appearances under concepts and relating them to the unity of the "I." This empirical imagination is the condition for the productive imagination through which we can reflect on these appearances and the concepts through which they are organized. As a student of Kant, Fichte rejected Kant's strictures on what can be known, and claimed that as the productive imagination enables us to comprehend mathematics, we can also comprehend the cognitive development of children. Concepts first form through active engagement of the child against resistance, Fichte argued. He also argued that cognition has a social dimension because the 'I' that accompanies all our perceptions and thoughts develops through mutual recognition. The child comes to experience itself from the perspective of the other as a subject among other subjects. Instead of transcendental deductions, Fichte argued that these developments are creative, involving new syntheses made possible through imagination. Cognition is dialectical.

Schelling began as a disciple of Fichte, but took his arguments further, synthesizing them with Herder's philosophy. He argued that through the productive imagination we can reconstruct not only the cognitive development of humans, which always takes place within a particular culture, but the evolution of nature. Advancing Herder's philosophy of nature, he called for a transformation of science and new forms of mathematics. In opposition to Kant, who identified science with mathematical theories, Schelling defended natural history by recognizing the primacy of narratives over abstract models in comprehending the world.

Nature was portrayed as developing through the limiting of activity. Humans, as essential socio-cultural beings, could still be seen as having evolved within nature, bringing nature to consciousness. Embracing the notion of mutual recognition as the foundation of ethics, he characterized the history of humanity as the advancement of freedom based on the development of recognition. While acknowledging that we first develop concepts of nature to overcome resistance to our will, transforming it in the process, the development of mutual recognition and self-understanding as participants in nature grants a place to the quest for justice and wisdom as far more than instruments to control the world to satisfy appetites. The quest for justice and wisdom also justifies respect for the intrinsic value not only of other people but also of nature. While the mechanistic view of nature of Descartes and Newton served the development of technology, it was clearly deficient because it could not account for the emergence of sentient life and human consciousness. While Kant had given a place to concepts in cognition and defended a particular set of concepts, Schelling argued that it is possible to advance beyond received concepts and develop new, more adequate concepts to understand the world and ourselves. To this end, it is also necessary to study and recover the wisdom of past ages, and of diverse cultures throughout the world. There can be scientific revolutions. There is no end point to these dialectical processes. The quest for proper recognition and for knowledge can go on indefinitely providing people are willing to take up the challenge of developing new conceptual frameworks and new forms of life to overcome current contradictions.

In the past, Schelling's speculations on the philosophy of nature were regarded as of little significance. Work on the history of science has since revealed that almost all major advances in science since Schelling have been influenced by his philosophical reflections, including the first law of thermodynamics, the development of field theories in physics and the notion of valency in chemistry, major advances in mathematics along with developments in biology where Schelling is seen as a precursor to complex systems theory and biosemiotics (Heuser-Kessler 1986; Gare 2013). These in turn facilitated the emergence of new disciplines, including ecology, strongly influenced by von Humboldt's work in geography, and cultural anthropology based on Herder's notion that there was a diversity of cultures (McIntosh 1985, 24; Tang 2008, 160ff.). In the twentieth century, human ecology emerged through the synthesis of geography, ecology, and cultural anthropology (Kormondy and Brown 1998).

The success of research traditions inspired by Schelling's philosophy of nature also vindicates Schelling's conception of science, which anticipated the arguments of the opponents of logical positivism, including Paul Feyerabend's argument that progress in science requires the proliferation of research programs and Alasdair MacIntyre's argument that narratives are required to defend major conceptual revolutions in science. That this achievement originated in Germany, refusing to accept the subordination of German culture to the intellectual centers of France and Britain, vindicated Herder's defense of national cultures and the need to resist the homogenization of national cultures with the dominant global culture. It revealed the value of diversity of cultures not only because the vitality of a nation derives from their specific cultures, but because it is the condition for the advancement of science itself. This vindicates von Humboldt's conclusion, influenced by Herder, Schleiermacher, and Schelling, that in the modern world, it is possible to overcome the fragmentation of knowledge and to reconcile the sciences and the humanities. And it vindicates their notion of education as *Bildung*, the development of culture and character through participating in the quest to advance our understanding of the world, not as Cartesian minds standing outside the world, but as situated as participants within nature and based in a particular culture in a particular place at a particular time in history.

19.4 Conclusion

The quest for biocultural conservation is an extension of the values promoted by these disciplines of ecology, cultural anthropology, and human ecology, as it has been proposed in the Ecology and Ethics book series (Rozzi et al. 2013, 2015, 2018). Ecology challenged the view that nature is nothing but matter to be transformed to serve human purposes and rejects the claim that evolution occurs through the competitive struggle for survival. Evolution occurs through organisms and species interacting with each other so as to augment the environments of each, and that to prosper, they must augment the health of ecosystems. This has generated ever more complex forms of symbiosis. Cultural anthropology countered the view of the social Darwinists that indigenous people are relics of the past, losers in the competitive struggle for survival, and doomed to extinction (Bowler 1989, p. 305). Human ecology integrated these perspectives, and it is this integration that justifies the call for biocultural conservation.

This has significant implications for education, because education is enculturation. Biocultural conservation implies that diverse cultures should be recognized, and that this needs to be recognized in education practices. This is an extension of Herder's argument that those formed by their cultures should not abandon them, but appreciate their uniqueness and strive to realize their full potential. Education should begin with students studying the world around them and mastering local knowledge developed in their particular habitats. This does not mean insulating students from all other cultures, but building on their own culture while learning from other cultures. What is required is not multiculturalism, but transculturalism. This should be recognized to be the case even with science, where in the past, advances in science were often due to new ways of thinking injected by scientists with different cultural backgrounds influenced by different practices and ways of living.

Today, however, the condition described by Ortega y Gasset, domination of intellectual life by overspecialized barbarians, the "mass-scientists" as opposed to broadly educated scientists, and the consequent corrosion of culture and diversity, is corrupting our education systems, research organizations and science itself as never before. With the transformation of educational institutions into business enterprises, the highest values have been devalued. The quest for wisdom is dismissed. Everything is defined in terms of economics, with the sole end being the growth of the economy as defined and measured by economists. Knowledge is only valued insofar as it serves this growth. The new supposedly more efficient universities run by managers are producing more graduates than ever before, graduates who are not only uneducated in the traditional sense but very often unemployable (Newfield 2008, p. 4). If we look at science itself, there have never been more scientists or greater production of scientific papers, but in almost every field, including physics, most papers are frequently shoddy, and science has stagnated (Smolin 2007; Charlton 2012). Climate science, developed against strong opposition, is an exception. Efficiency has produced stagnation, not unlike the stagnation of the late Roman Empire.

The triumph and proliferation of these "neo-barbarians" have collided with the reality that humanity is on a trajectory to disaster through global ecological destruction, and as with other major problems, ineffective responses associated with tunnel vision are leaving us on the same trajectory. This has opened a space where we can question the claimed authority of these anointed hyper-specialists. These "mass-scientists" in Ortega y Gasset's sense are nothing like the great scientists of the past. As Robert Root-Bernstein (2015, p. 204) pointed out, great scientists typically are "endowed with an abundance of restless imagination, spend their energy in pursuit of literature, art, philosophy, and all the recreations of mind and body." And as Joseph Ben-David observed, the sites of great intellectual achievements have been characterized by decentralization, facilitating the survival of a great diversity of schools of thought contending with each other, as in Ancient Greece, Renaissance Italy, and Nineteenth-Century Germany and the Anglophone world in the Twentieth Century. Efforts to control such creativity has inevitably destroyed it. Epstein (1995, p. 201) claimed that the greatest creativity occurs on the borders of cultures which challenge each other while retaining their integrity. The time is ripe for the recovery of genuine science.

In the spirit of Schelling, the theoretical ecologist Robert Ulanowicz (1997, p. 6) now argues that developments in ecology could provide the forms of thinking needed to break through the logjams afflicting physics. Such arguments are part of the recent revival of natural philosophy (Gare 2018). This could advance all other disciplines, bridging the gap between science and the humanities, supporting and reinvigorating history, the humanistic forms of the human sciences, particularly anthropology, and transforming our understanding of our place in nature (Rozzi 2019). It gives a place to human ecology, and through this, appreciation of the diversity of Indigenous peoples and non-European civilizations with cultures adapted to their often-unique ecosystems (Mackey and Claudie 2015). This development provides the means for such people to defend their own cultures against the homogenizing expansion of the culture of modernity. In line with these developments, and to effectively confront the crises of civilization, most importantly, the ecological crisis, we need a radical redirection of education. This required direction has already been indicated by Vico, Herder, Schelling, and von Humboldt, an education aimed at wisdom, fostering imagination, and the quest for a comprehensive understanding of the world, ourselves, and our place within it, cultivating humanity, and acknowledging and appreciating diversity and the significance of all life. Wisdom, including appreciation by people of their unique habitats and cultures, is now required more than ever to overcome the global ecological crisis.

References

Arnold G, Kloocke K, Menze EA (2009) Herder's reception and influence. In: Hans Adler H, Koepke W (eds) A companion to the works of Johann Gottfried Herder. Camden House, Rochester, pp 391–420

Ben-David J (1971) The scientist's role in society: a comparative study. Prentice Hall, Englewood Cliffs

Berlin I (1976) Vico and Herder: Two studies in the history of ideas. Viking, New York

Berlin I (1980) Note on alleged relativism in eighteenth century European thought. J Eighteenth Century Stud 3(2):89–106

Bowler PJ (1989) Evolution: the history of an idea, revised edition. University of California Press, Berkeley

Charlton BG (2012) Not even trying: the corruption of real science. University of Buckingham Press, Buckingham

Epstein M (1995) After the future (trans: Miller-Pogacar A). University of Massachusetts Press, Amherst

Gare A (1998) McIntyre, narratives and environmental ethics. Environ Ethics 20:3–21

Gare A (2006) Postmodernism and the environmental crisis. Routledge, Abingdon

Gare A (2013) Overcoming the Newtonian paradigm: The unfinished project of theoretical biology from a Schellingian perspective. Prog Biophys Mol Biol 113:5–24

Gare A (2018) Natural philosophy and the sciences: challenging science's tunnel vision. Philosophies 3(33):1–29

Harvard Working Group Arts on the and Humanities (2013) Mapping the future: the teaching of the arts and humanities at Harvard College. https://scholar.harvard.edu/files/jamessimpson/files/mapping_the_future.pdf. Accessed 20 Aug 2020

Heuser-Kessler M (1986) Die Produktivität der Natur: Schellings Naturphilosophie und das neue Paradigma der Selbsorganization in den Naturwissenschaften. Duncker & Humblot, Berlin
Hobbes T (1651) Leviathan. 1985 edition. Penguin, Harmondsworth
Jacob MC (2003) The radical enlightenment: pantheists, free-masons and republicans, 2nd edn. Temple Publishers, Greensboro, NC
Johnson M (1987) The body in the mind: the bodily basis of meaning, imagination, and reason. University of Chicago Press, Chicago
Kormondy EJ, Brown DE (1998) Fundamentals of human ecology. Prentice Hall, Upper Saddle River
Lakoff G, Núñez RE (2000) Where mathematics comes from. Basic Books, New York
Mackey B, Claudie D (2015) Points of contact: integrating traditional and scientific knowledge for biocultural conservation. Environ Ethics 37:341–357
May Jr RH (2023) Liberation philosophy and biocultural education. A Latin American journey. In: Rozzi R, Tauro A, Wright T, Avriel-Avni N, May Jr RH (eds) Field environmental philosophy: education for biocultural conservation. Ecology and ethics, vol 5. Springer, Dordrecht, pp 203–228
McIntosh RP (1985) The background of ecology: concept and theory. Cambridge University Press, Cambridge
Mirowski P (1989) More heat than light: economics as social physics, physics as nature's economics. Cambridge University Press, Cambridge
Müller-Michaels H (2010) Herder in office: his duties as superintendent of schools. In: Adler H, Koepke W (eds) A companion to the works of Johann Gottfried Herder. Camden House, Rochester, pp 373–390
Newfield C (2008) The unmaking of the public university: the forty-year assault on the middle-class. Harvard University Press, Cambridge
Nisbet HB (1970) Herder and the philosophy and history of science. The Modern Humanities Research Association, Cambridge
Ortega y Gasset (1961) The revolt of the masses. Unwin Books, London
Readings B (1996) The university in ruins. Harvard University Press, Cambridge
Root-Bernstein R (2015) Arts and crafts as adjuncts to STEM education to foster creativity in gifted and talented students. Asia Pacific Educ Rev 16:203–212
Rozzi R (2012) Biocultural ethics: the vital links between the inhabitants, their habits and regional habitats. Environ Ethics 34:27–50
Rozzi R (2019) Taxonomic Chauvinism, no more! Antidotes from Hume, Darwin, and biocultural ethics. Environ Ethics 41:249–282
Rozzi R (2020) A biocultural ethic for sustainable geographies. In: Sarmiento FO, Frohlich LM (eds) The Elgar companion to geography, transdisciplinarity and sustainability. Edward Elgar, Northampton, pp 172–189
Rozzi R, Pickett STA, Palmer C, Armesto JJ, Callicott JB (eds) (2013) Linking ecology and ethics for a changing world: values, philosophy, and action, Ecology and ethics, vol 1. Springer, Dordrecht
Rozzi R, Chapin FS III, Callicott JB, Pickett STA, Power ME, Armesto JJ, May RH Jr (eds) (2015) Earth stewardship: linking ecology and ethics in theory and practice, Ecology and ethics, vol 2. Springer, Dordrecht
Rozzi R, May RH Jr, Chapin FS III et al (eds) (2018) From biocultural homogenization to biocultural conservation. Ecology and ethics, vol 3. Springer, Dordrecht, pp 21–47
Schelling FWJ (1803) On university studies. Guterman N (ed), (trans: Morgan ES). 1966 edition. Ohio University Press, Athens
Sikka S (2011) Herder on humanity and cultural difference. Cambridge University Press, Cambridge
Singer DW (1950) Giordano Bruno: his life and thought. Henry Schuman, New York
Skinner Q (1998) Liberty before liberalism. Cambridge University Press, Cambridge

Smolin L (2007) The trouble with physics: the rise of string theory, the fall of science and what comes next. Houghton Mifflin, Boston
Tang C (2008) The geographic imagination of modernity: geography, literature, and philosophy in German Romanticism. Stanford University Press, Stanford
Toulmin S (1990) Cosmopolis: the hidden agenda of modernity. University of Chicago Press, Chicago
Ulanowicz RE (1997) Ecology: the ascendent perspective. Columbia University Press, New York
Vico G (1982) Vico: selected writings (trans, ed: Pompa L). Cambridge University Press, Cambridge
Vico G (1984) The new science of Giambattista Vico (trans: Bergin TG, Fisch MH). Cornell University Press, Ithaca, NY
Vico G (1990) On the study methods of our times (trans, ed: Gianturco E). Cornell University Press, Ithaca
Vico G (1993) On humanistic education (trans: Pinton GA, Shippee, AW). Cornell University Press, Ithaca
von Herder JG (1968) Reflections on the philosophy of the history of mankind. University of Chicago Press, Chicago
von Herder JG (2002) Philosophical writings (ed, trans: Forster MN). Cambridge University Press, Cambridge
von Herder JG (2004) Another philosophy of history and selected political writings (trans: Evrugenia ID, Pellerin D). Hackett, Indianapolis
von Humboldt W (1963) Humanist without portfolio (trans: Cowan M). Wayne State University Press, Detroit
Watson P (2010) The German renaissance. Simon & Schuster, London
Wiborg S (2000) Political and cultural nationalism in education: the ideas of Rousseau and Herder concerning national education. Comp Educ 36(2):235–243

Part III
Biocultural Reconnection: Recovering the Sense of Community Through Education

Chapter 20
Introduction to Part III. Biocultural Reconnection. Recovering the Sense of Community Through Education

Alejandra Tauro

Abstract This chapter introduces the part about educational alternatives towards biocultural conservation in a broad sense. The part's main goal is to present and to examine a diversity of ongoing initiatives that propose complementary approaches to FEP (see Part I) to overcome the prevalence of monocultural and "universal" approaches to formal, and even non-formal (see Part II) education. The reader is introduced to novel educational methodologies and experiences developed in different regions of the world with cases in Portugal, Japan, the United States of America, Venezuela, Mexico, and South Africa. This chapter focuses on reconnecting with the small, the silent, and the invisible in childhood and other co-inhabitants too often ignored by the prevalent view of education, and in this way recover the sense of community involving children and adults in different initiatives through philosophical dialogue. Also, it is proposed to reconnect society with biocultural community by incorporating ecotourism as an effective practice of non-formal education that can generate economic alternatives based on social solidarity for local communities. The chapter also shares experiences of collaborative actions and education facilitated by technological tools. These cases, that incorporate local development, territory defense, and traditional local knowledge, offer a broader and contextual understanding of education as a tool that reconnects people with biocultural community. These biocultural reconnections weave together local initiatives and global movements that seek to defend and create conditions that make possible the continuance of life on this planet.

Keywords Interwoven alternatives · Biocultural ethics · Ecotourism · Philosophy for children · Intergenerational perspective

---

A. Tauro (✉)
El Colegio de Puebla A.C., Puebla, Mexico

Cape Horn International Center (CHIC), Universidad de Magallanes, Puerto Williams, Chile

R. Rozzi et al. (eds.), *Field Environmental Philosophy*, Ecology and Ethics 5,
https://doi.org/10.1007/978-3-031-23368-5_20

20.1 Introduction

Faced with the accelerated destruction of the diversity of habitats and lifeways, educational alternatives and biocultural conservation initiatives are emerging that value life as the articulating axis for transformations toward sustainability (Rozzi 2013; Kothari 2020). In different parts of the world, we can observe formal and non-formal education alternatives that resist the forces of destruction and provide guidance for reconnection among different co-inhabitants and habitats. Reconnecting means connecting again, putting back in communication, and recovering linkages; etymologically it is associated with the verb to bind. Recovering the links between humans, today mainly city dwellers, and the diversity of co-inhabitants leads us to a re-understanding of what it means to live together in a matrix of diverse cultures, living beings, and ecosystems. This reconnection resonates with the original meaning of the term religion which, unlike any ecclesiastical institution, leads us to re-link ourselves with "the whole" or that portion of life that animates us. In other words, reconnecting refers to recovering the soul—"animus" or "spirit"—an allusion to breathing, a vital physiological function that connects life with the material world we co-inhabit (Rozzi 2013). Reconnection as recovery of the spirit or soul imbues the practices of the different co-inhabitants with spirituality that has been lost or weakened by the rationality of modernity. Therefore, practices of reconnections return linkages and sense of community to different social spaces.

In this book, we focus, on the one hand, on deconstructing teaching-learning spaces that have been weakened by the one-dimensionality of an education model centered on the employment market (Prakash and Esteva 2008). On the other hand, we highlight a diversity of alternative educational initiatives, programs, and systems that are deployed in communities both outside and inside schools. One of these alternative practices is Field Environmental Philosophy (FEP) whose essential purpose is to reconnect the spheres of the sciences (truth), the arts (aesthetics-beauty), and ethics (justice).

FEP has a transdisciplinary root that generates transformative actions based on knowledge that is commonly shared among the diversity of co-existing knowledges in a habitat. Its methodology motivates biocultural reconnection through direct and experiential encounters between co-inhabitants in a habitat. Then, by generating alternative views from analogical thinking based on art and poetic creation, the reconnection is deepened by fine-tuning the human senses to broaden the perception of co-inhabitants and their habits. Finally, reconnection experiences are encouraged to continue over time, generating structural changes in the political, social, and material infrastructure that lead to the long-term conservation of biocultural linkages (Rozzi et al. 2010a, also see Part IV).

In synergy with FEP, a "tapestry of alternatives" is spreading around the world (Kothari 2020). Among these alternatives are "forest schools" in parts of Europe that provide children with hands-on learning amid nature; the autonomous Zapatista schools in Mexico that teach about diverse cultures and social struggles; and the Ecoversities Alliance of educational centers worldwide that broaden disciplinary academic knowledge (Kothari 2021). These alternatives highlight the pedagogical dimension of transformational change. In this chapter, we present ongoing alternative initiatives that overcome the prevalence of monocultures and "universal" approaches to education. That education is a substantive part of these transformational changes is recognized as an indirect driver (Chan et al. 2020). However, we show that transformative change requires awareness of the power of education in establishing reconnections and opening paths of resistance, and not limiting it to capacity building or reducing it to homogenizing systems.

In this Part III we present different cases of creative and innovative initiatives in heterogeneous biocultural contexts, that vindicate socio-environmental struggles, and that emphasize underestimated dimensions of the amazing diversity of life. With an eye on the transforming power of education, we seek to integrate the lost and forgotten dimensions of human being with their co-inhabitants. Thus, biocultural reconnection leads to community recovery and revitalization.

A central motivation of the chapters that make up this part of the book is to show the virtues that co-exist around the world, that are built amid oppressive scenarios. This sharing of community biocultural reconnection could be an "antidote" to alleviate the gaps of inequities and "chauvinism" that dominate the societies of the current century (Rozzi 2019). We develop these ideas in chapters that organize the following themes:

- Education with "children's eyes" to reconnect with the world
- Ecotourism as an educational platform to generate various (re)encounters with "eyes of the earth"
- Education and co-participatory methods through technologies to reinforce the community fabric with "fraternal eyes"

20.2 Education with "Children's Eyes" to Reconnect with the World

From a reductionist view, formal education focuses on preparing pupils (small or large) to compete and achieve goals in the labor market. This vision limits the opportunities to share and create worlds beyond an employment system. Costa Carvalho (2023) chapter, "Small, Silent, and (In) significant: Childhood as a Minoritarian Experience of Education" (Chap. 21), re-states the meaning of education by looking at childhood through three philosophical concepts: "the small, the silent, and the insignificant." We start school in childhood when everything is great in color and imagination, full of songs and silences of astonishment, that express all the senses of our bodies in constant movement creating new possibilities. Education that prepares for the market takes childhood away from us and imposes a "normalizing" expectation on how childhood should be educated by the family and by the school, as shown in "With Children's Eyes" by Franceso Tonucci (2012). In her chapter Costa Carvalho (2023) reconnects us with those eyes of childhood, problematizing the implications of Western rationality in early childhood education, and showing the results we see in today's societies. The author moves the implications of this reductionist rationality beyond humans, to the invisibility of other beings. In the author's words, "education might expand images (becoming a diffractive lens), sounds (becoming a babel-like amplifier), and senses and meanings (becoming playful)." This inclusive and decolonizing education makes it possible to reconnect the links that are lost or weakened by Westernized rational thought, thus returning their power to the communities of co-inhabitants.

Communities are defined by the interactions between different co-inhabitants and generations sharing their habitats and habits. Faced with the exacerbated individualism typical of the predominant utilitarian worldview, community ties have weakened and even disappeared in many cases. In these scenarios, the recovery and revitalization of community fabrics is an antidote (*sensu* Rozzi 2019). A biocultural restoration experience carried out in Japan shows that by applying the holistic and systemic approach of FEP it was possible to revitalize biocultural interactions through a participatory and intergenerational process that proposed a collaborative form of governance (Toyoda 2018). These community processes require fostering dialogue to generate the conditions that enable otherness (looking at one another through others) and thus strengthen biocultural community fabrics (Fleuri and Fleuri 2018).

In his chapter "Communities of Philosophy Inquiry for the Empowerment of Ecological Agency" (Chap. 22), Johnson et al. (2023) show how philosophical dialogue provides opportunities to look at diverse environmental problems from different perspectives in order to expand the horizons of *biocultural awareness*. The authors reflect on the tension that can arise in philosophical dialogue as a source that generates both diversity and conflict. Paulo Freire also recognized such tension and contradictions when motivating pedagogical dialogue within communities as ways to overcome colonial struggles (Fleuri and Fleuri 2018). Acknowledging tension and creating mechanisms to integrate it into dialogue in a positive way empowers the biocultural community. The authors offer us a little explored dimension in environmental issues such as the therapeutic service that educational space can provide. We might note that this type of practice also has been proposed by Eugenio Echeverria (2004) based on his own experience in Philosophy for Children in Mexico, and by Abadio Green-Stocel et al. (2013) in *Pedagogy of Mother Earth*.

This therapeutic space is discussed in the work of Santiago-Jiménez (2023, Chap. 23). From the perspective of psychological resilience, the author reflects on recovery from traumas in people or communities through practices, such as rural tourism, that can provide successful adaptation to negative life events (see next theme). These practices promote personal and communal dialogue in a way that is complementary to social and structural dialogues, all fundamental in the search for socio-environmental justice. The dialogue communities that are born from philosophy for children and with children, complementing FEP, encourage critical thinking "with the eyes of children." This kind of thought is necessary to develop democratic dispositions that support fairer societies based on subjects and not on manipulative partisan political systems. The need for new institutional infrastructures (see Part IV) is imperative to overcome biocultural homogenization and associated utilitarian ethics.

20.3 Ecotourism as an Educational Platform to (Re)Encounter the "Eyes of the Earth"

FEP's methodological approach integrates the social, economic, and environmental dimensions of sustainability by generating new links between biological and cultural diversity at different spatial and social scales. The practice of FEP, in line with other movements such as Global Tapestry of Alternatives (Kothari 2021), emphasizes that the generation of local economic alternatives (for example, ecotourism) and the consolidation of these alternatives over time, are fundamental for generating conditions enabling the maintenance of biocultural diversity (Tauro et al. 2021). Two key aspects in the generation of these alternative actions are (1) education that promotes re-learning habitat and (2) community economic policies that facilitate a decent standard of living. Economic alternatives based on the principles of social solidarity are inspirations to move into a different kind of world (Kothari 2021).

Ecotourism has been recognized as an effective practice of non-formal education that facilitates reconnecting society with a sense of community (Walter 2009; Breakey and Breakey 2023). The incorporation of "genuine and correct" ecotourism practices in biocultural programs and initiatives generates opportunities that counteract migration in the absence of job opportunities, promotes local conservation, and improves conditions of community inclusion and well-being (Torres-Villa and Barragán-López 2016; Walter et al. 2018; Singh 2021). Continuing with the metaphor provided by the previous chapters, ecotourism, as one part of economic solidarity alternatives, invites us to travel with the "eyes of the earth" and seeing others as peers.

In the chapter "Biocultural Resilience Through Educational Tourism in Cholula, Mexico" (Chap. 23), Evelinda Santiago-Jiménez (2023) presents a case study of an educational and cultural tourism experience in a rural community of San Andrés Cholula. This educational experience takes "tourist animation" as a pedagogical tool to facilitate social gatherings between peasants who open the doors of their homes and city visitors who wish to have enriching experiences. This initiative offers a strategy to resolve the social vulnerability of different rural and indigenous communities in Mexico. Drawing on the concept of psychological resilience, Santiago-Jiménez explains that this strategy helps to "imagine the construction of an inclusive tourism process, which also provides a holistic vision sufficient to not only consider and integrate individuals who are in situations of poverty, but also those who are in a latent state of vulnerability due to poor income or social deprivation." The author provides a vision of how, through educational tourism, the economies of rural families can be strengthened in a way that complements their various productive activities that are increasingly weakened by the concentration of agro-industrial production and land grabbing. In the context of oppression, rural and cultural tourism becomes an opportunity with a dual purpose: (1) to generate economic income supported by community organization and (2) to educate tourists through culinary cultural experiences rooted in regional biodiversity. Ecotourism and other forms of local and biocultural tourism are linked to changes that are occurring in tourism in general. Current approaches open the door to developments focused on nature and local communities, where the populations themselves exercise their power to offer tourism based on the attractions of their biocultural heritage.

The worldwide tourism boom is leading to the generation of new forms of tourism, especially nature tourism (Sheldon 2020; Ateljevic 2020; Ioannides and Gyimóthy 2020). While the sustainability of the industry is questionable, in relative terms it does not compare with extractivist industries (Honey 2008; Devine and Ojeda 2017). Any initiative that goes beyond the still prevailing model of *over-tourism* or *mass-tourism*, and *fast-tourism* helps to enhance the benefits of this fast-growing economic sector. Some experiences from Costa Rica, Galapagos (Ecuador), and Cape Horn (Chile) show us some challenges to achieve a genuine ecotourism according to defining and establishing tourism boundaries, strengthening alliances with local people and institutions, and demanding corporate responsibilities. (Rozzi et al. 2010b). Biocultural reconnection is enhanced from the perspective of tourism in those destinations that are betting on a slow, genuine, community, creative type of tourism. In these destinations, imagination and connection with the cultural roots and the environment are opening a universe of experiences. An example is the *Creative Tourism* initiative, a new form of tourism that involves visitors and hosts in the joint creation of the tourism product (www.creativetourismnetwork.org). These creative tourism practices invite educational activities and artistic representations (such as exhibitions, concerts, and performances) in places where urban life habits are alienated or dis-enchanted from nature but can be re-enchanted through education with a biocultural ethical base (Tauro et al. 2021). Guided by specific didactic methodologies, education through the practice of tourism, especially ecotourism, can promote biocultural reconnection.

20.4 Education and Co-participatory Methods Through Technologies That Reinforce the Community Fabric with "Fraternal Eyes"

Forty years ago, personal computers and the internet in civil society were science fiction. The use of technological tools with satellite connection such as cell phones or GPS was inaccessible to ordinary people. When Orlando Fals Borda and his colleagues in Colombia proposed another way of doing social science to understand and respond to the demands for justice by the country's forgotten and invisible groups (Fals Borda 1993; Villasante 2010), the various current technologies did not exist. Today, action-participatory methodologies in scientific inquiry have spread among the social, environmental, and human sciences. Likewise, technological tools have emerged that provide new elements not only for discussion towards change, but for action as well. In addition to these advances, the demands for justice and visibility remain, and participatory methodologies combined with current technologies, have deepened the role of the academia in society for contributing to socio-environmental justice. These efforts of educational praxis and academic reflection trace a path to reconnect community fabrics with "fraternal eyes," where communities of co-inhabitants are joined together.

The three following chapters in Part III are based on participatory methodologies unique to their contexts and historical realities and that evolved into transdisciplinary collaboration using technological tools. Zent and Zent (2023, Chap. 24) employed Collaborative Action Research (CAR) between academics, leaders of religious institutions, local indigenous populations, and members of government agencies to shape a territorial project. This project, through CAR, documents more than 20 years of work with the Jotï communities of Venezuela. Documentation of the process is recorded in an auto-ethnographic book and in the Traditional Ecological Knowledge (TEK) assessment. The same book is also an educational resource for the children and future generations of the Jotï nation, as well as a source of information about the people and their "ecogonic" philosophy of *jkyo jkwainï* "to love/respect/care for the environment." During this collaborative work process, people learned to use computers and other technological means to document their territory. Different participatory mapping techniques that have been developed demonstrate decolonizing potential, that is, they allow different land use policies to be negotiated over the territory. In the voice of the authors, it is argued that indigenous communities "are effectively appropriating a non-traditional resource (the written word [and mapping technologies]) to protect a traditional resource (the land they occupy) within the surrounding power structure (the national constitution and subsidiary laws" (Zent and Zent 2023).

Meanwhile, O'Donoghue and Sandolval-Rivera (2023, Chap. 25) present a methodological proposal for student-led active learning based on the ethics of care and respect that includes both humans and other than-humans, twinned together in the community fabric. The authors apply the methodology of HAND Print-Care in Mexico and South Africa. The meetings between academics, teachers, and the local community around food systems and health, facilitated the creation of teaching materials rooted in local needs. This methodology has the power to recover local singularities in the face of problems generalized by institutional educational models where "externalizing of knowledge has produced abnormalities of exclusion that are not easily resolved in curriculum settings" (O'Donoghue and Sandolval-Rivera 2023, pp. xx). The HAND Print-Care methodology addresses a historical cultural perspective in education that allows the inclusion of intergenerational heritage, thus recovering the knowledge of generations excluded by a homogenizing educational system. These experiences lead to recovering school education that is linked to the realities of local communities.

Finally, Lepheana and Taylor (2023, Chap. 26) present a case in which technologies mediate community empowerment. Facing the Covid-19 pandemic in 2020, the authors and the South African Enviro-Champs movement innovated community environmental work methodologies using communication technologies. The virtual training programs allowed active learning to continue and opened new reflections in the field of job training and sustainability practices. This chapter reflects on some of their learning practices in relation to FEP by focusing on learning experiences in the field that led to discovering the intimate relationships between human life and other life that is manifested in the biosphere. This chapter also shares with the previous ones the recognition that indigenous knowledge practices are necessary to make learning meaningful and connected for the participants in the different environmental training programs.

20.5 Biocultural Reconnection: Learnings

The different initiatives presented in the chapters of this part teach us the importance of reconnecting with communities and with the land. The reconnection of people with the earth, according to Kirsti Luke, executive director of the Tuhoe Te Uru Taumatua organization in New Zealand, is related to the restoration of our reciprocal relationships and the renewal of our sense of purpose, as well as our sense of community and of the kindness between us (Crimmel and Goeckeritz 2020). The experiences of these chapters invite us to redefine relationships with the land and with diversity in different educational contexts, to learn to observe the small and the silent, and how to learn from intercultural dialogue recognizing tensions. Also, they teach us to learn from the opportunities offered by biocultural tourism experiences to see oneself reflected in the other and learn from the lives of others (humans and other-than-humans) with whom we decide to share the journey, to learn from the knowledge of local practices hosted by indigenous, peasant, and urban communities to rethink teaching, school culture, and education in general. They help us to seek education that is significant for communities and that contributes to the strengthening of community life of human and other-than-human cohabitants.

Acknowledgment Alejandra Tauro thanks Roy May and Ricardo Rozzi for his valuable comments on the manuscript. Roy May translated the introduction to English version. AT receives basal support from the Cape Horn International Center (CHIC) - ANID PIA/BASAL (PFB210018).

References

Ateljevic I (2020) Transforming the (tourism) world for good and (re)generating the potential 'new normal'. Tourism Geogr 22:467–475

Breakey H, Breakey NM (2023) Leopold's cultural harvest, biocultural tourism and field environmental philosophy. In: Rozzi R, Tauro A, Wright T, Avriel-Avni N, May Jr. RH (eds) Field environmental philosophy: education for biocultural conservation. Ecology and ethics, vol 5. Springer, Dordrecht, pp 281–294

Chan KM, Boyd DR, Gould RK, Jetzkowitz J, Liu J, Muraca B et al (2020) Levers and leverage points for pathways to sustainability. People Nat 2(3):693–717
Costa-Carvalho M (2023) Small, silent and (in)significant: childhood as an experience of education. In: Rozzi R, Tauro A, Wright T, Avriel-Avni N, May Jr. RH (eds) Field environmental philosophy: education for biocultural conservation. Ecology and ethics, vol 5. Springer, Dordrecht, pp 339–358
Crimmel H, Goeckeritz I (2020) The rights of nature in New Zealand: Conversations with Kirsti Luke and Christopher Finlayson. ISLE Interdiscip Stud Lit Environ 27(3):563–577
Devine J, Ojeda D (2017) Violence and dispossession in tourism development: a critical geographical approach. J Sustain Tourism 25(5):605–617
Echeverria E (2004) Filosofía para niños. Ediciones SM, México
Fals Borda O y otros (1993) Investigación-Acción-Participativa. En Documentación Social, n° 92. Madrid
Fleuri RM, Fleuri LJ (2018) Learning from Brazilian indigenous peoples: towards a decolonial education. Aust J Indigenous Educ 47(1):8–18
Green-Stocel A, Ramírez SYS, Pimienta ALR (2013) Licenciatura en pedagogía de la Madre Tierra: una apuesta política, cultural y académica desde la educación superior y las comunidades ancestrales. Relaciones Interculturales en la Diversidad, Cátedra Intercultural, pp 85–94
Honey M (2008) Ecotourism and sustainable development. Who owns paradise? Island Press, Washington, DC
Ioannides D, Gyimóthy S (2020) The COVID-19 crisis as an opportunity for escaping the unsustainable global tourism path. Tourism Geogr 22:624–632
Johnson B, Tsuji R, Lukey B, Toyoda M (2023) Communities of philosophical inquiry for the empowerment of ecological agency. In: Rozzi R, Tauro A, Wright T, Avriel-Avni N, May Jr. RH (eds) Field environmental philosophy: education for biocultural conservation. Ecology and ethics, vol 5. Springer, Dordrecht, pp 359–378
Kothari A (2020) Earth Vikalp Sangam: proposal for a global tapestry of alternatives. Globalizations 17(2):245–249. https://doi.org/10.1080/14747731.2019.1670955
Kothari A (2021) These alternative economies are inspirations for a sustainable world. https://www.scientificamerican.com/article/these-alternative-economies-are-inspirations-for-a-sustainable-world/
Lepheana A, Taylor J (2023) The enviro-champs movement: co-researching transformation through training processes in a post COVID world. In: Rozzi R, Tauro A, Wright T, Avriel-Avni N, May Jr. RH (eds) Field environmental philosophy: education for biocultural conservation. Ecology and ethics, vol 5. Springer, Dordrecht, pp 427–446
O'Donoghue R, Sandolval-Rivera JC (2023) Hand-Print CARE: intergenerational and plural knowledge in schools. In: Rozzi R, Tauro A, Wright T, Avriel-Avni N, May Jr. RH (eds) Field environmental philosophy: education for biocultural conservation. Ecology and ethics, vol 5. Springer, Dordrecht, pp 415–426
Prakash MS, Esteva G (2008) Escaping education: living as learning in grassroots cultures, 2nd edn. Peter Lang, New York
Rozzi R (2013) Biocultural ethics: from biocultural homogenization toward biocultural conservation. In: Linking ecology and ethics for a changing world. Springer, Dordrecht, pp 9–32
Rozzi R (2019) Taxonomic Chauvinism, no more! Antidotes from Hume, Darwin, and biocultural ethics. Environ Ethics 41(3):249–282
Rozzi R, Anderson C, Pizarro JC, Massardo F, Medina Y, Mansilla AO, Kennedy JH et al (2010a) Filosofía ambiental de campo y conservación biocultural en el Parque Etnobotánico Omora: Aproximaciones metodológicas para ampliar los modos de integrar el componente social ("S") en Sitios de Estudios Socio-Ecológicos a Largo Plazo (SESELP). Revista Chilena de Historia Natural 8:27–68
Rozzi R, Massardo F, Cruz F, Grenier C, Muñoz A, Mueller E (2010b) Galapagos and Cape Horn: ecotourism or greenwashing in two emblematic Latin American archipelagoes? Environ Philos 7(2):1–32

Santiago-Jiménez ME (2023) Rebuilding a life project through educational tourism in Cholula, Mexico. In: Rozzi R, Tauro A, Wright T, Avriel-Avni N, May Jr. RH (eds) Field environmental philosophy: education for biocultural conservation. Ecology and ethics, vol 5. Springer, Dordrecht, pp 379–390

Sheldon PJ (2020) Designing tourism experiences for inner transformation. Ann Tour Res 83: 102935

Singh A (2021) How Sarmoli became a poster child of ecotourism in Uttarakhand. Mongabay-India. Retrieved on 20 July 2021. https://india.mongabay.com/2021/06/how-sarmoli-became-a-poster-child-of-ecotourism-in-uttarakhand/

Tauro A, Ojeda J, Caviness T, Moses KP, Moreno-Terrazas R, Wright T et al (2021) Field environmental philosophy: a biocultural ethic approach to education and ecotourism for sustainability. Sustainability 13(4526)

Tonucci F (2012) Con Ojos de Niños. ELE Editorial Laboratorio Educativo

Torres-Villa R, Barragán-López E (2016) Hábitat de la cultura ranchera en la sierra de Jalisco y Michoacán, México. Potencial para el aprovechamiento de un turismo biocultural. Int J Sci Manag Tourism 2(4):281–301

Toyoda M (2018) Revitalizing local commons: a democratic approach to collective management. In: Rozzi R, May RH Jr, Stuart Chapin F III, Massardo F, Gavin MC, Klaver IJ, Pauchard A, Nuñez MA, Simberloff D (eds) From biocultural homogenization to biocultural conservation, Ecology and ethics, vol 3. Springer, Dordrecht, pp 443–457

Villasante TR (2010) Historias y enfoques de una articulación metodológica participativa. Cuadernos CIMAS. Available in: https://www.redcimas.org/wordpress/wp-content/uploads/2012/08/m_TVillasante_HISTORIAS.pdf

Walter P (2009) Local knowledge and adult learning in environmental adult education: community-based ecotourism in southern Thailand. Int J Lifelong Educ 28(4):513–532

Walter P, Regmi KD, Khanal PR (2018) Host learning in community-based ecotourism in Nepal: The case of Sirubari and Ghalegaun homestays. Tourism Manag Perspect 26:49–58

Zent S, Zent E (2023) Collaborative action research with the Jotï in Venezuela: experiences in autoethnography and TEK vitality assessment. In: Rozzi R, Tauro A, Wright T, Avriel-Avni N, May Jr. RH (eds) Field environmental philosophy: education for biocultural conservation. Ecology and ethics, vol 5. Springer, Dordrecht, pp 229–246

Chapter 21
Small, Silent, and (In)Significant: Childhood as a Minoritarian Experience of Education

Magda Costa-Carvalho

Eu acho que isso é a filosofia em si:
pega numa ideia simples e torna-a especial (I think that this is philosophy: take a simple idea and make it special. English translation by Ricardo Rozzi and Roy May).
Afonso Junco (12 anos)
Il existe d'autres cas où, pendant un temps plus ou moins long, une catégorie a réussi à en dominer absolument une autre. C'est souvent l'inégalité numérique qui confère ce privilège: la majorité impose sa loi à la minorité ou la persécute. Mais les femmes ne sont pas comme les Noirs d'Amérique, comme les Juifs, une minorité: il y a autant de femmes que d'hommes sur terre (There are other cases where, for a shorter or longer time, one category has managed to dominate another absolutely. It is often numerical inequality that confers this privilege: the majority imposes its law on or persecutes the minority. But women are not a minority like American blacks, or like Jews: there are as many women as men on the earth." English Translation: Beauvoir, S. (2011). The Second Sex, *Vintage Books, New York, p. 27).*
Simone de Beauvoir, Le Deuxième Sexe, *p. 22*
No pretende la Ética de la Libération *ser una filosofía crítica para minorías, ni para épocas excepcionales de conflicto o revolución. Se trata de una* ética *cotidiana, desde y en favor de las* inmensas mayorías de la humanidad excluidas de la globalización, *en la 'normalidad' histórica vigente presente. Las éticas filosóficas más de moda, las* standards *y aun las que tienen algún sentido crítico, con pretensión de ser pos-convencionales, son éticas de minorías (claro que minorías hegemónicas, dominantes, las que tienen los recursos, la palabra, los argumentos, el capital, los ejércitos) que frecuentemente pueden* cínicamente *ignorar a las víctimas, a los dominados y afectados-excluidos de las "mesas de negociaciones" del sistema vigente, de las comunidades de comunicación dominantes... (The* Ethics of Liberation *does not pretend to be a critical philosophy for minorities, nor for exceptional times of conflict or revolution. It is an everyday* ethic, *from and in favor of the* vast majority of humanity excluded from globalization, *in the current historical 'normality' in force today. The most fashionable*

R. Rozzi et al. (eds.), *Field Environmental Philosophy*, Ecology and Ethics 5,
https://doi.org/10.1007/978-3-031-23368-5_21

philosophical ethics, the standards *and even those that have some critical sense, pretending to be post-conventional, are minority ethics (of course hegemonic, dominant minorities, those who have the resources, the word, the arguments, capital, armies) that can frequently* cynically *ignore the victims, the dominated and affected-excluded from the 'tables of negotiations' of the current system, and of the dominant communication media.... English translation by Ricardo Rozzi and Roy May).*
Enrique Dussel, Ética de la Liberación, *p. 15*
Les minorités et les majorités ne se distinguent pas par le nombre. Une minorité peut être plus nombreuse qu'une majorité. Ce qui définit la majorité, c'est un modèle auquel il faut être conforme: par exemple l'Européen moyen adulte mâle habitant des villes... Tandis qu'une minorité n'a pas de modèle, c'est un devenir, un processus. On peut dire que la majorité, ce n'est personne. Tout le monde, sous un aspect ou un autre, est pris dans un devenir minoritaire qui l'entraînerait dans des voies inconnues s'il décidait à le suivre (The difference between minorities and majorities isn't their size. A minority may be bigger than a majority. What defines the majority is a model you have to conform to: the average European adult male city-dweller, for example... A minority, on the other hand, has no model, it's a becoming, a process. One might say the majority is nobody. Everybody's caught, one way or the other, in a minority becoming that would lead them into unknown paths if they opted to follow it through. English Translation: Deleuze, G. (1995) Negotiations, *Columbia University Press, New York, p. 173).*
Gilles Deleuze, Pourparlers, *p. 235*

Abstract It is possible to question and deconstruct the roots of some educational conceptions of childhood through approaches that find different ways of thinking about children, childhood, and even to experience education. Philosophy has contributed to some of these approaches by deconstructing and reconstructing the Western dominant ideas of rationality and its uses in education. In this chapter, I will approach these critical movements through disruptive provocations grounded in three concepts: *small*, *silent*, and *(in)significant*. These concepts are commonly associated with descriptive and normative images of childhood: what it is and is expected of children. By retaking these concepts, I will problematize the impact of Western rationality on the ways education has been thought and practiced.

Keywords Philosophy · Childhood · Minority · Political Agency · Voice

M. Costa-Carvalho (✉)
Faculty of Human and Social Sciences, University of the Azores, Ponta Delgada, Portugal

Institute of Philosophy, University of Porto, Porto, Portugal
e-mail: magda.ep.teixeira@uac.pt

21.1 Introduction

This reflection begins with three concepts—small, silent, and (in)significant—to reconfigure childhood from the margins, and also to question what they might say, on the one hand, about the social stigmatization and political oppression of childhood and, on the other hand, as affirmative powers of different educational practices. The hermeneutical use of small, silent, and (in)significant helps us reevaluate the importance of disregarded voices in educational experiences.

Additionally, I present three projects as differentiated readings of the potentialities present in educational spaces and times. The three educational practices presented in this chapter aim, then, to simultaneously illustrate the lines of analysis that we defend, as well as to give visibility to emancipating and challenging projects. Practices of disturbance, destabilization, and disruption (Murris 2016, p. 200) are ways to resist the macro dynamics of public policies and common institutional discourses. I present these practices as possible examples among others to inspire educators to engage in distinctive ways of understanding childhood education. I have chosen, then, three differentiated examples:

- The first is entitled *Em Caxias a Filosofia Em-Caixa?* [Does Philosophy Fit in Caxias?] (2014) and consists of a university extension project in a school on the outskirts of Rio de Janeiro, Brazil.
- The second—field environmental philosophy—covers an environmental education project in a research, education, and conservation center in the Cape Horn Biosphere Reserve, Chile.
- The third refers to a set of practices carried out by the educational service *Fábrica das Artes*, part of the Centro Cultural de Belém, in Lisbon, Portugal.

The geographical, thematic, and institutional heterogeneity of these experiences offers a diversity of proposals that cross through education, childhood, and philosophy. These proposals resist social and political hegemony that excludes oppressed and underappreciated "minorities." Likewise, I propose a reconfiguration of the small, the silent, and the (in)significant things and events in our daily life as educators and, most of all, as readers and learners of the worlds in which we live.

Is it, then, that to be in education in a *childlike* manner means to be small, silent, and insignificant? Or is it that by being attentive to what is small, silent, and insignificant can make us sensitive to the rhizomatic "minority fields" of difference?

21.2 Minority as Affront: The Childhood of Questioning

What can emerge from decolonizing educational encounters, which reshape the experience of childhood according to difference and uniqueness? Above all, what emerges is embodying education driven by dynamic relationships rather than those ruled by efficiency, success, productivity, or profit. Grounded on these dynamic relationships, educational places and times can be rethought as moments for the small, or what is (almost) not seen; for the silent, or that which is (almost) not heard; and for the insignificant, or that which (almost) does not matter. At another scale, education might expand images (becoming a diffractive lens), sounds (becoming a babel-like amplifier), and senses and meanings (becoming playful).

What can our educational encounters become if diversity and singularity are enhanced over homogeneity and universality? How important is the reconfiguration of minority processes, a theme we will retake later, which are generally little considered in schools and in other educational contexts? What forces do we need in order for these processes to assert themselves as places and times of forgotten marginalities? What changes are required for education to welcome everyone, especially those from the peripheries, in effective ways and not through mere empty abstractions?

An education born from the affirmation of childhood as a small, silent, and (in)significant force can be envisioned as a form of life that is distinct and that suspends adult normativity (Kennedy 2020). This is a vision from the reverse side of a colonizing process of thought that (since the beginning of Western civilization and in different geographical points) has imposed a logocentric and adult-centric model of reason. A masculine, adult, white, heterosexual, European, and urban rationality (the dominant standard of humanity of the so-called modern industrial and Euro-Western societies) has been imposed as normative (Mallory 2013; Gare 2022).

I begin the chapter with three thinkers: Simone de Beauvoir, Enrique Dussel, and Gilles Deleuze. These authors provoke us to rethink the notions of majority and minority beyond mere numerical or mathematical indicators. Rather, majority and minority are defined as political relationships with diverse people and contexts. Beauvoir, Dussel, and Deleuze argue that the concepts of "majority" and "minority" are masculine criteria used for segregating individuals into the category of "other." Moreover, these criteria are turned into social and political norms or patterns of relationships that the "other" must conform to.

Politically, majority is the entity that imposes a model. Retroactively, majority is that which is produced by the same model. It is a self-immune and exclusionary model. Therefore, the majority that imposes the model does not need to be more numerous. It is enough for it to be the normative pattern.

By denouncing the underlying meaninglessness of masculinity as majority, Beauvoir (1976) highlights the political nature of the dualistic division between those who have the power of speech and those who do not. In turn, Dussel (1998, p. 15) refers to the power relations involved in the hegemonic legitimization of certain groups and perspectives, inverting the use of concepts and referring *to the vast majorities of humanity excluded* from the discourses and practices in force in the globalization era. Complementarily, Deleuze (1977) relocates the problem and discovers how minorities can take advantage of a multiplicity of ways to challenge the system. By redefining themselves as "becoming-other" as dynamic transformers (*devenir*) or kinks in the system, minorities challenge normative models. From Beauvoir to Deleuze, we witness a decisive reversal in the use of these concepts, moving from denouncing the colonizing excesses of the majority to reconstructing minorities as potentially decolonizing dynamics.

Beauvoir's liberation from patriarchal models to re-situate women could also be applied to pedagogical and political educational models. Traditionally, one does not walk with children by following their steps. Never. The dominant majority (which may be the numerical minority) designs and implements the rule. It marches ahead of the children. This majority is the constant parameter against which one evaluates, serving as a normalizing device (Deleuze and Guattari 2007, pp. 145–146). The model installs itself. It governs. It dictates. It founds hierarchy and, noiselessly or silently, subjects and excludes. The *white heterosexual European city-dwelling, standard-language-speaking adult male* is the diffusing center of the majority's dominant model (Deleuze 1988). A certain concept of *logos* (abstract analytic discursive reason) is added to this modeling pattern of being properly human.

Deleuze and Guattari (2007, p. 146) distinguish two meanings in the idea of "minority":

- As objectively definable individuated states, entities with concrete ghetto territories (opposed to the majority).
- As moving germs that set up uncontrollable movements ("becoming-other" or "kinks" that destabilize accepted norms).

Deleuze's distinction between minority as ghetto and minority as subversive movement is crucial if one's thinking avoids being booby-trapped: recognizing the political dimension of minority is distinct from wanting to replace the established majority with another. It is thus a matter of asserting a different way of life, of seeking to *become a minority*. Becoming a minority disturbance means affirming multiplicities and variabilities that disturb, and sometimes implode, standardized logic. Consequently, it is an affirmative underground movement, which introduces strangeness and fracturing into institutions. These minority "disturbers" catch up with us like subterranean snakes, furtive, and not always visible, dynamically spreading between the cracks of the established order, lurking on the unsuspected margins of the models, and moving potentialities that at every moment crack the established domain.

What might happen in education if this were the challenge? What worlds would be born if we exchange fixed, unquestioned states for processes that affront? What if we allow education to be invaded by subterranean times and spaces of childhood as affirmation, difference, and questioning? Becoming a minority? Thinking from the margins?

21.3 Education as *Smallnessing*

Thinking about the use of words shows the immensity of meanings that run through them, the ambivalences in which they are constructed, but also the colonizing tendencies of the subterranean processes of signification. Beauvoir, Dussel, and

Deleuze suggested meanings of minority that bring it closer to an idea of childhood conceived as a resistant force of the voiceless (in Portuguese, *in-fante*; from the Latin *in* [= not], *fari* [=speaking], plus the suffix *nte* [=agent]), of those on the periphery.

Let us now consider the first conceptual intensity that we relate to childhood: *small*. Small is currently attributed to children. To it we can add expressions such as "underage" or "juvenile." These meanings establish a magnetic field (again with quantitative roots, but extending beyond it) that distinguishes the *larger* (in size, intensity, or duration) from the *smaller* (underprivileged, insufficient, flawed, weak, limited). The larger occupies more space or attention; it imposes itself. The smaller is disregarded, relegated to non-places, or forgotten places. Larger and smaller are exclusive binary concepts (one is either larger or smaller).

An analogy between children and so-called savage human beings echoes distant cultural imaginaries (Rollo 2016). Children and Indigenous people belong to the category of the "not yet" of the city and the civility of laws, viewed as "ungoverned savagery and backwardness." These people are assigned to a life in darkness, in unprepared obscurity. Rollo presents an analogy between children and other individuals without political agency who must be redeemed by the "benevolent modern age." The modern age is an "age" in the literal chronological sense of the moment of acquisition of speech, or rationalized mode of speech, in which children and Indigenous people enter into a process of proper humanization. The modern age views them as products that need "to be formatted" by a model that aims to insert them into fora of social and political participation.

However, modernity was also the arrival of industrialized exploitation, of macroscale production, of exclusionary democratic ideals that privileged only certain voices and disabled others (Rollo 2020). Therefore, a critical movement is to recognize "invisible and unviable smallnesses" as affirmative powers, which differentiate and make educational experiences concrete. This perspective lives with fissure, with rupture, with difference; it is not tempted to create a comprehensive and tension-free narrative. Argentine social critic Walter Mignolo refers to the notion of deconstruction forged by French philosopher Jacques Derrida. In the context of decolonial cultural studies, he proposes the "colonial difference" as a critical stance before the violence generated by hierarchies of privilege over subaltern discourses (Mignolo 2003). This effort entails overcoming the normative and exclusionary dualism of dominant majorities:

savage—civilized
nature—culture
emotion—reason
ignorance—knowledge
child—adult
little—big

Along these lines, Westernized rationalist humanist thinking has contributed to instituting the inferiorization or colonization of childhood (Murris and Haynes 2018). Unsettling the binaries of dominant educational conceptions allows rethinking childhood from non-quantitative, non-rationalist, and non-adult logics.

Children as the small, marginalized other, treated as an adult's property, as the economically disenfranchised of society, as the epistemically incomplete, or even as the cultural outsider of political discourses, is no longer the only available picture (Kennedy 2010). Seen as the stranger, the foreign, the outsider that transgresses the majorities, the child can assume a privileged epistemological place: "Like women, persons of color, or others marginalized by Eurocentric and patriarchal personal, interpersonal and social constructs, the child's location in the social and natural world affords her an 'epistemic privilege'" (Kennedy 2010, p. 15).

From this "displacement," childhood opens up as a perspective beyond the chronological (Kohan 2011) that breaks the limits, brings unsuspected questions, breaks with the dualism child–adult, educator–educated, small–large, by uninstalling the forces of the dominant majorities. Borders no longer make sense as markers of the norm and a non-instrumental experience of education is made possible, a *minor education* (Gallo 2002).

21.3.1 A Critical Educational Project in Brazil: Does Philosophy Fit in Caxias?

A philosophy project conducted in different Brazilian schools that resists the ideal that education formation of children is to create future citizens allows us to revisit the concepts outlined above (Kohan 2011, 2014). Walter Kohan, past president of the International Council of Philosophical Inquiry with Children and professor at the Center for the Study of Philosophies and Childhood, State University of Rio de Janeiro, Brazil (NEFI/UERJ 2021), has led the project *Em Caxias a Filosofia En-Caixa?* [Does Philosophy Fit in Caxias?][1] (Kohan 2014). Since 2007, he has coordinated a team of NEFI/UERJ students, researchers, and teachers from the Municipal School Joaquim da Silva Peçanha, in the municipality of Duque de Caxias in Rio de Janeiro, to practice everyday philosophy with children and resist the hierarchical standardization of educational times and spaces (Omelczuk 2016; Gomes 2019) (Fig. 21.1). The aim is to promote thinking that differs from the dominant social, cultural, and political forces (Kohan and Olarieta 2012).

Inhabiting the place of "not knowing" within an institution whose social recognition is marked by the authority of knowledge is not an easy task. Learning to be comfortable in this place is one of the greatest challenges and values of this Brazilian project (Omelczuk 2016, p. 239). Students have "the possibility of getting in touch with childhood thought that teaches them that knowledge is always partial, therefore, subject to being questioned" (Kohan and Olarieta 2012, p. 245). *Em Caxias a Filosofia En-Caixa?* focuses on the promotion of thought encounters in unforeseen

[1]The title is a play on words and the name of the municipality where the project occurs. Caixa in English is "box," but since the name of the municipality where the project takes place is Caixa, the literal translation would be: "In [the Municipality of] Boxes, does philosophy fit in boxes?"

Fig. 21.1 Children and teachers practicing philosophy at the Municipal School Joaquim da Silva Peçanha, in the municipality of Duque de Caxias in Rio de Janeiro (NEFI Archive). Photographs by NEFI/ Vanise Dutra Gomes

and unpredictable relationships with what is known and what is not known. With no distinctions inhibiting speech, everyone begins by unlearning the habits that prevent this shared and egalitarian experience of thought. It is about presupposing a condition of equality amid difference, between the small and the less small, about offering opportunities so that everything that concerns us can be problematized, even—and above all—that for which there seem to be no viable alternatives (Gomes 2016[2]). To educate is to expect the impossible, the unexpected. It is not being paternalistic to the little ones, to the minors, to the minorities nor wanting to integrate them, or model them, according to the established status quo of the institutions. It is about accepting multiple paths from philosophical experiences and thoughts as a practice leading to liberating, creative transformations. This is especially for those who are placed on the margins: Of course, it is important that the poor and the dispossessed, those who have been silenced for centuries, embrace their fundamental right to speak and to express themselves. But if we are really interested in their speaking in their own voices, and not in the voices of the ventriloquists of popular culture, the corporate media, the political, academic, or therapeutic elites, or, even more directly, in the voices of market or capital, the issue of enabling the voices of the marginalized becomes more complicated (Kohan 2011, p. 348).

This educational *smallness* is the smallness of creative impossibilities, in the adversities and restlessness of marginal and, at times, indigent and inhuman daily lives. However, more than an additive and integrative solution for the "small," which would still be hostage to an exclusivist and institutionally homogeneous background, the proposal is that of education as experiencing *smallness*. Or education as a movement of *smallnessing*.

21.4 Education as *Silencing*

Refusing exclusionary binaries allowed us to rethink the meaning of small. The aim is not to advocate that children are also capable of abstract reasoning, behaving just like adults, or maintaining dialogues that can be deemed philosophical. Instead, the aim is to replace a colonialist paradigm of childhood education with a decolonial paradigm of childhood education.

In recent decades, the approach to schooling as an instrument at the service of a certain conception of efficiency has become the majority. As instruments of a society driven by production, school programs and curricula are often evaluated according to their potential impacts in the production machinery. Currently, education is not a place of emancipation but of homogenous instruction in which fewer and fewer students are educated to be critical protagonists of their own lives (Biesta 2020), or even to hear the sounds of their own voices. Nevertheless, as with our exploration of the concept of "small," can the idea of silencing conceal other challenges for education?

The absence of speech that childhood carries from its etymology (*in-fans*) can be read in at least two ways.

---

[2]This and many other books related to the project "Does Philosophy Fit in Caxias?" are available on NEFI Edições website, free of charges: http://filoeduc.org/nefiedicoes/editora.php

First, childhood silence can be understood as a defect or a lack, reinforcing the idea that children are on the way to full human development, where (and only where) they are ready to acquire speech privileges. The silencing of childhood as a form of pedagogical bio-power shakes the established parameters of collective existence. What voice is muted when only one voice is allowed to be heard?

Second, the silence of childhood can be read as a presence, affirming resistance to the dominant rhythms that, within education, impose a productive conception of learning. These silences are marks of resistance in school spaces and times. When we remain in the primordial soup of unspoken potentialities, how can educational instances be transformed?

This question reminds me of the character of the nineteenth-century short story by Herman Melville *Bartleby, the Scrivener: A Story of Wall Street.* Bartleby is a young man who was hired as a clerk by a Wall Street lawyer to copy documents. After a time, with an affirmative silence, Bartleby escapes into the powers of imperturbable suspension, avoiding work by replying, "I would prefer not to" (Melville 2016).

The former question also reminds me of Mozambican writer António Emílio Leite Couto, better known as Mia Couto, who refers to childhood in terms similar to the idea of going back to the unheard, to a first language, a language of chaos, an unspeakable language where everything might have any name:

> Childhood/Infancy is not a time, it is not an age, a collection of memories. Childhood/Infancy is when it is not too late. It is when we are available to be surprised, to let ourselves be enchanted. Almost everything is acquired in that time when we learn the very feeling of Time. The truth is that we maintain a relationship with the child/infant as if s/he were a minor, lost in a precarious state. However, childhood/infancy is not just a stage towards maturity. It is a window that, whether closed or open, remains alive within us. (Couto 2009, p. 110)[3]

In educational encounters, when silence interrupts the expected and anticipated continuous flows of speaking and learning, each one—child and adult, regardless of age—is able to return to themselves. In their uniqueness and difference, they return to their freedom. Philosopher Igor Jasinski understands moments of silence as possibilities for education to be experiences that create unpredictability, rather than submission driven by the "productive and producer" model. He emphasizes the importance in the classroom of maintaining moments of intense silence—times that diverge from linear, sequenced, and productive communication flows (Jasinski 2016). The experience of philosophy in education limits the suspension of inoperability in the relationship between adults and children (Jasinski and Lewis 2015). It is this inoperability that frees education from the logic of results by replacing it with the practice of unpredicted potentialities.

The silences that settle after certain questions are asked and the questions that spring from certain silences are moments of uncertainty and expectation that are fundamental in education. These moments resonate with the production of these silences in the unspeakable language of Mia Couto or in the irreducibility of

[3]Translation by Ricardo Rozzi and Roy May.

Bartleby. More than filling educational mediation with productive sounds, it is choosing *to fine-tune silences* (Couto 2016).

Silencing does not mean refusing voice nor giving adults the privilege of choosing who speaks. This understanding also departs from the so-called silent pedagogy, which is a strategy that cooperates with the "adultization" and "learnification" of education by proposing a fixed and deliberate use of silence defined by teachers (Ollin 2008). On the contrary, the silence I refer to is an egalitarian invitation to suspension, to a halt as resistance to the dominant and shaping institutional logics and dominations.

Both the small and the silent offer opportunities to deconstruct the universal presuppositions implicit in the scientific knowledge of dominant rationalist-logocentric adult education. Instead, it opens opportunities to rediscover the singularities of ecosystems and the beings that inhabit them. The logic of efficiency and the discourse of modernization and economic growth marginalize and impose the globalizing structures of oppressive majorities. It is, therefore, important to reconfigure categories such as small and silent, promoting encounters with multiple ways—children's ways—of experiencing and building the world.

21.4.1 The Miniature Forests of Cape Horn, Chile: Experiencing the Small and Silent

In different educational environments, the small and the silent can be the center of refugee experiences and the affirmative power of different ways of experiencing ourselves in relation to others. In my second example, we encounter the small and the silent for an inversion of the common rhythms of education: the miniature forests of Cape Horn. A project conducted by the Universidad de Magallanes in Chile and the University of North Texas in the USA invites us to encounter biodiversity at the scale of tiny living beings (Rozzi et al. 2006). The project is based on an ethical biocultural approach (Rozzi 2013). Its *field environmental philosophy* methodology has generated novel activities such as "ecotourism with a hand lens." This activity was born out of a borderline experiment after, in 2000, Chilean biologist and philosopher Ricardo Rozzi got lost and inadvertently fell into a peat bog. The experience earned him a few hours of direct contact (although fearing for his own survival) with tiny plants in all their diversity and exuberance (Rozzi 2012). The episode was decisive for his idea of "change of lenses" for conservation and the proposal for valuing the non-human other through the creation of the UNESCO Cape Horn Biosphere Reserve (Rozzi et al. 2008a).

The need for direct encounters with living beings co-inhabiting their habitats at scales quite different from the specifically human macroscale environment inspired the creation of educational programs at the Omora Ethnobotanical Park. This park, at the southern end of Chile, includes *outdoor educational experiences* for observing tiny plants and other organisms in their "miniature forests" (Tauro et al. 2021). In

Fig. 21.2 Preschool and school children visiting the Omora Park for experiencing "ecotourism with a hand lens." This involves observations with a magnifying glass, drawings, composing metaphors, designing interpretive signs, and celebrating with the community (Rozzi et al. 2010a). Photographs by Yanet Medina and Sergio González Ayala (Omora Park Archive)

local workshops, the participation of children from local schools in Cape Horn has been fundamental in the implementation of ecotourism with a hand lens (Fig. 21.2). Children, teachers, and community Indigenous elders gave colloquial meaning to the image of the "miniature forests of Cape Horn" (Rozzi et al. 2010b). The slow rhythm of the educational experience with these small and silent "forests":

> gives us a hand lens that broadens our mental, perceptual and affective image about nature and about our relationship with it. The observation of the diversity and the unity of life ethical, aesthetical, and ecological dimensions that complement the narrow economic perspective that currently prevails in the relationship of contemporary society with nature. (Rozzi 2012, p. 23)

The Omora Ethnobotanical Park is based on *field environmental philosophy*, an educational approach that rejects the imposition of global development models that marginalize concrete singular realities, promoting the diversity of life(s) and ecological knowledge in local and regional communities, human, non-human, and other-than-human (Rozzi et al. 2008b). In this context, *field environmental philosophy* proposes an encounter with biocultural diversity as a direct experience of co-habitation with beings in ecosystem interaction, often overlooked by current educational models and decision-making processes in the public sphere (Rozzi et al. 2010b). Over two decades, this transdisciplinary project has been able to enhance crossovers between academic research and actors in the public sphere such as policy makers, in a collaborative way that involves people and institutions. Through collaboration between diverse school contexts, Indigenous communities, educators, researchers, and artists, the Omora Park has brought together partners of these children from preschool to community elders around the protection of biological and cultural diversity (Rozzi et al. 2008b).

The slow and silent experience of ecotourism with a hand lens contrasts with the current rapid worldwide depletion of biodiversity. In recent decades, we have witnessed a massive extinction of humanity's cultural heritage linked to an "extinction of experience" (Maffi 2001). The protection of biodiversity will be utopian unless it is combined with the preservation of cultural diversity. It is paradigmatic that of the approximately 6000 languages that currently exist around the world, half are spoken by communities with less than 10,000 inhabitants and that half of that half is spoken by communities with 1000 inhabitants or less (Rozzi 2013). In other words, most of the world's linguistic diversity belongs to small communities of numerically minority peoples. But how can these minorities affirm themselves as *becomings* that destabilize certain logics? Or as subversive movements that call the attention for different ways of life? It is necessary to recognize that, in *field environmental philosophy* perspective, the loss of *natural biodiversity* is not the core of the contemporary environmental crisis, but rather the loss of *biocultural diversity*. This refers to the multiplicity and deep interrelation of the manifestations of life on Earth, understanding life in its biological and cultural dimensions, within the framework of a complex socio-ecological system in constant co-adaptation (Maffi and Woodley 2010).

The extinction of experience is even more drastic due to the pervasive process of biocultural homogenization and its influence on education (Rozzi 2013, 2018). What biological and cultural microcosms do we not want to ignore due to the prevalence of majority and homogenizing biocultural monocultures? Which types of education do we need to foster biocultural conservation? What childhood or minority actors do we not want to exclude from broader pictures of education? Are adultism, futurism, and formation ways of homogenizing educational experiences? Are these also ways of suppressing childlike fashions of living and experiencing the world (through small, silent, and insignificant educational experiences)?

21.5 Education as (In)Significant Living

The small and the silent can be marks of a return to childhood that focus resistance against the global, homogeneous, universal, and totalizing rhetoric of the dominant pedagogies (adult, futurist, and formalist). But what about meaning? To what do we regularly attribute meaning? What do we consent to in our educational practices? The perspectives we are dealing with now converge to the importance of re-signifying educational times and spaces beyond the reproductive, homogeneous, and hierarchical logic that runs through school institutions.

The insignificant is another category usually attributed to childhood; children and their environments are often understood as without value or interest. Childhood is, in this context, understood as the "not yet" to someone who is not given a properly significant place or is seen as a lesser producer of meanings; as smaller and almost expendable senses; as senses that register only to mark the beginning of the journey toward the perfecting represented by the adult world.

However, what can happen if we graphically mark a tension within the very word "significant"? (insignificant [un-meaningful]). What can become (in)significant in infancy (in-fancy)? In Portuguese, the prefix "in"—inherited from the Latin language—can mean negation (as *indispensable* or *illogical*), but it can also mean an inward movement (as in the words *introduce* or *induce*). Could we then look for other semantic possibilities in the word (in)significant? What if, in relation to childhood senses, the prefix "in" (significant) meant an inward movement, and not a negation? What could be significant within childhood itself? To what things and dynamics do children attribute meaning? Is there any specific aspect to the meaning of the experience made by the children? Is childhood itself its own way of meaning? Or, then, is there a privileged *childlike* relationship with the meanings of the world? What relationship would that be? Who would be allowed? What will be the childlike (in)meanings? What will a childlike relationship with meanings be?

In the presence of the small, the silent, and the (in)significant, what challenges can be opened up in different educational spaces and times? The same effort of attention, of course. A slowdown of the step. The small, the silent, and the (in)significant ask us to pay attention to what cannot be found in the global, in the macro, in the noisy, in chronological efficiency, in the certainty of already defined meanings. The small, the silent, and the (in)significant request an interruption. This interruption requires different modes of experience in addition to the dominant signification processes. The rhetoric of modernization and economic growth, universalized in educational discourses and practices, produces and installs certain pre-established meanings (Rozzi 2013). However, despite the homogenization that crosses our institutional spaces, it is possible to find fractures that pave the way for childhood to assert other forces, minority actors, variations, and seismic movements.

21.5.1 *Fábrica das Artes in Lisbon, Portugal: Creation of Meaning*

The educational example that we chose to illustrate a possible configuration of this childhood relationship with meanings is perhaps one of these seismic movements. Within the scope of the Cultural Center of Belém (CCB 2021), in Lisbon, Portugal, several transversal projects have been developed in recent years aimed at "all childhoods" (Wallenstein 2016, p. 3). The provocation begins with the idea that childhood is plural, multiple, and that it can belong to everyone (no matter what is his or her age). The cultural and educational program of Fábrica das Artes at CCB focuses on the intersection of knowledge and actions from the performing arts (shows and workshops) that bring together different audiences: children and adults, schools and families, educators, scientists, philosophers, actors, and artists (Fig. 21.3). They all intersect as childhood, around specific themes (artistic thought and contemplation, curiosity and neuroscience, childhood), seeking to build what a collaborator with one of the projects, Portuguese philosopher José Gil (2014), designated as *a*

Fig. 21.3 Children and teenagers in different activities at Fábrica das Artes, Centro Cultural de Belém, in Lisbon (Fábrica das Artes-CCB Archive). Photographs by "CCB/ Manuel Ruas Moreira"

consistency plan, that is, moments when different actors connect through the cross-fertilization between philosophical questions and artistic creations, scientific experiments, and performances.

Committed to the democratization of the processes of artistic creation, as well as to the events of philosophical thinking or scientific production, the activities of Fábrica das Artes take place not only in the Centro Cultural de Belém building, but also in schools and other educational institutions. Children are involved from the beginning of the programming design, mixing the different moments of work with their new and challenging look, but not just children, chronological childhood. Other childhoods are also protagonists in the re-significant work of Fábrica das Artes, a work of recovering tenuous voices and their transformation into propellants of ideas:

> The natural relationship between childhood, art and philosophy, the childhood of philosophers and artists, their curiosity, their way of looking at things, of seeing the dimension of the sensitive, of positioning before them, of *play* with elements from life and the world, from the game of variables, in short, the approximation between the childhood experience and the experience of artists and philosophers can be aggregated through the idea of "becoming-child." (Wallenstein 2014, p. 17)[4]

Searching in the small, the almost invisible, the silent, the (in)significant of sensitive and delicate things, that which is childhood in all of us, the adultist and hegemonic representations of time, of adulthood within education, are rethought. Through research and transversal experiments, Fábrica das Artes places itself at the interface between educational and cultural institutions, resisting the ghettoization of childhood through transforming experiences.

As with the initiatives *Does Philosophy Fit in Caxias?* and field environmental philosophy at the Omora Park, the Fábrica das Artes *c*hooses to build another kind of school. It is a "school to come," which emerges as a critical counterpoint to the standardization by prevailing hierarchical education (Ramos do 2016). The school to come, the school of minor childhood—small, silent, and (in)significant—is, above all, an invitation to refuse reproductions and recognitions that kill questions, to refuse the main paths that erase unpredictability, and to create experience and knowledge (instead of just replicating them). It is an invitation to escape, to ease into the idea that nothing is completed, that everything is to be done, that everything is to come, and that everything is in the making.

I have finished writing. However, a smallness attracts my attention. So far, I have left this chapter's first epigraph in silence. Almost invisible by the age of its author, the opening sentence remained silent: *I think this is philosophy: take a simple idea and make it special.* I placed this affirmation by Afonso Junco, perhaps the smallest of voices, the most distant and almost inaudible, before the philosophers consecrated by historical chronology. He is a 12-year-old boy who lives in the Azores, one of the nine islands of this archipelago in the European periphery: a childhood periphery in a peripheral geography. In a short, almost silent, and apparently (in)significant sentence, I begin and I end the chapter. If philosophy is the movement of taking a simple idea and making it special, just like Afonso teaches us, then one's encounter with childhood could be this space–time of taking the smallest and turning it into an affirmative minority: an opportunity of receiving what happens and making it

[4]Translation by Ricardo Rozzi and Roy May.

diverge, of resisting indifference, and of allowing education to be a true creation of meanings. Is this not after all the only authentically possible gesture in education?

Acknowledgement The author would like to express her gratitude to Ricardo Rozzi, Alejandra Tauro, and Roy May, for their caring and insightful comments and editing to previous versions of this chapter.

References

Beauvoir de S (1976) Le Deuxième Sexe. Éditions Gallimard, Paris

Biesta G (2020) What constitutes the good of education? Reflections on the possibility of educational critique. Educ Philos Theory 52(10):1023–1027. https://doi.org/10.1080/00131857.2020.1723468

CCB (2021) Centro Cultural de Belém. Available via https://www.ccb.pt/en/about-ccb/

Couto M (2009) E se Obama fosse Africano? Editorial Caminho, Lisboa

Couto M (2016) Jerusalém. Editorial Caminho, Lisboa

Deleuze G (1977) Pourparlers. Les Editions de Minuit, Paris

Deleuze G (1988) Abécédaire. G como Gauche. Available via https://www.youtube.com/watch?v=c2r-HjICFJM

Deleuze G, Guattari F (2007) *Mil Planaltos. Capitalismo e Esquizofrenia 2*. Lisboa: Assírio & Alvim

Dussel E (1998) Ética de la Libéración em la Edad de la Globalización y de la Exclusión. Editorial Trotta, Madrid

Gallo S (2002) Em torno de uma Educação Menor. Educação & Realidade 27(2):169–178

Gare A (2022) Challenging the dominant grand narrative in global education and culture. In: Rozzi R, Tauro A, Wright T, Avriel-Avni N, May RH Jr (eds) Field environmental philosophy: education for biocultural conservation, vol 5. Springer, Dordrecht

Gil J (2014) Prefácio: A Contaminação. In: Wallenstein M (ed) Se não havia nada, como é que surgiu alguma coisa? Lisboa, Edição Fundação Centro Cultural de Belém, pp 8–11

Gomes VCAD (2016) Dialogar, conversar e experienciar o filosofar na escola pública: encontros e desencontro. NEFI Edições, Rio de Janeiro

Gomes VCAD (2019) Em Caxias a filosofia en-caixa? e as vozes infantis da periferia. Praxis & Saber 10(23):19–37

Jasinski I (2016) The passion of (not) Teaching: An Agambenian Meditation on the Value of Philosophy with Children. Philos Educ Soc, pp 486–493

Jasinski I, Lewis T (2015) Community of infancy: suspending the sovereignty of the teacher's voice. J Philos Educ 0:1–16

Kennedy D (2010) Philosophical dialogue with children: essays on theory and practice. The Edwin Mellen Press Lewiston, New York

Kennedy D (2020) Becoming child. Wild being and the post-human. In: Kohan WO, Weber B (eds) Thinking, childhood, and time. Contemporary perspectives on the politics of education. Lexington Books, London, pp 119–129

Kohan WO (2011) Childhood, education and philosophy: notes on deterritorialization. J Philos Educ 45(2):339–357

Kohan WO (2014) Philosophy and childhood. Critical perspectives and affirmative practices. Palgrave Macmillan, New York

Kohan WO, Olarieta BF (eds) (2012) A Escola Pública aposta no Pensamento. Editora Guttenberg, S. Paulo

Maffi L (ed) (2001) On biocultural diversity. Linking language, knowledge, and the environment. Smithsonian Institution Press, Washington/London

Maffi L, Woodley E (eds) (2010) Biocultural diversity conservation: a global sourcebook. Routledge, London
Mallory C (2013) Environmental justice, ecofeminism, and power. In: Rozzi R, Pickett STA, Palmer C, Armesto JJ, Callicott JB (eds) Linking ecology and ethics for a changing world: values, philosophy, and action, Ecology and ethics, vol 1. Springer, Dordrecht, pp 251–260
Melville H (2016) Bartleby, the Scrivener. CreateSpace Independent Publishing Platform, London
Mignolo W (2003) Histórias Locais, Projetos Globais. Colonialidade, saberes subalternos e pensamento limitar. Editora UFMG, Belo Horizonte, p 454
Murris K (2016) The post-human child. Educational transformation through philosophy with picturebooks. Routledge, London/New York
Murris K, Haynes J (2018) Philosophy for children: a postdevelopmental relationality. In: Murris K, Haynes J (eds) Literacies, literature and learning: reading classrooms differently. Routledge, London, pp 50–63
NEFI/UERJ (2021) Center for the Study of Philosophies and Childhood at the State University of Rio de Janeiro. Available via http://filoeduc.org/nefiedicoes/editora.php
Ollin R (2008) Silent pedagogy and rethinking classroom practice: structuring teaching through silence rather than talk. Camb J Educ 38(2):265–280
Omelczuk FW (2016) Uma aposta no invisível: o fazer filosofia com crianças. Revista Contemporânea de Educação 11(21):236–240
Ramos do ÓJ (2016) "Sobre o tradicional cerco à inventividade infanto-huvenil e o amanhã da escola como uma comunidade de iguais", em *Nós pensamos todos em nós*, coord. Madalena Wallenstein. Edição Fundação Centro Cultural de Belém, Lisboa, pp 216–223
Rollo T (2016) Feral children: settler colonialism, progress, and the figure of the child. Settl Colon Stud 8(1):60–79
Rollo T (2020) Democratic child's play. natality, responsible education, and decolonial praxis. In: Kohan WO, Weber B (eds) Thinking, childhood, and time. Contemporary perspectives on the politics of education. Lexington Books, London, pp 145–160
Rozzi R (2012) Serendipity in the origin of ecotourism with a hand lens. In: Goffinet B, Rozzi R, Buck LLW, Massardo F (eds) The miniature forests of cape horn: eco-tourism with a hand-lens. UNT Press, Denton, TX, pp 11–26
Rozzi R (2013) Biocultural ethics: from biocultural homogenization toward biocultural conservation. In: Rozzi R, Pickett STA, Palmer C, Armesto JJ, Callicott JB (eds) Linking ecology and ethics for a changing world: values, philosophy, and action, ecology and ethics, vol 1. Springer, Dordrecht, pp 9–32
Rozzi R (2018) Biocultural homogenization: a wicked problem in the Anthropocene. In: Rozzi R, May RH Jr, Chapin FS III, Massardo F, Gavin M, Klaver I, Pauchard A, Nuñez MA, Simberloff D (eds) From biocultural homogenization to biocultural conservation, Ecology and ethics, vol 3. Springer, Dordrecht, pp 21–47
Rozzi R, Massardo F, Anderson C, Heidinger K, Silander J Jr (2006) Ten principles for biocultural conservation at the southern tip of the Americas: the approach of the Omora Ethnobotanical Park. Ecol Soc 11(1):43
Rozzi R, Armesto JJ, Goffinet B et al (2008a) Changing lenses to assess biodiversity: patterns of species richness in sub-Antarctic plants and implications for global conservation. Front Ecol Environ 6:131–137
Rozzi R, Arango X, Massardo F, Anderson C, Heidinger K, Moses K (2008b) Field environmental philosophy and biocultural conservation: the Omora Ethnobotanical Park educational program. Environ Ethics 30(3):325–336
Rozzi R, Massardo F, Medina Y, Moses K, Caballero P, Camelio E, Morales V, Lewis L, Goffinet B, Mendez M, Cavieres L, Chávez J, Russell S (2010a) Ecoturismo con lupa: integración de las ciencias ecológicas y la ética ambiental. Ambiente y Desarrollo 24(2):30–37

Rozzi R, Anderson CB, Pizarro JC, Massardo F, Medina Y, Mansilla A, Kennedy JH, Ojeda J, Contador T, Morales V, Moses K, Poole A, Armesto JJ, Kalin MT (2010b) Field environmental philosophy and biocultural conservation at the Omora Ethnobotanical Park: methodological approaches to broaden the ways of integrating the social component ("S") in long-term socioecological research (LTSER) sites. Revista Chilena de Historia Nat 83:27–68

Tauro A, Ojeda J, Caviness T, Mosses K, Moreno R, Wright T, Zhu D, Poole A, Massardo F, Rozzi R (2021) Field environmental philosophy: a biocultural ethic approach to education and ecotourism for sustainability. Sustainability 13(8):4526. https://doi.org/10.3390/su13084526

Wallenstein M (ed) (2014) Se não havia nada, como é que surgiu alguma coisa? Edição Fundação Centro Cultural de Belém, Lisboa

Wallenstein M (ed) (2016) Nós pensamos todos em nós. Edição Fundação Centro Cultural de Belém, Lisboa

Chapter 22
Communities of Philosophical Inquiry for the Empowerment of Ecological Agency

Benn Johnson, Rika Tsuji, Benjamin Lukey, and Mitsuyo Toyoda

Abstract Using three case studies of philosophical dialogues in environmental education and environmental decision-making from the USA and Japan, this chapter explores how such dialogues contribute to advancing biocultural conservation by enhancing the voices of multi-stakeholders. This provides opportunities for examining ecological issues from different perspectives, and thus expanding horizons of biocultural awareness. Multi-stakeholder processes are not easy to implement due to "tension" that exists among stakeholders. Tension may be derived from differing senses of place and life habits, or assumptions regarding gender, race, ethnicity, class, age, political identities, and experiences. This tension significantly affects our interpretations of the world. While it is the source of diversity, it also has a possibility to generate narrow perspectives and unresolvable conflicts. The purpose of this chapter is to illustrate how philosophical dialogues can be designed to transform negative tension into constructive, creative, and caring relationships in light of three basic principles of such dialogues: building intellectual safety, appreciating the diversity of viewpoints, and connecting dialogues to empowerment and self-transformation. Based on the examination of case studies from these three principles, we argue that philosophical dialogue complements field environmental philosophy through transformatively fostering ecological agency in biocultural conservation.

B. Johnson · R. Tsuji
Kagawa University, Kagawa, Japan
e-mail: benjamin.johnson@kagawa-u.ac.jp; tsuji.rika@kagawa-u.ac.jp

B. Lukey
The Uehiro Academy for Philosophy and Ethics in Education, University of Hawaii, Honolulu, HI, USA
e-mail: lukey@hawaii.edu

M. Toyoda (✉)
Sado Island Center for Ecological Sustainability, Niigata University, Niigata, Japan
e-mail: toyoda@cc.niigata-u.ac.jp

R. Rozzi et al. (eds.), *Field Environmental Philosophy*, Ecology and Ethics 5,
https://doi.org/10.1007/978-3-031-23368-5_22

Keywords Biocultural conservation · Education · Community of philosophical inquiry (CPI) · Philosophical dialogue · Philosophy for children (p4c) · Tension

22.1 Introduction

In situ conservation approaches, such as field environmental philosophy (FEP), require collaboration and incorporation of the voices of multiple stakeholders (Rozzi et al. 2010). Key international environmental agendas, such as Agenda 21, Convention on Biological Diversity, and the 2030 Agenda for Sustainable Development, emphasize the importance of involving multi-stakeholders (Poole 2018). Whereas participatory processes of multi-stakeholders are considered important in environmental decision-making, which significantly affect local life and industry, the collaboration among them does not necessarily grow easily because there is tension among stakeholders/participants, which may be derived from each person's sense of place and life habits, differing based upon gender, race, ethnicity, class, age, and experiences (Toyoda et al. 2020).

In this chapter, we examine the issue of *tension* in biocultural conservation education, which poses the threat of alienating participants, locking them into narrow perspectives, and preventing them from engaging actively in education, and thus from generating creative solutions for biocultural co-inhabitation (*sensu* Rozzi 2019, pp. 274–278). *Tension* discussed here might be intellectual, political, and/or psychological. For example, if two people have different ideas about something, a certain tension emerges between them. The act of conservation itself consists in a sort of tension, since it inevitably implies that something or someone is at risk of loss at another's expense. People often avoid seeking a multiplicity of perspectives, viewing them as destructive tensions that cause unfixable conflict. But this is a product of attitudes toward tension rather than a product of multiple viewpoints.

We can change our pedagogical understanding of tension from destructive to constructive (Adamian and Jayakumar 2018; Dong 2008) by accepting that multiplicity is the state of things, and diversity is an aspiration. Then, multiplicity does not necessarily entail conflict that prevents us from understanding others. In tension we can find the potential for opening up new horizons, exploring challenging perspectives, and generating transformative experiences. One of the challenges of biocultural conservation education is to be aware of the existence of tensions in people's views and voices and turn them into caring, creative relationships.

One possible approach to overcome this challenge is creating opportunities to engage in *philosophical dialogue* in environmental education and environmental decision-making (Toyoda 2008; Tsuji and Johnson 2020). These two fields tend to be influenced by tensions generated from differing expertise, social occupations, political, and historical concerns, interests, and backgrounds. Sometimes, under such tensions, hierarchical relationships are built or reinforced, for example, between students and educators, between the expert and the lay, and among different ethnic or socioeconomic groups. This inhibits the equal and meaningful exchange of voices and thus diminishes the effects of a biocultural education program. The problem is *not* that there *is* tension. Rather, a problem arises when tension remains unproductive (Dong 2008). Our approach demonstrated in this chapter is to transform tension into a productive mode by developing intellectually safe communities of inquiry.

The philosophical dialogue that we propose is a way to overcome undesirable intellectual hierarchy and to cultivate the soil of collaborative thinking. It encourages us to transform our habit of holding onto one's own views and values and enables us to deal with difficult and complex issues of ecological crisis by fostering the alternative habit of multi-perspectival thinking. "Dialogue" has become one of the key concepts in education since the late twentieth century (Stewart and Zediker 2000). Our conception of philosophy is not purely academic and theoretical; it is also dialogical, experiential, and transformational. In our pedagogical framework, informed by the Philosophy for Children movement (see Lipman 2003, Gregory et al. 2017), philosophical dialogues that are practiced with a view to unfolding various perspectives, participants are expected to challenge one's predominant thinking through the encounter with "the other" and to expand one's cognitive horizons for perceiving the world. Thus, we emphasize the notion of *community of philosophical inquiry* (CPI, Lipman 2003) in our pedagogy to draw attention to the power of collaborative thinking. We enhance this aspect of philosophical inquiry by acknowledging the importance of cultivating *an intellectually safe community* as the necessary foundation of meaningful dialogues (Jackson 2004). Thomas B. Yos (2012), one of the pioneering practitioners of the philosophy for children Hawai'i (p4cHI)[1] pedagogical framework emphasizes that the idea of community is often neglected in philosophical inquiry, especially in this day and age (p. 52). He claims that "we need to purposefully build loving places—with the same forethought that we give to other instructional strategies" (Yos 2012, pp. 55–56). Encouraging intellectual safety in a community is a way to transform tensions into positive sources of intellectual exploration.

In this chapter, we provide three case studies of *philosophical dialogues* from different geographic and cultural regions, and discuss the issue of tension that emerges in biocultural conservation education. In these cases, three core principles can be seen in the common approach to creating *communities of philosophical inquiry*:

1. *Intellectual safety* needs to be cultivated in communicative spaces and communities of inquiry to foster the inclusion of various stakeholders' voices.
2. *Diversity of viewpoints* should be valued as the source of ideas and inspirations.
3. *Dialogue* should be regarded as an invitational opportunity for *empowerment* (e.g., the self-discovery of empowerment in thinking for oneself and with one's community) and/or *self-transformation*.

After presenting the case studies that include a summer camp in Texas and a school program in Hawai'i as well as a community process in Japan, we will discuss the role

[1] Here, "philosophy" and "children" are purposely left lower-case. For a fuller explanation of p4cHI, visit www.p4chawaii.org

of these principles in cultivating ecological agency: the ability to respond to socio-environmental challenges as members of ecological communities. The authors of this chapter have all been educated in philosophy for children at the University of Hawai'i, and have a combined 44 years' experience in facilitating this pedagogy and supporting schoolteachers and community leaders. The pedagogical strategies we discuss here are thus drawn from practical experience in p4cHI pedagogy, which recognizes that communities of philosophical inquiry develop differently depending on environments, risks, and tensions of particular places. We are aware that each of these cases corresponds to specific cultural contexts that require careful consideration of cultural and political differences. Creating intellectually safe communities of inquiry must begin from the needs and particulars of a given community and will thus vary according to context (Lukey 2012). We recognize the threats involved in honest and genuine expression of perspectives and the persecution of environmental leaders in many regions of the world (May 2018). With respectful awareness of these constraints, we aim to highlight conceptual and pedagogical commitments that facilitate the empowerment of different perspectives within communities of philosophical inquiry.

22.2 Case Studies

22.2.1 Environmental Philosophy with Children: A Summer Camp Utilizing Philosophy for Children and FEP Practices and Values

In June of 2019, a group of university students led the Environmental Philosophy with Children (EPWC) summer camp, through interdisciplinary collaboration among philosophers, artists, local educators, and conservationists. [2] The summer camp was a 5-day place-based day camp in an urban and suburban part of north Texas (USA), which aimed to incorporate philosophical dialogue, ecology, and art activities for 24 students, ages six through ten.

The summer camp took place in a suburban area where habitats have been highly disturbed for ranching and urbanization. Participants, including the organizers, were from diverse cultural, educational, experiential, and linguistic backgrounds. Because of this, the summer camp was organized keeping three core attributes of the group in mind. First, not all participants would be long-term residents in the area. Second, participants' backgrounds would be diverse, and this diversity would need to be supported rather than suppressed. Third, the camp would need to ensure a community where those who were familiar and unfamiliar with the place would feel

[2] The reflection in this section is drawn from observation within the camp, and is not intended to be a description of long-term empirical analysis, as there was no long-term follow-up with participants after the initial five days of the camp in June 2019.

included, and where the community could discuss and explore existing and marginalized biocultures in the place.

The EPWC camp attempted to incorporate this urban and suburban context into its biocultural education by drawing from philosophy for children and field environmental philosophy (FEP, Rozzi et al. 2010) approaches, and aimed to create place-based environmental education that would integrate the community of philosophical inquiry. Although the EPWC project was informed by FEP and philosophy for children, the summer camp was not intended specifically to be an example of biocultural conservation, so it did not clearly map onto the FEP's 4-step cycle. For example, instead of implementing new conservation areas, as described in the FEP fourth step, the EPWC team worked with local conservationists and the nonprofit school to utilize existing conservation areas.

The purpose of *in situ* conservation for FEP is to redevelop a biocultural ethic grounded in the idea that ethics, or habits, are a product of the interactions among co-inhabitants in shared habitats (Rozzi 2019, p. 275). Habitats, or "places," are integral parts of our ethical development and lives (Rozzi et al. 2010). The summer camp aimed to provide a space for both facilitators and children to engage in and to explore their place critically, creatively, and caringly to integrate "place" into communal "ethical" inquiry. In this sense, the summer camp carried the spirit of biocultural conservation, without consistently carrying out its methodology. Even so, the team implemented the three other steps of FEP: composition of metaphors, interdisciplinary research, and design of field activities.

To recognize and work through tensions that commonly arise in multidisciplinary groups, the camp organizers implemented a community of philosophical inquiry as part of the camp goals and methods. Through open dialogue, the organizers shared what they thought was important for the camp, which included emphases such as place, narrative, embodiment, artistic representation, and underrepresented co-inhabitants. As a result of inquiry, "mapping" was developed as the primary metaphorical tool to translate these emphases into activities for exploring the local habitat.

Through the metaphor of mapping, camp facilitators and children explored what is cared about and not cared about as well as what is forgotten or neglected by our narratives about and representations of particular places. For example, in one field activity students created texture maps of a local conservation site following face-to-face encounters with locally vulnerable box turtles. Face-to-face encounters with other beings are an essential component of FEP (Rozzi et al. 2006), which in turn inspire the creation of metaphors and other creative compositions. In their texture maps, some children insisted on including hiding places for turtles, by adding twigs and grasses to their cardboard maps (Fig. 22.1). Through the critical, creative, and caring lens of maps, students were able to see themselves and other co-inhabitants, not purely as abstract geographic entities, but as embodied and situated co-inhabitants (Tsuji and Johnson 2020).

Drawing from p4cHI approach, the development and implementation of the camp were grounded in caring for one another and building an intellectually safe community (Jackson 2013). Forming an inclusive and intellectually safe community was

Fig. 22.1 Children making cardboard maps at the Environmental Philosophy with Children (EPWC) summer camp, June 2019. Photograph by Rika Tsuji

particularly important for addressing the potential tensions arising from implementing biocultural conservation and cosmopolitanism of suburban areas. As Rozzi (2013) discusses, more than half of the world population now lives in cities rather than rural areas, and this trend is increasing, meaning that fewer people have "face-to-face encounters" with biocultural diversity, creating a challenge for biocultural conservation. For example, biocultural conservation-focused education could provoke conflicts stemming from participants' newfound awareness of socio-political responsibility and the loss or marginalization of particular biocultures. It may lead to contradictory or under-processed feelings, such as shame over one's affiliation with colonialism, questioning of the dominant economic narrative, or even shallow excitement over the exoticism of seeing something unusual.

The principle of plurality inherent in the community of philosophical inquiry helped the camp community to navigate tensions by minimizing exclusion and ensuring the opportunity to explore questions about living well together in the same place. Incorporating the community of philosophical inquiry into biocultural education provided an intellectually safe community to discuss these experiences and to acknowledge that they were legitimate experiences of the community (instead of having ignored them or letting them remain latent and unspoken).

For example, in the camp, the facilitators shared a Kiowa people's story about the ecocide of bison due to US expansion and colonialism. In the story, the bison are members of the Kiowas' families, who fought against white settlers to protect their family members (Caduto and Bruchac 1997). After listening to the story, one student, Robert, 7 years old, questioned the moral implications of having ancestral connections with colonial practices, asking whether his ancestors who were in the army could be considered "bad guys." [3] Although it is not clear what exactly Robert felt in that moment, his response shows an internalization of the story. For Robert, although the Kiowa people's story was not his story *per se*, the intellectually safe space to question and approach the story from his own perspective seemed to allow for a greater sense of awareness of his position amid a broader historical biocultural community, without seeing that history as a discontinuous event. Moreover, the space to share his question spurred others in the community to see together the colonial realities of the place. In this case, intellectual safety allowed the camp community to explore difficult personal feelings without minimalizing nor condemning these feelings, or neglecting the complex relations from which they arose.

22.2.2 Sado Community Dialogue for Ecological Co-inhabitation

The observation of this section is based on a series of community dialogues that Mitsuyo Toyoda, since 2007, carried out on Sado Island, Niigata, Japan. This island is well known for re-introducing the bird *toki* (*Nipponia nippon*, the crested ibis), which once was extinct in Japan but recovered as a result of local and international efforts of captive breeding and ecological conservation (Fig. 22.2a). The crested ibis is used as a flagship species (see Zhu 2023) for biodiversity conservation in this region because its habitat extends to the mosaic of traditional agricultural landscapes known as *satoyama*. This mosaic includes forests, agricultural fields, grassland, and irrigation systems (Fig. 22.2b). Farmers on Sado Island engage in environmental farming by reducing the amount of agricultural chemicals and artificial fertilizers as well as by managing rice paddies in a way that provides suitable habitats for fish and insects. Since the first release of the bird into the natural environment in 2008, its population has been steadily growing and reached 450 in July 2020. The growth of the wild population indicates the success of environmental conservation practices.

The community dialogues are oriented toward finding community resources for sustainable ecological co-inhabitation. Toyoda began coordinating communal workshops in 2007 as a member of a research group led by Toshio Kuwako in order to create opportunities for local stakeholders to share both worries and hopes for co-inhabiting with the *toki* (Fig. 22.2c). Those workshops were particularly important because some local farmers had a negative image of this bird, seeing it as an agricultural pest that tramples rice seedlings while foraging for food in paddies. Most of the other residents were indifferent to the project. Before the re-introduction started, it was uncomfortable for most local residents to consider this issue with a

[3] Per institutional review board (IRB; 19-289), all student names used in this section are pseudonyms.

Fig. 22.2 (**a**) Bird *toki* (*Nipponia nippon*, the crested ibis) in a rice paddy. (**b**) Rice paddies in a satoyama landscape on Sado Island. (**c**) Workshop about toki habitats with local stakeholders on Sado Island. Photographs "**a**" by Hisashi Nagata, "**b**" and "**c**" by Mitsuyo Toyoda

sense of involvement. This tendency generated challenges because the ecological restoration of *satoyama*, the main habitat of *toki*, would be possible only with the cooperation of local residents.

By 2020, Toyoda had coordinated more than 60 meetings on this topic in light of the three key ideas of philosophical dialogues mentioned in the introduction to this chapter. Intellectual safety was particularly important for dialogues about co-inhabitation with the *toki*, given the fact that not all residents acknowledged the value of the project. Recognizing difficult tensions among stakeholders was vitally important to encourage participants to share their honest voices, both positive and negative perceptions. This enabled uncovering diverse issues that emerged with the re-introduction project as well as to create an atmosphere where participants could think together about their sensitive concerns regarding biocultural conservation.

Since many participants were indifferent to the project, they tended to remain quiet when asked to share their opinions. For this reason, at the beginning of each meeting, we began by asking participants to share *any* concerns about their communities. This resulted in the inclusion of concerns about aging society, children's education, social-welfare services, and local industry, among other concerns. Based on the variety of voices shared by the participants, they were encouraged to identify sustainability challenges emerging in their community and to consider how co-inhabitation with the *toki* might provide solutions to their important concerns. At these meetings, the coordinator group never tried to teach participants the value of biological diversity and the legitimacy of the re-introduction project. The primary aim of these meetings was not to teach environmental values but to create a dialogical space in which participants could think together based on the diverse voices they heard at the meetings, and to make connections between their everyday lives and ecological issues. Such a multi-perspectival dialogue possesses three important educational and transformative goals:

1. Expansion of one's epistemic framework through encounters with diverse viewpoints.
2. Growth of democratic culture in which each voice is valued as a unique perspective.
3. Empowerment of participants through engagement in creative problem-solving.

These goals are particularly important for the growth of ecological agency. In any environmental project, it is crucial to empower actors who are concerned about their environment, as well as to experiment with various actions for an adaptive improvement. These goals are not achieved simultaneously, but occur gradually. At the beginning of community dialogues, many participants might remain quiet, thinking that they have nothing to share. It is critical that coordinators of dialogues create an atmosphere in which people feel safe to share their ideas (anything that comes to their minds), and invite each of them to dialogical engagement to explore together their hopes and expectations. It is also critical to include multi-stakeholders in the dialogue, such as local residents, governmental workers, and specialists, from young to elder, in order to expose each participant to different viewpoints and to recognize the richness of voices in their community. Even though there are conflicting opinions among them, it is crucial to look at all ideas as the source of inspiration for environmental actions and to encourage participants to engage in creative thinking.

Community dialogues for co-inhabitation with the *toki* generated a variety of outcomes, including (1) the creation of a physical community space for the conservation of rice terrace landscapes and (2) the establishment of a watershed conservation project for ecological co-inhabitation (see Toyoda 2018). In both cases, many local stakeholders were initially indifferent or opposed to the project because they considered that their livelihood had not been prioritized compared to global missions such as the conservation of biodiversity. Dialogues resulted in a deeper public engagement in various projects for ecological conservation and community development, involving more participants from both inside and outside of each region. The extension of communities for biocultural conservation connected diverse knowledge, skills, and ideas from participants and generated positive outcomes for conservation challenges.

22.2.3 *SEEQS: The School for Examining Essential Questions of Sustainability*

The example of Environmental Philosophy with Children in Texas illustrated the importance of intellectual safety in opening up children's voices about sensitive biocultural issues. The value of intellectual safety was reaffirmed in actual environmental decision processes in the example of Sado Island. By committing to intellectual safety, communities began to be empowered to think together about possible actions toward ecologically sustainable development. How then can the power of philosophical dialogues be implemented in schooling, a pivotal system to foster responsible citizens in the future? The third example of School for Examining Essential Questions of Sustainability (SEEQS) shows a concrete pedagogical framework that applies the power of philosophical dialogue in biocultural conservation education.

SEEQS is a public charter middle school in Honolulu, Hawaii, founded in 2013 (SEEQS 2018). Its student body is ethnically, culturally, and socio-economically diverse. SEEQS envisions that its graduates will be stewards of planet Earth and healthy, effective citizens of the world (Fig. 22.3). The vision is realized through an approach to teaching and learning that builds from the metaphor of cultivating plants. The "soil" is the school community and much time is dedicated to its development. Within this soil "seeds" are planted. These seeds are equivalent to the content courses found in standard public schools, but here they are "watered" by the interdisciplinary Essential Questions of Sustainability (EQS) courses that all students must take—all students participate in a mixed-grade level, interdisciplinary, project-based, year-long pair of courses designed around examining sustainability questions. These courses are co-taught by a team of five or more teachers from different disciplines. They focus on interconnections between humans and the environment; create a real-world context to apply disciplinary knowledge; engage community members as partners in learning; empower students to choose the focus of their project; and motivate students through authentic exhibitions of learning (SEEQS 2021a). The combination of soil (i.e., community-building), seeds (i.e., skills development), and water (i.e., real-world application) helps students "sprout"

Fig. 22.3 Children at an outdoor classroom in the School for Examining Essential Questions of Sustainability (SEEQS), June 2019. Photograph by Buffy Cushman-Patz

the five sustainability skills that enable students to grow into "trees" and live out the SEEQS vision to become Stewards of Planet Earth. The five SEEQS sustainability skills are:

1. Reasoning analytically: making judgments based on reasons and evidence.
2. Managing effectively: finishing what you start, with effort, organization, and care for quality.
3. Communicating powerfully: expressing yourself so that others are able to understand.
4. Collaborating productively: working together toward a common goal.
5. Thinking systemically: seeing patterns, making connections, and designing solutions (SEEQS 2021b).

In order to graduate from SEEQS, students must prepare and publicly defend a portfolio that demonstrates their cultivation of these sustainability skills.

The University of Hawai'i at Manoa Uehiro Academy for Philosophy and Ethics in Education, the home of p4cHI, has been supporting and collaborating with SEEQS since its inception. Many of the elements of p4cHI pedagogy are woven through the SEEQS school model and curriculum. Uehiro Academy faculty have worked alongside SEEQS teachers as they developed curriculum, school routines, activities, and assessments. This collaboration has helped create a constructive tension, balancing the value and practice of sustainability curricula with philosophical inquiry. It has also provided teachers with guidance and practice in facilitating diverse, student-driven communities of philosophical inquiry. Thus, a visitor to SEEQS today will see not only many elements of p4cHI, but also notice the consistent emphasis on intellectually safe dialogue through most classes and activities.

SEEQS' approach is compared with FEP in Table 22.1. Though not part of the school design, it is possible to see how all four FEP steps are interconnected and expressed through the SEEQS's model. To compare FEP with SEEQS, it is most convenient to begin with Step 3 (Table 22.1).

Table 22.1 School for Examining Essential Questions of Sustainability (SEEQS) and field environmental philosophy (FEP) (Rozzi et al. 2010)

FEP Step	Expression at SEEQS
Step 3: Field activities with an ecological and ethical orientation	The five core principles that describe SEEQS's foundational educational philosophy describe aspects of bioculturally robust field activities: 1. Real-world situations and real-world contexts enable real-world learning 2. Learning occurs when learners take ownership of their learning 3. Everyone is a teacher, everyone is a learner, all of the time 4. A learning environment is composed of its community members, cultural values, and physical surroundings 5. Improvement of the organization requires consciously collaborative participation by community members (SEEQS 2021c) These principles acknowledge the importance of a culturally responsive, multi-perspectival community of philosophical inquiry that is engaged in real-world (i.e., field) actions. These actions are carried out daily through the EQS curriculum, and are punctuated with field experiences such as the day-long EQS camps and learning/activity trips (e.g., to a watershed restoration site, to the state legislature, or to a local farm). These activities culminate in a public semester-end presentation of learning where students explain their projects and practice their advocacy Starting with FEP's Step 3 as the focus for learning, the other three FEP steps become necessarily integrated
Step 1: Interdisciplinary ecological and philosophical research	Three distinct aspects important for younger learners/agents: 1. The "seeds" of *content knowledge* (humanities, arts, and sciences) nourish one aspect of creating an understanding of the world 2. The "water" of project-based learning through EQS adds understanding gained from *application* and from researching solutions to specific real-world problems 3. p4cHI *dialogue* with peers and adults is an inward activity of philosophical research whereby the participants co-create meaning for the information they are accumulating and the activities they are practicing

(continued)

Table 22.1 (continued)

FEP Step	Expression at SEEQS
Step 2: Communication	Asking and exploring questions is not only a process of research but also a *process of communication.* p4cHI dialogue allows participants to test provisional answers to questions within their community of inquiry. This then allows them to revise and refine their understanding and the meaning of these answers. Thus, when students are expected to demonstrate the sustainability skill of communicating powerfully, in their end-of-semester presentation of learning, their end-of-quarter Student-Led Conferences, or in their eighth grade portfolio defense, they have honed their communication skills (often via stories) through the multiple aspects of their research and their extensive field activities
Step 4: *In situ* conservation	Becoming an effective steward of the planet requires a sense of personal empowerment. As a middle school, SEEQS does not have a dedicated biocultural conservation site. However, it is educating students who are able to identify and act upon opportunities for *in situ* conservation. Successful SEEQers have a sense of what they are able to learn from others in shaping their own understanding of the world. They are able to communicate this understanding effectively to others by drawing upon what they have learned in class and what they have done in the field. Thus, they are empowered so that what they learn can have an impact on the planet, especially when teachers and learners are together. For example, the in situ conservation of SEEQers will be seen in one student's push to persuade restaurant owners to stop using plastic straws or another group of students' fundraising efforts for the Humane Society to humanely control the feral cat population decimating indigenous bird life

22.3 Discussion

The previous examples show how providing a framework for integrating tensions among stakeholders' viewpoints can be constructive for practices of field environmental philosophy (FEP), and by extension biocultural conservation. These examples illustrate three key ideas seen in a common approach to creating communities of philosophical inquiry: (1) an intellectually safe community, (2) valuing a diversity of viewpoints, and (3) empowerment and transformation through dialogue. They also illustrate how these key ideas were integrated successfully into an educational approach in three different situations. To understand how tension can become constructive in such cases we must unpack both the conceptual and pedagogical frameworks that are not limited to schooling but apply to ***communities of philosophical inquiry*** in multiple contexts.

22.3.1 Intellectual Safety and the Community of Philosophical Inquiry

Intellectual safety is an iterative/processual quality that requires collaborative effort within the community of inquiry. Educational spaces are not necessarily safe for everyone (Leonardo and Porter 2010); however, recognition of this fact allows for the communal development of intellectual safety as a regulative ideal. Although intellectual safety is never finished, this regulative ideal provides an asymptote for the community as the participants continually revisit their ***community of philosophical inquiry***. It also guides the modeling of the facilitator (and eventually the modeling of all participants) while doing the inherently risky work of examining tensions. As Thomas Jackson explains:

> In an intellectually safe place, there are no put-downs and no comments intended to belittle, undermine, negate, devalue, or ridicule. Within this place, the group accepts virtually any question or comment, *so long as it is respectful of the other members of the circle.* What develops is a growing trust among the participants and with it the courage to present one's own thoughts, however tentative initially, on complex and difficult issues (Jackson 2001, p. 460, emphasis added).

Intellectual safety should cultivate participation habits emphasizing "being listened to" and "listening to others." However, "safety" is not synonymous with comfort. Intellectual safety necessitates critical examination and intellectual discomfort, that is, the creation of "critical communities" (Butnor 2012). Such communities can thus help participants to develop the critical and dialectical tools necessary to recognize and critically address challenges such as misinformation and "fake news."

Thus, the concept of an intellectually safe community must be translatable into pedagogical practices. We identify the following three practices for facilitating intellectually safe communities of philosophical inquiry. First, a facilitator recognizes where the community is beginning from, observing the variety of interpersonal dynamics and power structures present among the participants that could contribute to a negative tension keeping participants from listening to others and being heard. These observations then guide how the facilitator models expectations in the community. The examples in this chapter all provided different starting points for the community of philosophical inquiry, which facilitators recognized and adapted their pedagogical practices to. Rika Tsuji (2020) has discussed the importance of actively identifying and removing destructive social tensions that prevent the inclusion of plural voices in the inquiry.

Second, destructive social tensions (e.g., bullying, put-downs, intimidation) undermine listening and being listened to. Listening to another just to find faults prevents others from being genuinely heard. Facilitators must discourage ridicule, intimidation, emphatic indifference, and other anti-social behaviors among participants. Social discrimination according to gender, race, and ability, for example, is often revealed both within school classrooms and outside of school and must be actively addressed by the facilitator (Lin and Sequeira 2017).

Third, it is difficult to model genuine listening while speaking all the time. The example from Sado Island illustrates that facilitators (even if they are qualified or trained specialists) must step back and let others speak. If we wish participants (e.g., students, farmers, educators) to be fully engaged biocultural agents, then we must resist the temptation to infringe upon or manipulate the participants' inquiry by doing the thinking for them. While the organizer of a dialogue may bring a predetermined purpose to the community, the purpose is always subject to reformulation by the community. Thus, a facilitator is constantly shuttling roles. Especially at the beginning stage, facilitators have to be pedagogically strong to stop behaviors that undermine intellectual safety, occasionally relying upon their "authority" for compliance with such orders, while also being philosophically self-effacing so that they can step back, invite others to speak, and model genuine listening.

22.3.2 Thinking with Multiple Viewpoints Within the Community of Philosophical Inquiry

An intellectually safe community enables a more engaging dialogue by including a greater diversity of viewpoints. This creates two related challenges, which then create a further tension. First, when I listen to others, how does what they share impact my thinking and inquiry, personally? Second, when people listen to one another, how do they think together, communally? These questions produce an obvious tension between what it means to think for oneself and what it means to think with others.

In the community of philosophical inquiry, we need to avoid the common practice of "specialists" doing the thinking for a participant and instead encourage all participants to actively think together about their concerns. As Paulo Freire (2014, p. 77) argued, "the teacher cannot think for her students, nor can she impose her thought on them," and any attempt to do so would remove the authenticity of the thinking process. The inquiry which begins to emerge within each of us (the Freirean authenticity of thinking) incorporates the viewpoints, experiences, and thinking of others. Resonating with Freire's desire for liberatory education, what results will not be an individual's thought but a co-constituted thought developed through dialogue (Kohan 2018). Perhaps paradoxically, engagement with a multiplicity of viewpoints, even though it initially looks divergent and even chaotic, helps cultivate ownership of the collaborative thinking processes.

Pedagogically, the non-hierarchical structure of participation (i.e., participants eventually becoming co-facilitators of philosophical inquiry, not dependent on a designated "facilitator") catalyzes authentic thinking amid a multiplicity of viewpoints. The right to speak and share does not need to go through the teacher/facilitator; participants should actively seek input from others in the community of philosophical inquiry. To further intellectual safety, however, this should be invitational and not intimidating. As the example in Sado illustrates, not all stakeholders are interested in participating in dialogue. Forcing participants to do so or "putting them in the spotlight" when they do not want to be in the spotlight is counterproductive to the aims of eliciting more perspectives. Invitations can be declined, especially when appropriately extended (another skill that the facilitator models).

22.3.3 Empowerment and Transformation Through Philosophical Dialogue

The decentralization of participation puts those with expertise in a different (and often uncomfortable) role within the community of philosophical inquiry. Jackson (2001, pp. 461–462) reminds us that, "No one, especially not the teacher, knows either 'the' answer to the inquiry or where it will lead. Any effort to guide an inquiry to a predetermined answer or outcome corrupts the process from the start." In philosophical dialogue, the facilitator rather needs to be "ignorant," that is, to be able to unlearn what one knows and believes (Kohan et al. 2017).

Instead of providing prefixed knowledge and answers, the facilitator focuses on unfolding a variety of viewpoints by eliminating epistemic hierarchy between experts and the public and inviting all participants to share their voices. The philosophical examination of the world advances when everyone is empowered to be a thinker, an explorer, and an experimenter.

To fully explain our argument, we need to unpack the understanding of the term "philosophical." "Philosophy" comes in many varieties, but all varieties involve content and activity. Our focus is on Jackson's "little p" philosophy, where "content refers to the set of beliefs we begin to acquire at birth that continue to inform our experience, becoming the framework with which we make sense of the world" (Jackson 2013, p. 108). These beliefs and the resulting framework are constantly challenged and revised (to differing degrees). Hence, "little p" activity "refers to the fact that we don't passively acquire this content but engage it almost from the beginning" (Jackson 2001, p. 109). Philosophical inquiry within the community consists of this form of engagement. Since the content of philosophy includes the beliefs and frameworks of participants, the community of philosophical inquiry naturally cultivates a sense of common ownership of the dialogue. Since participants listen to others and consider multiple viewpoints while revisiting their beliefs and frameworks to create meaning with others in the activity of philosophy, the process of philosophical inquiry naturally cultivates a sense of empowerment.

If one of the goals of FEP is to create new conceptual lenses for ecological awareness, then our understanding of philosophy recognizes that creating conceptual lenses is both a personal and active process for the whole community. The experience of communal inquiry involves a meandering exploration more than a teleological progression (Klaver 2018). This active engagement of the community prevents participants from being alienated from their experiences and ensures that nobody is left behind. Ownership is a co-constitutive transformative experience, not only personal but also communal. This communal sense of ownership helps participants recognize and exercise their agency, thus empowering them to act on behalf of and with the biocultural community of which they are a part. Participants do not simply become distinct selves, but realize themselves as part of an interrelated community, embodying the ethos of biocultural diversity.

22.3.4 *Implications of the Community of Philosophical Inquiry for FEP*

Field environmental philosophy was created at the Omora Ethnobotanical Park as an approach to integrate ecological sciences and environmental ethics into biocultural conservation and education in the Cape Horn Biosphere Reserve, Chile (Rozzi et al. 2008). FEP aims to foster intercultural dialogues and an inter-species sense of co-inhabitation (Rozzi 2013), and has been applied successfully on at least two levels. First, organizers have created programs and experiences that integrate a four-step cycle for biocultural research. Second, a large number of students and members of the public have participated in these programs and either directly or indirectly become involved in biocultural conservation. We propose that the pedagogy of p4cHI offers an educational approach that has the potential to enrich engagement activities with students and stakeholders. We advocate that the FEP model could be adapted to some educational and conservation initiatives to incorporate a philosophers' pedagogy (Makaiau and Miller 2012), an approach that is consistent with the founders of the worldwide philosophy for children movement who saw that education and philosophy were inseparable.

> At the heart of philosophy is ... dialogue; at the heart of this discipline is therefore what is essential to education. The craft of philosophy contains itself a pedagogy—the need for dialogue, the need for questioning and a method of inquiry—which are essential characteristics of education in general. This is why education cannot be divorced from philosophy and philosophy cannot be divorced from education (Lipman and Sharp 1978, p. 259).

FEP envisions research, arts, and education as inseparable: it is a way to integrate ethics into biocultural conservation through the 4-step cycle (Rozzi et al. 2010; Tauro et al. 2021). The philosophers' pedagogy we have presented provides another role for "philosophy" within FEP, namely, the participants' dialogical engagement with their own beliefs and experiences. Doing philosophy with others is not just about epistemology and ethics, but is about the way we live *with* others as we communicate and work through biocultural questions. As Ann Margaret Sharp (2017) explains, "the community of inquiry constitutes a *praxis*—reflective communal action—a way of acting in and on the world. It is a means of personal, moral transformation that inevitably leads to a shift in meanings and values that affect the daily judgements and actions of all participants" (p. 246). Through this dialogical model of inquiry, we disclose webs of interrelation within our biocultural communities.

This communal approach to philosophy embraces tensions within biocultural education, which offer opportunities for self-transformation when the community is open to exploring the diversity of perspectives "together with a sensitivity to the particularity of each situation" (Sharp 2017, p. 245). In light of biocultural education, this particularity needs to include biocultural experience and the historicity of the place, as shown in the Environmental Philosophy with Children summer camp. Forming a community of philosophical inquiry requires recognizing the perspectives of others without "othering" them or oneself. This recognition provides the basis for its synthesis with FEP, because it is through dialogical reflection in a diverse community that we can expand our horizons of biocultural awareness. Through such expanded awareness, many different tensions that arise within a biocultural community can become a constructive source for communal transformation and biocultural conservation.

Acknowledgements The EPWC summer camp (promoted as Environmental Explores Summer Camp) was funded by the Philosophy Learning and Teaching Organization (PLATO), the Onstead Institute, and the Department of Philosophy and Religion at the University of North Texas. The camp team sincerely appreciates the financial and institutional support from these institutions. We would also like to acknowledge and express deep gratitude to the rest of the student camp team, Shoshana McIntosh, Beatriz Galuban, T Wright, and Emily Hudson for their hard work and care. We would like to extend our gratitude to our faculty team, Tyson Lewis, Tran Templeton, Chris Moffett, and Adam Briggle for their strong support, including providing resources and supervision. The camp team greatly appreciates the collaborating institutions and their educators, Koan School, Lewisville Lake Environmental Learning Area (LLELA), and the Dallas Zoo for providing the camp with in-depth resources and educational opportunities. The camp was also made possible through each of the camper's thoughtful participation and their parental support. We appreciate the chance to form a community of philosophical inquiry with these students.

The authors would also like to thank SEEQS founder, Buffy Cushman-Patz, for providing permission, photographs, and valuable feedback for including SEEQS in this chapter, as well as all the participants of communal dialogues for the ecological co-inhabitation on Sado Island.

References

Adamian AS, Jayakumar UM (2018) Mutual engagement in spaces of tension: moving from dialogue toward action across multiple contexts. The Edu Forum 82(3):335–350

Butnor A (2012) Critical communities: intellectual safety and the power of disagreement. Edu Pers 44:29–31

Caduto MJ, Bruchac J (1997) Keepers of the animals: native American stories and wildlife activities for children. Fulcrum Publishing, Golden, CO

Dong YR (2008) Productive tensions: student teachers' handling of sociocognitive conflicts during the classroom discussion. Eng Ed 40:231–255

Freire P (2014) Pedagogy of the oppressed. 30th anniversary edition. Tras., Ramos MB. Bloomsbury, New York

Gregory MR, Haynes J, Murris K (eds) (2017) The Routledge international handbook of philosophy for children. Routledge, Oxford

Jackson TE (2001) The art and craft of "gently Socratic" inquiry. In: Costa AL (ed) Developing minds: a resource book for teaching thinking, 3rd edn. Association for Supervision and Curriculum Development, Alexandria, VA, pp 459–462

Jackson TE (2004) Philosophy for children Hawaiian style—"On not being in a rush...". Thinking 17:4–8
Jackson TE (2013) Philosophical rules of engagement. In: Goering S, Shudak N, Wartenberg T (eds) Philosophy in schools: an introduction for philosophers and teachers. Routledge, New York, pp 99–109
Klaver I (2018) Re-claiming rivers from homogenization: meandering and riverspheres. In: Rozzi R, May RH Jr, Chapin FS III, Massardo F, Gavin M, Klaver I, Pauchard A, Nuñez MA, Simberloff D (eds) From biocultural homogenization to biocultural conservation. Ecology and ethics, vol 3. Springer, Dordrecht, pp 49–69
Kohan WO (2018) Paulo Freire and philosophy for children: a critical dialogue. Stud Philos Educ 37:615–629
Kohan WO, Santi M, Wozniak JT (2017) Philosophy for teachers: between ignorance, invention and improvisation. In: Gregory MR, Haynes J, Murris K (eds) The Routledge international handbook of philosophy for children. Routledge, Oxford, pp 253–259
Leonardo Z, Porter RK (2010) Pedagogy of fear: toward a Fanonian theory of 'safety' in race dialogue. Race Ethn Edu 13(2):139–157
Lin CC, Sequeira L (eds) (2017) Inclusion, diversity, and intellectual dialogue in young people's philosophical inquiry. Sense Publishers, Rotterdam
Lipman M (2003) Thinking in education. Cambridge University Press, Cambridge
Lipman M, Sharp AM (1978) Introduction to Section IV: the philosophical education of the child. In: Lipman M, Sharp A (eds) Growing up with philosophy. Temple University Press, Philadelphia, pp 258–273
Lukey B (2012) Philosophy beyond boundaries: a new model of philosophy in high schools. In: Mohr Lone J, Israeloff R (eds) Philosophy and education. Cambridge Scholars Publishing, Newcastle upon Tyne
Makaiau AS, Miller C (2012) The philosopher's pedagogy. Educ Persp 44(1&2):8–19
May RH Jr (2018) Land grabbing and violence against environmentalists. In: Rozzi R, May RH Jr, Chapin FS III, Massardo F, Gavin M, Klaver I, Pauchard A, Nuñez MA, Simberloff D (eds) From biocultural homogenization to biocultural conservation. Ecology and ethics, vol 3. Springer, Dordrecht, pp 109–123
Poole AK (2018) The UN sustainable development goals and the biocultural heritage lacuna: where is goal number 18? In: Rozzi R, May RH Jr, Chapin FS III, Massardo F, Gavin M, Klaver I, Pauchard A, Nuñez MA, Simberloff D (eds) From biocultural homogenization to biocultural conservation. Ecology and ethics, vol 3. Springer, Dordrecht, pp 315–331
Rozzi R (2013) Biocultural ethics: from biocultural homogenization toward biocultural conservation. In: Rozzi R, Palmer C, Armesto JJ, Callicott JB, Pickett S (eds) Linking ecology and ethics for a changing world: values, philosophy, and action. Ecology and ethics, vol 1. Springer Science+Business Media, Dordrecht, pp 9–32
Rozzi R (2019) Taxonomic chauvinism, no more! antidotes from Hume, Darwin, and biocultural ethics. Environ Ethics 41(3):249–282
Rozzi R, Massardo F, Anderson C, Heidinger K, Silander J Jr (2006) Ten principles for biocultural conservation at the southern tip of the Americas: the approach of the Omora Ethnobotanical Park. Ecol Soc 11(1):43. http://www.ecologyandsociety.org/vol11/iss1/art43/
Rozzi R, Arango X, Massardo F, Anderson C, Heidinger K, Moses K (2008) Field environmental philosophy and biocultural conservation: the Omora Ethnobotanical Park educational program. Environ Ethics 30(3):325–336
Rozzi R, Anderson CB, Pizarro JC, Massardo F, Medina Y, Mansilla A, Kennedy JH, Ojeda J, Contador T, Morales V, Moses K, Poole A, Armesto JJ, Kalin MT (2010) Field environmental philosophy and biocultural conservation at the Omora Ethnobotanical Park: methodological approaches to broaden the ways of integrating the social component ("S") in long-term socio-ecological research (LTSER) sites. R chile de hist natl 83:27–68
SEEQS (2018) SEEQS strive HI performance report. Available via https://www.hawaiipublicschools.org/Reports/StriveHISEEQS18.pdf. Accessed on 31 May 2021

SEEQS (2021a) Essential questions of sustainability (EQS). http://www.seeqs.org/eqs-essential-question-ofsustainability.html. Accessed 23 Apr 2021
SEEQS (2021b) SEEQS sustainability skills. https://www.seeqs.org/sustainability-skills.html. Accessed 23 Apr 2021
SEEQS (2021c) Philosophy of learning. Available via https://www.seeqs.org/why-seeqs. Accessed on 28 May 2021
Sharp AM (2017) The community of inquiry: education for democracy. In: Gregory MR, Laverty MJ (eds) In community of inquiry with Ann Margaret Sharp: childhood, philosophy and education. Routledge, Florence, pp 241–250
Stewart J, Zediker K (2000) Dialogue as tensional, ethical practice. South J Commun 65(2-3): 224–242
Tauro A, Ojeda J, Caviness T, Moses K, Moreno R, Wright T, Zhu D, Poole A, Massardo F, Rozzi R (2021) Field environmental philosophy: a biocultural ethic approach to education and ecotourism for sustainability. Sustainability 13:4526. https://doi.org/10.3390/su13084526
Toyoda M (2008) Applying philosophy for children to workshop-style environmental Education. Proceedings of the XXII world congress of philosophy, vol 27, pp 101–109
Toyoda M (2018) Revitalizing local commons: a democratic approach to collective management. In: Rozzi R, May RH Jr, Stuart Chapin F III, Massardo F, Gavin MC, Klaver IJ, Pauchard A, Nuñez MA, Simberloff D (eds) From biocultural homogenization to biocultural conservation. Ecology and ethics, vol 3. Springer, Dordrecht, pp 443–457
Toyoda M, Takashima T, Kita A, Nakagawa K (2020) Consensus building processes for agricultural land management in the mountainous regions of Japan. Policy Pract Stud 6(2):255–266
Tsuji R (2020) Revisiting the community of philosophical inquiry through the lens of Arendt and Butler. Precoll Philos Public Pract 2:1–17
Tsuji R, Johnson B (2020) Philosophy meets place: creating an environmental philosophy summer camp. In: Katz CE (ed) Growing up with philosophy camp. Rowman & Littlefield, Lanham, MD, pp 99–109
Yos TB (2012) Raising the bar: love, the community of inquiry, and the flourishing life. Educ Perspect 44(1&2):52–57
Zhu D (2023) Bridge the channel, enhance the inclusivity: a comparison between flagship species-centered and moss-centered conservation in Chile and China. In: Rozzi R, Tauro A, Wright T, Avriel-Avni N, May RH Jr (eds) Field environmental philosophy: education for biocultural conservation. Ecology and ethics, vol 5. Springer, Dordrecht, pp 457–482

Chapter 23
Biocultural Resilience Through Educational Tourism in Cholula, Mexico

María Evelinda Santiago Jiménez

Abstract For decades, mass tourism activity has contributed, like other extractives economic activities, to environmental pollution and dislocation of cultures. This impact is associated with massive structures that tourism enclaves have made in countries with low *per capita* income. Tourism is an opportunity to reduce poverty; so metaphorically it becomes in those impoverished countries the "*manna* from heaven." However, much of that *manna* remains in only a few national and foreign hands, leaving crumbs for local inhabitants. This chapter proposes resilience and literacy as alternative foundations for tourism based on an example from a rural-urban community in the city of San Andrés, Cholula, Mexico. This example reveals that under the surface of the *glamour* of tourist destinations there is a cultural dynamism closely related to local biological biodiversity that has been cultivated by the inhabitants themselves. Building alternative tourism in Mexican urban-rural communities can bring well-being to local people and their culturally important biological species and ecosystems. This type of tourism requires education and dialogue among different human actors, and even with other-than-human co-inhabitants based on sharing and respect.

Keywords Environmental literacy · Intercultural dialogues · Responsible tourism · Traditional ecological knowledge

23.1 Introduction

Global social-environmental crisis is immersed in a deeper crisis that is cognitive, affective, and experiential crisis (Rozzi 2013). Global society has limited the world-view only to humans, and certain humans, ignoring the relationships with other-than-human co-inhabitants (Rozzi 2012). To solve this crisis and reconnect citizen with biological and cultural diversity, tourism can foster well-being to local people

---

M. E. Santiago Jiménez (✉)
Tecnológico Nacional de México/Instituto Tecnológico de Puebla, Puebla, Mexico
e-mail: evelinda.santiago@puebla.tecnm.mx

R. Rozzi et al. (eds.), *Field Environmental Philosophy*, Ecology and Ethics 5,
https://doi.org/10.1007/978-3-031-23368-5_23

and their culturally important biological species and ecosystems. This type of tourism requires education and dialogue among different human actors, and even with other-than-human co-inhabitants based on sharing and respect. This process will catalyze biocultural resilience and environmental literacy. Local communities reconstruct their own life projects, and tourists learn about different ways communities co-inhabit with nature. This experience contrasts the prevailing knowledge that global society has about nature today, negotiated by the media, and mediated by virtual realities. Alternative community tourism offers an alternative *in situ* educational experience that is cognitive, affective, and experiential. The meaning of being human is reconstructed in such a way that both local people and tourists can have a playful reunion with nature (Rozzi 2008).

In situ educational tourism leads to *socio-ecological literacy*. However, this literacy must be accompanied by rebuilding life projects of local communities, suggesting that this reconstruction is founded on resilience. There is no consensus on the definition of resilience. Spanish psychologist Cristina Villalba (2004) suggests that resilience implies (a) overcoming difficulties and succeeding in spite of being exposed to high-risk situations; (b) maintaining competence under pressure, meaning knowing how to successfully adapt to high risk; and (c) recovering from a trauma by successfully adjusting to negative life events. Resilience is a process that has to do with dynamic adaptations to risk events that threaten an individual or a group of people, such as the lack of means of consumption and production to build a dignified life project. Therefore,

> [. . .] although it requires an individual response, it is not an individual characteristic since it is conditioned by both individual and environmental factors, emerging from a great heterogeneity of ecological influences that converge to produce an exceptional reaction to a major threat (Villalba 2004, p. 3).

The concept of resilience helps to *imagine* the construction of an inclusive tourism process, which also provides a holistic vision sufficient to not only consider and integrate individuals who are in situations of poverty, but also those who are in a latent state of vulnerability due to poor income or social deprivation. In this way, resilience makes possible the vision to create strategies for solving the vulnerability of some individuals, and at the same time contribute, in a sustained way, to creating alliances among individuals, families, and societies. According to the above, the objective is to integrate the diversity of capacities so that they become elements for fighting vulnerability and poverty.

The concept of resilience is complemented by environmental and cultural literacy. From the perspective of reflexive, critical, and permanent learning, a person is provoked to stop judging cultures based exclusively on their own values. The person realizes that there are no absolute good nor absolute bad cultures (Levi-Strauss 2008). Rather each culture has carved its own history through sustained relationships with their ecosystems either in abrasive or respectful ways (Leff 2008), or perhaps in intermediate or culturally hybrid ways (García-Canclini 2005). These mixtures of abrasive and respectful relationships will be visible in the landscape that a society inhabits. In this sense, the concept of environmental literacy has evolved into a plural vision:

> Literacy has evolved beyond the simplistic notion that defines it as a set of: technical skills of reading, writing and arithmetic . . . to a pluralistic concept composed of a great diversity of meanings and dimensions encompassed by these skills. In the face of current economic, political, and social transformations, including globalization and advances in information and communication technologies (ICTs), this vision recognizes the existence of many forms of literacy that are embedded in diverse cultural processes, personal circumstances, and collective structures (UNESCO 2004, p. 6).

The critical point is to extend literacy toward cultural and environmental literacy. In this chapter, I illustrate the importance of resilience and literacy for building alternative tourism in rural communities. I call this process *resilient tourism.*

23.2 Tourist Animation

Resilient tourism can be implemented via tourist animation. Ramón Fernández (1992) defined tourist animation as a psycho-sociological activity, intimately linked to the motivations of visitors. Alternatively, Yannick Gallien (1993) defined it as creating a pleasant atmosphere that allows and facilitates exchanges between people. Ezequiel Ander-Egg (1989) offered a more comprehensive definition by explaining that tourist animation is a set of social techniques based on a participatory pedagogy that promotes voluntary practices and activities. These are developed with the active participation of people within a given group or community and are manifested in different areas of sociocultural activities that foster quality of life. According to this definition, tourist animation is conceived as an educational strategy of exchanges and learning with different cultures. Animation is a technique of social intervention that promotes popular culture. This requires an adequate form of organization to create processes in which the individual or group takes an active part, providing them with opportunities for physical and mental development. In this sense, the tool is a form of intervention that encourages citizens and tourists to participate in cultural life (Table 23.1).

Animation creates environments that allow local realities to be presented to tourists. It is a pedagogical strategy that brings tourists together with local inhabitants. It stimulates local people and tourists to have "a reencounter [. . .] with their own history [and with] the history of humanity" (Toledo 2007, p. 45). This delicate process of learning should be conducted by local people who through their identities and cultural roots can convoke bonds of solidarity, empathy, surprise, freedom, tolerance, and especially intercultural dialogues and exchanges of knowledge.

This process can be considered as literacy. However, it is a long-term process that involves human development and has different dimensions linked to cultural diversity and histories of places (Freire and Macedo 1989). Consequently, tourist animation can make the tourist literate. Through critical reflections, at the destination

Table 23.1 Tourist animation strategy for responsible tourists and travelers, combined with possible spaces of resilience. Adaptation from the brochure for the Tourist and Responsible Traveler (TRT n/d) of the Global Code of Ethics for Tourism of the World Tourism Organization (WTO 1999). Tourism of the World Tourism Organization (WTO 1999). Edited from Santiago-Jiménez and Morfín-Herrera (2016, p. 111)

Guidance for responsible tourists	Tourism animation strategy for socio-ecological education	Promotion of social and ecological resilience
Open your mind to other cultures and traditions to transform your experience, earn respect, and be welcomed by local people Be tolerant and respect diversity Observe social and cultural traditions and practices	Offer tourists a literacy experience of the local social and ecological reality to foster solidarity and respect An example of this strategy is the Mole Poblano workshop developed in Cholula, Puebla, Mexico	Recover or strengthen local identity and traditions Provide opportunities for multicultural awareness and dialogue between the local community and visitors sharing different forms of knowledge Generate interdisciplinary and multicultural projects that diversify the local economy Recognize and take advantage of the cultural, social, ecological, and economic capacities of the local community

tourists are able to "read" and be in solidarity with the complex reality of social and ecological diversity.

23.2.1 Educational Tourism

Tourist animation should combine recreation and responsibility. Thus, tourism provides opportunities to support the basic strengths of rural communities: solidarity, spirituality, dedication, effort, and commitment (De Miguel 1995). Tourists can select among offerings involving pedagogical recreation processes that lead them to be sensitive to social-ecological diversity of the place they are visiting. This catalyzes local collective actions with the participation of tourists who discover **identities** inscribed in cultural and environmental processes grounded in specific places and social topologies (Porto-Gonçalves 2002; García 2010: 78).

The destination is not just a tourist destination but a place embraced by local actors as their territory. Tourism animation designed as pedagogical tourism involves dialogue as a central component that weaves together both the stories of the local community and those brought by tourists. In other words, it provokes dialogical encounters. Unfortunately, massive tourism has depersonalized the territory objectifying it as a destination, which puts all the emphasis on pleasure experiences of tourists and does not value the experiences of local inhabitants. These local experiences are *folklorized* (or stereotyped) to delight visitors. The local culture becomes depersonalized and is affected by distorted meanings that often provoke social and ecological conflicts. On the contrary, when tourism is planned to be inclusive and plural, it generates alliances of knowledge and exchange relationships. This melting pot of tourist animation generates feelings of empathy with the social and ecological diversity of the destination/territory.

Animation, when used as a catalyst to endogenously induce the reconstruction of life projects, can also well serve to design schemes that make tourists socio-ecologically literate at a "destination," thus turning this destination into "a dynamic book" that facilitates the reading of reality and highlights the complexity of the place. The experience will give the tourist the possibility of internalizing knowledge, while the practice of solidarity will be a means of personal and collective development, provoking processes of self-education, mutual tolerance, and learning the legitimate differences between peoples and cultures and their diversity (WTO 1999).

23.2.2 A Case of Tourist Animation in Cholula, Mexico

Tourist animation is embodied in specific cases. Here, we present a case based on a gastronomic workshop that focused on the preparation of mole in its ecological and cultural contexts in San Andrés Cholula, state of Puebla, Mexico (Fig. 23.1). The mole as the central dish was selected based on four criteria:

(a) It represents the economic, social, and cultural complexity of the locality.
(b) It promotes the preservation and reinvention of local uses and customs.
(c) It is a key link in the relationship between local actors and their biological and cultural diversity.
(d) It is a platform for local ceremonies.

Evelinda Santiago-Jiménez and collaborators reflected on the sociocultural, economic, and ecological dimensions of two representative dishes of Cholula: *Chiles en nogada* and *mole* (Santiago-Jiménez and Morfín-Herrera 2016). *Chiles en nogada* is a Mexican dish of *poblano chiles*, pepper stuffed with shredded meat, fruits, and spices, and topped with a walnut cream sauce called *nogada*. Our analysis led to discarding this dish because it inherits the historical process of colonialism in Mexico. It was prepared as a tribute to Agustín de Iturbide, the general who won decisive battles to liberate Mexico from Spain. However, workshop participants pointed out that this independence was only legal and did not drive profound changes in social, cultural, political, or economic terms. Colonization remained under the rule of the independent creoles who continued subordinating Indigenous populations and cultures. Consequently, the "chile en nogada" is a dish that represents the passing of the Spanish colonial power to Creole colonial power. For this reason, we opted for the other representative dish, mole, because it represents the "tears and struggles" that the peasant and Indigenous peoples have had and maintain over time. However, when mole is consumed in restaurants it becomes a neutral product devoid of its historical ties, community networks, and ceremonial depth. In

Fig. 23.1 San Andrés Cholula Municipality in Puebla, Mexico. In the background, the "Nuestra Señora de los Remedios" Sanctuary that was built atop the Cholula Pyramid, which was considered by the pre-Columbian the Choluteca to be sacred to a female rain deity called Chiconahuiquiahuita (Nine Rains). Photograph by Evelinda Santiago-Jiménez

contrast, when mole is consumed in town celebrations, the dish acquires socio-ecological connotations.

Our innovative tourist animation activity strengthens the tradition of mole by reviving its preparation within the Cholula culture. Following the theoretical research, we designed a series of steps to organize the tourist animation workshop. The first field activity was to visit the local market to learn about the life of local people, and to buy from them the traditional ingredients for mole (Fig. 23.2). Then participants joined a local cook to learn with her how to prepare mole in a traditional clay pot. Cooks belong to a group of women called Cihuame, a Nahua word that means "working woman" (Fig. 23.3). The workshop is designed for groups of maximum ten tourists; hence, it is low impact. Finally, participants shared the mole during the re-creation of a wedding ceremony that customarily includes the mole dish for the wedding banquet.

The ecological component is based on the commitment of Cihuame women to safeguard ecosystems while carrying out their daily activities. For example, they separate the garbage, grow aromatic and medicinal herbs, trade regularly, and walk or use bicycles. This is a basis for including alternative technologies and to restore impacted ecosystems. Although these projects are potentially instrumental for

Fig. 23.2 At the local market ("Mercado Cosme del Razo") of San Pedro Cholula buying ingredients to prepare the mole and, in the same activity, showing life habit and teaching to appreciate local fruits, legumes, and seeds conserved there. Photograph by Evelinda Santiago Jimenez

Indigenous peoples to realize their dreams or to reconstruct their life projects, social complexity often causes an impasse and it seems that projects have failed. Stimulating everyone's interest is essential for the project to move forward. Local participants know that tourists will become spontaneously interested in local culture. Therefore, the challenge is to reach out to the fair-trade market where people are interested in culture, gastronomy, and meeting courageous women who are leaving a legacy for their community.

Tourism is still incipient because the number of visitors has been very low. Our workshop is conceived as a pilot for Cihuame women to train them marketing techniques and business planning, as well as ways of presenting themselves to visitors. Through open dialogue where the opinion of the women is the guideline that defines the course of action, our workshop has carefully prevented impacting their ways of life. Cihuame women's knowledge is very valuable regarding content as well as the way in which they present it to visitors and researchers. The complexity of this type of research requires long-term processes. The pace of this complexity may seem inappropriate for modern science that demands speed in generating data. However, dialogue across different forms of knowledge takes time because it is based on personal relationships.

Fig. 23.3 Cooking together. Photograph by Evelinda Santiago-Jiménez

In summary, this tourist animation revitalizes cultural expressions from the past that still have an important presence. This re-creation of the past connects with contemporary life through the participation the Cihuame women. Their gastronomic gifts enhance the appreciation of mole. In turn, their events raise money to finance the reconstruction of the San Diego de Alcalá Convent in Cholula. In this way, past heritage acquires contemporary relevance. The Cihuame women have decorated this historical monument and take care of figures and furniture that require repair. Through this reconstruction, the Cihuame collective contributes to Cholula's cultural attractions, which combine Indigenous and colonial elements in this tourism animation (Table 23.2).

The dialogue construction process of the Cihuame Project only managed to integrate some of the aspects included in Table 23.1. For example, with "local actors" it was only possible to implement "tourist information" and "cultural identity" (Table 23.1). Transdisciplinary dialogues not only involved Cihuame women but also professionals from different disciplines who developed interdisciplinary synergies. Moreover, to achieve transdisciplinary integrations, fortnightly meetings with the Cihuame collective allowed women to orient the project according to their worldviews and practices. Finally, socio-ecological literacy needs to take into account the actions of the responsible tourist indicated by the World Tourism Organization (WTO 1999).

The central question is: How to provoke the tourist to open his mind to other cultures and traditions? To address this question, a strategy based on the concept of sociocultural animation suitable for tourism was designed. It is important to note that

Table 23.2 Reconstruction of the cognitive, affective, and experiential values of the tourist destination. Adapted from Santiago-Jiménez and Morfín-Herrera (2016, pp. 108–109)

Social actors						
Variables	Local actors	Travel agencies and tour operators	Municipalities	Tourist services	Tour guides	Tourists
Tourism information	Define which components of the local culture should be shared with tourists	Promote respectful behaviors toward both nature and culture	Inform about the natural and cultural diversity and values of the place	Provide orientation about public transport and activities of interest	Organize activities that allow tourists to discover the place	Learn about local biocultural diversity, its values, and threats
A biocultural view	Recover the collective memory of the relationship between local society and nature	Tour guides must be trained about local biocultural diversity and its values	Use the territory in a way that respects its natural values and architectural heritage	Require environmental impact assessment and/or ecological/ architectonic restoration	Design activities that avoid negative impacts	Choose offers that do not cause negative environmental or cultural impacts
Fauna and flora	Provide training to recover the historical memory about the values of nature in the local culture	Do not offer tours to places containing vulnerable natural objects or that are culturally sensitive	Protect and mark vulnerable areas or species	Adapt buildings and gardens (e.g., restaurants and hotels) to encourage wildlife	Always avoid discomfort or damage	Respect animals and plants, and avoid buying them as souvenirs
Cultural identity	Identify factors causing erosion of cultural diversity, and seek ways to preserve its integrity	Guides should know deeply the culture of the host village	Enhance cultural identity so that tourism is beneficial to all local actors	Offer comfort appropriate to the local standard of living and provide traditional cuisine	Facilitate the integration of tourists into the local way of life	Reject tourism "ghettos," contact local people, and buy local handicraft
Energy	Training for using alternative energies and technologies	Offer nearby destinations and activities accessible via public transportation	Improve public transportation, promote non-motorized transport, and create pedestrian areas	Adopt energy-saving measures and create awareness programs	Avoid activities that consume fossil fuels	Use public transportation, walk, bike, or ride horseback

(continued)

Table 23.2 (continued)

Social actors						
Variables	Local actors	Travel agencies and tour operators	Municipalities	Tourist services	Tour guides	Tourists
Water	Participate in water use policy making to generate collective and mutual benefits	Do not offer destinations to places with water shortage and treatment problems	Build water treatment plants and promote water-saving measures	Implement savings mechanisms and engage customers in installing them	Avoid activities that overconsume water and/or impact aquatic environments	Save water and avoid wasteful activities
Solid waste (garbage)	Training to collectively dispose local solid waste	Use recycled paper in advertising brochures	Use separate waste for recycling, and create awareness about reuse	Reduce the use of non-recyclable packaging	Use equipment and materials that do not generate waste	Avoid non-recyclable packaging and dispose of waste according to designated containers

our tourist animation did not have the connotation of mere fun, as it is often the case in hotels where an animator entertains the audience with games in swimming pools or other spaces such as beaches. In contrast, sociocultural tourism animation is based on achieving literacy in the destination and catalyzing processes of resilience in the territory to enable the reconstruction of the life habits and projects of local actors in their own habitats and share them with visitors (Rozzi et al. 2010).

In this sense, tourism activity within a fair-trade context would make it possible to increase or create resilience in the localities, while providing literacy and recreational spaces that would not only make free time fun or entertainment, but also help show the diversity of visions and knowledge, and promote the eradication of all forms of domination, discrimination, and exclusion of cultural, social, and group identities.

23.3 Concluding Remarks

To overcome destructive practices of mass tourism, a pilot tourism activity is proposed as a dialogued pedagogical strategy based on concepts of critical pedagogy and social disciplines. This activity is consistent with the conceptual framework of biocultural ethics and the methodological approaches of field environmental philosophy, biocultural education, and conservation. The destination is recognized as a territory inhabited by stories together with unique biota and cultural practices. Our study provides an approach that incorporates two new variables: reflective literacy and resilience.

The specific case of the tourist animation "Mole, a Cholultec experience" set in San Andrés Cholula, Puebla, was carried out by a group of women called Señoras Cihuame San Andrés AC, who initially came together to recover their historical memory and restore a convent built in 1557. It is proposed that tourist activity can be a catalyst for the reconnection of members of global society with an integral meaning of being human rooted in biocultural territories. The trend toward overcrowd tourist destinations must change. There is an undeniable reality: we must catalyze an era change of era. Tourism animation contributes to this change by orienting global society to assume just relationships with the biological and cultural diversity housed in unique, and still little known and valued, corners of the planet.

References

Ander-Egg E (1989) La animación y los animadores: pautas de acción y formación. Narcea, Madrid

De Miguel S (1995) El perfil del animador sociocultural. Marcea, S.A, Madrid

Fernández R (1992) Desarrollo de la animación turística en la gerencia hotelera, Villa Coral. Trabajo de diploma, Camagüey

Freire P, Macedo D (1989) Alfabetización. Lectura de la palabra y lectura de la realidad. Paidós Ibérica, Madrid

Gallien Y (1993) Revitalización del espacio natural y urbano para la promoción de actividades turísticas y recreativas programadas. III Simposio Internacional de Turismo, Ecología y Municipio. SEDESOL/PNUMA/SECTUR/OMT/BTO-Gobierno de Sinaloa. Mazatlán, Sinaloa, Septiembre de 1993

García-Canclini N (2005) Culturas híbridas. Estrategias para entrar y salir de la modernidad. Grijalbo, México

García L (2010) Espacios de articulación, redes autogestivas e intercambios alternativos en la ciudad de Buenos Aires. Otra Economía - Volumen IV - No 6 – 1er semestre/2010, pp 68–82

Leff E (2008) Discursos sustentables. Siglo XXI, México

Lévi-Strauss C (2008) Antropología estructural: mito, sociedad, humanidades. México, Siglo XXI

Porto Gonçalves CW (2002) De geografia as geografias: um mundo em busca de novas territorialidades, en Ceceña y Sader (Coord.), La guerra infinita. Hegemonía y terror mundial, Buenos Aires, CLACSO

Rozzi R (2008) Educación para valorar la biodiversidad. Biodiversidad en la educación informal. Turismo con lupa en cabo de hornos. CONAMA. Biodiversidad de Chile. Patrimonio y Desafíos. Ocho libros editores, Santiago de Chile, pp 626–628

Rozzi R (2012) Biocultural ethics: the vital links between the inhabitants, their habits and regional habitats. Environ Ethics 34:27–50

Rozzi R (2013) Biocultural ethics: from biocultural homogenization toward biocultural conservation. In: Rozzi R, Pickett STA, Palmer C, Armesto JJ, Callicott JB (eds) Linking ecology and ethics for a changing world. Ecology and ethics, vol 1. Springer, Dordrecht, pp 9–32

Rozzi R, Massardo F, Cruz F, Grenier C, Muñoz A, Mueller E, Elbers J (2010) Galapagos and Cape Horn: ecotourism or greenwashing in two emblematic Latin American archipelagoes? Environ Philos 7(2):1–32

Santiago-Jiménez E, Morfín-Herrera MDC (eds) (2016) Animación Turística para el Diálogo de Saberes. Universidad de las Américas (UDLAP) & Instituto Tecnológico (ITP), Puebla

Toledo V (2007) Hábitat: del riesgo a la sustentabilidad. Fórum Universal de las Culturas Monterrey 2007. UNESCO, México, pp 19–59

unesco (2004) La pluralidad de la alfabetización y sus implicaciones en políticas y programas. UNESCO, Francia

Villalba C (2004) El concepto de resiliencia. Aplicaciones en la intervención social. Asociación para la promoción y desarrollo de la resiliencia. ADDIMA, Madrid. Texto completo disponible en http://www.addima.org/Articulos.htm. Consultado el 17 de julio de 2011

WTO (1999) The responsible tourist and traveller. Available via https://www.amandasettle.com/wpcontent/uploads/2019/07/responsibletouristbrochureen.pdf

Chapter 24
Collaborative Action Research with the Jotï in Venezuela: Experiences in Autoethnography and TEK Vitality Assessment

Stanford Zent and Egleé Zent

Abstract This chapter describes a relatively recent phase of long-term collaboration between local communities of the Jotï Amerindian group and researchers from the Venezuelan Institute for Scientific Research (IVIC). A sustained dialogue between the collaborating parties over ongoing threats to the Jotï population, environment, culture, and sovereignty coalesced into the realization of two overlapping and interlocking Collaborative Action Research projects: (1) an assessment of the current state of TEK vitality, through the measurement of patterns and levels of knowledge transmission across generations; and (2) an autoethnographic book, written in Spanish and Jotï, that would synthesize key aspects of their unique cultural heritage, cosmovision, environmental ethic, and resource management practices. The book project was intended to provide a documentary record and educational resource for the children and future generations of the Jotï nation, as well as a source of information about the people and their ecogonic philosophy of *jkyo jkwainï* "loving/respecting/caring for the environment" for the benefit of outsiders. The TEK vitality assessment was aimed at diagnosing the current state of health, resilience, and vulnerability of different aspects of Jotï TEK, as well as the contextual factors affecting this dynamic. The projects were conceived to run concurrently and go hand in hand. Thus, data collected on various TEK-related topics for the vitality assessment would augment the record of ethnographic material available for the book. The book in turn would turn the substantive data register compiled though the assessment into an attractive and accessible product that could be read by anyone in the community and thereby provide an immediate return on their investment. In this chapter, we recount the imbricated histories of the two projects: antecedents, justification, methodologies, data management, results, follow-up, and current status moving forward.

S. Zent (✉) · E. Zent
Laboratorio de Ecología Humana, Centro de Antropología, Instituto Venezolano de Investigaciones Cientificas (IVIC), Altos de Pipe, Miranda, Bolivarian Republic of Venezuela

R. Rozzi et al. (eds.), *Field Environmental Philosophy*, Ecology and Ethics 5,
https://doi.org/10.1007/978-3-031-23368-5_24

Keywords Collaborative action research · Jotï · Autoethnography · TEK · Loving/caring for the environment · Ecogony

24.1 Introduction

Indigenous peoples today face many challenges with respect to affirming their rights to land, language, health, and self-determination. Forging a healthy balance between the maintenance of cultural traditions and independence, on the one hand, and the innovation of adaptive strategies for coping with rapid environmental and social change, on the other hand, is crucial for meeting those goals. In this chapter, we describe our experiences with the Jotï people in southern Venezuela in dealing with these issues based on the strategy of Collaborative Action Research (Zent and Zent 2023). Early on in our relationship with the Jotï, they expressed a sense of vulnerability to losing their traditional culture, *jyedï jawa* "our things," and distinct identity, as *nï jodï* "true people," in view of the considerable changes going on and the general perception that the outside world was closing in around them. One of the points that came up time and again during our conversations was the crucial importance of the land for maintaining their culture and the culture for maintaining their land. This was not surprising given their holistic worldview. What was surprising however was their quick awareness that scientific research methods and modern information technology could help them in their quest to safeguard their cultural knowledge and habitat for future generations of Jotï growing up in a rapidly evolving situation. So, soon after receiving the news that the new national constitution, passed in 1999, granted land ownership rights to Indigenous peoples, the Jotï sought our help to demarcate their traditional territories. We eventually accepted their proposal and, from 2001 to 2006, worked together with them to plan and execute a land self-demarcation and community-based mapping project covering most of the tribal land area (see Zent and Zent 2023 for more information on this).

Besides producing impressively detailed maps and textured accounts of their land ethic and use patterns, perhaps the most valuable outcome of the demarcation project was intellectual growth of the Jotï themselves, as they became familiar with using computers and cameras, writing in their own language, recording data, and managing electronic databases. Upon witnessing the remarkable results of their newly acquired skills, they also began to grasp the potential benefits of teaming up with scientists to pursue other goals. We discussed future action research possibilities aimed at traditional ecological knowledge (TEK) documentation and transmission. In earlier fieldwork, much information had already been recorded with respect to landscape and data on plants, animals, fungi, and their habits and uses. However, there were still many knowledge domains or areas that had not been investigated in depth or at all. Those plans lingered in an indefinite state until 2009–10, when the march of surrounding events and another round of conversations brought them more into focus. The gist of those conversations can be grouped into two main threads corresponding to the distinct action projects described below, but in reality they were intertwined.

Fig. 24.1 Cover of the Jotï's community book. See the table of contents in Appendix of this chapter (cover photograph provided by Eglée Zent)

24.2 Two Interwoven Projects

The first project commenced with a call from the Jotï via short-wave radio. They asked our advice on how to write a book or, better yet, they wanted us to help them write a book about themselves (Fig. 24.1). As explained at the time, their main motive for writing the book was simply to educate people as to who they are, where they came from, where they live, what their customs are, how they take care of the

Fig. 24.2 (**a**) Forest and river landscape in the Jotï territory (left). (**b**) Family group navigating the Moyá River, Bolívar State, Venezuela (right) (Photographs by Eglée and Stanford Zent)

forest, how they make a living, what they make, what they celebrate, and what they know and think (Fig. 24.2). In other words, they aimed to do a complete ethnographic account of themselves and their history, territory, culture, ecology, economy, technology, ritual, wisdom, and philosophy. The intended audience was both external and internal. For other people, the *dododï* "clothed people/nonindigenous others," it could help to promote mutual understanding. In the context of expanding contacts with different actors of the national society—e.g., missionaries, military personnel, government officials, educators, healthcare practitioners, merchants, other Indigenous groups, scientists, law officers—they had encountered unhappy situations in which they felt that they were misunderstood and mistreated.

The book idea was hatched soon after a disturbing legal incident involving some young men from the Kayamá community who were attending school in the city of Ciudad Bolívar (see map in Fig. 1 in Zent and Zent 2023). They were accused of assaulting a woman and taken to a court of law to answer the charges. Fortunately, they were eventually acquitted on the grounds of mistaken identity. The judge in the case remarked something to the effect that it would help in the future if the government authorities and the general public were to have a better idea of who the Jotï are and where they come from in order to avoid such misunderstandings. While the legal incident could be considered a catalyst for the initial idea, it also became apparent throughout the long period of project design and execution that the Jotï had more transcendental motives for doing it. For their own people, particularly the younger and future generations, a book that summarized their immemorial wisdom would contribute to their continued socialization and persistence as *nï jodï*, considering all the time that children and adolescents spend in school nowadays and the fact that they are already well accustomed to book learning. They knew that we had spent years collecting precisely this type of information, but they also wanted the book to reflect their own words and point of view rather than our own.

To sum up the original justification of the book: it represented a strategy of survival for the Jotï people and culture moving forward in a brave new multicultural world. In effect, they were proposing to appropriate the power of the written word to defend their right to be different. Over the next few years, the matter of general objectives would be submitted to further discussions and reflections, which served to reaffirm, refine, and expand the initial ones. In sum, we are able to discern at least three major goals:

(A) To legitimize their right to exist as a culturally distinct entity within the larger national society. This act of legitimization would operate on two levels: from external and internal points of view.
(B) To create a permanent and comprehensive account of their knowledges, beliefs, and practices in order to facilitate the transmission of their most cherished cultural elements from older to younger generations.
(C) To assert their rights of ownership and use of their traditional territory and the resources on it to the exclusion of others. They would do this by creating a public record that defines the exact land area and boundaries, manifests everything of cultural significance within, and attests to their ancestral occupation and maintenance of it.

The second project or conversational thread took place in parallel and overlapped somewhat with the previous one. The research proposal involved creating a documentary record of major facets of Jotï TEK and assessing the vitality of TEK transmission. In this case, the idea for the project was originated by the researchers, but it also grew out of earlier research results and the reactions to it. The distribution of plant knowledge by age was explored during previous ethnobiological research. One of the general findings of that study was that there was a gap between younger and older individuals with respect to the ability to identify forest trees and their use values in Jotï communities subjected to greater degrees of acculturation by contact with outsiders (Zent and Zent 2004). These results were presented to the participating communities. At Kayamá, one of the schoolteachers responded by opening a class on ethnobotany which he then taught for a number of years.

The perceived problem of heritage endangerment was a recurrent topic of discussion throughout the auto-demarcation project and the idea of carrying out a more comprehensive study of TEK transmission was floated. Local receptivity to this idea was generally favorable, but nothing concrete was decided until July 2010 when we were summoned to Kayamá to discuss the state of progress (or lack thereof) of the Jotï land claim. Community leaders were peeved by the apparent lack of progress and the inability or unwillingness of the government to make good on their constitutional promise. The main conclusion coming out of those meetings was that they could no longer count on the government to guarantee their land security or cultural survival, and therefore, they themselves would have to be more proactive. The decision was made at that time to go ahead with a project focused on TEK documentation and transmission.

The methodology chosen for this project is called VITEK, which stands for "Vitality Indicator for Traditional Environmental Knowledge." The VITEK is an experimental indicator designed to measure the state and trends of TEK transmission in diverse biocultural settings. It consists of a standardized yet flexible methodology for collecting empirical data and calculating statistics on the rate of intergenerational retention/change of TEK. The indicator consists of three interrelated measures:

1. Intergenerational rate of retention (comparing the similarities/differences between adjacent age groups).
2. Cumulative rate of retention (comparing oldest and youngest age groups).
3. Annual rate of change (annualized versions of the previous statistics).

The VITEK was developed as part of a larger effort to create reliable indicators for biocultural diversity conservation (Zent and Maffi 2008). Distinct from macro-scale indicators that have been created for global or national policy initiatives, the VITEK indicator is intended for primary application and use at the local level. It can help subject populations diagnose the general state of health of their traditional knowledge, identify which knowledge domains are more/less threatened by erosion, gauge how fast changes may be occurring, and tell which sociodemographic groups are most affected. This type of information can be used to prioritize remedial actions to areas where they are most needed. It can also be employed to evaluate the impact of planned interventions, by way of before and after measurements, and to monitor the changing state of TEK transmission over an extended period, through periodic applications at regular intervals.

The VITEK design includes eight basic features.

(a) *Direct measurement*—the dynamic states and trends of TEK itself are measured rather than using proxy variables.
(b) *Quantitative measurement*—these trends are represented in numeric form.
(c) *Replicability*—the indicator is based on a standardized protocol of data collection and analysis that can be replicated in more than one place, enabling comparison.
(d) *Representative sampling*—the assessment is based on representative sampling as much as possible.
(e) *Scalability*—statistical results from more than one community or population can be aggregated to produce higher-scale measures.
(f) *Global applicability*—the indicator assessment can be applied in any biocultural context.
(g) *Local appropriateness*—it incorporates local criteria, categories, and meaning in all phases.
(h) *Gender sensitivity*—it takes into account potential gender differences.

24.3 Community Research Sites

From 2008 to 2013, a pilot study of the indicator methodology was conducted by the Human Ecology Laboratory at the *Instituto Venezolano de Investigaciones Científicas* (Venezuelan Institute for Scientific Research) (HE-IVIC) in two locations in northern Venezuela: (1) Chuao, an Afro-descendant creole community on the Caribbean coast in Aragua State and (2) Páramo los Conejos, a mixed-race (Euro-Amerindian), peasant society in the Andean cordillera of Mérida State. IVIC researchers were keen on expanding it to other biocultural contexts, especially among Amerindian populations in the south. A presentation on the work being done in Chuao and the Páramo was given for the people at Kayamá, and they agreed to participate in a forthcoming second phase of the pilot study.

The community of San José de Kayamá (*Jwabekï Jkyo*) was the logical place to carry out the projects described here. The inspiration and aspiration to do the book had originated with the people of that community. Due to time and budgetary constraints, and because the activities and subject matters overlapped, it made sense to do both projects at the same time and place to the extent that was feasible. Kayamá is the largest and most acculturated Jotï community, with a population of 640 persons in 2012. It has a comparatively large number of persons who are bilingual, literate, and computer savvy. Thus, there were many capable people who could fulfill active roles. Another important consideration was that they are well organized. They have a civil association, Jodena U̲, with elected officials, an effective decision-making process, and wide support among the population. There are also additional managerial structures that complement the work of the association, including a communal council, a healthcare team, and the school (administrators and teachers). As demonstrated during the autodemarcation project, they are efficient at getting things done. Finally, besides the Jotï there is a fairly large Eñepa population (~450 persons in 2012) living in the vicinity of the Kayamá mission. Pending their approval, we would also be able to incorporate them into the VITEK pilot study with little additional cost.

With oral agreements between the collaborating parties in hand, written research proposals were prepared to formalize the plans and seek funding. The investigation into TEK transmission and vitality was included in a multi-site, multicultural pilot study of the VITEK indicator. The indicator methodology and statistical analysis would be applied among three indigenous groups of Venezuela: the Jotï and Eñepa populations at Kayamá (Bolivar state), and the Huottöja (Piaroa) community of Betania de Topocho (Amazonas state) (see map in Fig. 1 in Zent and Zent 2023). This proposal was awarded a grant by the National Fund for Science, Technology and Innovation (Fonacit) of the Ministry of the Popular Power for Science and Technology of Venezuela. These funds covered the purchase of equipment for data recording and storage (computer, printer-scanner, camera, voice recorder, and memory sticks), travel expenses, and modest stipends for active local participants (data collectors, processors, translators, informants, photographers). The book project received generous financial support from the Ian Andrew Porter Foundation through a sponsorship by the NGO Acaté (both based in the USA). This funding was used for equipment and supplies for the project activities (computer, GPS, video camcorder, portable memories, batteries, battery charger, drawing materials) and the community (microscope, video projector, school supplies, medicines, subsistence work tools) and food and travel costs to carry out the fieldwork. It also paid for a large portion of the costs associated with the publication itself, such as the layout, art design, and print job. A second grant, from the Global Diversity Foundation, covered the cost of printing additional copies. All of the equipment was retained in the community after the conclusion of the respective projects, as they will continue to be used for building and accessing the information stored in the community TEK database.

Formal discussions and planning for the projects got underway with a series of well-attended public assemblies at Kayamá in November 2011 (Fig. 24.3). The

Fig. 24.3 Public meeting to discuss collaborative projects at Kayamá, November 2011 (Photograph by Eglée and Stanford Zent)

purpose of those meetings was to define and disseminate the precise objectives, subject matter, methods, and anticipated products of the proposed activities. Another purpose was to secure the free and prior informed consent for the projects in a manner consistent with the norms and procedures dictated by our sponsoring institution, IVIC, which meant written on paper. Both the book and VITEK projects were discussed side by side. It was agreed that work on both projects would go hand in hand, as they had overlapping aims. The vitality assessment method entails the documentation of a substantial amount of TEK data and corresponding creation of a local TEK database. The contents of the database would be used to provide material for the book. The book in turn would turn the substantive results of documentation into an attractive and accessible product that could be read by anyone in the community. Thus, both the plan and the formal agreement to collaborate on the VITEK assessment and auto-ethnography book projects came together at the same time.

24.4 VITEK Assessment

24.4.1 Methods

The Jotï module of the VITEK assessment was carried out at Kayamá from November 2011 to July 2017. It formed part of a pilot study which involved the

Table 24.1 Traditional ecological knowledge (TEK) domains studied in two components for the VITEK indicator. Each domain is expressed in English and Jotï nomenclature

Component	Domain	Jotï nomenclature
Conceptual knowledge	Plants	*Jtau*
	Animals	*Jađï*
	Interspecific relationships	*Jodena me jtuna jawa jkwa̲ï̲*
	Ecotopes	*Jkyoda jawa*
	Soils	*Ne*
	Climate	*Jkyejo jawa*
	Landscape	*Jkyodä majawana*
Practical skills	Agriculture	*Lïđï dekawa*
	Animal husbandry	*Ja̲ bajkibï dekawa*
	Hunting	*Jkyo balebï ja̲nï ju, janï dojkwe dewa*
	Fishing	*Mojto jlai dewa, me dewa*
	Collection	*Jkyo balebï jawa ulï*
	Food processing and preparation	*Jawa jkwa̲ï̲ jawa akï ja̲ï̲ đïđï dekawa*
	Ethnomedicine	*Yu dekawa*
	Handicrafts	*Jawa ja̲ï̲ bebï madä*
	Housebuilding	*Nuwe̲ ja̲ï̲ bebï madä*

experimental application of the indicator methodology in different biocultural contexts. The methodology was designed to fulfill the criteria of being universally applicable yet also locally appropriate. This means that it began with a general protocol for data collection and measurement but then was tailored to fit the particular criteria and value judgments of the local collaborators in terms of the way it was implemented. The paramount goal of the implementation process was not merely to deliver a statistical result to the participating communities but also to capacitate the local partners in all phases of the method so that they would be capable of carrying out the assessment on their own at some future date if they choose to do so. Therefore, the operational plan for implementing the method was developed in direct consultation with the community from the outset and throughout the different stages of its execution.

The first phase of the indicator method was focused mainly on the construction of a TEK database that would then serve as baseline data for the assessment. The first thing that had to be done was to define the domains of cultural knowledge that would be included. In this case, we started with a predetermined set of “cosmopolitan” TEK domains (i.e., relevant cross-culturally) and then went through a process of confirming them, eliminating them, or adapting them to the nearest local equivalent on the basis of input from the community. This technique was used because it would permit direct comparisons across the cultural groups where the indicator was applied. A total of 16 domains were chosen for the entire assessment, seven corresponding to conceptual knowledge and nine corresponding to practical skills (Table 24.1). After the domains were set, the next step was to elaborate an inventory of categorical items for each domain (e.g., plant taxa, soil types, gardening techniques or crops, medical treatments) through free listing or similar unbiased elicitation technique. This was carried out with individuals as well as small focus groups. For each category thus discovered, semi-structured interviews were conducted with cultural experts (i.e., local knowledgeable individuals) pertaining to cultural and ecological data. The interviews were structured in the sense that a set of questions was formulated for each domain. Ready-made data sheets showing the questions, written in Spanish and Jotï languages, and space for writing down the answers were prepared. They were open-ended to the extent that the interviewees were free to expand on their answers and encouraged to do so via follow-up questions.

Much of the work realized during this phase was performed by local participants. The training of the local team included at least nine basic operations.

(a) Using and caring for the equipment (computer, printer-scanner, voice recorder, digital camera, portable memory devices).
(b) Conducting the interviews with cultural experts and filling out the data sheets.
(c) Recording the interviews with the voice recorder.
(d) Scanning the data sheets and storing them on a computer.
(e) Transcribing the interview texts and notes onto Excel worksheets in the computer.
(f) Translating the texts written in Jotï into Spanish.
(g) Operating a digital camera to take photographs and short videos of relevance to the local TEK system.
(h) Backing up all of the digital files - of texts, recordings, photos - onto flash drives.
(i) Keeping a log of personnel, dates, and time spent working on the project.

The basic data collection and elaboration of the database proceeded at a convenient Jotï pace throughout the first three years of the project. Approximately 800 interviews were conducted and the cognitive and behavioral facts corresponding to 1500 distinct categorical segregates had been recorded on paper and transcribed onto Excel worksheets. Only about half of these entries were translated into Spanish. The database also had >400 digital photographs and >100 interview recordings.

The next phase of the indicator development concerned the preparation of a TEK aptitude test (TEKAT) (Table 24.2). This test would be administered to a sample of persons of different generation groups and the results of the respective groups compared to determine rates of retention/change across generations. The first step was to determine the test composition by asking community focus groups to assign relative importance weights to each domain. Thus, the number of questions pertaining to a particular domain reflected the relative weight assigned to it. A subset of categorical items was randomly selected from the total inventory of items in each domain, and the questions drawn up for the test were based on those selections. In some cases, stimulus materials such as photographs and drawings were used for the test. Several versions of the TEKAT were elaborated to vary test exposure and hence preserve independence between test takers. Every test version was written in both Spanish and Jotï languages.

Table 24.2 Complementary methodological approaches and phases of doing Collaborative Action Research, CAR, among the Jotï (Preliminary CAR; Vitality Indicator for Traditional Environmental Knowledge, VITEK; Autoethnography)

	Methodological approach		
Phase	I. Preliminary CAR	II. VITEK	III. Autoethnography
Research/ Data collection	a. Population census	a. Definition/identification of local, culturally specific TEK semantic or behavioral domains (e.g., plants, animals, soils, ethnomedicine, food preparation)	a. Selective incorporation of data from previous research projects (ethnobiology, behavioral ecology, community mapping, landscape construction, ethnomedicine, language study).
	b. Participant observation in daily activities	b. Free listing of categorical items pertaining to each domain	b. Selective incorporation of data from TEK memory bank
	c. Casual conversations and unstructured interviews	c. Semi-structured interviews with cultural experts pertaining to cultural and ecological data about each categorical item	c. Personal histories on specific topics (first interethnic contact, residential history, health experiences, rituals, etc.)
	d. Plant and fungal specimen collections (~3400)	d. Domain weighting exercises to determine relative importance value of each domain	d. Construction of album of culturally significant videos and photographs
	e. Structured ethnobotanical interviews within forest plots (169)	e. Community census updates	e. Student hand drawings of cultural themes and scenes
	f. Resource harvest accounting (1074 days)	f. Reproductive histories to determine population-specific generation interval	f. Updating GPS data points
	g. Focal follow observations (90)	g. Questionnaire on relative age at intellectual maturation	g. Revision of native language text portions
	h. Cultivated plant inventories in gardens (125 inventories)	h. Collection or preparation of stimulus materials (e.g., specimen samples, drawings, photographs, sound recordings) for TEK aptitude test (TEKAT)	h. Revision of contents with schoolteachers and local cultural experts
	i. Geographic coordinate registration of culturally significant places and spaces (>5000 records)	i. Stratified sample selection of TEKAT test subjects by age group and gender	i. Final draft review and approval
	j. Mental maps of landscape	j. Random sample of categorical items for making up questions on TEKAT	

(continued)

Table 24.2 (continued)

Phase	Methodological approach I. Preliminary CAR	II. VITEK	III. Autoethnography
		k. TEKAT test administration	
Training of local collaborators	a. Consensus alphabet learning	a. Interviewing cultural experts (voice recording and writing answers on data sheets)	Photo and drawing selection for the book
	b. Local language literacy	b. Transcribing and translating interview data on computer	Book manuscript line editing and copyediting
	c. Basic computer use	c. Use of electronic equipment (basic operation, data backup and scanning, digital data storage)	
	d. Computer program use (Windows OS, Word, Excel, GIS)	d. Physical specimen collecting	
	e. GPS use	e. Photography	
	f. Filling out data sheets	f. TEKAT test administration	
		g. VITEK statistical calculations	

The test administration was accomplished in the first quarter of 2016. About 90% of the tests were administered by local participants while the IVIC-HE team members directly oversaw the entire operation. Before the tests were administered, each and every question was confirmed for comprehension and accuracy by an exhaustive review process involving the combined efforts of literate, bilingual assistants, and cultural experts. The population of test takers was determined by drawing a stratified sample (grouping variables: gender, age group) from the community census. A total of 96 subjects successfully completed the test. For analytical purposes, we divided this population into eight sociodemographic groups: female and male adolescents, young adults, mature adults, and elderly. The figuring of descriptive statistics based on the test results (e.g., means, standard deviations) was sufficient to derive the VITEK statistics and make a general assessment of the major trends of TEK transmission in the community.

24.4.2 Results

Eleven main results were found. (1) At the level of the total test, average TEK competencies are declining steadily across successive generations but at a very mild rate (10–20%/generation). (2) Even though Kayamá is one of the most acculturated Jotï communities, the rather mild level of overall decline is most likely a consequence of their continuing self-sufficiency in subsistence matters. (3) Competencies are more stable (i.e., higher retention) among females. (4) Young adult males in particular exhibit the largest gap in comparative competencies. (5) This gap is most evident with respect to practical knowledge domains (i.e., activity-based skills). (6) Young male performance is mildly lagging in the traditional subsistence domains of cultivation, hunting, and gathering. (7) The performance of this demographic group is on par with their older counterparts with regard to fishing and animal husbandry, which are subsistence activities marked by recent innovation. (8) Diminished participation and time spent in traditional subsistence activities is probably a major factor in this result. (9) Dedication of young males to school and related extracurricular activities (e.g., sports games, mission infrastructural works) is another conditioning factor. (10) Potential TEK conservation actions should target young males and prioritize expanding opportunities or incentives for acquiring practical knowledge. (11) If possible, it makes sense to try and integrate such actions into the school setting.

The complete results, including all of the VITEK statistical measures domain by domain, were presented to the community in 2017. Digital copies of the final version of the database were also handed over at that time. A lengthy discussion was held regarding what the next move would be, and it was decided that the first line of action would be completion and delivery of the autoethnography book. We also talked about eventually preparing study materials for use in the local school. However, about the same time as the book came out in print, the COVID pandemic hit and the community school shut down, and is still closed at the present time.

24.5 Building an Autoethnography and Writing a Book

The scope and contents of the book were defined for the most part by the Jotï themselves with some input from the authors of this chapter. Like most decisions taken by the Jotï at Kayamá in regard to collective matters, this one was arrived at by consensus through open discussion during public gatherings. At one of the planning meetings, a whiteboard was set up and the tentative table of contents was discussed and written down, chapter by chapter (Appendix). The initial outline underwent a few modifications as time went by and the work unfolded, but the same basic *modus operandi* of collective consultation was maintained throughout the process. The main modification to the original table of contents was the decision to add a chapter on territory and the autodemarcation project. Other key decisions made were the following.

(a) The book would be written mostly in Spanish but with certain parts in Jotï. The Jotï astutely recognized that it was necessary for the book to be written in Spanish to reach an external audience and therefore achieve its educational

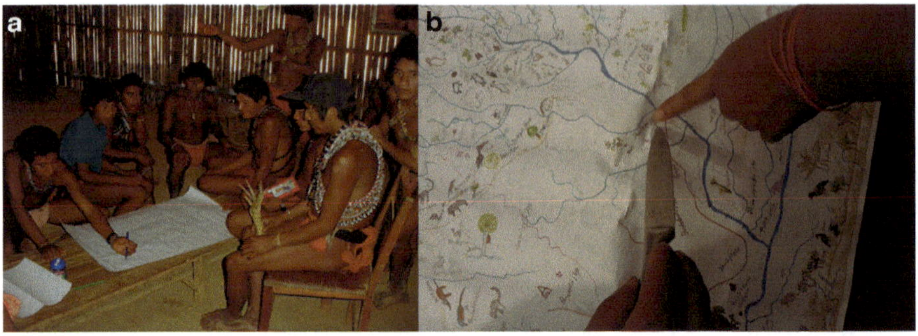

Fig. 24.4 (**a**) Group of Jotï working on a mental map for the auto-demarcation project (left). (**b**) Detail of the community map showing landscape features (right) (Photographs by Eglée and Stanford Zent)

mission. They also understood that this would pose a constraint for them. For these reasons, they requested the researchers' literary assistance.

(b) Egleé Zent would be the lead author.
(c) Stanford Zent would work with bilingual schoolteachers to review and edit the Jotï language content.
(d) Alirio Juae Molo was designated as the community coordinator and liaison for the project.
(e) Community elders and others would be available to provide testimony and information about selected topics if needed.
(f) Data and information collected previously as well as during the realization of the TEK vitality and documentation project could be included in the book.
(g) Schoolchildren would make drawings of cultural themes for inclusion in the book.
(h) Photographs and illustrations would be selected through a joint review process.
(i) Maps showing land uses and borders would be updated to reflect any significant changes that had occurred since the end of the auto-demarcation project (Fig. 24.4).

The first draft of the manuscript was reviewed word by word with members of the community at Kayamá in July 2017. Subsequent drafts were also relayed by various means to the community for their revision and correction. The final product was a tome of 525 pages, published in 2019 by Ediciones IVIC under the title *Nï jotï aiye: jkyo jkwainï / Libro comunitario jotï: historia, territorio y vida* (Zent et al. 2019). It contains 223 pictures, 31 hand-drawn illustrations, 15 tables, 3 maps, and 6 appendices. Besides the two authors of this chapter, there are 206 Jotï co-authors whose names appear after the title page, giving credit only to individuals who played a substantial role in its making, either as direct sources of testimonies or information or as reviewers during the editing process. A large number of other individuals also contributed with illustrations, photographs, or translation skills. Considerable emphasis was placed on letting the Jotï voice(s) be heard as much as possible. The main text includes 1389 direct or paraphrased narratives or quotes by many different Jotï speakers. The book is divided into nine chapters: introduction, population, territory, history, subsistence, ritual, medicine, environmental ethic, and contemporary situation and future outlook (Appendix). A large portion of the printed books were destined for distribution to Jotï communities for their internal use and for re-distribution to whomever they decide. Due to severe travel restrictions imposed by the government when the pandemic broke out, it was not possible to deliver on this commitment right away. The travel situation eased in 2022 and we were then able to hand out 150 copies to individuals from at least a dozen Jotï communities in Amazonas and Bolívar states.

By virtue of its collaborative makeup, the book could be considered an unconventional autoethnography or hybrid ethnography, reflecting, on one hand, Jotï culture, ecology, and history from a native perspective and, on the other hand, striving to maintain a high academic standard for cultural description. This was achieved by interspersing verbatim phrases or narratives by local collaborators with synthetic statements and interpretations by the academic coauthors. The range of material appearing in the book was drawn from different research projects, of both the conventional and the collaborative kind, carried out among the Jotï during the last 25 years. The principal data collection or compilation methods as well as training phases that went into this effort are summarized in Table 24.2.

Three aspects of the broader significance of the book for the Jotï and others deserve special mention. The first is the emphasis on the concept of *jkyo jkwainï*, which refers to "respecting/loving/caring for the environment." This idea constitutes a fundamental principle of the Jotï ethos and underlies different aspects of their social and ecological behavior. It implies that the health, harmony, prosperity, and survival of people and everything else in the world depend on the exercise of proper attitudes and actions toward others, including not only human beings but also nonhuman living beings, nonliving entities, and spiritual beings. The basis for proper treatment in turn is the assumed truth that all beings are connected. Such connectivity is considered to be polyvalent in nature: physical, spiritual, social, and emotional (Box 24.1). The intimately spiritual dimension of *jkyo jkwainï* is amply expressed in a written statement on the basic rules and principles of Jotï territoriality drafted and signed by Jodena U officials and other community leaders in April 2010. An excerpt from this document states:

> These jungles, mountains, rivers, and rocks are the souls in which we live, they are the heart of our community. It is the principal point, the living soul. If this heart dies, we will all die (Zent et al. 2019, p. 468; our translation).

This is in effect a succinct statement of biocultural conservation.

The second aspect is that the codification of traditional oral knowledge in a nontraditional written form represents the appropriation of foreign technologies of information for purposes of expanding native enculturation possibilities. The use of a book as an educational resource for younger generations of Jotï is meant not to replace but rather to supplement customary means of cultural knowledge transmission. This act of conversion adds value to local knowledge precisely for that sector of the population—schoolchildren and literate adults—which is more likely to consider the knowledge forms of the dominant society, such as print and digital media, to be more prestigious. In a similar fashion, coauthoring a book also confers a literary status on the Jotï in the eyes of the national society at large. This is an example of the adaptive capacity of TEK and how it can evolve to survive.

The third thing is the decision to devote an entire chapter to the territorial autodemarcation process, including the publication of the maps and land boundaries for the first time. In the absence of official land rights recognition by the government, publishing the results of the demarcation work (e.g., the map and description of land use and natural resource management by the Jotï) effectively brings it into the public domain and thereby confers some measure of legitimacy to their claim. A detailed description of the autodemarcation project in all of its phases and methods is also provided to be completely transparent about how the map was produced and who did the work. As a published document that is potentially available to a national and international public readership, it acquires a quasi-legal status that explains why they effectively have consuetudinary rights to occupy those territories as well as to defend them from potential threats by outsiders. This reflects the significance of knowledge as a tool for political agency (Fig. 24.5).

Box 24.1 Part of the Jotï Hunter's Craft: Connecting with the *Aemo* "Guardian Spirits" of Game Animals

Hunting is a main subsistence activity of the Jotï (Fig. 24.6). A hunter who loses their ability to shoot or capture a certain animal type is thought to have lost their amicable connection with the animal's *aemo* "guardian spirits." This could have been caused by a failure to follow certain norms and procedures when the animal was last hunted (e.g., not kill more than one needs, maintain a serious frame of mind, say words of blessing, deposit discarded organs properly, put skull on display after consumption). The animal *aemo* responds to the transgression by cutting off the supply of their wards to the hunter.

To recuperate their connection and hunting ability, the hunter must take an infusion made from the inner bark of several tree species. This material act has an effect on a spiritual plane: the botanical *aemo* are family or friends of the animal *aemo* and they intercede on behalf of the hunter's *namodï* "spiritual essence" to recover the good relationship. Spiritual recovery translates into material recovery and the hunter is once again able to kill the prey species.

24.6 Discussion

We presented two interrelated projects focused on the problematic of TEK transmission: the VITEK indicator assessment and the community autoethnography *Jkyo jkwainï*. The first project entailed the measurement of intergenerational rates of TEK retention/change among a stratified sample of subjects for the purpose of diagnosing

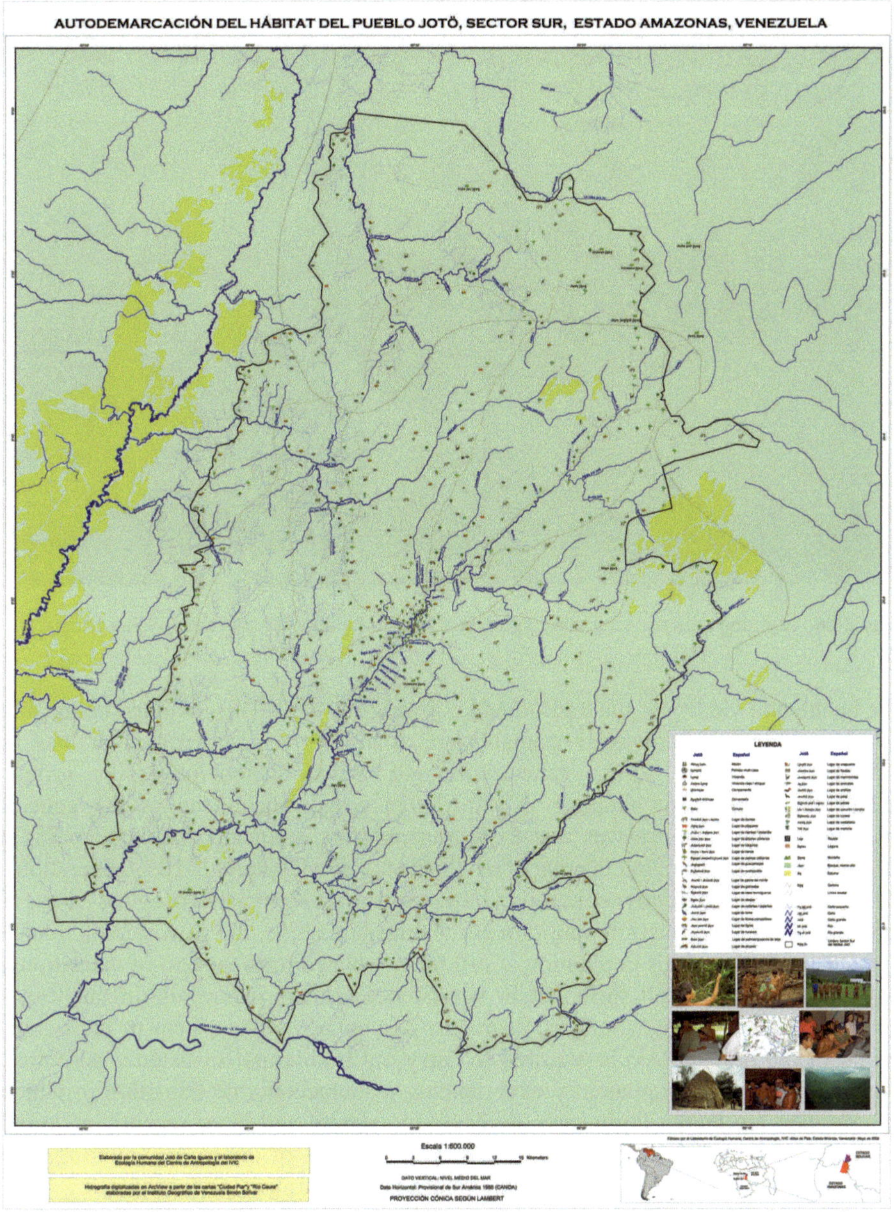

Fig. 24.5 Autodemarcation map of the Jotï habitat (northern sector) elaborated by local communities together with the Human Ecology Laboratory at the Instituto Venezolano de Investigaciones Científicas (HE-IVIC, Venezuelan Institute for Scientific Research)

Fig. 24.6 A young man carrying a blowgun, ready to go hunting (Photograph by Jan Dungel)

the health and vitality of the traditional knowledge system. This included being able to identify which knowledge domains are at highest risk of erosion or loss of information content across generations and which sociodemographic groups manifested significant slippage in knowledge competence. One of the general findings of this exercise was that TEK transmission at Kayamá is experiencing a broad trend of erosion, but the rate of decline is nonetheless mild (~10%/generation). This result is apparently conditioned by some of the social-ecological changes associated with the missionizing process. At the same time, the slow rate of change reflects the fact that this population is still fairly isolated from the rest of the national society and mostly self-sufficient in subsistence matters. More significant losses were registered among young adult males, particularly with respect to their command of the practical skills required to carry out traditional subsistence activities. This demographic is apparently experiencing a socioeconomic transition in which young men are increasingly more dedicated to school and other recently acquired pastimes and less dedicated to the customary pursuits of gardening, hunting, and gathering. Based on this assessment, one possible strategy for reinforcing TEK transmission where it is most needed would be to enhance learning opportunities in the traditional occupations through nontraditional means (e.g., incorporation of traditional vocational subjects into the school curriculum, virtual lessons, more flexible school schedules).

Beyond producing measures of the state and trends of TEK vitality at a given moment, other positive outcomes of the process included TEK documentation and local capacity-building. The creation of a digital database of 1500 TEK info bites preserves in readily accessible form fragments of the wisdom of community cultural experts and elders that might otherwise be lost upon their passing. Putting this information in writing and then onto a computer stands to benefit young people in particular since they are the subgroup most likely to use and learn from this new technology and also manifest greatest risk of TEK decline. A large cross section of community members participated actively in most phases of the project development. The only activities in which local team members did not participate was with the mathematical calculations used to figure the indicator statistics and the creation of graphic images for viewing the results. This gap in the training will be addressed in the future, pending clear indications from the community that they want to continue with this process.

Given the collaborative design, the execution of the work plan required significant investment in the training component and added significantly to the overall length in time from start to finish. However, one of the primary goals of the project was precisely to build local community capacity for conducting their own vitality assessments, or similar type of inquiry, in the future. The only real possibility for carrying out this type of diagnostic assessment recurrently or over the long term, for example to measure the impact of documentation and transmission reinforcement initiatives, is for the people themselves to take the lead or do it independently.

The autoethnography project involved the making of an illustrated book, available in digital and print form, which tells the story of their philosophy and lifeway from a distinctively Jotï point of view. It amounts to the creation of a new information product from the traditional knowledge passed down and evolved through countless generations of forebears. Material for this literary work came from the database compiled for the indicator as well as other data and information recorded during previous research and collaborative endeavors with the Jotï. It is destined for two main user groups:

(a) The Jotï of Kayamá and other communities
(b) The non-Jotï population within the country and beyond

For the former, the proximate goal for the book is to educate the children and future generations of the Jotï community about their biocultural heritage. This represents a conscious attempt to enhance traditional knowledge transmission by leveraging social changes—literacy and bilingualism—to foster cultural survival and environmental sustainability. For the latter, the intention is to inform the general public in regard to Jotï culture, ecology, and ethics, and thereby promote better communication and comprehension in their dealings with outsiders.

The inclusion of detailed information about their territory, the autodemarcation work, and their persistent efforts to gain legal recognition of their land claim in a published document represents a public declaration of their rights. They are effectively appropriating a nontraditional resource (the written word) to protect a traditional resource (the land they occupy) within the surrounding power structure (the national constitution and subsidiary laws). The book, viewed as both a process and a product, brings the enterprise full circle. Moreover, the collaboration between community members and academics inverted the traditional roles. The local partners came up with the original idea and made the major decisions in regard to the content, use, and distribution. The researchers served as advisors, providers of information, and facilitators of communication to another world. While the description of this case study will hopefully serve as an example that others can learn from, we are not suggesting that it be replicated. Collaborative action research by nature strives to be attentive to the particularities of the people and their environment.

Appendix: Table of Contents of the Book Nï jotï aiye: jkyo jkwainï/Libro comunitario jotï: historia, territorio y vida.

1. ***Ñamuliyë*/**Prologue (how and why the book was created).
2. **¿Jwaidï nï jotï da?/**Who are the Jotï, Jodï, or Jotö?
    - *Jibï ayadï da/*How many are there?
    - *Jteböna mee̱ dekadï da/*Where do they live?
    - Contact period, 1970–1990.
        - *Jkąyǫ Ijkwąlą* /Caño Iguana, Amazonas State.
        - *Jwabekï jkyo/*Kayamá, Bolívar State.
        - *Bǫlojkojto/*Río Asita, Amazonas State.
3. ***Jkyo*/**Territory or environment.
    - Territorial auto-demarcation: process and progress, 2000–2017.
4. ***Jo baede jawa - ñamuliyë jotï*/**Narratives about the ancestors and first people.
    - *Baede jadï jkyo jawa/*Ancient history.
    - Cosmogony.
        - *Ne/*Earth, the physical setting of life.

          *Jkyo/*The Forest, living environment.
          *Jkyo a̱e̱mo/*Guardian spirits.
        - *Au/*Water, the sap of life running through the world.
        - *Jkyö/*Air, like the guardian spirits, we cannot see it but cannot live without it.
        - Jkulë/Fire, for cooking food and warmth at night.
        - *Jkyo bëjkya/*Cosmos.
    - Anthropogony.
        - *Jti̱nëwa̱/*Sun, the person energy of the earth.
        - *Yowale/*Opossum, humanity from the inner earth.
        - *Jlae, jkyomalïdëkä/*Divine creators and many more, different but equals.

  - *A̱mia̱*/Moon, nocturnal light.
  - *Jo*/Human person.

    Ijkwöju/Heart, soul.
    Inëja̱/Physical body, what the person looks like.
    Namodï/Spirits.

- Ecogony.
  - *Jawa*/Food of humankind.
  - *A̱jta̱i*/Palms, ancestral and current food.
  - *Jkyo jtau ïdï*/Cultivating wild trees.
  - *Jkwë̱ jtawï jka̱jka̱*/Primordial tree of food.
  - *Ja̱nïka̱ baloka*/Game and gardens.
  - *Ijtïedema jadï*/Sharing/generous people.

5. ***Jkyo balebï, jawa balaijkyë***/Daily sustenance, hunting, gathering, horticulture.
   - *Au wai, abuwe wë, jobe*/To drink, dream, and sing as life strategies.
     - *Au wai*/To drink water, liquids.
     - *Au ibu, au dïlï*/To bathe, wash oneself.
     - *Nana duwi*/To paint body with resins.
     - *Abuwe wë*/To dream.
     - *Jobe, o̱nemai̱, jtajwibo*/To sing and play flute.
     - *U̱dïjte*/To teach, socialize.
   - *Jkyo balebï ja̱nï ju ja̱nï dojkwe dekae bada*/To hunt with blowgun or lance.
     - *Jwa̱na*/Blowgun.
     - *Jwaiyë-alëjtä*/Darts.
     - *Malawa*/Curare.
     - *Lujkuwi*/Lance.
     - *Yaya-alëjtä*/Snare traps.
     - *Jnimëla*/Bow and arrow.
     - *Yëwi*/Dog.
   - *Ja̱nï ju jtobaibï, ja̱nï dojkwe jtobaibï*/Hunting with blowgun and lance.
     - *Waña*/Gall bladder.
   - *Ju dekae*/Hunting with blowgun.
   - *Jkyo balebï*, *Jawa jamu jtoba*/Gathering.
     - *Jtau*/Plants.
     - *Badebodï*/Palm larvae (Coleopteran).
     - *A̱mië̱dï*/Seed larvae (Coleopteran).
     - *Mae̱na*/Bee honey.
     - *Ñajki̱*/Fungi.
     - *Abajladï jadï*/Other animals.

- *Jkyo balebï, mojto̲ jlai*/Fishing.
- *Jawa balaijkye*/Horticulture.
  - *Jkyo balo, mana balo, mana ajkunë lïdï dekae*/Trail garden, forest garden.
  - *Nuwë̲kï jwëla majae balaijkyë*/Dooryard gardens.
  - *Balo*/Swidden.
- *Baede balo u̲dïdewa*/Ancient narratives about cultivation.
  - *Ñamuliyë uli yëwi*/Jaguar, the first horticulturist.
  - *Uli jkajwiyë*/ peccary, the horticulturist.
  - *Jkyomalïdëkä*/The divine creator as horticulturist.
  - *Jkawai*/Tobacco.
- *Balo ja̲ï, balo bake, Balo bake u̲ wë*/To make a new garden, select a plot.
  - *Balo na̲ï, jtau jwëu*/To slash and clean the garden area.
  - *Balo bude*/To burn the garden.
  - *Balo majawa lïdë*/To plant the crops.
  - *Balo wae*/To weed the garden.
  - *Imëe̲*/To harvest the crops.

6. ***Jkye wibï majkyö***/Ceremonies, ancestral rituals, and celebrations
   - *I̲nimonï jkye wae majkyö*/Rituals for infants and children.
     - *Jtido jadï*/Infants and small children.
     - *Jtido-ja̲nï au ijte*/To bathe the newborn child.
     - *Jtidoa̲ ikyunëka-jojo*/To put a belt on the baby.
     - *I̲ni jkwanakadä ijtï*/To put necklaces on the child.
     - *O̲neka-jelë ilïjto ikye*/To pierce the ears.
     - *Dodo jkajkyëu jtï dekae*/To put on the loincloth.
   - *Bae jodï jkye wibï majkyö*/Rituals for adolescents.
     - *Ñonakalë*/To pierce the septum and insert nasal stick.
     - *Jkawai jkajtïjto ikye*/To give (first) tobacco.
   - *Bëkyadï jotï jkye wibï majkyö*/Rituals for everyone.
     - *Ja̲nï ju jtoba ibï, ja̲nï jkejto ikye*/Hunting ritual, calling the animals' guardian spirits.
     - *Miñelinïdë jadï*/Mortuary ritual.
     - *Miñeli*/Journey to the afterlife.
     - *Budejkï yalu jadï ñajadï iledaibïjteibï jkyewayakï*/Mourning ritual.
7. Yudeka jawa, jibidï deka jawa/Traditional Medicine
   - *Jti jadï*/Healthy, beautiful, and good.
     - *Ñaja̲ baebï dëwa*/Origin and cause of diseases.

- *Ña jadï*/Predators, bad people who eat, harm, or make others sick.
- *Aweladï*/Malevolent beings.
  - *Ajkukë*/Spider.
  - *Ijti*/Scorpion.
  - *Ejko*/Snake.
  - *Yëwidï*/Jaguars.
  - *Abaladï ña jadï*/Other evil agents (*Awela, Bulujadï, Jkïmekaja, Jkodo boma ja, Uli ja ñëdoja, Jkwañuwë o Yajkeba, Iye, Jkyo Aweladï, Jtujeladï jadï, Jidï jadï, Jkwa, Lojkoi o Jwaidï, Manajadï, Mëjna jtojwali, Ñëdoja, Uli Bolë, Uli Jkwayo jawabo, Uli Jtujkuli, Waijlo*).
- *Jti jadï*/Benevolent beings, stewards, builders of goodness.
  - *Ajkuli*/Agouti.
  - *Jkajojadï*/Enlightened and wise persons.
  - *Jkajwiyë mowali*/The peccary chiefs.
  - *Jkyomalïdëkä*/Eternal, good, divine being of the sun (house).
  - *Jkyo ae o Jlae*/Eternal, good, divine being of the mountain (house).
  - *Jkyo aemo*/Guardian spirits of the forest.
  - *Ñowali bujkajadï*/Benevolent teachers from the beginning of time.
  - *Uli Jkonojtodï*/Giant oriole.
- *Ñajade bae ibï ikye*/Being ill.
- *Namodï luwëde bae ibï*/Means by which the spirits leave the body.
  - *Baebï dëwa jtowekawa*/Uncontrollable or unmanageable causes.
  - *Baebï dëwa bajkewa*/Controllable or avoidable causes.

    Ñëdojae/Poison.
    Yu/Blessing of food or drink.
    Nuwe jkwa/Taking shelter in the house.
  - *Baebï dëwa anuwëdïdekawa*/Causes for which the person can learn to control or avoid.
- *Yudeka jawa*/Traditional medicine.
- *Jkwai-wekï, Udï bebï, Jodaikë maï*/To prevent, favor, correlate (ontogeny).
  - *Jwai*/Birth.
  - *Ini*/Childhood.
  - *Bae jadï*/Adolescence and young adulthood.
  - *Miñelinïdë-ja*/Death.
- *Jtijadï baebï yu*/Healing.
  - *Yudeka jadï*/Healers.
- *Yudeka jawa*/Medicines.

- *Baede ñaja̱baebï*/Ancestral diseases.
- *Jti̱doa̱ jwai miñeli*/Newborn death, stillborn.
- *Jkwa̱junëebï ño dulu*/Cough and flu.
- *Ja̱ni diyënï ma, jawa jkwa̱ï̱*/Food taboos during menstruation.
- *Jti̱doa̱ ñaja̱ jkelau*/Upset and crying baby.
- *Jkulë jkye̱jo̱*/Fever and chills.
- *Jtu julëe*/Headache.
- *Ekyo jkwa julu*/Stomach ache.
- *Ñëdo̱wa*/Poisoning.
- *I̱ni ñaja̱*/Sick child.
- *Jkwa̱ julu*/General pain, stings, and bites.
- *Duwëwe jae*/Erysipelas.

8. ***Jkyo jkwainï***/Caring, loving, respecting, and conserving the environment.

   - *Jtija̱ mee̱, joja̱ mee̱*/Living well, doing good, staying healthy, being beautiful.
   - *Jkïmañë*/Sacred sites and spaces.
   - Decalogue of territoriality and ethics for the good life.

9. ***Bae anuwëdï majkyö, jta i̱nï***/Today, the future, and farewell message.

   - *Bae anuwëdï majkyö*/Contemporary life and future.
   - *Jwaï dekae nuwe*/The school.
   - *Jti jadï baebï dekae nuwe̱*/The health clinic.
   - *A̱e̱modï, jlaedï, ja̱ï̱ jibidï*/Wise persons, leaders, social organization.
   - *Jta i̱nï*/Farewell message.

References

Zent S, Maffi L (2008) Final report on indicator no. 2: methodology for developing a vitality index of traditional environmental knowledge (VITEK) for the project 'global indicators of the status and trends of linguistic diversity and traditional knowledge.' Terralingua. Available at: https://ivic.academia.edu/Zent

Zent S, Zent E (2004) Ethnobotanical convergence, divergence, and change among the Hoti. Adv Econ Bot 15:37–78

Zent S, Zent E (2023) Collaborative action research for biocultural heritage conservation. In: Rozzi R, Tauro A, Wright T, Avriel-Avni N, May RH Jr (eds) Field environmental philosophy: education for biocultural conservation. Ecology and ethics, vol 5. Springer, Dordrecht, pp 229–246

Zent E, Zent S, Jotï N, Jodena U (2019) Nï jotï aiye: jkyo jkwainï/Libro comunitario jotï: historia, territorio y vida. Ediciones IVIC, Caracas

Chapter 25
Hand-Print CARE: Intergenerational and Plural Knowledge in Schools

Rob O'Donoghue and Juan Carlos A. Sandoval-Rivera

But, in the end, knowledge can fulfil its orientation function only if the voice, and thus its communication function, is brought back to life. (Elias 2011: 158)

Abstract This chapter is a critical review of Hand-Print CARE toward the inclusion of intergenerational knowledge and ethics in contemporary curriculum settings as plural knowledge mediation environments. The Hand-Print concept emerged as a proposition for learner-led action learning in the Centre for Environment Education, India. "CARE" emerged as an acronym reflecting an ethic of inclusive respect through ***C**oncern* for others, being ***A**ttentive* to needs, showing ***R**espect* for each other, and being ***E**ngaged* in learning actions for the common good. Hand-Print CARE was thus activated as a co-engaged mediation process toward "Learning to look after others to best care for ourselves and the surroundings we all share." The Hand-Print CARE program informed a South African setting of course-supported teacher education developed in collaboration with a community-engaged research program in Mexico. The co-engaged research process was used to explore the inclusion of Indigenous knowledge practices so as to enhance learning in school subject disciplines. Working from a cultural historical perspective in environment and sustainability education, a process of immanent critique is used to review Hand-Print CARE materials and practices of course-supported design research collaboration and the inclusion of intergenerational heritage in locally relevant themes for teaching in conventional subject disciplines.

Keywords Intergenerational knowledge · Indigenous knowledge · Sustainable living · Inclusion

R. O'Donoghue
Environmental Learning Research Centre, Rhodes University, Grahamstown, South Africa
e-mail: r.odonoghue@ru.ac.za

J. C. A. Sandoval-Rivera (✉)
Instituto de Investigaciones en Educación, Universidad Veracruzana, Xalapa, Mexico
e-mail: csandoval@uv.mx

R. Rozzi et al. (eds.), *Field Environmental Philosophy*, Ecology and Ethics 5,
https://doi.org/10.1007/978-3-031-23368-5_25

25.1 Introductory Overview

The Hand-Print concept emerged as a proposition for learner-led action learning in the Centre for Environment Education, India. "CARE" emerged as an acronym reflecting an ethic of inclusive respect through ***Concern*** for others, being ***Attentive*** to needs, showing ***Respect*** for each other, and being ***Engaged*** in learning actions for the common good. This brief sketch scopes the terrain on which Hand-Print CARE developed to integrate diverse knowledge for Indigenous livelihood contexts and for clarifying sustainable living in educational settings.

The study presents case studies of Hand-Print CARE from Latin America and Southern Africa with local Indigenous peoples where knowledge practices closely align with habits and habitats. In the southern Africa context of this study, much of the natural science knowledge related to local landscapes was appropriated into the colonial institutions (Shava 2008). The situation is similar in the Mexican context (Jiménez-Naranjo and Mendoza-Zuany 2016) regarding the exclusion of Indigenous knowledge from schools due to education systems and colonial policies that persist in Indigenous education. The histories of colonial and emancipatory struggle in Africa, Latin America, and elsewhere that have produced modern school curricula have neglected three intermeshed things:

- Abstract knowledge of modern subject disciplines has been constituted in object congruent but exclusionary ways.
- Barriers to abstract knowledge generate exclusions of local knowledge for young learners today.
- Institutional patterns of abstraction that constituted more "object congruent" knowledge exclude cultural histories related to local habitats, habits, and associated sustainability practices.

This analysis reveals that accompanying the exclusionary injustices of colonial domination and institutional appropriation is a disjunction between knowledge providing orientation for sustainable living and the communicative processes to achieve this. Noel Gough (1999) reminds us that the curriculum is nothing more than the stories we tell the children. Here, it is important to underline the powerful nature of the stories that are represented in the curriculum (Lotz-Sisitka et al. 2013), as well as those that are excluded. The epistemic problem is especially significant among Indigenous communities who have suffered the disjunctures of colonial abjection and the associated exclusions while having to be successful in a modern education system dominated by the learning of abstract propositions that are difficult to relate to local habitats, let alone use as a means of orientation in a changing world. Incorporating local knowledge into the school curriculum resolves this problem by taking seriously the ancestral wisdom that has developed in Indigenous contexts from the knowledge of their own social and ecological environment, of the strategies of adaptation to the biocultural environment, and more recently, of the processes of innovation and adaptation to the effects of climate change.

Four locally relevant Hand-Print CARE topics are reviewed to scope some of the methodological and ethical dimensions in a more inclusive and integrating approach to Indigenous knowledge and modern scientific knowledge related to:

1. The traditional milpa crops and nutrition in *Nahua* Indigenous communities in Mexico.
2. *Nguni* cattle and herding for sustainable pasture management in South Africa.
3. Intergenerational knowledge practices related to regional climate change among the *Xhosa* in South Africa.
4. Cultural practices of hand-washing to avoid 2020 COVID-19 among the *Nguni*, South Africa.

25.2 Background on Hand-Print CARE and the Epistemic Paradox in Modern Schooling

Within these histories of exclusion, Hand-Print CARE has centered on the development of locally relevant themes in learning materials to foster the inclusion of intergenerational heritage in school curriculum settings. Clearly, subject content knowledge is part of this heritage. However, this is seldom communicated as, and with, an intergenerational heritage that links local habitats and life habits in ways that are relevant to Indigenous and rural children learning scientific knowledge at school. The pedagogical challenge is, on the one hand, the restoration of voice and legitimate heritage in Indigenous knowledge within curriculum settings. On the other hand, the challenge is to create emancipatory objectives in which education "foster [s] in learners/participants a critical stance toward the world and oneself by promoting discourse, debate and reflection" (Wals and Dillon 2013, p. 254). O'Donoghue (2019) explored how the curriculum delivery of abstract knowledge was a barrier to the acquisition of knowledge related to everyday cultural life in meaningful ways. This was especially true for Indigenous children who found it difficult to appropriate abstract curriculum knowledge in a culturally relevant way so as to inform them how they might best thrive within natural habits and habitats with the dispositions and practices necessary for their communities to thrive.

Both in southern Africa and Latin America, during the last century we find that intergenerational knowledge and production among Indigenous people has been accompanied by the appropriation and recontextualization of explanatory abstractions in scientific and administrative institutional settings. The latter are reflected in the curriculum and little attention has been given to the former so that young African and Latin American children are confronted by knowledge about their surroundings that they cannot recognize as their own (Odora-Hoppers 2002). In this sense, we agree with Rozzi (2013) in the search to counteract processes of biocultural homogenization. We need to make it evident that knowledge is diverse. For this reason, it is important to retake knowledge that has been subjugated or marginalized in the

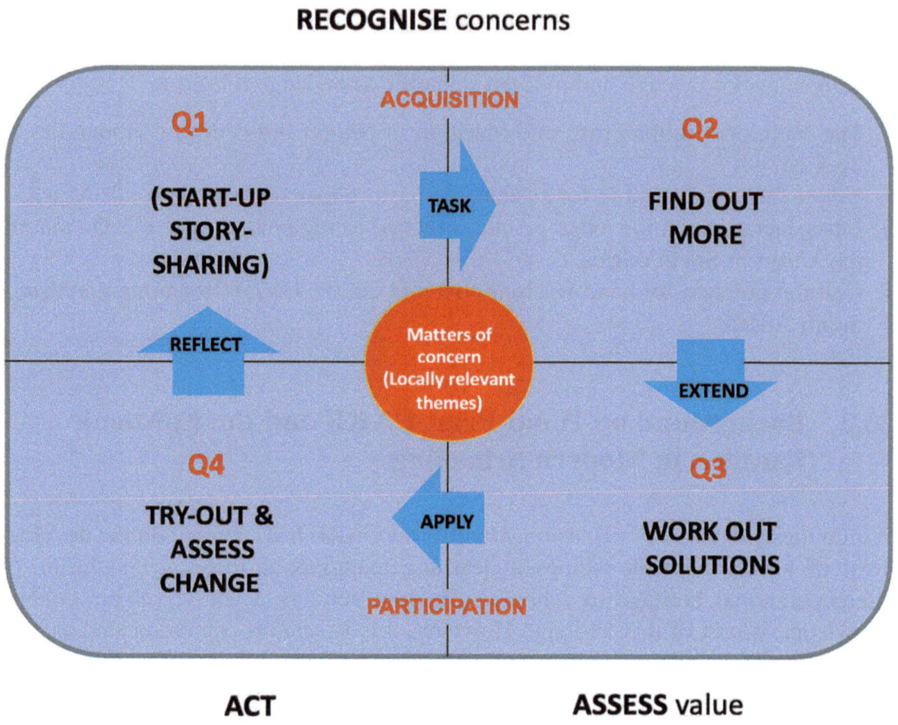

Fig. 25.1 A framework developed to integrate locally relevant themes into start-up story sharing and inquiry within a four-quadrant expansive learning process around local matters of concern. Source: Authors

academy. Doing this will help us to better understand the socio-ecological challenges that we confront and, in collaboration with schools, to face them.

The double challenge of epistemic abstraction and the relating of insights to habits and habitat were resolved for Hand-Print CARE through the development of locally relevant themes (Östman et al. 2013). A library of locally themed materials was developed as start-up true stories for teachers to work with on locally relevant themes (Fig. 25.1). A "start-up story" approach enables learners to contribute their life experiences and intergenerational knowledge through a co-engaged knowledge acquisition process of action learning using a four-quadrant task sequencing schema after Edwards (2014) through a sharing of their own stories that both resonate with and open up the learning trajectory specified for the teacher to mediate in meaningful ways.

What follow are a series of illustrative vignettes on the inclusion of plural knowledge heritage in South Africa and Mexico. The vignettes illustrate how teachers are navigating some of the learning challenges across intergenerational local knowledge practices and disciplinary concepts that commonly are present as inaccessible abstractions in school subject disciplines.

25.2.1 Milpa and Nutrition (Nahua, Mexico)

The milpa is the ancestral food production system that originated in Mesoamerica and consists of the associative planting of maize, squash, tomatoes, and various legumes (Sandoval-Rivera and Mendoza-Zuany 2021). Some studies (Molina-Anzures et al. 2016) have reported up to 50 different species, whether cultivated, sponsored, or tolerated. The depth of cultural knowledge practices embedded in milpa seldom articulates with diverse themes in the curriculum. This is most notable in rural multigrade schools for children aged 6 to 11 (grades 1–6) where milpa is still evident in daily life and the cultural knowledge is almost totally absent in urban environments. It is worth mentioning that the milpa is a traditional crop that has existed in the peasant culture since ancient times because community life is sustained by it. The milpa is the main means of subsistence for peasant families because from there they obtain the basic products for their food, as well as economic income when there is any surplus in production that can be traded (Sandoval-Rivera and Mendoza-Zuany 2021). Production in the milpa is relevant to families, even involving children. In Hand-Print CARE work on this traditional crop, in collaboration with teachers, we developed an integrating learning sequence using several themes of the curriculum, notably:

- Caring for Health by teaching about the different parts of the body, good hygiene practices, healthy eating habits, and the risks of consuming alcohol and smoking and how to avoid them.
- Caring for the Environment by showing that some actions can affect nature and that we can help to take care of it.
- Appreciating ecosystems by identifying air, water, and soil as essential resources for living beings.

Engaging these topics from the vantage point of milpa as a cultural practice disrupted the normal curriculum treatment of the above topics as the delivery of facts (abstract propositions) for healthy modern living. For example, diet is important in teaching about health. However, the diet shown in the curriculum belongs to the European diet, which is culturally unknown because it includes foods that are not easily found in rural regions or are not even known by the children. Participating teachers and students were challenged, on the one hand, to develop explanatory reasoning and critical insights on the health risks in modern diets that were not present in earlier milpa times. On the other hand, teachers explored with students the potential of the milpa to provide balanced nutrition that is culturally and environmentally situated to satisfy the nutritional requirements of rural families (Fig. 25.2).

The main learning for teachers is that the curriculum is dynamic and ceases to be the static and linear delivery of health information. The topics are no longer addressed according to the way they are presented in the educational curriculum. Instead, they are organized according to the productive and natural cycles known to children and people in the communities. In this way, the curriculum is directly related to real life and the challenges that people face (such as climate change and

Fig. 25.2 Photographs illustrating traditional milpa at different stages of growth: (**a**) planting corn and associated crops, (**b**) corn and other plants maturing, (**c**) corn ready to harvest (Source: Authors)

lack of water). The topics are addressed based on the potential they have to connect to real situations, such as the milpa.

25.2.2 Nguni Cattle and Pasture Management (Zulu, South Africa)

Masuku (2018) worked with young cattle herders in Mpembeni to clarify the acquisition of knowledge and to explore conflicts in modern schooling. A key process she uncovered was the central role of story and practical learning interactions in mediating care of cattle with insights into the knowledge practices undertaken at a village level among young children. Working from case evidence, she developed a Hand-Print CARE start-up story to encourage learning through open processes of story sharing and practical exchanges.

Local storytelling by elders shares intergenerational knowledge with children, especially those in urban schools few of whom herd cattle today, about cattle herding in their childhood. In the rural areas where Sibongile Masuku (2018) worked with young herders there are still tick-dipping day conflicts with schooling. This animal hygiene practice is not considered by school calendars and throughout the country little of the cultural knowledge on herding is included in the curriculum. The Hand-Print CARE module included an historical overview of the origin of Nguni cattle husbandry and a task to investigate cattle care in a local area. The Nguni cattle materials invited students to share knowledge in their mother tongue as well as to explore how Nguni cattle herding practices are being researched to resolve the loss of Indigenous pasture biodiversity owing to the colonial practice of fencing pastures in drier areas. One of the key observations that teachers made during a course-supported process of design research intervention was that the learners related differently to knowledge. Instead of learning being a process of trying to remember certain facts, they were learning about something that related to a past cultural world of herding practices that carried a solution for a modern problem of biodiversity loss.

25.2.3 Stories on Climate Change (Xhosa, South Africa)

Climate change is approached as a problem of the modern world where pupils are taught the concepts of global warming and about increasing drought conditions in the southern African region. No curriculum attention had been given to how the Nguni had long successfully lived on the Trankei grasslands of eastern southern Africa as an area of high climate variability driven by the El Nino/La Nina seasonal oscillations in the southern Pacific Ocean.

Hand-Print CARE materials were developed to introduce long-forgotten seasonal practices to capture soil moisture for the next season. In the process of preparing cropland for the next planting season, cattle are taken to the fields to eat the remaining stalks of the summer crops and also to fertilize the lands for the next season. Then the lands could be cleaned. This cultural Xhosa practice is called *Galesha*. In drought years, the rains were usually late and scattered so the water captured in the soil over the winter season enabled weakened cattle to plow for a late planting of the summer crops. This module had an interesting link to the Nguni cattle materials since another key knowledge practice was the Zuurveld migrations to the nutritious southern pastures of the region. Zuurveld is the name that the Boer settlers gave to the region of South Africa that is known today as Albany. This area belonged to the Xhosa populations before the arrival of the European colonists. The practice of annually moving cattle to other pasturage is environmentally sound because it allows grass cover to restore before being grazed in subsequent years.

Once again, the teachers found high interest among the students as they were engaged in deliberative learning that generated greater explanatory depth as the knowledge, although an artifact of Nguni history, shed light on the changing climatic processes in the region.

25.2.4 Hand-Washing to Avoid 2020 COVID-19 (Nguni, South Africa)

Hand-washing took on new significance with the advent of the COVID-19 viral disease. Schools were closed and under lockdown to slow down the spread of the disease in the region. This was new and frightening for pupils doing their studies over the Internet. However, their interests were sparked when they learned how previously the Nguni adopted a process of self-isolating using hand-washing to restrict the spread of cholera.

Colonial medicine saw cholera as a water-borne disease, but the Nguni treated it as a hand-to-hand and hand-to-mouth disease. To keep the disease away everyone would wash hands when returning to the village or before preparing food. The manner of washing (pour-washing) was important so as not to contaminate water. This changed with the advent of enamel and plastic bowls because instead of freshly pouring water for each other and drip-drying their hands after washing, visitors were offered a bowl of water, soap, and a towel. Whereas cholera is easily carried away in a stream of water, a virus needs washing with soap for twenty seconds. A village woodworker picked up on the original Nguni practices and the emerging knowledge about the new virus to produce a soap dispenser in combination with a foot-operated trickle of water for pour-washing. This became a good start-up story for use with a video on "Hand-Washing and Health."

25.3 Some Curriculum Strategies for Fostering Plural Knowledge Inclusion

These four case experiences involved a series of significant shifts in thinking and practice. We noted during fieldwork that these shifts were relatively easy for teachers with Indigenous roots but were not easy for institutional curriculum professionals. Institutional professionals found it difficult to expand their perspectives beyond the inclusion of "prior knowledge" as a route to learning abstract curriculum concepts. In our work on inclusion (O'Donoghue and Roncevic 2020), key examples that assisted us to move to more authentic plural knowledge exchanges were cases where Indigenous knowledge practices could be shown to be equal and better than institutional narratives. Across this Hand-Print CARE work with teachers on course-supported design research intervention (O'Donoghue et al. 2019), we identified the following strategies to expand conventional subject discipline methodology:

- **Storytelling**
  This was relatively easy for teachers to take up and commonly included inviting community members into a classroom for contextualizing knowledge.
- **Active village knowledge practices**
  Practical demonstration was combined with more work on videos to activate start-up deliberations in urban school settings.
- **Story-sharing deliberations**
  Hand-Print CARE start-up stories with photographs proved to be one of the best ways to activate story-sharing deliberations among students.
- **Deliberation circles and demonstration learning**
  The democratizing of knowledge in deliberation circles was important to activate inter-subjective learning.
- **Inquiry in community**
  Early on in learning sequences, depth inquiry in community settings is effective.
- **Practical games**
  Practical play activities, for example, comparative hand-washing and "smarty grazing" simulations, emerged as important for exploring the embedded knowledge in intergenerational knowledge practices in community settings.
- **Mother tongue inclusion**
  Exploring plural language environments has probably been one of the most significant interventions to foster inclusive learning environments. Mother tongue with simultaneous translation was particularly notable in urban settings where there are students from diverse language communities in classroom settings.
- **Professional learning communities**
  Finally, recent approaches to professional learning among teachers have disrupted an assumption that expertise must come from the outside experts. School inspectors have now become subject advisors and teachers are being supported to work as "professional learning communities." The step to the

classroom as a plural knowledge environment of dialectic learning interaction is being explored and achieved in increasingly inclusive ways to mediate many of the contradictions in modern schooling.

It has not been possible to narrate all of these more inclusive strategies. What is involved here is a loosening up of the assumption that community knowledge is inferior or backward. Here, the key is a shift toward an inclusive approach to curriculum.

25.4 Re-Framing the Curriculum as a Plural Knowledge Environment of Dialectic Inquiry

Hand-Print CARE insights on the interplay of Indigenous knowledge practices and abstract concepts in the modern curriculum point to how learning in school subjects might best be approached as dialectical processes engaging plural knowledge systems. The institutional externalizing of knowledge has produced anomalies of exclusion that are not easily resolved in curriculum settings. The strategy explored in Hand-Print CARE has had two dimensions:

- A story-sharing inclusion of life experience, intergenerational knowledge practices, and local matters of concern
- An inquiry-centered engagement with curriculum knowledge that blends these (heritage and concerns) in a Transformative Model of Education for Sustainable Development (TMESD—Chikamori et al. 2019) so as to activate critical knowledge acquisition and participatory action learning across habitus, habitats, and sustainable habits of mind.

Through case evidence in work on the inclusion of Indigenous knowledge, we have been able to conclude that learning is a collaborative process of expansive learning and transformative action.

Curriculum settings have not often been effective at including and successfully integrating diverse cultural capital in learning transaction. This has deepened the exclusion of generations of Indigenous peoples, often shaping as disjunctures across intergenerational epistemic processes of learning to live together in more sustainable ways. Contrary to this, our work is finding that there are ways to integrate knowledge and contextualize learning as situated and intergenerational processes in plural knowledge environments.

25.5 Final Remarks

Although the pedagogy we have been using in collaboration with teachers has been very well received by them, questions remain. On the one hand, the Edwards model explained above allows a pendular pedagogical strategy that navigates between the skills of the teacher and the previous knowledge of the students, passing through the materials and their contents that are elaborated in dialogue with people from the local communities. This triangle pedagogical dialectic allows us to establish a permanent dialogue between different systems of reason. However, in order to understand issues related to sustainability and socio-ecological challenges in the local context and in order to articulate them to the subjects that must be addressed from the curriculum, teachers require certain technical or specialized knowledge in ecological or biophysical terms, which they do not master due to their professional training in pedagogy, education, or social sciences. This represents a challenge in terms of the training and follow-up that teachers require on these topics.

Another important challenge that is emerging in the interaction between Indigenous knowledge and scientific knowledge is that the school context is related to the one previously mentioned. Sometimes, there are practices and beliefs rooted in the local culture that can contradict scientific knowledge because they are generated in their own cultural systems or belief systems different from the systems or paradigms from which scientific knowledge is constructed. This situation causes a tension that is difficult to deal with in the classroom because teachers have, on the one hand, little scientific background and, on the other hand, little knowledge of local culture and context or the ontological and epistemological implications of cultural practices and knowledge production. In this sense, it is necessary to explore in more detail how to take advantage of these tensions between knowledge systems, since this can help to understand the complexity of the inclusion of Indigenous knowledge in the school context and to explore new pedagogical possibilities aimed at sustainability.

Some open questions in the experience that we are sharing in this chapter are that although the content addressed from the mediation between curriculum and Indigenous knowledge is relevant for both teachers and students, the main objective of the project is not that there is only an appropriation of the curricular contents by the students from a cultural mediation with what they know or are familiar with, but rather that the scientific knowledge that the school offers can be articulated in actions that contribute to improving living conditions and to face the socio-ecological challenges. Or to better understand that the socio-environmental situations that they face, they have historically faced, or the one they are facing currently in the context of climate change can have some other explanations. In other words, the evidence that the appropriation of scientific knowledge from sociocultural mediation processes can be articulated with local practices and knowledge to solve concrete challenges in the communities is still lacking. This requires more time to see if the articulation of school knowledge with Indigenous knowledge really allows a better understanding and use of concepts that can be applied in daily life to improve living conditions in line with the paradigm of sustainability.

The insight emerging in this study is that it is possible to re-establish links between the orientation and communication functions of our symbolic systems in curriculum processes that work across past, present, and future.

References

Chikamori K, Tanimura C, Ueno M (2019) Transformational model of education for sustainable development (TMESD) as a learning process of socialization. J Crit Realism. https://doi.org/10.1080/14767430.2019.1667090

Edwards A (2014) Designing tasks which engage learners with knowledge. In: Thompson I (ed) Designing tasks in secondary education. Routledge, London, pp 13–27

Elias N (2011) The symbol Theory. Collected works

Gough N (1999) Surpassing our own histories: autobiographical methods for environmental education research. Environ Educ Res 5(4):407–418

Jiménez-Naranjo Y, Mendoza-Zuany RG (2016) La educación indígena en México: una evaluación de política pública integral, cualitativa y participativa. LiminaR 14(1):60–72

Lotz-Sisitka H, Fien J, Ketlhoilwe M (2013) Traditions and new niches. International handbook of research in environmental education. Routledge, New York, pp 194–205

Masuku LS (2018) In-betweenness: a postcolonial exploration of sociocultural intergenerational learning through cattle as a medium of cultural expression in Mpembeni, kwaZulu-Natal. Unpublished PhD thesis, Makhanda/Grahamstown, Rhodes University

Molina-Anzures M, Chávez J, Gil A et al (2016) Eficiencias productivas de asociaciones de maíz, frijol y calabaza (Curcurbita pepo L.), intercaladas con árboles frutales. Phyton 85(1):36–50

O'Donoghue R, Roncevic K (2020) The development of education for sustainable development-materials for inclusive education in South African curriculum settings. ZEP: Zeitschrift für internationale Bildungsforschung und Entwicklungspädagogik 43(1):20–26

O'Donoghue RB, Kibuka-Sebitosi E, Tshiningayamwe S, Palmer C (2019) Navigating non-sense by exemplifying situated life experience and intergenerational heritage knowledge in Education for Sustainable Development learning spaces. S Afr J Environ Educ 35. https://doi.org/10.4314/sajee.v35i1.8

Odora-Hoppers C (ed) (2002) Indigenous knowledge and the integration of knowledge systems: towards a philosophy of articulation. New Africa Books, Cape Town

Östman L, Svanberg S, Aaro Östman E (2013) From vision to lesson: education for sustainable development in practice. WWF, Stockholm

Rozzi R (2013) Biocultural ethics: from biocultural homogenization toward biocultural conservation. In: Linking ecology and ethics for a changing World. Springer, Dordrecht, pp 9–32

Sandoval-Rivera JCA, Mendoza-Zuany RG (2021) The milpa goes to school: teacher training in climate change adaptation from a situated learning perspective. Sustain Clim Change 14(1):42–46

Shava S (2008) Indigenous knowledges: a genealogy of representation and applications in developing contexts of environmental education and development in southern Africa. Unpublished PhD Thesis, Rhodes University, South Africa

Wals AE, Dillon J (2013) Conventional and emerging learning theories. International handbook of research on environmental education. Routledge, London, pp 253–261

Chapter 26
The Enviro-Champs Movement: Co-Researching Transformation Through Training Processes in a Post-COVID World

Ayanda Lepheana and Jim Taylor

Abstract This chapter presents a research and educational process investigating the Enviro-Champs movement. This South African movement works with local communities to address complex social-environmental problems. The effects of COVID-19 in 2020 interpenetrated every facet of human life and, for us, working with the Enviro-Champs movement the effects were enormous. Due to the impacts of COVID, our training programs had to be re-shaped from face-to-face learning to online learning. This led to some remarkable innovations that are developed and explained below. In addition to online learning, our work draws on field environmental philosophy (FEP), a pedagogy that takes students outdoors to investigate the environment around them. Together with the communities, we discovered that the environment is alive and that our lives are linked to the larger life of the biosphere, which supports all life on Earth. The purpose of our study is to determine the most effective form of training and capacity building in a post-COVID period. In addition, we examine how to apply this to support job skills training and sustainability practices. We applied action learning processes to structure and guide the learning. We found that indigenous knowledge practices can serve to situate learning and make it more meaningful and connected to the participants.

Keywords Adult-based learning · Collaborative research learning · Indigenous knowledge · Online learning · Sustainability

A. Lepheana
GroundTruth Associate, Hilton, South Africa
e-mail: ayanda@groundtruth.co.za

J. Taylor (✉)
Centre for Water Resources Research, University of KwaZulu-Natal, Pietermaritzburg, South Africa

R. Rozzi et al. (eds.), *Field Environmental Philosophy*, Ecology and Ethics 5,
https://doi.org/10.1007/978-3-031-23368-5_26

26.1 Background

Ecological infrastructure refers to naturally functioning ecosystems that produce and deliver services that are of value to society—fresh water, climate regulation, soil formation, and disaster risk reduction (Cumming et al. 2017). Ecological infrastructure is the nature-based equivalent of the built infrastructure. Arguably, ecological infrastructure is more important for providing services and underpinning socio-economic development than the built infrastructure. In South Africa, unsustainable development is eroding ecological infrastructure at an alarming rate (Department of Forestry, Fisheries and the Environment 2021).

Efforts to work against the tide and support work that is strengthening ecological infrastructure are to be welcomed and encouraged. In this regard, the Enviro-Champs movement in townships in various parts of South Africa is a breath of fresh air in a steadily declining atmosphere of unsustainable development (Department of Forestry, Fisheries and the Environment 2021). Environmental degradation is especially visible and evident in the streams and rivers that supply our cities and peri-urban areas. Most urban or peri-urban catchments in South Africa are degraded to a significant, alarming extent (Hay 2017). Of grave concern is that most of the catchments are continuing to degrade, over time, at a worsening rate.

26.1.1 Co-engaged Research

We joined hands with the Enviro-Champs. The Enviro-Champs movement was started in Mpophomeni Township in KwaZulu-Natal South Africa in 2009. The movement commenced in response to problematic environmental degradation such as inadequate sewerage provision and management, water pollution, and solid waste dumping. The movement is largely independent although affiliations to the Duzi uMngeni Conservation Trust (DUCT—www.duct.org.za) are commonplace when the Enviro-Champs undertake funded projects. Members of this movement were trained in stream ecology. However, we, as co-researchers, were very much part of the learning. This helped make the research more engaging and more meaningful to us all. This understanding that we were training and discovering together emerged through the various joint deliberation processes that were built into the research. Through such processes, we enhanced our ability to better design, scaffold, and engage in adult-based learning and transformative change.

In the decade of the 2000s, the Enviro-Champs developed as a modest, yet powerful, social movement with gathering momentum. Known by different names across South Africa the work of the Enviro-Champs was singled out by the Presidential Skills Summit Framework Agreement as an example of best practice in community-based catchment and water management (Republic of South Africa 2018). This social movement manifests in different forms and with different names in various parts of the country to represent a groundswell of people committed to sustainability and the common good.

26.1.2 Background to the Enviro-Champs Concept

The Enviro-Champs project aims to mobilize communities to understand and take action to address water quality, quantity, and access (equity) issues. Solid waste issues are also a priority in a consumer-driven society where solid waste dumping and over-spilling dump sites, known as land-fills, are commonplace. The project also addresses sanitation and other pressing township issues such as environmental management and air pollution. The Enviro-Champs project is guided by the UN Sustainable Development Goals (SDGs), particularly SDG 4 (Education) and SDG 6 (Water), to develop skills and to apply these capabilities to the above environmental issues. The project contributes to the implementation of South Africa's Department of Environment Affairs' National Environmental Education and Training Strategy and Action Plan (2015–2024) (AEETAP 2021). It is also part of the Education for Sustainable Development network based in South Africa, India, Mexico, and Germany (ESD 2021).

26.1.3 Connecting Through Technology

Since face-to-face training was not possible, due to the COVID lockdown, other learning methods had to be explored. At first, we thought Internet-based programs like web conferencing platforms would be the answer, but none of the participants had laptops. They did have access to smart-phones, however, and these could be used through text and voice messaging apps such as "WhatsApp." This proved to be a major breakthrough. "WhatsApp" chat is a trending tool suited to informal, adult, learning. "WhatsApp" is a popular instant messaging tool, which can record voice notes, share photographs and videos, and enable video calls.

26.1.4 Indigenous Knowledge Practices

In the research processes, we explored how indigenous or local knowledge practices could enhance the ability of participants to connect with the study topics. This proved to be helpful in that participants could meaningfully integrate their prior knowledge and understanding into the learning context. Indigenous knowledge practices are a body of knowledge, know-how, and practices maintained and developed by people, usually in rural or semi-rural areas, who have long histories of interaction with the natural environment. In addition to practices, these knowledge systems include values, beliefs, and worldviews (Mandikonza 2019). These forms of knowledge practices are handed down orally, from generation to generation (Boven and Morohashi 2002). Unfortunately, modernizing and colonizing processes are silencing such histories, and this leads to disjointed learning or learning that does not connect well with the participants' lived reality.

26.2 Our Studies with the Enviro-Champs

Our study aimed to determine the most effective form of training in a post-COVID period and how this training might be applied to support the Enviro-Champs Job Skills project. It continues to explore how indigenous knowledge practices can be mobilized as part of adult-based learning and change. Supported by Transforming Education for Sustainable Futures (TESF), we embarked on a research process to better understand the project, strengthen the way it is managed, and see where we can develop learnings or principles of social change, at a catchment level, which may be applied elsewhere. The project is managed in a collaborative manner and common goals are addressed with participants. This collaboration includes effective communication, actively listening to others, being accountable, and respecting the diversity of others' information and ideas. Toward this aim we have adapted the pedagogical model developed by Jim Taylor and collaborators (2018) called "5Ts:" "tuning in, talk, touch, think, and take action" (Fig. 26.1). The 5T's do not have to be undertaken in any particular order (Graham and Taylor 2018).

26.2.1 The Research Methodology

The research methodology may be described as a mixed-method study. The monitoring and evaluation processes, the various reports, and the learning for monthly reporting, reflection, and semi-structured interviews were applied to evaluate the project. This included data on participant observation and opinions. These tools are useful in qualitative research as they offer systematic processes and reveal hidden insights. Semi-structured interviews were used to record participants' learning experiences and opinions about the program in relation to its purpose. Participants also completed a semi-structured interview to establish the Enviro-Champs' satisfaction toward the use of "WhatsApp" as an effective form of training in a post-COVID period. The research also explored the acceptance of the WhatsApp tool for Enviro-Champs training and a chat conversation register to analyze the different types of interactions.

Using indigenous knowledge and practices in social learning enables indigenous people and local communities to actively participate and become meaningfully part of the learning processes. They are powerful resources held by rural people.

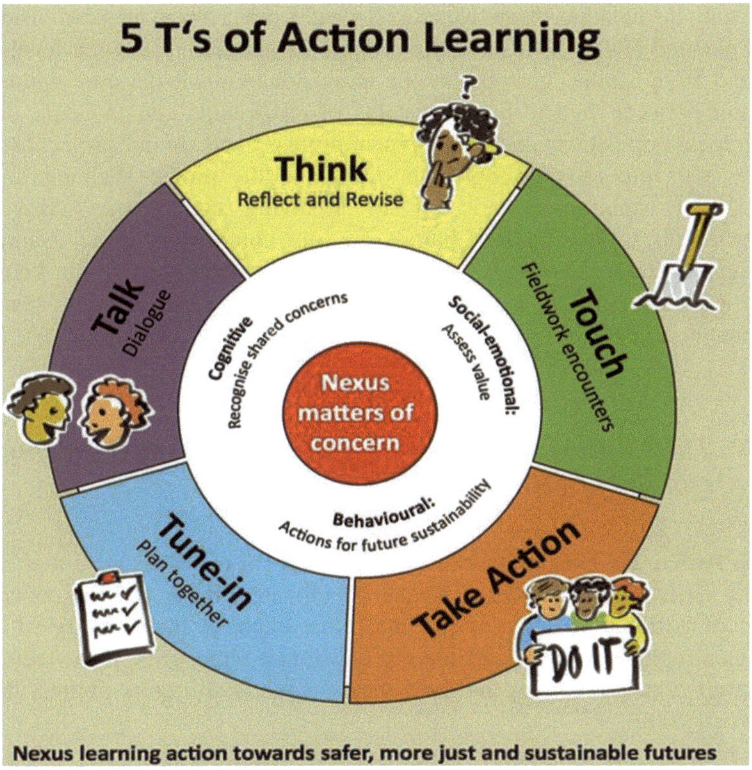

Fig. 26.1 The 5T's of action learning. The five T's are tuning in, talk, think, touch, and take action. Figure adapted by Rob O'Donoghue from Taylor, O'Donoghue, and Venter (2018, p. 117)

Consequently, they are a key element in the fight against poverty and social exclusion for many rural communities worldwide (Boven and Morohashi 2002).

26.2.2 Modern Science and Indigenous Knowledge: Enriching Our Practices Through Dialogue

We noticed that the different interactions between the Enviro-Champs and facilitators during the learning should be valued not only for the relationships of knowledge construction but also for the social and interdependent experiences that took place. Such learning processes are usually not taken into account. We also found that by articulating indigenous knowledge practices with modern techniques the richness and mix of the dialogue proved more valuable than either one on its own. The indigenous knowledge practice of only collecting water where you can hear it, for example, alludes to the fact that moving water is likely to have a higher oxygen content and be healthier than water collected from a stagnant pool. Biological oxygen demand (BOD) is a scientific process for measuring oxygen levels in the water, and BOD studies corroborate the indigenous knowledge observations. The interaction between two different systems of knowledge can facilitate dialogue between local populations and development professionals. In our project as well as in others, this proved meaningful for designing the project, helping to reflect people's real aspirations, and actively involving communities (Boven and Morohashi 2002). The Enviro-Champs training courses are good examples of including indigenous knowledge practices to successfully involve the knowledge and participation of local people and communities for purposes of sustainable development.

26.2.3 The Context: Governance Failures and Deteriorating Quality of Life

In South Africa, governance failures are the origin of many resource management problems. These are exacerbated further by climate change and the concomitant increase of extreme weather events that have exposed the inability of current governance regimes to deal with present and future challenges. Knowledge about resource governance regimes and how they change is still quite limited, however (Pahl-Wostl 2009).

A conceptual framework is required to address the dynamics and adaptive capacity of resource governance regimes as multi-level learning processes. We identified major structural characteristics of governance regimes that need to be transformed, including the influence of formal and informal institutions, the role of state and non-state actors, the nature of multi-level interactions, and the relative importance of bureaucratic hierarchies, markets, and networks. Change is conceptualized as social and societal learning that proceeds in a stepwise fashion moving from single to double to triple loop learning (Pahl-Wostl 2009).

The Enviro-Champs adopt the conceptual framework of Education for Sustainable Development (ESD). It also adapts citizen science tools, such as the Stream Assessment Scoring System (www.minisass.org), as well as the application of various other monitoring tools. For example, these include a clarity tube for measuring water turbidity as well as a velocity plank for measuring velocity and volume. These tools proved very useful, and coincide with the emphasis that field environmental philosophy (FEP) gives to "direct encounters" with co-inhabitants from multiple cultures and multiple species (Rozzi et al. 2006, 2008).

Table 26.1 Sources for research insights in the Enviro-Champs project

Data set	Data source	Purpose of data source	Recording method
1	External moderator group/ consultation group	This group consisted of key people who guide the project, work with the project leader, and deliberate issues as well as support decision making	Record of deliberations and how decisions are arrived at
2	Code of Conduct for working together as well as a "WhatsApp" code of conduct	The code of conduct was designed through deliberative processes with the participants. It proved helpful in guiding transparent and respectful interactions	Written document
3	Introduction to Aquatic Ecology course	This short field course was offered through a workshop that involved a practical question-guided learning strategy, and an exploration of indigenous knowledge practices	Course outline and support notes
4	Fieldwork adapting field environmental philosophy (FEP)	A fieldtrip to a local stream supported action learning processes and different forms of knowledge	Report with photos
5	"WhatsApp" group chat training sessions—online group dialogue and co-engagement	Discussions were tracked, including agreements and group dynamics. The 5T's of action learning (Taylor et al. 2018) provided a structure for the learning	"WhatsApp" group chats Chart recording digital interactions—voice, text, and video
6	Semi-structured interviews	Interviews recorded and explored participants' learning experiences and opinions about the program in relation to its purpose	Recorded, transcribed, and analyzed interviews

26.2.4 *Document Analysis and Data Sources*

The data which we used to co-research how the Enviro-Champs were learning are tabulated in Table 26.1.

26.2.4.1 The External Moderator and Consultation Group

There were many key people who guided the project by helping and orienting the project leader. They gave support in planning for workshops and discussing challenges and opportunities that engaged the project. By having a moderating group, the project leaders' subjectivity could be tempered. This group proved to be a useful sounding board when the project leader was uncertain or was looking for guidance.

26.2.4.2 The Code of Conduct

The code of conduct was a jointly compiled collection of the rules, principles, values, and Enviro-Champs expectations. The code of conduct outlines expected behavior toward each other as well as the relationships among partners and members of the public. What made this code of conduct special was that it was jointly drafted, and agreed upon, through the active participation of all members of the Enviro-Champs. A similar participatory process was followed to develop the WhatsApp code of conduct. These codes of conduct for the Enviro-Champs were something they could continuously reflect on, look back to, and draw on where necessary.

26.2.4.3 An Introduction to Aquatic Ecology

The field course "Introduction to Aquatic Ecology" consisted of a workshop that involved a practical question-guided learning strategy and an exploration of indigenous knowledge practices. The course was situated in the reality of the lives and work of the Enviro-Champs participants. The curriculum was organized through key questions to the participants. It included indigenous knowledge processes to ensure that the wisdom of the past was not neglected and could strengthen and complement the wisdom of the present. The participants identified with many of the indigenous knowledge practices. They also shared examples from their experiences, which strengthened the course curriculum. Applying an action learning framework guided the learning and provided an orientating framework (Taylor et al. 2018). This enabled participants to "tune in" to the training by mobilizing their prior knowledge and understanding. In this way, their life world experiences were very much part of the learning process. This approach contrasts with academic courses where information is conveyed to participants in a top-down manner.

26.2.4.4 Fieldwork

The fieldtrip and action learning included discovery, experimentation, learning about, and connecting to the natural world (Fig. 26.2). The fieldwork enabled the participants to engage with the real world and identify problem areas that need solving. Our fieldwork approach could add to the field environmental philosophy (FEP) methodology by emphasizing participatory action. In turn, it could benefit from adopting the FEP's 4-step cycle by including the composition of metaphors and/or narratives (Tauro et al. 2021).

Fig. 26.2 Enviro Champs working in the field. (Photograph by Liz Taylor)

26.2.4.5 The Training Session

The Enviro-Champs from the Palmiet area of Durban attended 40 days of training between July and December 2020. The training included empowering Enviro-Champs with knowledge they could share during door-to-door activities. This knowledge will be invaluable beyond the finite period of the project. The training included the following ten core subjects:

(a) Riparian Health Audit tool
(b) Grassland study
(c) MinSASS and citizen science tools
(d) Snake handling
(e) Water testing in the Palmiet Catchment
(f) Names of indigenous trees in the field
(g) Production of videos on environmental problems in the community
(h) Names of invasive alien plants
(i) Layout of sewage pipes and sewage systems in the field
(j) Geographical Open Data Kit (GeoODK), a computerized process for collecting and organizing data.

26.2.4.6 The "WhatsApp" Group

"WhatsApp" chat is an inexpensive trending tool that is being used in informal and formal education. It is an instant messaging tool, voice notes recording, and video calls. It can be used to track discussions, agreements, and group dynamics in relation to the themes of action learning. During the learning sessions, with "WhatsApp" we recorded the types of interactions that occur among the Enviro-Champs.

26.2.4.7 Semi-Structured Interviews

We applied semi-structured interviews to elicit participants' learning experiences and opinions about the program. Only six Enviro-Champs participated in the detailed semi-structured interviews. However, the non-structured dialogues while conducting the interviews disclosed key findings (Box 26.1).

Box 26.1 Key Findings for Collaborative Research and Education

(a) **Everyone could talk and share.** During the training processes everyone could talk and share rather than just receive input. This allowed the Enviro-Champs to experience teaching and learning about their local environmental challenges in an interactive manner that also allowed them to incorporate their own knowledge. This was facilitated through practicing co-engaged inquiry (Jaworski 2010). The development of knowledge through inquiry enhanced the participant's learning about socio-environmental issues.

(b) **Confidence and development of a sense of agency.** Enviro-champs participants felt they were developing a sense of agency and confidence. This encouraged them to believe in themselves and to be able to talk and present in front of people as well as to conduct door-to-door education. As noted by one Enviro-Champ, a young woman who speaks Zulu:

> Last time I gave the presentation was when I was at high school 7 years ago. I was nervous to stand in front of the group and give presentations. The project made me feel like an important member and I have gained knowledge. I feel I now understand the issues better and I can talk about and share with my community.

The importance of such an exercise is to confront context and to find a sense of relevance. Confidence is self-assurance of one's own abilities or qualities (Bäck et al. 2017). By investing time in this effort, their attitudes and motivation were positively influenced.

(c) **Kindness and generosity are appreciated.** To serve one's community requires special qualities that include a public-spirited outlook and a commitment to the common good. When these qualities are informed by

(continued)

Box 26.1 (continued)

kindness and generosity, these virtues become powerful transformative characteristics.

(d) **Connecting with indigenous knowledge practices is powerful and develops situated meaning.** In one debate on land in South Africa, one of the panelists, a Khoisan chief, spoke about custodianship and stewardship in the following way: when one uses the land, they become accountable, and that implies leaving the land in a better state for those who will inherit it (Tladi 2019). Indigenous knowledge practices, such as these, are a critical factor for effective sustainable development.

The empowerment of local communities is a prerequisite for the integration of indigenous knowledge in development processes. The integration of appropriate indigenous knowledge practices into development programs has already contributed to efficiency, effectiveness, and sustainable development impact and training. Indigenous knowledge, like any other knowledge, needs to be constantly practiced and further adapted to evolving local contexts. Supporting local and regional networks of traditional practitioner and community exchanges can help disseminate relevant indigenous knowledge and enable communities to participate more actively in the development processes. While innovative mechanisms for the protection of indigenous intellectual property rights need to be developed, many indigenous knowledge practices can, at the same time, be integrated into local, national, regional, or global development efforts.

Experience has shown, however, that this cannot be done by one institution alone. Partnerships between community-based organizations, nongovernmental organizations, academia, the private sector, research, government, and donor institutions are needed to enhance the use of indigenous knowledge practices for development. Numerous examples are given of the successful implementation of indigenous knowledge in the areas of health care, agriculture, education, and women's issues by communities and governments in Africa, India, and South America. This course process included indigenous knowledge processes to ensure that the wisdom of the past is not neglected and can indeed strengthen and complement the wisdom of the present. The participants identified with many of the indigenous knowledge ideas that were shared. They also shared examples from their experiences, which strengthened the course curriculum.

Mobilizing indigenous knowledge practices in the Enviro-Champs training processes enabled them, along with local communities, to actively participate in the social learning processes (Wals 2007). One young man, who grew up in the rural region of the Eastern Cape Province, noted, for example:

(continued)

Box 26.1 (continued)

Back at home in rural areas water is clean and the people have a connection with the streams, rivers and springs because they collect water from these water bodies and it is everyone's duty to keep them clean. Here in Durban, we have lost that connection and practice of keeping water bodies clean, because we now think that water comes from a tap and that is one of the reasons why people are polluting our rivers

Indigenous knowledge practices are powerful resources for rural people and are a key element in the fight against poverty and social exclusion for many communities (Boven and Morohashi 2002). It has been encouraging and interesting to observe the growing interest from the Enviro-Champs in participating in Education for Sustainable Development through the role of indigenous knowledge practices.

(e) **Action learning was greatly beneficial especially the 5Ts.** An action learning framework was used to structure the learning and the evaluation of the issues and risks that participants faced. Our "action learning" framework adapts the approach developed by Jim Taylor et al. (2018). Action competence and other methodologies that ground democratic learning processes (Taylor et al. 2018) are important dimensions of action learning. These processes engage the participants in decision making, rather than assuming that they should simply implement externally derived solutions. Such pedagogical models are therefore potentially powerful ways of enhancing "agency," understood as the ability of people to develop their capacities.

Action learning, as an open-process framework, offers participants a structured engagement in fieldwork activities. This framework includes the "5T's model which includes tuning in, talk, touch, think, and take action (Fig. 26.1). Central in the model is the "Nexus or Matter of Concern" which is the focus or issue which is being addressed. This enabled all participants to "tune in" to the training by mobilizing their prior knowledge and understanding. In this way, their lived world experiences were very much part of the learning.

Question-guided learning was a key component of the learning processes that were appreciated by the participants. They noted how it enabled them to figure things out, and to ask more informed questions about the issues and risks they were facing. Question-guided learning activated prior knowledge. Participants were encouraged to mobilize knowledge they had about rivers and streams in their local water resources management context.

Practical sessions were greatly appreciated. Practical sessions at the river proved useful in many ways. People were amazed by what they discovered

(continued)

Box 26.1 (continued)
and enjoyed showing it to each other. This practical experience generated rich dialogues, and some participants remembered other streams, sometimes from their youth.

26.3 Learning Exchange Workshop

To strengthen the learning an exchange workshop was organized with Enviro-Champs from the Mpophomeni region of KwaZulu-Natal. This enabled social learning processes among different Enviro-Champs to observe and assess their discussions for evidence of social learning. Participants experienced the work of the Mpophomeni Enviro-Champs first-hand, understanding the role they play and the tools they use. The workshop generated joint discussions, identifying learning opportunities for all Enviro-Champs, particularly how different tools and approaches may be adapted for use in other areas.

26.3.1 Feedback from a Participant in Her Own Words

One participant noted the following:

> Thank you to Mpophomeni Enviro-Champs for giving us this learning opportunity. I am 44 years old and not working. I thought my time has passed and now I am waiting for pension grant. But when I saw a 69-year-old (Baba Cele) helping his community, I realize that there so much I can still offer my community, and beyond, through the Palmiet Project. My friend and I always think that it is the role of the municipality keep our environment clean. Through the learning exchange workshop, I have learned how I can be eyes and ears of the municipality. Now I am going to have a chance to teach my friend about the role we can play, together, to assist the municipality.

26.3.2 Ownership and Trust

For applying a social learning process (*sensu* Wals 2007), it was important to co-develop the workshop to foster a sense of ownership and trust. As expected in social learning processes, there were many instances where it was necessary to accommodate unanticipated gaps in the "competencies" of individuals in the group. This was not in the original design; ownership and trust were essential to adapt to changes.

26.4 What Kind of Scaffolding Is Needed to Enable People to Participate in Enviro-Champs's Activities?

Scaffolding is a negotiated, two-way process of learning implying that participants and facilitators of learning are active co-participants (Maybin et al. 1992). "Scaffolding" is a metaphor used in educational theory to describe ways of supporting students in their learning processes. To succeed participants need to develop an understanding of the end goal, even if it is not reached. Enviro-Champs constantly remind each on the "WhatsApp" group about the goal of their work, and what is needed to achieve it. Learning is supported by negotiated scaffolding, much more than the structured, rigid form of scaffolding which takes place in formal teaching. This kind of negotiated scaffolding allows participants to situate themselves within their context and make meaning of their learning as it relates to their own reality.

26.4.1 How Does One Build Confidence or Agency in Ourselves and in the People We Work With?

It is necessary to create a sense of belonging and to build confidence. In contrast to prevailing formal education, the Enviro-Champs approach is not classroom centered and overcomes the distinction between teachers and learners. By relying on things that participants are familiar with (talking to each other about their culture and youth), every individual manages to participate, present, and contribute significantly to the program.

Enviro-Champs is developing as a Community of Practice (Wenger-Trayner and Wenger-Trayner 2015). This term refers to a group of people or organization that develops a process of collective learning and that meets regularly to achieve a desired or unintended goal. Such a community is usually governed by a defined culture of practice (Brown 2013). Communities of practice are diverse, including sports club, professional organizations, or even groups seeking new ways of learning (Wenger-Trayner and Wenger-Trayner 2015).

26.4.2 Ways in Which Enviro Champs Has Developed as a Community of Practice

Working with the Enviro-Champs and sharing the tools of science, participants have developed confidence and understanding from their own perspective rather than simply being the recipients of other experts telling them what they know, in a top-down manner. These are exciting issues to research because they are central to the development of agency among the participants and there are many ways this can be done. We are finding that learning is strengthened when people are involved and co-engaged in the project processes. This helps them develop self-reliance and improves their attitudes toward learning.

26.5 Interim Conclusions

The conclusions we are discovering together are coming through the various joint deliberation processes that this learning experience and research process is making possible. Through such processes we are enhancing our ability to design, scaffold, and engage in adult-based learning and transformative change. The sharing of tools of science and encouraging people to use them helps participants develop confidence and understanding from their own perspective rather than simply being the recipients of other experts telling them what they know in a top-down manner. These are exciting issues to research because they are central to the development of agency among the participants, and there are many ways this can be done. As mentioned above, the Enviro-Champs approach could add to the field environmental philosophy (FEP) methodology by emphasizing participatory actions. In turn, it could benefit from adopting the FEP's 4-step cycle by including the composition of metaphors and/or narratives (Tauro et al. 2021).

Learning is strengthened when people are involved and co-engaged in the project processes. This helps them develop self-reliance and improves their attitudes toward learning. WhatsApp chat is an inexpensive tool that can be effectively used in informal and formal education. As an instant messaging tool, by recording voice notes and video calls WhatsApp can be used to track discussions, agreements, and group dynamics in relation to the 5T's of action learning. According to participants, the WhatsApp training using "group chat" and other linked features such as sharing videos and photographs was effective and rewarding. The Enviro-Champs have thus contributed significantly to their own enrichment and training and that of their fellow candidates. This was made possible by the training approach adopted which included question-based learning, cultural, and indigenous practices, as well as action learning and knowledge sharing.

As a social movement the Enviro-Champs may take different forms and have different names in various parts of the country, but they have common ways of working. The training program promoted these and included indigenous knowledge practices. This enabled the Enviro-Champs to connect with what they were learning and to be able to address and understand challenges facing their surrounding communities. Using holistic and community-based indigenous knowledge practices helped to mobilize the Enviro-Champs through storytelling, local language teaching, sharing, and learning. This puts knowledge production and sharing at the forefront of the Enviro-Champs work. Learning thus functions at a community level, in its cultural context, by involving people as community knowledge holders. Mobilizing indigenous knowledge practices, research, teaching, and learning through participatory processes is an important component of good quality learning and training.

Acknowledgments This training and research could not have been completed without the help and support of many people. We greatly appreciated the dialogues that Injairu Kulundu-Bolus set up and managed as part of the transformative research processes. We learned so much there. The meetings enabled us to share our ideas with increasing confidence as the research developed. We would also like to thank Prof. Rob O'Donoghue for insights on research methodology and educational ideology. Catherine Meyer, from GroundTruth, was an enormous help despite many challenges that the project faced. Further support from Mark Graham, Patsy Hampson, Sindiswa Mthalane, and Smiso Bhengu is also greatly appreciated. They supported the project and guided it. Of most importance was the capacity building and financial support from TESF, the Development Bank of Southern Africa, eThekwini Metro, and the Presidential Employment Stimulus of the Department of Science and Innovation. We are grateful for their trust in us. All the Enviro-Champs were a great pleasure to work with and they too shared freely their insights, challenges, and knowledge. It was a pleasure to co-research this project with them.

References

AEETAP (2021) Africa environmental education and training action plan 2015–2024: strengthening sustainable development in Africa. Available via https://www.unep.org/resources/publication/africa-environmental-education-and-training-action-plan-2015-2024

Bäck L, Hildingsson I, Sjöqvist C, Karlström A (2017) Developing competence and confidence in midwifery-focus groups with Swedish midwives. Women Birth 30(1):e32–e38. https://doi.org/10.1016/j.wombi.2016.08.004

Boven K, Morohashi J (2002) Best practices using indigenous knowledge. Nuffic, The Hague, and UNESCO/MOST, Paris

Brown T (2013) T10 situated learning theory of Lave and Wenger. Retrieved, September 9, 2019, from https://s3-eu-west-1.amazonaws.com/s3-euw1-ap-pe-ws4-cws-documents.ri-prod/9780415829496/T10.pdf

Cumming TL, Shackleton RT, Förster J, Dini J, Khan A, Gumula M, Kubiszewski I (2017) Achieving the national development agenda and the Sustainable Development Goals (SDGs) through investment in ecological infrastructure: a case study of South Africa. Ecosyst Serv 27: 253–260

Department of Forestry, Fisheries and the Environment (2021) Annual performance plan. Government Printers, Pretoria 2021/222021.2022annualperformanceplan.pdf

ESD (2021) Our history. Education for Sustainable Development (ESD) Expert Net. Available via: https://esd-expert.net/our-history.html

Graham M, Taylor J (2018) Development of citizen science water resource monitoring tools and communities of practice for South Africa, Africa and the World. Water Research Commission

Hay D (2017) 'Our Water, Our Future' Securing the water resources of the uMngeni River Basin. Institute of natural Resources, Pietermaritzburg

Jaworski B (2010) How we teach: inquiry in teaching and learning mathematics, vol 1(2). Loughborough University, pp 105–121

Mandikonza C (2019) Integrating indigenous knowledge practices as context and concepts for the learning of curriculum science: a methodological exploration. Southern African. J Environ Educ 35. https://doi.org/10.4314/sajee.v35i1.13

Maybin J, Mercer N, Stierer B (1992) Scaffolding learning in the classroom. In: Maybin J, Mercer N, Stierer B (eds) Thinking voices: The work of the National Oracy Project. Hodder & Stoughton, Seven Oaks, Kent, pp 186–195

Pahl-Wostl C (2009) A conceptual framework for analysing adaptive capacity and multi-level learning processes in resource governance regimes. Institute for Environmental Systems

Research, University of Osnabru¨ck, Barbarastrasse, Revised June 2 2009; from journal homepage: www.elsevier.com/locate/gloenvcha

Republic of South Africa (2018) Presidential jobs summit, framework agreement. Government Printers, Pretoria

Rozzi R, Massardo F, Anderson C, Heidinger C, Silander JA Jr (2006) Ten principles for biocultural conservation at the southern tip of the Americas: the approach of the Omora Ethnobotanical Park. Ecol Soc 11(1):43. http://www.ecologyandsociety.org/vol11/iss1/art43

Rozzi R, Arango X, Massardo F, Anderson C, Heidinger K, Moses K (2008) Field environmental philosophy and biocultural conservation: the Omora Ethnobotanical Park educational program. Environ Ethics 30(3):325–336

Tauro A, Ojeda J, Caviness T et al (2021) Field environmental philosophy: a biocultural ethic approach to education and ecotourism for sustainability. Sustainability 13(8):4526

Taylor J, O'Donoghue R, Venter V (2018) How are learning and training environments transforming with education for sustainable development? In: Leicht A, Heiss J, Byun W (eds) Issues and trends in education for sustainable development. UNESCO, Paris. http://unesdoc.unesco.org/images/0026/002614/261445E.pdf

Tladi R (2019) Episode 1, The big debate: land without compensation: 10 March 2019. Interview with panellists (Video). SABC Digital News. Retrieved 12 March 2019, from https://www.youtube.com/watch?v=olrU3bRfk4U&t=878s

Wals AEJ (2007) Social learning towards a sustainable world: principles, perspectives, and praxis. Wageningen Academic Publishers, Wageningen

Wenger-Trayner E, Wenger-Trayner B (2015) Introduction to communities of practice: a brief overview of the concept and its uses. Retrieved 13 February 2021, from https://wenger-trayner.com/introduction-to-communities-of-practice/

Part IV
Introduction to Platforms for Integrating the Sciences, Arts, and Humanities into Participatory Education

Chapter 27
Introduction to Part IV: Platforms for Integrating the Sciences, Arts, and Humanities into Participatory Education

Noa Avriel-Avni

Abstract Long-term ecological (or socio-ecological) research (LTER or LTSER) sites, as well as biological stations, botanical gardens, and other platforms, enable the integration of diverse disciplines and cultural forms of knowledge rooted in specific localities to conduct biocultural educational programs. Consequently, these platforms provide valuable sites to foster intercultural dialogues and co-production of knowledge, as well as integrating sciences, arts, and philosophy in a site-based frame of reference.

Keywords Biosphere reserves · Botanical gardens · Long-term ecological research sites · Science education centers · Teachers college

27.1 Introduction

One of the greatest challenges to biocultural conservation derives from the Cartesian dichotomy between human and nature, and between body and mind (Schultz 2001; Vining et al. 2008; Rozzi 2013). In the age of globalization and consumer capitalism, the Cartesian dualism is exacerbated (Blühdorn 2017). On the one hand, a geographical detachment between consumers and the ecosystems that supply their resources (Costanza et al. 2014) obscures the damage caused to these systems and is therefore often overlooked (Massey 1997). On the other hand, lack of reliance on local resources may trigger irresponsible behavior toward local ecosystems. Traditional knowledge and practices developed by local cultures over the years, as co-inhabitants, may lose their significance (Rozzi et al. 2008; Gould et al. 2010; Mackey and Claudie 2015). The combination of these processes generates a positive feedback which encourages the creation of a global culture that displaces local cultures (Rozzi et al. 2018).

N. Avriel-Avni (✉)
Dead Sea and Arava Science Center, Mitzpe Ramon, Israel
e-mail: noa@adssc.org

R. Rozzi et al. (eds.), *Field Environmental Philosophy*, Ecology and Ethics 5,
https://doi.org/10.1007/978-3-031-23368-5_27

An alternate approach, the field environmental philosophy (FEP), develops a perception of the environment as a habitat shared by humans from different cultures, other living beings, and other ecosystem components. Alongside scientific knowledge, the cultivation of the arts and values' awareness are considered legitimate ways of understanding the environment (Rozzi et al. 2010). Thus, FEP contributes to the abolition of the Cartesian distinction between human society and its ecosystems, and between body and mind. Instead, it perceives human beings as integral components of ecosystems (McDonnell and Pickett 1993), and proposes an epistemology that incorporates awareness of biological and cultural diversity in all dimensions of consciousness and learning processes (Tauro et al. 2021).

From the perspective of environmental education, eliminating the distinction between humans and nature is a deep and long-term process (Fletcher 2017; Leopold 2004). This complex task requires emotionally and conceptually re-embedding people in their place, and reweaving the loose threads that previously tied humans to their ecosystem (Capra 1996). It is a challenging endeavor because it is opposed to the firm promise of current policy makers that consumer capitalism supports sustainability and can actually be reconciled with values of social justice, political equality, and ecological integrity (Blühdorn 2017). Environmental education that works according to this approach perceives the environment as a commodity that nevertheless enables sustainable development. Students are expected to develop personal environmental responsibility without challenging prevailing economic and political policies regarding the environment (Hursh et al. 2015). Sometimes, even educational approaches that call for reconnection with nature, in fact, preserve the conceptual separation between humans and nature (Fletcher 2017). Hence, new environmental education approaches are needed to prevent this blind imposition of global models and to appreciate singular biological and cultural realities (Tauro et al. 2021).

In Part IV of the book, we explore long-term platforms and approaches to encourage this necessary paradigmatic transformation. Having uncovered in the previous parts of the book the challenges and possibilities for preserving biocultural diversity through environmental education (EE), Part IV reviews various organizations that can serve as platforms for such education. Each chapter presents different platforms and different versions of environmental education. Together they present a portrait of EE based on participatory education and a combination of sciences, arts, and humanities.

27.2 Questions Environmental Education Needs to Address

In order to conserve biocultural diversity, FEP nurtures perceptions of the environment as habitats shared by humans from different cultures, other living creatures, and other ecosystem components (Rozzi et al. 2010). This paradigm shift poses a variety of questions and challenges to EE, which are raised in the various chapters of this section. Part of them are due to the Cartesian dichotomy between human and nature and between body and mind, which are repeatedly embedded in education systems and globalized Western thinking (Schultz 2001; Vining et al. 2008), and are exacerbated in consumer capitalism (Blühdorn 2017). In light of this, EE programs need to address various questions.

First, how to nurture reconnection of humans with nature, as a basis for developing awareness and caring for biocultural diversity? UNESCO set biodiversity conservation as an international mission in the 1970s (Ishwaran et al. 2008). However, the slow progress that has been achieved shows that this task is not simple (Ardoin et al. 2020), partly because of the distancing of modern societies from the ecosystems that support their existence and the abundance of natural substitutes that technological culture provides them. To reconnect society and nature, Joy Zhu (2023) focuses in her chapter on "flag species" for biological conservation, including among these flagship species, little ones such as mosses. Flagship species are used in botanical gardens, protected areas, and other places to attract the public back to nature, and to invite people to care for biodiversity conservation (Zhu 2023). This approach may have some success, but EE still needs to address a second question: How to raise awareness of the vitality of species that are not so attractive to ecosystem integrity?

When adding the cultural aspect to biodiversity conservation, global cultural homogenization must be considered (Rozzi et al. 2018). This poses another challenge to education for the preservation of not only biodiversity but rather of biocultural diversity throughout the world. In places where there is or has been in the past (which is in fact the majority of the modern world) an intercultural geopolitical conflict, there is a tendency to actively forget and eliminate cultures that compete with the dominant culture (Maffi 2001). Consequently, EE programs need to address a third question: How to develop an awareness of the importance of biocultural diversity in this complex reality?

In this complex reality, global transportation systems interconnecting distant ecosystems and societies pose another challenge today (Collins et al. 2011). The global socioeconomic system causes the separation of people from their local ecosystems that used to support their lives. This separation poses two new challenging questions for EE:

(4) How to develop awareness and responsibility toward the local ecosystem and biocultural diversity?

(5) How to develop awareness and responsibility toward remote life-supporting ecosystems, which are out of sight and hence their interdependence with them is not perceived?

In the light of the rapid development of technology and "screen culture," which may further alienate people from natural environments, now, EE needs to face further questions: (6) How to attract students to stay away from the screens for a while? Or alternatively, (7) how to harness the screens in favor of quality and meaningful environmental education? And (8) how to incorporate local and international collaboration as an essential value of EE?

Understanding learning as a holistic process that combines body and mind (Dussler et al. 2023), another challenging question for EE is: (9) How to encourage diverse ways of knowing ecosystems through not only sciences but also arts and humanities to develop a sense of place and responsibility toward conservation of biocultural diversity?

The significant paradigm shift that FEP outlines as a basis for maintaining biocultural diversity also places demands on the nature of EE platforms. They need to provide in-depth knowledge and understanding of the social-ecological systems (Folke 2006). Since the ecosystem is conceptualized as a co-product of both human cultural and biophysical components of the environment, a long-term familiarity with the ecosystem is required. The need for knowledge based on long-term research and monitoring is reinforced by understanding ecosystems as open and dynamic (Collins et al. 2011; Folke et al. 2010).

27.3 Platforms for Fostering Biocultural Conservation

Part IV presents a variety of platforms for long-term environmental research and a variety of associated environmental education programs. Each of them fosters biocultural conservation under different theoretical and practical approaches.

27.3.1 Botanical Gardens

Joy Zhu (2023) describes botanical gardens as platforms that display plant species and their habitats. Botanical gardens are also used for long-term research and ex situ conservation of local species, and they connect the public with biodiversity through recreation and participatory educational activities. Exposure of visitors to scientific and cultural knowledge and direct experiences with biocultural diversity play a fundamental role in cultivating "field environmental philosophy" (Rozzi et al. 2006). The chapter includes a comparison between botanical gardens in China and Chile, thereby raising the possibility of international collaborations for understanding heterogenous ecosystems and cultures.

27.3.2 Biosphere Reserves

Elke Schüttler et al. (in this book) develop a comparative approach to examine the role of biosphere reserves as platforms for cultivating conservation of biocultural diversity in the spirit of FEP. Biosphere reserves were conceptualized and today are supervised by the UNESCO Man and Biosphere Programme (MaB). In contrast to preservationist approaches to conservation, biosphere reserves aim to integrate human needs and conservation practices (Schaaf 2003). These protected areas are understood as social-ecological systems, and their research platforms work in this vein over time. Schüttler et al. (in this book) show how countries highly committed to the biosphere reserve concept allow implementing sustainability in a wider array of policies. Similar biomes have similar conservation challenges and should learn from each other, highlighting the necessity to extend partnerships among biosphere reserves globally. The network of biosphere reserves offers an opportunity to essay FEP and suitable curricula in a broad spatial scale while deepening awareness of biocultural diversity.

27.3.3 Science Education Centers

Rob Dussler et al. (2023) introduce Science Education Centers as another platform for long-term environmental education. They describe the Meadows Center for Water and the Environment at Texas State University as a vital hub for understanding socio-ecological systems through mindfulness and interdisciplinary perspectives and approaches to cultivate sensitivity and awareness. Using FEP approach, this science education center focuses on a freshwater aquifer spring that nourishes the San Marcos River and has been used by humans since their arrival in North America. This rich biocultural setting creates unique opportunities for understanding the vital linkages between human and the ecosystem and for cultivating environmental responsibility.

27.3.4 Teachers College

Nimrod Aloni et al. (in this book) demonstrate how pre-service and in-service teacher training curricula can serve for introducing the new paradigm of eco-humanist education and field environment philosophy. The curriculum of the Kibbutzim College of Education in Tel Aviv, Israel, is based on interdisciplinary collaboration between science and humanist lecturers. It is also a good platform for the long-term monitoring of the effects of the curricula on paradigm shifts in teachers and students. In this way, this College of Education becomes a distribution point for a new educational paradigm that promotes integrative sustainability to foster the quality of life of the individual, the community, and the natural environment.

27.3.5 LTER and LTSER

Long-term ecological research sites (LTER) are dedicated to understanding the structure and functioning of a complex and delicate ecosystem at their locations (Holzer et al. 2018). Understanding of the vital linkages of the inhabitants, habitats, and habits is achieved through building long-term and interdisciplinary research and collaboration with other LTER stations in the global network (Dick et al. 2018). Therefore, LTER stations by their very nature educate for collaboration (between disciplines, geographical areas, and generations), ecological literacy, and scalar thinking (in time and space).

Frederick J. Swanson (in this book) presents the Andrews Experimental Forest LTER, Oregon, USA. The experimental Forest site was established as a collaboration between US Forest Service system and scientists, in order to support forest exploitation. Since 1980, Andrews Experimental Forest has become an LTER site and gradually established strong partnerships between scientists, forest managers, artists, and ethicists. Education programs and public tours, developed jointly by the LTER scientists and the Art Center at the site, have brought about a shift in residents' perception of this native forest, from their utilitarian values to their intrinsic value. The chapter describes this perceptual change and its causes through the lens of biocultural ethics. The sharing of programs with other LTER sites, such as the Omora Ethnobotanical Park in Chile, has led to mutual inspiration and refinement of the educational approach. This process highlights the importance of collaboration between LTER stations worldwide in order to adapt the educational programs to the specific set of biocultural conditions at each site.

A detailed description of one of the curricula developed in Andrews Experimental Forest LTER can be found in Julie Markiewicz et al. (in this book). The EE program is based on the FEP approach and combines scientific knowledge of the forest, along with exposure to art that was inspired by the forest. The curriculum also includes hands-on experience in science and the arts. This combination of science and art serves as an original and inspiring way of fostering conservation of biocultural diversity. The chapter presents a study that examines the impact of learning on the environment in a multi-sensory approach and, using diverse ways of knowing, on the reconnection of people to their habitat and fostering the value of biocultural conservation.

Noa Avriel-Avni et al. (in this book) present the long-term social-ecological research platform (Mirtl et al. 2018) for reframing sense of place. Adding the social dimension to LTER incorporates research on perceptions of ecosystems, now understood as social-ecological systems. This perspective gives legitimacy to diverse ways of understanding the linkages of the inhabitants, habitats, and habits. LTSER platforms typically span geographic areas that include several social-ecological systems and therefore challenge the concepts of environmental responsibility (Rozzi et al. 2012). The chapter is based on the LTSER platform in the Negev Highlands, Israel. It proposes an understanding of "place" as co-produced by humans and nonhumans through long-term processes and multiple spatial scales (Massey 1997). In this way, the sense of place concept also changes and becomes a suitable basis for local and global environmental education.

Despite the differences among the environmental education platforms that are presented in the various chapters of Part IV, they share three common characteristics: (a) understanding ecosystems as dynamic and open social-ecological systems; (b) integrating long-term research and monitoring of changes in ecosystems and education; and (c) welcoming different ways of knowing the social-ecological system.

27.4 Insights for Environmental Education

EE curricula outlined in this part of the book encourage paradigmatic transformations. Some explicitly implement FEP and others share FEP's theoretical framework and use similar educational approaches. Moreover, some chapters suggest expanding FEP. Three paradigmatic shift questions proposed by Goldman et al. (2018) help describe these main outlines.

1. *The ontology question: What the world is?*
   Environmental education (EE) programs in Part IV seek to foster a perception of the world:

   - As a social-ecological system, which is open, dynamic, and complex.
   - As the outcome of co-production by human and non-human co-inhabitants over the long term.
   - As social-ecological systems that are perceived differently, depending on the culture.

In order to change the perception of the world, EE uses a variety of notions and methods:

- Exposure of students to the biological and cultural diversity within the SES.
- Focus on little perceived undervalued biota, including as "flagship species" little organisms.
- Emphasis on integrative and locally centered educational approaches.
- Practices at multiple scales in time and space.

2. *The epistemological question: How do we know the world?*
   EE programs included in this part of the book operate with the understanding of and respect for multiple ways of knowing and co-inhabiting the world. Hence, they expose students to social-ecological systems and biocultural diversity using:

   - Contemplative environmental education
   - A holistic approach
   - A combination of scientific literacy alongside art, music, and storytelling
   - Exposure of students to different ways of understanding a place from the perspective of diverse cultures that co-inhabit ecosystems.

The constructivist approach underlying EE bases learning on the students' personal experience using:

- Outdoor education
- Active learning, whether it is scientific research learning or socio-ecological action research.

3. *The ethics question: What are the desired relationships between humans and world?*

   The understanding that paradigmatic changes also include ethical changes leads EE programs to:

   - Expose students to biological and cultural diversity and emphasize that we share our habitats with other co-inhabitants.
   - Emphasize the value of collaboration among humans (from different cultures) for conservation of biocultural diversity.
   - Highlight the effects of our life habits on ecosystems and the importance of taking responsibility for those habits.
   - Foster biocultural conservation.

Each of the chapters that comprise Part IV responds, in its own way, to these questions. In so doing, they present a holistic vision integrating the sciences, arts, and humanities into participatory environmental education.

References

Ardoin NM, Bowers AW, Gaillard E (2020) Environmental education outcomes for conservation: a systematic review. Biol Conserv 241:108224

Blühdorn I (2017) Post-capitalism, post-growth, post-consumerism? Eco-political hopes beyond sustainability. Glob Discourse 7(1):42–61

Capra F (1996) The web of life: a new synthesis of mind and matter. HarperCollins, New York

Collins SL, Carpenter SR, Swinton SM, Orenstein DE, Childers DL, Gragson TL, Whitmer AC (2011) An integrated conceptual framework for long-term social ecological research. Front Ecol Environ 9(6):351–357

Costanza R, de Groot R, Sutton P, van der Ploeg S, Anderson SJ, Kubiszewski I, Farber S, Turner RK (2014) Changes in the global value of ecosystem services. Glob Environ Chang 26(1):152–158

Dick J, Orenstein DE, Holzer JM, Wohner C, Achard AL, Andrews C et al (2018) What is socio-ecological research delivering? A literature survey across 25 international LTSER platforms. Sci Total Environ 622:1225–1240

Dussler R, Williams J, Massey S (2023) Mindfulness and reconnection with freshwater ecosystems. In: Rozzi R, Tauro A, Wright T, Avriel-Avni N, May RH Jr (eds) Field environmental philosophy: education for biocultural conservation. Springer, Dordrecht

Fletcher R (2017) Connection with nature is an oxymoron: a political ecology of "nature-deficit disorder". J Environ Educ 48(4):226–233

Folke C (2006) Resilience: the emergence of a perspective for social–ecological systems analyses. Glob Environ Chang 16(3):253–267

Folke C, Carpenter SR, Walker B, Scheffer M, Chapin T, Rockström J (2010) Resilience thinking: integrating resilience, adaptability and transformability. Ecol Soc 15(4) http://www.ecologyandsociety.org/vol15/iss4/art20/

Goldman MJ, Turner MD, Daly M (2018) A critical political ecology of human dimensions of climate change: epistemology, ontology, and ethics. Wiley Interdiscip Rev Clim Chang 9(4):e526

Gould WA, González G, Walker DA, Ping CL (2010) Commentary. Integrating research, education, and traditional knowledge in ecology: a case study of biocomplexity in arctic ecosystems. Arct Antarct Alp Res 42(4):379–384

Holzer JM, Adamescu MC, Bonet-García FJ, Díaz-Delgado R, Dick J, Grove JM et al (2018) Negotiating local versus global needs in the International Long Term Ecological Research Network's socio-ecological research agenda. Environ Res Lett 13(10):105003

Hursh D, Henderson J, Greenwood D (2015) Environmental education in a neoliberal climate. Environ Educ Res 21(3):299–318

Ishwaran N, Persic A, Tri NH (2008) Concept and practice: the case of UNESCO biosphere reserves. Int J Environ Sustain Dev 7(2):118–131

Leopold AC (2004) Living with the land ethic. Bioscience 54(2):149–154

Mackey B, Claudie D (2015) Points of contact: integrating traditional and scientific knowledge for biocultural conservation. Environ Ethics 37(3):341–357

Maffi L (2001) On biocultural diversity: linking language, knowledge, and the environment. Smithsonian Institution Press, Washington, DC

Massey D (1997) A global sense of place. In: Gray A, McGuigan J (eds) Studying culture. Edward Arnold, London, pp 232–240

McDonnell MJ, Pickett STA (eds) (1993) Humans as components of ecosystems: the ecology of subtle human effects and populated areas. Springer, New York

Mirtl M, Borer ET, Djukic I, Forsius M, Haubold H, Hugo W, Jourdane J, Lindenmayer D, McDowell HM, Muraoka H, Orenstein DE, Pauw JC, Peterseil J, Shibata H, Wohner C, Yu X, Haaseel P (2018) Genesis, goals and achievements of long-term ecological research at the global scale: a critical review of ILTER and future directions. Sci Total Environ 626:1439–1462

Rozzi R (2013) Biocultural ethics: From biocultural homogenization toward biocultural conservation. In: Rozzi R, Pickett STA, Palmer C, Armesto JJ, Callicott JB (eds) Linking ecology and ethics for a changing world: values, philosophy, and action. Ecology and Ethics, vol 1. Springer, Dordrecht, pp 9–32

Rozzi R, Massardo F, Anderson C, Heidinger K, Silander J (2006) Ten principles for biocultural conservation at the southern tip of the Americas: the approach of the Omora Ethnobotanical Park. Ecol Soc 11(1):43. http://www.ecologyandsociety.org/vol11/iss1/art43/

Rozzi R, Arango X, Massardo F, Anderson C, Heidinger K, Moses K (2008) Field environmental philosophy and biocultural conservation: the Omora Ethnobotanical Park educational program. Environ Ethics 30(3):325–336

Rozzi R, Anderson CB, Pizarro JC, Massardo F, Medina Y, Mansilla A et al (2010) Field environmental philosophy and biocultural conservation at the Omora Ethnobotanical Park: methodological approaches to broaden the ways of integrating the social component (“S”) in long-term socio-ecological research (LTSER) sites. Rev Chil Hist Nat 83(1):27–68

Rozzi R, Armesto JJ, Gutiérrez J, Massardo F, Likens G, Anderson CB, Poole A, Moses K, Hargrove G, Mansilla A, Kennedy JH, Willson M, Jax K, Jones C, Callicott JB, Kalin MT (2012) Integrating ecology and environmental ethics: earth stewardship in the southern end of the Americas. Bioscience 62(3):226–236

Rozzi R, Roy H Jr, Chapin FS III, Massardo F, Gavin MC, Klaver IJ, Pauchard A, Nuñez MA, Simberloff D (eds) (2018) From biocultural homogenization to biocultural conservation. Springer International Publishing, Dordrecht

Schaaf T (2003) Biosphere reserves. In: Harmon D, Putney AD (eds) The full value of parks: from economics to the intangible. Rowman & Littlefield, Lanham, MD, pp 185–196

Schultz PW (2001) The structure of environmental concern: concern for self, other people, and the biosphere. J Environ Psychol 21(4):327–339

Tauro A, Ojeda J, Caviness T, Mosses K, Moreno R, Wright T, Zhu D, Poole A, Massardo F, Rozzi R (2021) Field environmental philosophy: a biocultural ethic approach to education and ecotourism for sustainability. Sustainability 13:4526. https://doi.org/10.3390/su13084526

Vining J, Merrick MS, Price EA (2008) The distinction between humans and nature: human perceptions of connectedness to nature and elements of the natural and unnatural. Hum Ecol Rev 15(1):1–11

Zhu D (2023) Bridge the channel, enhance the inclusivity: a comparison between flagship species-centered and moss-centered conservation in Chile and China. In: Rozzi R, Tauro A, Wright T, Avriel-Avni N, May RH Jr (eds) Field environmental philosophy: education for biocultural conservation. Ecology and ethics, vol 5. Springer, Dordrecht

Chapter 28
Bridge the Channel, Enhance the Inclusivity: A Comparison Between Flagship Species-Centered and Moss-Centered Conservation in Chile and China

Danqiong Zhu

Abstract I contrast flagship species-centered approach and moss-centered approach in biological conservation. The context I conduct this research is the status quo of contemporary ecological sciences and communication. Like other science fields, ecological science has been inundated by massive and diverse sources of information and data. The affluence of ecological studies is valuable to address pressing socio-environmental issues, but it is not sufficient because big data are available and usable almost exclusively for scientists. The channel between the academia and the field needs a bridge. Multiple communication approaches, such as paper tools, participation, education, ecotourism, and social medias, are also essential to socially incorporate ecological information like a bridge of arches, decks, and cables. In terms of conservation, flagship species offer an effective way to engage the public and attract the attention of policy makers. However, there are two limitations. First, flagship species almost exclusively have included vertebrate animals and vascular plants. Second, citizens access quite incomplete ecological information in a flagship species-centered conservation and communication. To overcome the first limitation, I compare the effectiveness of flagship species based on vertebrates with novel approaches that include nonvascular plants. Regarding the second limitation, I examine how ecological interrelations with other co-inhabitants are understood by citizens in educational and tourism programs associated with bryophytes' conservation in the Omora Ethnobotanical Garden, Chile, and the Shenzhen Fairy Lake Botanical Garden, Shenzhen, China. We conclude that the bridge between ecological science and citizens functions more effectively once the inclusivity of ecological sciences is guaranteed.

D. Zhu (✉)
Department of Philosophy and Religion, Subantarctic Biocultural Conservation Program, University of North Texas, Denton, TX, USA

Xidian University, Xi'an, Shaanxi, China Cape Horn International Center (CHIC), Universidad de Magallanes, Puerto Williams, Chile
e-mail: Danqiong.Zhu@unt.edu

R. Rozzi et al. (eds.), *Field Environmental Philosophy*, Ecology and Ethics 5,
https://doi.org/10.1007/978-3-031-23368-5_28

Keywords Biocultural conservation · Flagship species · Inclusivity · Moss-centered · Taxonomic bias

28.1 Introduction

For reconnecting society and nature, botanical gardens can play a fundamental role in cultivating "field environmental philosophy" and other forms of environmental education (Rozzi et al. 2006; Kono 2018). These gardens connect the public with biodiversity through recreation and participatory educational activities (He and Chen 2012; Nadkarni 2013; Poole et al. 2013). Additionally, they have a part in research and *ex situ* conservation, as well as in sharing scientific information with citizens (Chen and Sun 2018; Faraji and Karimi 2020). In this chapter, I will elaborate on how FEP could be relevant to strengthen the role that botanical gardens and protected areas play to enhance social awareness about the severity of the contemporary global socio-environmental crisis in the Anthropocene. We need an awareness about the gravity of the crisis as well as how the majority of living beings are underperceived and undervalued due to taxonomic biases in conservation efforts to save planetary biodiversity (Rozzi 2019). In global conservation and communication, there are two main reasons for exclusion. The first reason is the *exclusion of the majority of citizens* due to limitations of access to information and data produced by scientists (Dawson 2014). Without citizens' participation, ecological restoration and conservation are tepid (Rozzi et al. 2006); hence, science research and education need to be more broadly communicated and shared. The second reason is the *exclusion of the majority of living beings or "co-inhabitants"* due to a lack of consideration of tiny organisms as target species for conservation or educational materials to inform citizens (Rozzi 2019).

Regarding the first problem of exclusion, in global society the attitude toward conservation is lukewarm, with little identifiable success. For example, Lera Miles and collaborators (2006) found that tropical dry forests are critically threatened, but they are not in the forefront of public concerns. John Croxall and his team (2012) discovered that commercial fisheries and pollution, alien invasive predators, habitat degradation, and human disturbance continually and increasingly threaten seabird conservation. In many protected areas and biodiversity hotspots, human activities continue to hazard ecosystems and biological conservation. For example, Megha Verma and collaborators (2020) used the most recent human footprint dataset to demonstrate increasing impacts of crops, pastureland, roads, and infrastructures on biodiversity in Southeast Asia, which in turn reduces the efficacy of conservation.

Efforts by citizens are essential for biological conservation to succeed. Hence, it is necessary that the public has access to data regarding human pressure and how it continually endangers global and local biodiversity. Even if big data are accessible to citizens, how to communicate such information meaningfully is a problem. Inaccessibility to ecological information and limitations of science communications hinder the commitment of citizens to conservation. As illustrated by pioneering environmentalists like Aldo Leopold (1949) and Rachel Carson (1962), it is essential to motivate citizens by purposely writing for the general public. Their books *A Sand County Almanac* and *Silent Spring* continue to inspire young generations. These are pertinent models for communicating with public today.

Regarding the second problem of exclusion, many conservation researchers and organizations have focused on flagship species (Walpole and Leader-Williams 2002; McGowan et al. 2020). The organisms chosen as flagship species commonly are "charming species," which are privileged because they trigger greater concerns than "not-so-charming" species. Ricardo Rozzi (2002) has criticized that taxonomically speaking, flagship species are usually vertebrates like mammals and birds, or vascular plants like gymnosperms and angiosperms. This taxonomic bias leads to the exclusion of small animals and plants, which is deeply associated with global biocultural homogenization and a "taxonomic chauvinism" (Rozzi 2019). Barbara Clucas and her team (2008) expressed alarm that mammals and bird species—rather than invertebrates, fish, and amphibians—are prioritized for conservation, thus exposing the public to only a few flagship species. Gregor Kalinkat and his collaborators (2016) are also concerned that aquatic biodiversity is overlooked in conservation. In other words, numerous invertebrates and nonvascular plants are invisible to the public. Therefore, it is crucial to educate citizens to "open their eyes" to see these marginalized organisms.

The exclusion of the majority of citizens and that of the diversity of living beings in conservation efforts compel me to undertake the dual task of:

1. Bridging the scientific academy and citizens who are situated in different local cultures and contexts
2. Including a more comprehensive taxonomic diversity of the co-inhabitants in conservation.

To contribute to solving these two forms of exclusions in conservation, in this chapter, I will examine traditional charismatic flagship species-centered and lesser-known moss-centered conservation initiatives in Chile and China. Based on this analysis, I explore promising ways to bridge the channel between ecological sciences and citizens, and enhance taxonomic inclusivity. In Chile and China, moss-centered conservation and communication with citizens have succeeded through implementing multiple activities and projects that display the possibility and necessity of including invertebrates and nonvascular plants in conservation and education. These initiatives highlighted the ecological, aesthetic, cultural, and ethical values of those marginalized organisms, and stated that an understanding of these values is essential to foster biocultural conservation in order to secure the future of the planet (Rozzi et al. 2008a, 2012a).

28.2 Flagship Species-Centered Approach

In conservation biology, flagship species usually represent those charming species that can attract public attention. The World Wide Fund for Nature (WWF), formerly World Wildlife Fund, defines the interrelated concepts of flagship, keystone, and indicator species. Flagship species "act as an ambassador, icon or symbol for a defined habitat, issue, campaign or environmental cause; are usually relatively large, and considered to be 'charismatic' in western cultures; may or may not be keystone species and may or may not be good indicators of biological process" (WWF 2020).

From the angle of ethics, it is odd that the definition of flagship species is both anthropocentric (as humans decide which species are charismatic) and non-anthropocentric (as other-than-human species are valued). This definition is not ecocentric (as flagship species might or might not be keystone species) but biocentric (as one particular species becomes the conservation target). In terms of values, the focus is on the "charming" features of the species, and therefore, aesthetic or socioeconomic values outweigh ecological and ethical values. In this section, I will briefly review the use of flagship species as conservation targets in Chile and China, particularly with regard to China's Giant Panda (*Ailuropoda melanoleuca*) and Chile's Magellanic Woodpecker (*Campephilus magellanicus*). From the perspective of ecosystem management, Daniel Simberloff (1998) cautioned about the risks of focusing on single-species management, and using conservation strategies based on flagship, umbrella, and/or endangered species, because they overlook ecological interactions and the relevance of conserving biotic communities. From a philosophical perspective, the implications of a narrow focus on flagship species are probably greater than the ecological ones.

28.2.1 Case Studies of Flagship Species-Centered Approach

28.2.1.1 The Magellanic Woodpecker Conservation in Cape Horn Biosphere Reserve, Chile

To create the Cape Horn Biosphere Reserve, it was not coincidental that the Magellanic Woodpecker (Fig. 28.1) was chosen as the flagship species. The creation of this biosphere reserve was led by the interdisciplinary research team of the Omora Ethnobotanical Park, a biocultural garden established in 2000. It covers an area of 1100 ha near Puerto Williams, the capital of the Chilean Antarctic Province (Rozzi et al. 2006). A lot of research and efforts disclosed the ecological and cultural significance of this woodpecker. Early in the 1870s, a Yahgan-English dictionary compiled by the Anglican missionary Thomas Bridges included the word *lana* as the Yahgan name for the Magellanic Woodpecker (Rozzi et al. 2010a). Úrsula Calderón, one of the last two Yahgan people who at that time spoke her native language, explained to Ricardo Rozzi (2002) that "*lan*" in Yahgan means "tongue." Hence, the

Fig. 28.1 Magellanic Woodpecker in a forest close to Puerto Williams, Chile. Photo by Danqiong Zhu during her field environmental philosophy practice at Omora Park

bird's name *lana* could be associated with the skillful habit of extracting larvae from tree trunks (Rozzi et al. 2010a). It is notable that the scientific name *Campephilus* means "caterpillar lover" (Rozzi and Massardo 2011).

A woodpecker of the genus *Campephilus* is found only in the Americas (Arango et al. 2014). The close phylogenetic relationship between the Magellanic Woodpecker and the North American Ivory-billed and Imperial Woodpeckers (*Campephilus principalis and C. imperialis*) attracted US ornithologist Lester L. Short to study Magellanic Woodpeckers in the 1970s. His research disclosed valuable information regarding distribution, foraging behavior, sounds produced by bills, vocalizations, and morphology of the Magellanic Woodpecker, and compared it with other woodpeckers of the genus *Campephilus* (Short 1970).

Before the establishment of the Cape Horn Biosphere Reserve, researchers investigated the ecology of Magellanic Woodpeckers, discovering a close relationship for foraging and cavity-nesting with trees of the genus *Nothofagus* (i.e., the beeches of the "South") (Rozzi et al. 2010a; Duron et al. 2018). Magellanic Woodpeckers may play a key ecological role because in the harsh climatic conditions of Cape Horn their cavities are used by other birds (Vergara and Schlatter 2004). In 2005, to implement sustainable economic activities, and avoid negative environmental and social impacts, Omora Park researchers suggested identifying a flagship or charismatic species in order to enhance the participation of the local community. Questionnaires and interviews of members of the indigenous *Yahgan* community, teachers, students, Navy personnel, and authorities showed a marked preference for the Magellanic Woodpecker (Arango et al. 2014). Omora Park study

Fig. 28.2 In April 2007, in coordination with the Municipality of Cape Horn, the Omora Park team celebrated the Earth Day at the local school of Puerto Williams by distributing reusable cloth bags printed with the images of the Magellanic Woodpecker drawn by three students that had won a contest called "The Magellanic Woodpecker and its habitat." The cloth bags were given to local families to deter the use of single-use plastic bags and promote the protection of the woodpeckers' habitats in the Cape Horn Biosphere Reserve, Chile. Photograph by Ricardo Rozzi

included not only the preference of local communities but also ecological attributes for conservation, ethical values, endemic geographical distribution, conservation status, and biocultural or ethnobiological values. This research provided a theoretical foundation for designing a practical program to implement the Magellanic Woodpecker as a flagship species. Copious activities with the local community raised awareness about the need to conserve this species and its forest habitats (Arango et al. 2014, also Fig. 28.2).

The recognition of the Magellanic Woodpecker as a flagship species, fueled by theoretical contribution and practical efforts, eventually had a profound regional and international influence. At the theoretical level, the "3Hs" (habitats, habits, and co-in-habitants) framework of the biocultural ethic linked this species with its ecological roles and value for the conservation of *Nothofagus* forests (Rozzi 2018, pp. 33–34). At the practical level the methodology of "field environmental philosophy" (FEP) oriented the creation of conservation narratives and the design of participatory field activities (Rozzi et al. 2010b). Combined, these theoretical and practical frameworks exhibited their potential for both education and conservation. Additionally, a multi-scale institutional structure brought together local, national, and international private and public organizations and universities (Rozzi et al. 2006). The set of charismatic features, the endangered conservation status, the key ecological role, and rich traditional ecological knowledge linked to the Magellanic Woodpecker successfully grabbed the public's attention (Arango et al. 2014).

28.2.1.2 Panda Conservation in Foping National Natural Reserve, China

Pandas enjoyed their reputation first in Europe, then in North America, and lastly in its native land, China. It was discovered in 1869 by a French missionary and naturalist, Armand David, who sent specimens to the Muséum National d'Histoire Naturelle in Paris (Holland 2016). The first scientific description of the species was published in the early 1870s. In 1936, US fashion designer Ruth Harkness captured a baby panda and took it to Chicago's Brookfield Zoo, which attracted 53,000 visitors on the first day of its exhibition (Croke 2005, p. 1). At the same time, Chinese people were enmeshed in a web of continual domestic wars, the intrusion of Japan, unstable politics, and poverty without attending to the panda's reputation abroad. Even in 1961, when a panda was sent to the London Zoo and the newly founded WWF adopted the image of a panda as the logo, Chinese citizens seemed to be absent from this Western panda zeal. Chinese public appreciation was stimulated in 1972 when former China Premier Zhou Enlai presented two giant pandas as a gift to the USA to initiate diplomatic relations (Buckingham et al. 2013). The US–China relationship had a huge domestic and diplomatic impact on the public, and the establishment of Foping National Nature Reserve followed these events.

In the 1960s, Giant pandas in the Foping region attracted the attention of biologists after a group of ornithologists conducted fieldwork on birds, and accidentally found incomplete skulls of giant pandas on the south side of Qinling Mountain. In 1964, ornithologists Guangmei Zheng and Xu Pingyu published their discovery. In 1973, the Shaanxi Biological Resources Survey team conducted large-scale research on Qinling Mountain. They suspected that the population and density of giant pandas in Foping County were the highest in China. In 1978, China's State Council approved the establishment of Shaanxi Foping National Nature Reserve, covering a total area of 350 km^2 (app. 86,487 acres) to conserve giant pandas and their habitats. Roughly during that time, the WWF signed an agreement with China to cooperate in the conservation of giant pandas.

Since the 1970s, China has conducted four national surveys of giant pandas. The most recent survey included quantitative information about population density, genetic diversity, isolated population, habitat types, conservation management, and the status of captive populations (China State Forestry Administration 2015). The highest wild population was detected in the Foping National Nature Reserve.

Later in the 1980s, several ecological and biodiversity studies were conducted in Qinling Mountain, and fortunately in the Foping region researchers observed wild giant pandas. Later, researchers began studying pandas without interfering their life habits. Eventually, a female giant panda became accustomed to the presence of people (Schaller 1994, p. 119).

In the 1990s, new technologies such as GIS and radiotracking facilitated the ecological surveys, which enhanced investigation about the demography, geographical distribution, seasonal movement, and the relationship between seasonal movement and habitat parameters. From 1991 to 1995, researchers mapped the home range patterns of six radio-collared giant pandas, including three males and three females in Foping National Nature Reserve. Their results were published years later (Yong et al. 2004, pp. 158–169). Regarding the seasonal vertical movement of the giant pandas and their activity range, the radiotracking data show that the pandas live in their winter habitats for almost three quarters of the year, and in the summer habitats for only one quarter of the year (Liu et al. 2002). The seasonal migration of pandas called attention to their movements for habitat management, particularly to estimate their required protected areas. Pandas select their preferred habitats based on different bamboo species as food. In winter, giant pandas select *Bashania fargesii*, a bamboo with short and dense culms distributed in middle and southwest China, and in summer *Fargesia qinlingensis* with tall and sparse culms (Liu et al. 2005). Research on food preferences provided insights for restoration of panda habitats. This limited economic activities in some local areas, but also stimulated a paradigm shift in conservation that focused on the nutritional basis of food selection (Nie et al. 2015). Differences of seasonal movements between Foping Pandas and Wolong Pandas demonstrated that local habitat conditions play an important role in their behaviors (Liu et al. 2015). Hence, reintroducing pandas to wilderness needs to adopt different strategies based on local habitat conditions.

Five decades of research have enriched ecological information, panda conservation, and public education. The scientific, ecological, and cultural values of giant pandas have become common knowledge among educated Chinese (Fig. 28.3). The panda is also an umbrella species for the protection of habitats and other endangered co-inhabitants (Li and Pimm 2016; Li et al. 2020). Pandas also became economically important, and in 2010 visits to their reserves yielded US $2.6–6.9 billion per year (Wei et al. 2018). Chinese citizens were soon informed through multiple channels about this economic success of panda conservation (Fig. 28.4). Based on their ecological and aesthetic values, and contributions to local economy, the panda was designated as one of the "four treasures of Qinling Mountains," along with Crested Ibis (*Nipponia nippon*), Snub-nosed Monkey (*Rhinopithecus roxellana qinlingensis*), and Golden Takin (*Budorcas taxicolor bedfordi*). The concept of four treasures or "Sibao" (四宝, Fig. 28.5) in Chinese refers to the rareness of those animals and their multiple values, such as aesthetic value and their kinship with humanity, not exclusively economic value. The Chinese word "bao" etymologically refers to "jade," generated as the quintessence of heaven and earth by the power of nature. In modern everyday language, used as either a noun or an adjective, "bao" can convey meanings such as "precious," "treasure," "kinship," "distinguished," or "dearie."

Fig. 28.3 Exhibition in Foping Giant Panda Valley Scenic Area that introduces citizens to the domestic distribution and overseas journeys of the pandas as icons of international friendship. Photo by Meng Jia

Fig. 28.4 This panda in Foping Giant Panda Valley Scenic Area was named as "*Thinking Panda-Life*" by Dr. Zhu's Chinese student. Photo by Chao Gao

Fig. 28.5 Left: "Four Treasures of Qinling" inspired animations that released their theme song "The Spirit of the Heaven and Earth" on April 21, 2020. The series tells stories about their habits, unique life habits that include surviving skills, fighting for their habitats, struggles in everyday life, and their will to thrive. Source: *Sanqin Daily*. Right: "Four treasures of Qinling" were designed as the mascot of China's 14 National Games which were held in Xi'an, Shaanxi, in 2021. Photo by Meng Jia

28.3 Mosses-Centered Approach

Globally the awareness of bryophytes' conservation is novel. Despite the fact that the British Bryological Society, the first organization for bryological research and conservation, was founded in 1896, and that the International Association of Bryologists (IAB) was established in 1969, there is still little public awareness about mosses (Hallingbäck and Hodgetts 2000). Nevertheless, earlier taxonomic identification of bryophytes was not exclusive to professional bryologists, but also amateurs were very active in the field (Lawley 2019).

Some non-European cultures enjoy ancient traditions appreciating the beauty of mosses. For instance, in Chinese classical literature, mosses comprise a unique topic for generations of writers composing poetry, eulogy, and verses. In Japan, *Saihō-ji*, monks at the vintage Buddhist temple collected more than 120 species of bryophytes.

The exclusion of mosses in biological conservation might be interpreted as a twentieth-century expression of what Rozzi (2012a) has called *global biocultural homogenization*. Mosses receive much less attention than megafauna and vascular plants in conservation as well as in formal and non-formal education. Consequently, people's perspectives on the taxonomic breath of biodiversity are reduced (Rozzi 2013). To counteract this trend and to invigorate "moss-centered conservation," in this section I will review two cases studies. The Omora Ethnobotanical Park in Chile and Shenzhen Fairy Lake Botanical Garden in China exemplify successful moss-centered approaches. An analysis of the concepts, methodologies, and activities conducted at these botanical gardens can disclose multiple epistemic, ecological, cultural, and ethical values of mosses.

28.3.1 Case Studies of Moss-Centered Approaches to Conservation

28.3.1.1 "Ecotourism with a Hand Lens" at the Omora Ethnobotanical Park, Chile

The Omora Ethnobotanical Park, a botanical garden, is the research, education, and conservation center that has led the creation and implementation of the Cape Horn Biosphere Reserve (CHBR) (Rozzi et al. 2006; Moreira-Muñoz et al. 2020). The historical process of Omora Park and the creation of ecotourism with a hand lens (EHL) have been summarized by Rozzi and collaborators (2020).

In 2001, the former president of Chile, Ricardo Lagos, visited Omora Park and experienced the early stage of EHL. As a result, President Lagos supported not only the Omora Park but also the creation of the CHBR and stimulated the establishment of the Cape Horn Center to be a world observatory for climate change and more broadly global socio-environmental change. The mission of the center would be to focus on the "micro-world" represented by mosses and other small organisms, but scaling up to the global biosphere, thereby contributing to planetary sustainability (Rozzi et al. 2020).

In subsequent years, Omora Park researchers discovered a unique and endemic richness of moss species, other bryophytes, and lichens. Cape Horn was identified as a world biodiversity hotspot for bryophytes and therefore deserved consideration *per se* in conservation (Rozzi et al. 2008b). For the first time in history, a protected area was created based on the diversity of little "non-charismatic" organisms. This changed the perspective that small organisms can be effectively protected only as a consequence of protecting large charming "umbrella species" (Rozzi 2012b).

For the former achievement, networking and applying the four-step cycle of FEP methodology was essential. FEP includes not only research but also communication, participatory creation of metaphors, and narratives. Rozzi (1999, p. 917) has affirmed that metaphors become strong "cultural messengers from science to ethics" and culture. FEP also includes field activities guided with an ecological and an ethical orientation to appreciate the biophysical and cultural diversity embedded in ecosystems. Finally, FEP ask participants to undertake an *in situ* conservation action to foster a sense of responsibility among them regarding their links with other co-inhabitants (Rozzi et al. 2012b). Through FEP's four-step cycle, researchers were themselves stunned by the diverse cornucopia of formerly "invisible" nonvascular plants. Omora Park researchers investigated not only biological diversity but also the cultural diversity embedded in vernacular names, as well as the etymology of scientific names. Visitors, especially schoolchildren, propose names for moss species that do not have common names. Together with Omora Park

Fig. 28.6 Photo taken using a hand lens to observe the feather-like gametophyte and bean-like capsule of *Hypnum skottsbergii* at one of the stations along the Miniature Forests Interpretative Trail at the Omora Ethnobotanical Park. Photo by Danqiong Zhu

researchers, visitors also compose metaphors, including the "Miniature Forests of Cape Horn" (Rozzi 2012b). This metaphor effectively communicates the richness of tiny organisms that co-inhabit with mosses.

During the following period, EHL provided new opportunities for multiple visitors, including tourists, public authorities, philosophers, artists, teachers, and students, to have "face-to-face encounters" with mosses, liverworts, hornworts, and lichens, and to appreciate their diverse life habits and habitats (Rozzi et al. 2012b). Participants are awed by their direct encounters with nonvascular plants (Fig. 28.6). EHL is a guided field activity that connects people with other co-inhabitants and opens them to unrivaled ecological, aesthetic, and ethical values. These values have remained invisible due to biocultural homogenization (Rozzi et al. 2014). Eventually, *in situ* conservation led to the creation of the Miniature Forests Interpretive Trail at Omora Park. The trail, in turn, became a platform for research, education, communication, and conservation activities, including new forms of ecotourism (Rozzi 2012b).

The Chilean government supported this kind of diversification of tourism (Fig. 28.7). It catalyzed a series of workshops and published bilingual books in Spanish and English, which included field guides for the bryophytes and lichens of the Miniature Forests of Cape Horn (Goffinet et al. 2006, 2012) and introduced the EHL activity to the general public and tour guides (Rozzi et al. 2006, 2012a, b; Rozzi 2018). Rozzi (2012c, p. 45) highlighted that "small is diverse, small is beautiful, and small is essential."

Fig. 28.7 Local people decorate souvenirs with lichens and mosses. Photo by Danqiong Zhu

In recognition of the importance of bryophyte research, and the creation of the protected area in Cape Horn, and a novel ecotourism activity, the International Association of Bryologists (IAB) held its 2015 IAB Conference at the Omora Ethnobotanical Park, Chile. Participants came from all around the world. A strong delegation from China included researchers from the Shenzhen Fairy Lake Botanical Garden, China, who remained in contact with Omora Park researchers.

28.3.1.2 "The Miniature Angels in the Plant Kingdom" at the Shenzhen Fairy Lake Botanical Garden, China

The Shenzhen Fairy Lake Botanical Garden (SFLBG) is located in the Luohu District, Shenzhen, Guangdong province of southeast China. Founded in 1983, it covers an area of 546 hectares. Renowned as "An Emerald Embedded in the Metropolis," it combines plant collections, scientific research, popular science, recreation, and tourism (Li et al. 2018). A bryophyte research team was formed in 2006 with the initial goals of conducting bryology-related research, particularly focusing on taxonomy, diversity cataloging, genealogy, conservation, and horticulture. Zhang Li (2020) has provided a historical account of the bryological work at SFLBG.

In August 2012, bryologists from SFLBG participated in an expedition to Tibet. In a mountain village at an altitude of 4000 m, a researcher rediscovered a cluster of bryophytes that they had never observed. After 2 years of searching records and evidence from the US National Herbarium, they finally identified this species as

Fig. 28.8 *Impression, Eastern Himalayas* (30″ × 40″, acrylic on canvas) painted by Shihua Li, the artist from the Shenzhen Fairy Lake Botanical Garden. It reflects bryophytes' diversity in the Eastern Himalayas. The painting includes various species of mosses (31 species), liverworts (5 species), and even a hornwort. This work of art pays tribute to Ernest Häckel. It also encourages the audience to play a treasure hunting game to identify species ("the hidden angels"), and it offers an opportunity for the audience to observe morphological characteristics and habitats of bryophytes. Source: Shenzhen Fairy Botanical Garden

Brachymeniopsis gymnostoma, the only moss species that had been previously announced as extinct. In January 2015, Zhang Li and collaborators reported this rediscovery at the International Association of Bryologists (IAB) Conference held at the Omora Ethnobotanical Park, Chile (https://chile.unt.edu/iab2015/en-resumenin-brief).

The team at SFLBG includes bryologists and artists. Science and art play significant roles in conveying scientific information to citizens (Fig. 28.8). Compelling photographs, paintings, and calligraphies are combined with textual representation to convey scientific information to the public. With the biocultural ethic's 3Hs conceptual framework, we can understand the diverse life habits of mosses as well as their various habitats. This achievement eventually developed into a publication in 2019, titled as *The Magic and Enchantment of Bryophytes* (Zhang et al. 2019). This brilliant achievement is a result of more than 10 years of efforts to bridge the academy and the public.

Bridging scientific information and citizens was catalyzed by the collaboration between SFLBG team of bryologists and the Department of Gardens and Green Areas (Civic and Municipal Affairs Bureau of Macau Special Administration Region) to investigate Macau's diversity of bryophytes. This built a foundation for bridging academic research and Macau citizens. The team not only completed a catalogue of the diversity of bryophytes but also shaped how citizens began to think about mosses. A public exhibition entitled "The Miniature Angels in the Plant Kingdom" was held for Macau citizens in 2007, and in 2009, the same titled book was published bilingually in Chinese and English (Zhang et al. 2015, second edition). Macau citizens welcomed the book because it introduced them to bryophytes. In 2015, a second edition reached beyond Macau Island to the mainland, hence larger audience. Furthermore, in 2016, the book *Field Guide to Wild Plants of China: Bryophytes* (《中国野外植物手册:苔藓卷》) included around 300 species and introduced the method of using a hand lens to observe bryophytes. This book was rated as one of the outstanding popular science books by the Chinese Academy of Science.

To introduce knowledge about bryophytes to the general public, SFLBG also held exhibitions to communicate via expressive metaphors. In addition to "miniature angels," other metaphors such as "Adventure to the Green Mini-Cosmos" (绿色小宇宙之奇妙历程) and "The Plant Elves Who Lighten-up a Barren Land: Bryophytes" (点亮荒芜的植物小精灵:苔藓) have been constructed. These successful exhibitions have not only been opened to local citizens but also to those of other cities. For instance, in 2017 and 2018 the exhibition was held at the Shanghai Botanical Festival to show the beauty of the mini-landscape of ferns, sorrel plants, and bryophytes (Fig. 28.9). Moreover, the international audience was impressed by the exhibition conducted during the 19th International Botanical Congress in 2017. Around 6000 domestic and international researchers were exposed to the 80 paintings of bryophytes and moss flora. In 2019, SFLBG researchers were invited to the Beijing EXPO to co-organize "The Beauty of Mosses" (苔藓之美) that presented bryology, moss art, photographs, and miniature entities to reveal the aesthetical values of bryophytes (Fig. 28.10). The development of meaningful metaphors encourages the general public to understand the ecosystem services that bryophytes provide to the functioning of the biosphere. Meaningful metaphors successfully connect the imagination of citizens and open them to a wonderland of mosses, thus echoing the long-standing Chinese tradition of mosses appreciation.

28.4 Inclusivity: The Epistemic and Ethical Entanglement

In the introduction, I presented two senses of exclusion in contemporary conservation: (1) exclusion of the general public from the plethora of ecological information and (2) exclusion of "not-so-charming" species from conservation. In this section, I address the entanglement between epistemic biases and ethical practices in contemporary conservation.

As shown in the above case studies, to overcome the limitations of science communication and taxonomic biases, flagship species have strengths and limitations. Flagship species-centered conservation approaches focusing on large charismatic organisms display huge advantages in raising the conservation awareness of

Fig. 28.9 Citizens visit the mosses corridor that displays miniature mosses landscapes as well as moss science paintings, specimens, photographs, poetry, and other cultural representations of mosses at the 2018 Exhibition of Shade Plants, a theme section of Shanghai Botanical Festival. This exhibition was co-organized by the Shanghai Botanical Garden, Shenzhen Fairy Lake Botanical Garden, and Shanghai Normal University. Source: Shanghai Botanical Garden, photos by Li Zhang

citizens and facilitating communication with them. The sociocultural values of flagship species not only bridge the human community and other-than-human co-inhabitants but also connect diversified cultures. In our case studies, Magellanic Woodpeckers function both as a flagship species and as a biocultural keystone species. In addition, it associates Western ecological science with Yahgan ethno-ornithology. In the case of Giant Panda, its use as the logo of the WWF has

Fig. 28.10 Artists from the Shenzhen Fairy Lake Botanical Garden instruct children drawing mosses during the 2019 Beijing World Horticultural Exposition. Photos by Li Zhang

positioned this mammal as a global conservation icon that grabs attention for wildlife preservation. Giant Panda reserves also enlightened Chinese citizens about conservation needs. In terms of scientific communication, flagship species such as the Magellanic Woodpecker and the Giant Panda are powerful and productive. In southern Chile, the Magellanic Woodpecker's habit of knocking on woods producing drumming sounds characterizes its ecological function that rescues native forests from beetle larvae outbreaks. In Foping National Nature Reserve, the Giant Panda's *in situ* conservation encourages a national upsurge of wildlife conservation that fosters the mandate for an ecological civilization. The efficacy of communication in flagship species conservation does not exclude citizens from understanding the significance of those species, their rareness, their threatened habitats, and their interactions with other co-inhabitants. There is correspondingly strong proof between direct exposure to charismatic species and citizens' pro-conservation attitudes (Skibins et al. 2013). However, if citizens are only exposed to those charismatic species, cognitive bias might exclude not-so-charming species.

The charisma of biological species comprises ecological, aesthetic, and corporeal attributes. These attributes have important ontological, epistemological, and ethical implications (Lorimer 2007), which motivate citizens to participate in biodiversity conservation. Thus, exposed solely to charming species produces epistemic bias. Processes of how flagship species are identified in different regions disclose only the tip of the iceberg of epistemic bias. Research in Switzerland demonstrates that arbitrary elements are present in the identification of a flagship species (Schlegel and Rupf 2010). In this research, 415 students participated in questionnaires about their affinity for or antipathy to 27 indigenous wildlife species, including vertebrates and insects. Respondents indicated higher preference for species that they know rather than species unknown to them. At the species level, respondents showed higher affinity for butterflies, birds, and most mammals than for reptiles and other non-butterfly insects. In other words, the limited ecological information that people receive plays an important role in selecting flagship species. If critical biological or ecological information regarding conservation is absent in citizens' cognition, they seldom will pay attention to unknown species and their habitats. Based on "willingness-to-pay" (WTP) to investigate people's preference for species conservation, another study proved that participants commit higher attention to charismatic species rather on their conservation status (Colléony et al. 2016). People are constrained by what they know. If they merely know a global homogenized biological reality but do not know about regional biocultural diversity, then their cognition, attitude, and habits will be correspondingly limited. The absence of regionally endemic biodiversity and its inclusivity in our ecological communication drives the exclusive focus on large flagship species. Consequently, a large portion of the globe's biodiversity is excluded, particularly invertebrates and nonvascular plants.

To overcome the former exclusion, it is necessary to adjust scientific communication and increase the spectrum of species, for example by including invertebrates and nonvascular plants. Several reasons compel us to correct epistemic biases and enhance the taxonomic inclusivity in ecological communication and conservation.

First, generally, scientific communication can be hindered by taxonomic bias and misrepresentation. "Vertebratism" and "taxonomic chauvinism" (*sensu* Rozzi 2019) are symptoms and drivers of biocultural homogenization in modern philosophy, contemporary sciences, and global cultural representation that have excluded the most diverse groups of animals, i.e., invertebrates (Rozzi 2019). By "symptoms," Rozzi points to the phenomena that our mental images of animals are mostly represented by mammals or vertebrates, thus making invisible the majority of animals. He proposes that this taxonomic bias is "a driver" of global biocultural homogenization, because citizens feel more connected to a few non-local, iconic mammals (i.e., panda or Mickey Mouse) rather than to their own local biodiversity. In biocultural homogenization, these symptoms and drivers form vicious feedbacks.

Taxonomic chauvinism also permeates conservation terminology. Maan Barua (2011) investigated how misrepresentation of conservation terminology and taxonomic bias can hinder public conservation literacy. His analysis shows that terms such as flagship species, keystone species, and umbrella species are used differently by academics than by non-academics. Non-academic communication is less precise than technical academic terminology. This might lead to a stereotype fixation on mammals and birds. This misunderstanding can lead to ethical issues. To counterbalance fixation with vertebrates, Barua and collaborators (2012, p. 1458) called for promoting "invertebrate flagship as an invertebrate species or group that resonates with a target audience and stimulates awareness, funding, research and policy support for the conservation of invertebrate diversity." In the sphere of plant conservation, nonvascular plants like mosses are also underrepresented. Conservation terminology with rich communication about invertebrates and nonvascular plants will bridge the communication channel and improve citizens' conservation literacy.

Second, the limited accessibility of citizens to ecological information might fall into the "availability heuristic" (Tversky and Kahneman 1973) and provoke the occurrence of "neglected beauty" or "unattended charisma." Hence, the attracting features of the available species known to citizens are overestimated and those unknown species are disregarded. The inaccessibility to regional endemic species, such as sub-Antarctic mosses, and the accessibility to homogenized species, such as roses and apple, in turn exacerbate biocultural homogenization (Rozzi 2012a, 2013). Additionally, in the original Greek "charisma" (χάρισμα) means "divine gift." To determine which species are "divine gifts" is arbitrary. Thus, "neglected beauty" or "unattended charisma" is a cognitive bias, an epistemic defect, implying an absence of justice. To a very large extent, our notion of charisma is socially and culturally constructed. It is notable that the term charisma was applied to animals only in recent decades. Even Giant Pandas were unknown to the Chinese as well as the rest of the world. Therefore, the long journey to recognize the charisma of many neglected species has just begun. Work in the Omora Ethnobotanical Garden and the Shenzhen Fairy Lake Botanical Garden represents early stages of correcting these epistemic biases. This work is enhancing the inclusivity of ecological communication and conservation terminology. Due to these mosses-centered researches, communication, and conservation, the charisma of bryophytes has been impressed on the public's mindsets. These successes in disclosing the "neglected beauty" of bryophytes indicate the possibility of discovering the charisma of neglected species, enlarging the sphere of ecological and evolutionary understanding, thus enhancing the inclusivity of conservation.

Third, flagship species tend to focus the public's attention on the symbolic and representational value of large charismatic species. This tendency risks overlooking interactions among co-inhabitants in the ecosystem, thereby reducing ecological complexity to a few flagship species. Nabhan (1995) has cautioned about the danger of reductionism via the shades of meaning between endangered species and biodiversity. The ethical implication here is that taxonomic spectrums ought to be broadened for more inclusive conservation and ethical consideration. Moreover, to avoid extreme reductionism, citizens not only ought to know which entities (individual, species, community, etc.) are present; they also need to understand the interactions among them. These interactions are vital links among co-habitants, their life habits, and shared habitats (Rozzi et al. 2014). Consequently, the Magellanic Woodpecker and the Giant Panda should not be regarded only as flagship species. They also should be understood embedded in their complex ecological interactions and valued as umbrella species.

Notably, the mosses-centered conservation initiatives at the Omora Ethnobotanical Garden and the Shenzhen Fairy Botanical Garden have emphasized ecological complexity from the beginning. In the case of Omora, citizens attend to the integrity of the habits, habitats, and interactions of bryophytes with other co-inhabitants through ecotourism with a hand lens (Rozzi et al. 2012a). In the case of Shenzhen Fairy Lake, citizens are exposed to the diverse habits and delicate structures of bryophytes, their life circles, photosynthesis, participation in geochemical cycles, and their capacity to mold the landscape (Zhang et al. 2015). Since its creation, Omora Park researchers innovated by integrating environmental philosophy, the arts, and sciences because they considered that an ethical reflection on the relationships of co-inhabitation was essential for a systems and contextual approach to orient biological and cultural conservation (Rozzi et al. 2008a). Communication needs to include these dimensions and convey them to the general public and decision-makers. Conservation cannot achieve its goals if it excessively focuses on single flagship species in isolation from their ecological and cultural contexts.

Fourth, the epistemic bias that favors compartmentalization and specialization among disciplines impedes adequate ecological communication. A foundational principle of the biocultural ethic's "3Hs" framework is that conservation takes place in social-cultural-political contexts which are influenced by local ecological and cultural particular conditions (Rozzi 2015). This biocultural heterogeneity clashes with the prevailing narrative of universal, homogeneous, and linear progress (Gare 1998). This is analogous to the tension between flagship species- and moss-centered conservation approaches. The first expects single-species-centered communication to motivate citizens to biodiversity conservation. The second communicates ecological interactions among often overlooked organisms, which are essential components of biodiversity. Saroj Chawla (1991) examines the close associations among languages, worldviews, and the ways humans interact with their natural environments. Val Plumwood (2002) attributes the ultimate causes of our contemporary environmental crisis to citizens' failure to identify themselves as ecological beings due to a culture that separates humans from nature. Further, global citizens subordinate nature, because it is merely understood as an object for exploitation. Rozzi (2018) identifies "biocultural homogenization" to be a wicked problem because it involves feedbacks between the homogenization of biota and cultures.

Biocultural homogenization is a fundamental driver of rapid global change. In the case of panda conservation, the international attractiveness of the panda as an icon of Kungfu or WWF logo did contribute to saving pandas and their habitats in China.

However, the panda concentrated more than 50% of conservation research resources at the expense of less charismatic species (Wang et al. 2021). For global culture of connecting panda with people has facilitated local conservation and associated local environmental awareness. On the one hand, it has obscured the need for conservation of less conspicuous species. Additionally, the diversity of society–nature connections among various local communities has been often overlooked in biodiversity conservation. FEP addresses these problems by situating people in their regional and local biocultural complex contexts, and by enhancing communication and enriching people's understanding of native biocultural diversity.

28.5 Concluding Remarks

FEP can be effectively implemented at botanical gardens and protected areas for creating more inclusive conservation programs. Entanglements between epistemic biases and ethical practices generate complex problems linked to (1) taxonomic biases, (2) popular interpretation of biodiversity, (3) overlooked ecological interactions, and (4) homogenization of society–nature connections. FEP can counter epistemic biases and foster meaningful ethical practices through its four-step cycle (Rozzi et al. 2010b).

- To counter taxonomic bias, the case studies of moss-centered approaches at the Omora Park in Chile and the Shenzhen Fairy Lake Botanical Garden in China show that field investigation makes possible face-to-face encounters with a broad variety of other-than-human co-inhabitants. People not only encounter nonvascular plants but also invertebrates, including freshwater and marine invertebrates as illustrated by FEP's snorkeling activity designed by Chilean biologist Jaime Ojeda, "open your eyes to dive" (Rozzi 2019; Tauro et al. 2021). In this way, FEP methodology allows people to immerse in the local biodiversity rather than focusing attention on a few mammals or other vertebrates.
- In terms of popular interpretation of biodiversity, FEP broadens the spectrum of organisms co-inhabiting ecosystem that also ought to be conservation targets. An urgent task is to make endemic organisms known to local communities. FEP can incorporate the active participation of local communities in observing, drawing, documenting, and protecting their endemic biodiversity (Tauro et al. 2021). The local citizen's involvement generates locally grounded ecological communication that can scale up to regional, national, and global levels and inform them with the richness of particular ecosystems and cultures.
- To address overlooked ecological interactions among co-inhabitants of the ecosystem, FEP links ecology and ethics in educational field activities. By locating the biocultural ethic's framework of 3Hs (habitat, life habits, co-inhabitants) in the local ecosystems, FEP commits citizens to appreciate the complex interactions in local ecosystems (Rozzi 2012a). In turn, citizen can understand that local ecosystems are connected to global climate change and more broadly to social-

environmental degradation. This fosters a culture based on the concept that humans are co-inhabitants and ecological beings. In this chapter, I have focused on how "moss-centered conservation" invites citizens to value their co-inhabitation with little plants and lichens, and to understand the impact that global problems such as air pollution and acid rain have on them (Rozzi 2012c). This illustrates how a locally grounded experience can convey messages for ethical dwelling at the global scale.
- Regarding homogenization of society–nature connections, FEP opens opportunities for participants to understand how local biological diversity is embedded in local cultural contexts. This insight of existing diversified biocultural links at local and regional scales helps to overcome epistemic bias that drives biocultural homogenization by overlooking local forms of ecological knowledge (Rozzi 2018). The case studies on moss-centered approaches acknowledge that they are embedded in social–cultural–political matrixes. This acknowledgment captures the complex relationships among philosophy, language, cultural tradition, ecological science, and politics that FEP aims to integrate into conservation programs. In the Shenzhen Fairy Lake Botanical Garden, the "moss team" highlights Chinese tradition in classical literature and calligraphy through their appreciation of the charisma of mosses. China's contemporary political agenda of Ecological Civilization Construction plays a significant role in encouraging urban citizens to participate in exhibitions, nature experiences, and other types of communication activities. Both successful cases of the Omora Park and Shenzhen Fairy Lake Botanical Garden have constructed multiple metaphors embedded in their particular social-cultural-political backgrounds to communicate with citizens. Furthermore, FEP's openness to biocultural diversity and contextualization also inspires and embraces traditional ecological knowledge (Rozzi et al. 2008a), decolonial environmentalism (Gallegos de Castillo 2015), and empowerment of local communities to create novel conservation strategies (Toyoda 2018).

In conclusion, to counter the two senses of exclusion that I presented in the introduction (exclusion of general citizens from the plethora of ecological information and exclusion of many invertebrates and nonvascular plants from biological conservation), FEP contributes to enhancing inclusion through valuing the complexity and singularity of local biological and socio-cultural-political contexts. As shown in our case studies, to reconnect global citizens with "invisible" organisms through ecological communication and conservation, nonvascular plants like mosses can provide a miniature but powerful bridge. Botanical gardens and protected areas offer ideal places for citizens to connect with local co-inhabitants and native biocultural diversity. In this chapter, I appeal to botanical gardens, protected areas, environmental organizations, and other institutions to attend to FEP and its merits for more inclusive ecological communication, comprehensive taxonomic diversity, and biocultural conservation.

Acknowledgments The author thanks Ricardo Rozzi and Roy May for their tremendous work on English revision and comments. The author also appreciates the invitation of the Institute of

Ecology and Biodiversity (IEB) and the Cape Horn International Center (CHIC) through their Basal Funding projects AFB170008 and FB210018 (Agencia Nacional de Investigación de Chile, ANID), which made her fieldwork in Chile possible. Ricardo Rozzi, Juan Armesto, Javi Malebran, Terrance Caviness, Javier NE Vergara, Kelli Moses, and Omar Barroso helped with the field trip. The author's students Jia Meng and Gao Chao helped with some of the photographs. Special thanks to Li Zhang at the Shenzhen Fairy Lake Botanical Garden who suggested valuable references and allowed the author to use his photographs in high resolution.

References

Arango X, Rozzi R, Massardo F, Anderson C, Ibarra T (2014) Discovery and implementation of the Magellanic Woodpecker (Campephilus magellanicus) as a charismatic species: a biocultural approach for conservation in the Cape Horn Biosphere Reserve. In: Rozzi R, Jiménez JE (eds) Magellanic Sub-Antarctic Ornithology: first decade of long-term bird studies at the Omora Ethnobotanical Park, Cape Horn Biosphere Reserve. Chile University of North Texas-University of Magallanes Press, pp 302–316

Barua M (2011) Mobilizing metaphors: the popular use of keystone, flagship and umbrella species concepts. Biodivers Conserv 20:1427–1440

Barua M, Gurdak DJ, Ahmed RA, Tamuly J (2012) Selecting flagships for invertebrate conservation. Biodivers Conserv 21:1457–1476

Buckingham K, Neil J, David W, Jepson P (2013) Diplomats and refugees: panda diplomacy, soft "cuddly" power, and the new trajectory in panda conservation. Environ Pract 15(3):262–270

Carson R (1962) Silent spring. Fawcett World Library, New York

Chawla S (1991) Linguistic and philosophical roots of our environmental crisis. Environ Ethics 13: 253–262

Chen G, Sun W (2018) The role of botanical gardens in scientific research, conservation, and citizen science. Plant Divers 40:181–188

China State Forestry Administration (2015) The Giant Pandas of China: Status Quo. Chinese-English version of the report can be accessed online through: http://www.forestry.gov.cn/main/69/content-743562.html

Clucas B, McHugh K, Caro T (2008) Flagship species on covers of US conservation and nature magazines. Biodivers Conserv 17:1517–1528

Colléony A, Clayton S, Couvet D, Jalme MS, Prévot AC (2016) Human preferences for species conservation: animal charisma trumps endangered status. Biological conservation. Elsevier. https://doi.org/10.1016/j.biocon.2016.11.035

Croke V (2005) The lady and the panda: the true adventures of the first American explorer to bring back China's most exotic animal. Random House, New York

Croxall JP, Butchart SHM, Lascelles B et al (2012) Seabird conservation status, threats and priority actions: a global assessment. Bird Conserv Int 22:1–34

Dawson E (2014) Reframing social exclusion from science communication: moving away from 'barriers' towards a more complex perspective. J Sci Commun 13(2):1–5

Duron Q, Jiménez J, Vergara P, Soto G, Lizama M, Rozzi R (2018) Intersexual segregation in foraging microhabitat use by Magellanic Woodpeckers (*Campephilus magellanicus*): seasonal and habitat effects at the world's southernmost forests. Austral Ecol 43:25–34

Faraji L, Karimi M (2020) Botanical gardens as valuable resources in plant sciences. Biodivers Conserv 29. https://doi.org/10.1007/s10531-019-01926-1

Gallegos de Castillo L (2015) Sketch of a decolonial environmentalism: challenging the colonial conception of nature through the biocultural perspective. Inter-Am J Philos 6(1):32–47

Gare A (1998) McIntyre, narratives and environmental ethics. Environ Ethics 20:3–21

Goffinet B, Buck W, Massardo F, Rozzi R (2006) The Miniature forests of Cape Horn. Bilingual English-Spanish edition. Ediciones Universidad de Magallanes, Punta Arenas
Goffinet B, Rozzi R, Lewis L, Buck W, Massardo F (2012) The Miniature forests of Cape Horn: eco-tourism with a hand-lens ("Los bosques en miniatura del Cabo de Hornos: ecoturismo con lupa"). Bilingual English-Spanish edition. University of North Texas Press, Denton TX
Hallingbäck T, Hodgetts N (2000) Mosses, liverworts, and hornworts: status survey and conservation action plan for bryophytes. IUCN/SSC Bryophyte Specialist Group
He H, Chen J (2012) Educational and enjoyment benefits of visitor education centers at botanical gardens. Biol Conserv 149:103–112
Holland JS (2016) Who discovered the panda? National Geographic. Accessed through https://www.nationalgeographic.com/magazine/2016/08/explore-panda-mania-history/ on November 15, 2020
Kalinkat G, Cabral JS et al (2016) Flagship umbrella species needed for the conservation of overlooked aquatic biodiversity. Conserv Biol 31:481–485
Kono T (2018) The garden as a representation of nature: a space to overcome biocultural homogenization? In: Rozzi R, May RH Jr, Chapin FS III, Massardo F, Gavin M, Klaver I, Pauchard A, Nuñez MA, Simberloff D (eds) From biocultural homogenization to biocultural conservation. Ecology and ethics, vol 3. Springer, Dordrecht, pp 459–474
Lawley M (2019) A history of British and Irish bryology and bryologists. The Bryologist. https://websites.rbge.org.uk/bbs/Learning/Bryohistory/Bygone%20Bryologists/Bryohistory%20book_front%20matter.pdf
Leopold A (1949) A sand county almanac and sketches here and there. Oxford University Press, New York
Li B, Pimm S (2016) China's endemic vertebrates sheltering under the protective umbrella of the giant panda. Conserv Biol 30:329–339
Li S, Wang Q, Wang T (2018) The localization of UK experience in environmental education in Fairy Lake Botanical Garden, Shenzhen, China. J Agric Sci Technol A 8:55–58
Li S, McShea WJ, Wang D et al (2020) Retreat of large carnivores across the giant panda distribution range. Nat Ecol Evol 4:1327–1331
Liu X, Skidmore AK, Wang T et al (2002) Giant panda movements in Foping Nature Reserve. J Wildl Manag 66:1179–1188
Liu X, Toxopeus AG, Skidmore AK et al (2005) Giant panda habitat selection in Foping Nature Reserve, China. J Wildl Manag 69(4):1623–1632
Liu X, Wang T, Wang T et al (2015) How do two giant panda populations adapt to their habitats in the Qinling and Qionglai Mountains, China. Environ Sci Pollut Res 22:1175–1185
Lorimer J (2007). Nonhuman charisma. *Environment and Planning D: Society and Space* 25(5): 911–932
McGowan J, Beaumont LJ, Smith RJ et al (2020) Conservation prioritization can resolve the flagship species conundrum. Nat Commun 11(1):1–7
Miles L, Newton AC, DeFries RS et al (2006) A global overview of the conservation status of tropical dry forests. J Biogeogr 33:491–505
Moreira-Muñoz A, Carvajal F, Elórtegui S, Rozzi R (2020) The Chilean Biosphere Reserves network as a model for sustainability?: Challenges towards regenerative development, education, biocultural ethics and eco-social peace. In: Reed MG, Price MF (eds) UNESCO biosphere reserves: supporting biocultural diversity, sustainability and society. Earthscan studies in natural resource management. Routledge, New York, pp 61–75
Nabhan GP (1995) The danger of reductionism in biodiversity conservation. Conserv Biol 9:479–481
Nadkarni NM (2013) Not such strange bedfellows: underserved public audiences as collaborators for ecologists. In: Rozzi R, Pickett STA, Palmer C et al (eds) Linking ecology and ethics for a changing world: values, philosophy, and action, ecology and ethics. Ecology and ethics, vol 1. Springer, Dordrecht, pp 333–342

Nie Y, Zhang Z, Raubenheimer D et al (2015) Obligate herbivory in an ancestrally carnivorous lineage: the giant panda and bamboo from the perspective of nutritional geometry. Funct Ecol 29:26–34
Plumwood V (2002) Environmental culture: the ecological crisis of reason. Routledge, New York
Poole AK, Hargrove E, Day P et al (2013) A call for ethics literacy in environmental education. In: Rozzi R, Pickett STA, Palmer C et al (eds) Linking ecology and ethics for a changing world: values, philosophy, and action, ecology and ethics. Ecology and ethics, vol 1. Springer, Dordrecht, pp 349–372
Rozzi R (1999) The reciprocal links between evolutionary-ecological sciences and environmental ethics. Bioscience 49:911–921
Rozzi R (2002) Biological and cultural conservation in the archipelago forest ecosystems of Southern Chile. Ph.D. Dissertation, Department of Ecology and Evolutionary Biology, University of Connecticut
Rozzi R (2012a) Biocultural ethics: the vital links between the inhabitants, their habits and regional habitats. Environ Ethics 34:27–50
Rozzi R (2012b) Serendipity in the origin of ecotourism with a hand lens. In: Goffinet B, Rozzi R, Lewis L, Buck W, Massardo F (eds) The miniature forests of cape horn: eco-tourism with a hand-lens. University of North Texas Press, Denton, TX, pp 11–26
Rozzi R (2012c) Introduction to ecotourism with a hand lens. In: Goffinet B, Rozzi R, Lewis L, Buck W, Massardo F (eds) The miniature forests of cape horn: eco-tourism with a hand-lens. University of North Texas Press, Denton, TX, pp 28–79
Rozzi R (2013) Biocultural ethics: from biocultural homogenization toward biocultural conservation. In: Rozzi R, Pickett STA, Palmer C, Armesto JJ, Callicott JB (eds) Linking ecology and ethics for a changing world. Ecology and ethics, vol 1. Springer, Dordrecht, pp 9–32
Rozzi R (2015) Earth stewardship and biocultural ethics: Latin American perspectives. In: Rozzi R, Chapin FS III, Callicott JB et al (eds) Earth stewardship: linking ecology and ethics in theory and practice, Ecology and ethics, vol 2. Springer, Dordrecht, pp 87–112
Rozzi R (2018) Biocultural homogenization: a wicked problem in the Anthropocene. In: Rozzi R, May RH Jr, Chapin FS III et al (eds) From biocultural homogenization to biocultural conservation. Ecology and ethics, vol 3. Springer, Dordrecht, pp 21–47
Rozzi R (2019) Taxonomic chauvinism, no more! Antidotes from Hume, Darwin, and biocultural ethics. Environmental Ethics 41(3):249–282
Rozzi R, Massardo F (2011) The road to biocultural ethics. Front Ecol Environ 9:246–247
Rozzi R, Massardo F, Anderson C, Heidinger K, Silander J Jr (2006) Ten principles for biocultural conservation at the southern tip of the Americas: the approach of the Omora Ethnobotanical Park. Ecol Soc 11(1):43. http://www.ecologyandsociety.org/vol11/iss1/art43/
Rozzi R, Arango X, Massardo F et al (2008a) Field environmental philosophy and biocultural conservation: the Omora Ethnobotanical Park educational program. Environ Ethics 30:325–336
Rozzi R, Armesto J, Goffinet B et al (2008b) Changing lenses to assess biodiversity: patterns of species richness in sub-Antarctic plants and implications for global conservation. Front Ecol Environ 6(3):131–137
Rozzi R, Massardo F, Anderson C, McGehee S, Clark G, Egli G, Ramilo E, Calder6n U, Calder6n C, Aillapan L, Zárraga C (2010a) Multi-ethnic bird guide of the sub-Antarctic forests of South America. University of North Texas Press, Denton, TX
Rozzi R, Anderson C, Pizarro C et al (2010b) Field environmental philosophy and biocultural conservation at the Omora Ethnobotanical Park: Methodological approaches to broaden the ways of integrating the social component ("S") in Long-Term Socio-Ecological Research (LTSER) Sites. Rev Chil Hist Nat 83:27–68
Rozzi R, Lewis L, Massardo F et al (2012a) Ecotourism with a Hand-Lens at Omora Park. Ediciones Universidad de Magallanes, Punta Arenas
Rozzi R, Armesto JJ, Gutiérrez J et al (2012b) Integrating ecology and environmental ethics: earth stewardship in the southern end of the Americas. Bioscience 62(3):226–236

Rozzi R, Draguicevic JM, Arango X et al (2014) From science towards conservation: the education and environmental ethics program of the mora Ethnobotanical Park. In: Rozzi R, Jiménez JE (eds) Magellanic Sub-Antarctic Ornithology: first decade of long-term bird studies at the Omora Ethnobotanical Park, Cape Horn Biosphere Reserve, Chile. University of North Texas Press, Denton, TX, pp 295–302

Rozzi R, La Valle MT, Russell R, Goffinet B, Massardo F (2020) Ecotourism with a hand-lens: a field environmental philosophy experience from the South of the World. In: Brister E, Frodeman R (eds) Philosophy for the real world: an introduction to field philosophy. Taylor & Francis/ Routledge, New York, pp 222–239

Schaller G (1994) The last panda. The University of Chicago Press, Chicago, IL

Schlegel J, Rupf R (2010) Attitudes towards potential animal flagship species in nature conservation: a survey among students of different educational institutions. J Nat Conserv 18(4):278–290

Short LL (1970) The habits and relationships of the magellanic woodpecker. The Wilson Bulletin 82:115–129

Simberloff D (1998) Flagships, umbrellas, and keystones: is single-species management passé in the landscape era? Biol Conserv 83(3):247–257

Skibins JC, Powell RB, Hallo JC (2013) Charisma and conservation: charismatic megafauna's influence on safari and zoo tourists' pro-conservation behaviors. Biodivers Conserv 22:959–982

Tauro A, Ojeda J, Caviness T, Moses K, Moreno R, Wright T, Zhu D, Poole A, Massardo F, Rozzi R (2021) Field environmental philosophy: a biocultural ethic approach to education and ecotourism for Sustainability. Sustainability 13:4526. https://doi.org/10.3390/su13084526

Toyoda M (2018) Revitalizing local commons: a democratic approach to collective management. In: Rozzi R et al (eds) From biocultural homogenization to biocultural conservation. Springer, Dordrecht, pp 443–457

Tversky A, Kahneman D (1973) Availability: a heuristic for judging frequency and probability. Cogn Psychol 5:207–232

Vergara P, Schlatter RP (2004) Magellanic woodpecker (*Campephilus magellanicus*) abundance and foraging in Tierra del Fuego, Chile. J Ornithol 145:343–351

Verma M, Jones KR, Watson JEM et al (2020) Severe human pressures in the Sundaland biodiversity hotspot. Conserv Sci Pract 2:e169. https://doi.org/10.1111/csp2.169

Walpole MJ, Leader-Williams N (2002) Tourism and flagship species in conservation. Biodivers Conserv 11:543–547

Wang ZN, Yang L, Fan PF et al (2021) Species bias and spillover effects in scientific research on Carnivora in China. Zool Res 42(3):354

Wei F, Costanza R, Dai Q et al (2018) The value of ecosystem services from giant panda reserves. Curr Biol 28:2174–2180

WWF (2020). https://wwf.panda.org/our_work/wildlife/flagship_keystone_indicator_definition/, accessed on July 13, 2020

Yong Y et al (2004) Giant pandas migration and habitat utilization. In: Lindburg D, Baragona K (eds) Giant pandas: biology and conservation. University of California Press, Berkeley, CA

Zhang L (2020) How to conduct unpopular science education: case studies from Shenzhen Fairy Lake Botanical Garden. In Education on biodiversity conservation: Proceedings of the 4th Luosuo River Forum of Science Education, Xishuangbanna, Yunnan Province, China, pp 145–151. Available via http://www.luosuoriverforum.com/qfycontent/uploads/2020/03/43134bc2cd5d9aafea4f25cca0678303.pdf

Zhang L, Zuo Q, Hong B (2015) The miniature angels in the plant kingdom (《植物王国的小矮人》). Department of Gardens and Green Areas, Civic and Municipal Affairs of Bureau of Macao Special Administrative Region, Macao

Zhang L, Zuo Q, Mao L (2019) The Magic and Enchantment of Bryophytes (《苔藓之美》), extended edition. Phoenix Science Press, Nanjing, Jiangsu

Zheng G, Xu P (1964) Giant pandas found at the Southern Foot of the Qingling Mountain. Chin J Zool 1(in Chinese):3

Chapter 29
Biocultural Conservation in Biosphere Reserves in Temperate Regions of Chile, Estonia, Germany, and Sweden

Elke Schüttler, Roy Mackenzie, and Lucas Muñoz-Petersen

Abstract UNESCO Biosphere Reserves are protected areas aiming to reconcile conservation with human well-being. In 2021, there are 714 biosphere reserves worldwide located in a variety of biomes. However, in the temperate south of South America, between latitudes 45 and 60°S, there are only three biosphere reserves. Among them, the Cape Horn Biosphere Reserve (CHBR, 55°S) in southernmost Chile implements a biocultural conservation approach. In contrast, in Europe, at latitudes 45–60°N, there are 70 biosphere reserves in different biophysical, cultural, and institutional-socio-political contexts. Our study compares approaches to biocultural education and conservation of the CHBR with three European biosphere reserves. We found exemplary conservation initiatives in biosphere reserves in Estonia, Germany, and Sweden, which have similarities with their southern equivalent. First, in all these reserves education approaches focus on under-perceived and under-valued biota, particularly small organisms. Second, research includes an important education component. Third, conservation actions take place giving consideration for cultural aspects. From our case studies, we derive three

E. Schüttler (✉)
Sub-Antarctic Biocultural Conservation Program, Universidad de Magallanes, Puerto Williams, Chile

Cape Horn International Center (CHIC), Puerto Williams, Chile
e-mail: elke.schuttler@umag.cl

R. Mackenzie
Sub-Antarctic Biocultural Conservation Program, Universidad de Magallanes, Puerto Williams, Chile

Cape Horn International Center (CHIC), Puerto Williams, Chile

Millennium Institute Biodiversity of Antarctic and Subantarctic Ecosystems (BASE), Santiago, Chile
e-mail: roy.mackenzie@umag.cl

L. Muñoz-Petersen
Master of Science Program in Biosphere Reserves Management, Eberswalde University for Sustainable Development, Eberswalde, Germany
e-mail: lpe027@hnee.de

R. Rozzi et al. (eds.), *Field Environmental Philosophy*, Ecology and Ethics 5,
https://doi.org/10.1007/978-3-031-23368-5_29

lessons: (1) holistic conservation can be achieved, even in the absence of great funding schemes, when approaches are simple, integrative, and locally centered; (2) countries highly committed to the biosphere reserve concept more widely implement local innovation for sustainable activities; and (3) similar biomes have similar conservation challenges. This highlights the necessity to establish partnerships among biosphere reserves located in similar biomes.

Keywords Biodiversity · Biocultural ethics · Environmental education · Protected areas

29.1 Introduction

Biosphere reserves are learning sites for sustainable development and biocultural conservation in terrestrial, maritime, and coastal ecosystems. They are designated by UNESCO's Man and the Biosphere (MaB) Programme, an intergovernmental scientific program that aims to establish a scientific basis for enhancing the relationship between people and their environments. MaB has worked since the 1970s to promote the conservation of biodiversity and human well-being through the establishment of biosphere reserves. These protected areas provide opportunities for conservation, sustainable socio-economic development, and logistic support based on local community efforts and science. Hence, biosphere reserves are a practical arena for countries to gain experience in biodiversity management, conflict prevention, and the promotion of sustainable economic activities that are socially and culturally appropriate and environmentally sustainable (UNESCO 2017).

Biosphere reserves use a territorial planning strategy based on three types of zones (Guevara and Laborde 2008). First, *core zones* for strict preservation that must align with the country's protected areas, often national parks. Second, *buffer zones*, adjacent to core zones that can be used for activities that strengthen scientific research, monitoring, and education. Third, *transition zones* surrounding buffer zones, allowing human settlements and sustainable activities in harmony with the conservation of wild areas, semi-natural and cultural landscapes, and seascapes. The number of biospheres is constantly changing. In April 2021, the world network of UNESCO biosphere reserves encompassed 714 sites in 129 countries worldwide, including 21 transboundary sites (UNESCO 2021). South America has 60 biosphere reserves in 10 countries, while Europe has 256 in 38 countries. The legal recognition of biosphere reserves is given through national legislation, making their implementation dependent on the commitment of governmental, public, and private stakeholders (Elbakidze et al. 2013; Cuong et al. 2017). Notably, environmental education has been identified as one of the most important factors influencing the success of biosphere reserves (Stoll-Kleemann and Welp 2008).

With the aim of contributing to understanding the role of education and participatory approaches, in this chapter we compare biocultural conservation initiatives in biosphere reserves located in temperate latitudes of northern Europe with the methodology of the field environmental philosophy (FEP) developed in the Cape

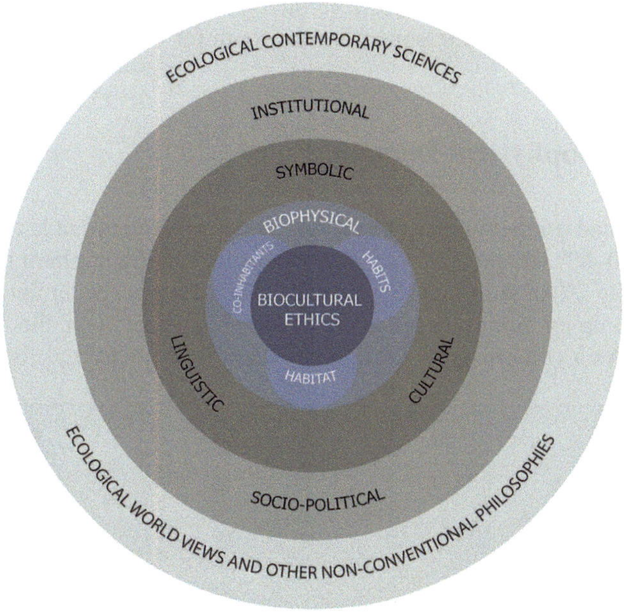

Fig. 29.1 Habitats, habits, and co-inhabitants and their interrelationships with the biophysical, institutional-socio-political, and symbolic-linguistic-cultural domains (modified from Rozzi 2016). The bullseye organization of biocultural domains in the form of expanding layers (spheres), locates the biocultural ethics at the center: a basic ethical conceptual framework for the ecological analysis of global society and its economic and cultural development policies affecting the relationship of human and non-human species. The co-inhabitants and their life habits that modify the habitat belong to the first level: the biophysical level. The following sphere is immaterial and is informed by the interpretation of the biophysical world. This symbolic-linguistic-cultural sphere represents the interpretation of the biophysical realm forged by different human groups. The third sphere, the institutional socio-political sphere, enables and regulates life habits. The biocultural ethic aims to maintain and foster the plurality of relationships between the biophysical and symbolic-linguistic-cultural dimensions. The scheme differs from the original publication as it points to an ecocentric socio-political turn, which puts humans and non-humans as the central axis of the development strategy

Horn Biosphere Reserve (CHBR), Chile (Rozzi et al. 2010). To accomplish this objective, we categorized the information into the three dimensions of **h**abitats, **h**abits, and co-in**h**abitants or the 3 “Hs” (Rozzi 2013). Each of the 3 “Hs” comprises biophysical, institutional-socio-political, and symbolic-linguistic-cultural domains which, along with their interrelationships, compose the conceptual framework of biocultural ethics (Fig. 29.1). Following Rozzi (2013, p. 10), biocultural ethics is a “new eco-philosophical paradigm that transforms prevailing ethics” as it considers all beings in their relationship with their environment, including their habits. These ethical dimensions are the basis of the FEP methodology that promotes the interaction of research, communication, in situ conservation, and field activities. The comparison will allow us to structure and identify the different conditions by which southern and northern biosphere reserves work and whether elements of

FEP can be found beyond Cape Horn in three European biosphere reserves. From this analysis, we can derive lessons from the south for the north and vice versa.

29.2 The Biophysical Domain

The northern and southern hemispheres have a major difference in land: ocean ratios at latitudes 40–60°. While land mass represents 54% in the northern hemisphere, it represents only 2% in the south (Rozzi et al. 2012). Europe, being much larger than the southern tip of South America, is in direct contact with the Arctic Circle (Fig. 29.2). As a consequence, Europe's climate is characterized by the interplay

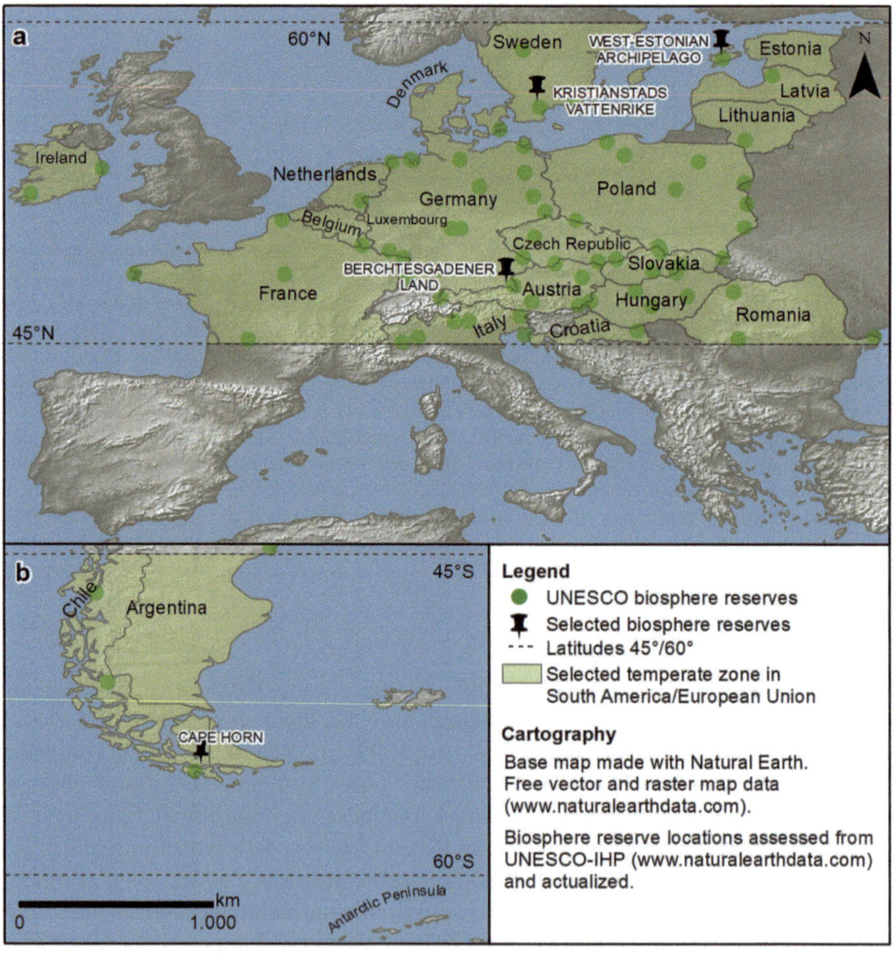

Fig. 29.2 Biosphere reserves located in the temperate zone between (**a**) latitudes 45–60°N in countries of the European Union and (**b**) latitudes 45–60°S in South America. Green circles indicate only the central location; they do not represent their geographical size

of the Atlantic Ocean, polar, and continental air masses which produce extreme conditions of temperature and humidity (Berentsen et al. 2021). In contrast, the climate in southern South America is strongly moderated by oceanic conditions (Donghi et al. 2021).

The South American temperate forests biome includes the sub-Antarctic Magellanic ecoregion that extends from the Gulf of Penas (47°S) to Cape Horn (56°S) (Rozzi et al. 2012). This ecoregion still has >90% of its surface classified as low human impact area considering land cover and human population density, livestock density, forest change, and illumination at night (Jacobson et al. 2019, but see the relevance of fauna intactness in Plumptre et al. 2021). This austral ecoregion has large wetland areas and a high endemism among fauna and flora with up to 90% of woody species (Arroyo et al. 1996) and 50% of bryophyte species (Villagrán Moraga 2020) being endemic. For Europe, over 200 terrestrial habitat types have been described (EUNIS 2017), but in contrast to the austral equivalent, pristine habitats are scarce. For example, Europe's old-growth forests only occur in 0.7% of the forested area and are concentrated in boreal and alpine regions (Sabatini et al. 2018). Moreover, as many as 37% of the habitats of the EU member states were categorized as critically endangered (2%), endangered (11%), and vulnerable (24%) on the European red list of habitats with grasslands, freshwater, and coastal habitats being the most threatened (Janssen et al. 2016).

29.3 The Symbolic-Linguistic-Cultural Domain

The Inuit of Greenland and the Sámi of Scandinavia and Russia are the only indigenous people living partly in the territory of member states of the European Union (European Commission 2016). However, those territories are located above 60°N; hence, there are no recognized indigenous people between latitudes 45 and 60°N. Although it is uncommon to apply the concept of "indigenous people" to Europeans, who began the colonization of oversea territories in the fifteenth century, some authors speak of "Europe's largely forgotten indigenous peoples "(Grote 2006/07, p. 425) because colonization took place internally, too. A written declaration by the European Parliament (2015) stated that over 60 regional or minority languages are spoken by more than 40 million people in the European Union. Among them, 49 languages are severely endangered. In Germany, for example, the Sorbs, a West Slavic people that have shared their land with Germans for over 1000 years, still maintain their own language and habits, but have the status of "minority" (Grote 2006/07).

The languages and cultures of indigenous people in the southern cone of South America are highly threatened as well. Before the European colonization, the "ethnic Fuegian complex" (Kawéskar, Selknam or Ona, Aónikenk or Tehuelches, and the Yagan) inhabited Patagonia and Tierra del Fuego as terrestrial or maritime hunter-gatherers (Borrero et al. 2011). Today, Kawéskar and Yagan descendants struggle to protect their territories, cultures, and languages spoken by fewer than ten people

(Rozzi et al. 2012). In contrast with the majority of Latin American countries, Chile does not contemplate any mention of indigenous peoples in its constitution (Aguilar et al. 2010), but recognizes them in Law Nr. 19253 (Ministry of Planification and Cooperation 1993). At latitudes 45–60°S, only the southernmost people, the Yagan, live in a biosphere reserve (CHBR).

29.4 The Institutional Social-Political Domain

Of the 13 South American countries, only Chile and Argentina extend their territories to include areas located between 45 and 60°S. Of the 27 member states in the European Union, 19 include territories located on the equivalent northern latitudes. Regarding investments in sustainable development, in 2017 and 2018 western Europeans spent 106 billion dollars to address mitigation and adaptation to climate change, while South America and the Caribbean only spent 28 billion dollars (CPI 2019). The European Union has a common strategy to address climate and environmental challenges through the "EU Biodiversity Strategy for 2030," with a budget of 1.8 trillion euros for 2021–2027 to recover from the corona crisis and to build a greener, more resilient, and digital post-COVID-19 Europe (European Union 2021). In South America, no comparable transnational strategy has been implemented so far. Although the Mercosur Agreement contains references to the Paris Agreement and the Convention of Biological Diversity, Mercosur lacks a legal framework for environmental protection (Heyl et al. 2021).

While Europe has 158,450 protected areas (13.1% of the continental surface), South America and the Caribbean have fewer protected areas (9962), but much larger ones (24.2%, UNEP-WCMC 2021; note that this database is not exhaustive). Southern Chile even protects 79.5% of land within the latitudes 45–56°; Argentina though protects only 10.5% and has fewer protected areas in its south (UNEP-WCMC 2021). Biosphere reserves in South America below 45° southern latitude include the Laguna San Rafael Biosphere Reserve (47°S), Torres del Paine (50°S), and Cape Horn (55°S), all situated in Chile (Fig. 29.2). In comparison, the same latitudinal range in the European continent (45°–60°N) harbors 60 national and 10 transboundary biosphere reserves, while seven are even located in higher latitudes (>60°N), in countries close to the Arctic circle (not considering the Russian Federation, UNESCO 2021).

29.5 The Cape Horn Biosphere Reserve (Chile)

The Cape Horn Biosphere Reserve (CHBR, 55°S) is one of ten biosphere reserves in Chile. Designated in 2005, it has an extension of 20,700 km^2 of land and 29,600 km^2 of sea. The archipelago is composed of southern beech forests, Magellanic tundra, high-Andean habitats, fjords, channels, and glaciers (Rozzi et al. 2012). Currently, no more than 2100 people are permanently living in the reserve, concentrated in the town of Puerto Williams on Navarino Island, with artisanal fishery, tourism, and livestock farming being the principal activities. The remaining islands are mostly uninhabited sub-Antarctic wilderness. The biosphere reserve is rich in culture, with the outstanding presence of sea nomads, the Yagan, for over 6000 years (Orquera and Piana 2009). Research and education have been the strength of the reserve since its creation in 2005 (Rozzi and Schüttler 2015). These activities have been following a biocultural approach with the FEP at its heart. In the next section, we will present three biosphere reserves in Europe focusing on biocultural aspects while drawing comparisons to activities in the CHBR. We first present a short overview of how biosphere reserves are managed in each country, followed by a description of the particular biosphere reserve and two conservation initiatives (box).

29.6 Three Examples from Europe

29.6.1 Estonia

Estonia has only one biosphere reserve, the West-Estonian Archipelago Biosphere Programme Area (BPA), designated in 1990 (Box 29.1). The Environmental Board of the Ministry of the Environment, in charge of the protected areas' management and the supervision of the national environmental policy, operates an office in the biosphere reserve, and one position of the 26 full-time specialists is reserved for an adviser dedicated exclusively to the BPA (Jallon 2019). The Environmental Board works with local communities, businesses, and representatives of educational institutions on the implementation of the five-year Sustainable Development Programme 2014–2020, while the West-Estonian Archipelago BPA Council is the advisory body to ensure the participation of all relevant stakeholders (Environmental Board 2014). The major focus of the development program is to lead the BPA into a pilot area for green economy, particularly through partnerships that encourage sustainability.

Box 29.1 West-Estonian Archipelago Biosphere Programme Area (Estonia)

The West-Estonian Archipelago Biosphere Programme Area (WABPA) is located on latitude 58°N (Fig. 29.2) and covers 4000 km^2 of islands and islets and 11,600 km^2 of sea. It has similar characteristics to those in the sub-Antarctic Magellanic ecoregion: a mosaic of marine and terrestrial habitats such as peat bogs and an ancient cultural heritage. For example, several dialects evolved in the BPA, being the insular and the Võru dialects, the ones most spoken (WABPA 2013). The BPA has a population of 40,500

(continued)

Box 29.1 (continued)

inhabitants (approximately 3% of the country's population), living mainly in the biosphere reserve transition zone. Since the mid-2000s, environmental education and local tourism have increased significantly, and infrastructure has been implemented to foster environmental education. The WABPA has five strategic objectives: (1) to become a pilot area for sustainable economy and use of natural resources; (2) to preserve biodiversity; (3) to preserve and exhibit the islands' cultural heritage; (4) to become a research, monitoring, and training center that supports green economy; and (5) to promote cooperation among stakeholders.

Restoration of Estonian Alvar Grasslands

The project Life to Alvars (2014–2019) was a successful restoration project of alvar grasslands of semi-natural origin. It was a winner of the Natura 2000 Award in the category of socio-economic benefits (2018). Alvars are calcareous grasslands with a thin layer of soil (<20 cm) on a limestone plain, mainly distributed on the islands of the Baltic Sea, harboring high species richness (Pärtel et al. 1999). Estonian alvar grasslands represent one-third of Europe's alvars and have high conservation value. They developed and are maintained thanks to traditional practices by the islanders, such as cutting shrubs or trees and allowing constant grazing. Over the past century, most alvars have become overgrown by shrubs and trees due to the changes in traditional land use and the introduction of pine plantations (Pärtel et al. 1999). Project Life to Alvars was implemented by governmental and non-governmental organizations, the Estonian University of Life Sciences, and the University of Tartu. They restored 2500 ha of the most valuable but overgrown alvar grasslands, including private and public land. Besides the cutting of shrubs and trees and their utilization as wood chips or fence posts, local farmers were also encouraged to sustainably manage the alvars through traditional grazing patterns by sheep, horses, and cattle (Fig. 29.3). The project success was guaranteed by emphasizing cooperation among livestock keepers, landowners, government agencies, and experts (ESCCA 2019).

Adult Education in Biosphere Reserves—Nordic Learning Network

The Biosphere Reserves in Adult Education Network (BRAEN) aimes to establish a Nordic-Baltic educational region creating a network among educational and scientific institutions in biosphere reserves in Estonia, Latvia, and Sweden. The network's functions include data collection, knowledge analysis, organization of summer/winter universities, and exchange of experiences on different topics, issues, or challenges shared by involved actors. BRAEN promotes adult education as a tool for raising awareness about environmental problems. The Hiiumaa Vocational School, for example, offers courses about horticulture, sustainable construction, and transportation; the Kuressaare Training Center offers advanced training on sustainable tourism, cooking, mechanics, and handicrafts, among other themes (https://braen.voog.com/en) (Fig. 29.3).

Fig. 29.3 Examples of similar cultural and educational activities for adults from the WABPA, Estonia, and the CHBR, Chile. (**a**) Traditional grazing on alvars on Hiiumaa Island, Estonia (photograph by Annely Holm), and (**b**) Julia González collecting reeds for artisanal Yagan basketry on Navarino Island, CHBR, Chile (photograph by Sandra Vallejo in Massardo and Rozzi 2006). (**c**) Gastronomy courses on seaweeds conducted by the CHBR educational team in Punta Arenas, Chile (photograph by Andrés Mansilla), and (**d**) workshop on "Bioeconomy in the kitchen – local food, low-waste, reasonable solutions in catering" of the Nordic Learning Network, Hiiumaa, Estonia (photograph by Jane Üksik)

29.6.2 Germany

Biosphere reserves are anchored in German law (§ 25, BNatSchG 2021) and are part of the German Sustainability Strategy. They are managed by the governmental administration of the respective federal states. The MaB National Committee has the role to continuously enhance the concepts of sustainable development and education of the UNESCO MaB strategy. It is currently composed of 17 members of science, administration, and non-governmental organizations, chaired by the Federal Agency for Nature Conservation. There are 16 biosphere reserves located throughout the diverse habitats of the country, often including cultural landscapes. They are important destinations for sustainable tourism and are hotspots of organic farming. Local businesses are also crucial players in the economy (UNESCO 2021). All German biosphere reserves cooperate closely, and their managers have regular meetings, as illustrated by the case of the Berchtesgadener Land Biosphere Reserve (Box 29.2).

Box 29.2 Berchtesgadener Land Biosphere Reserve (Germany)

The Berchtesgadener Land Biosphere Reserve (BLBR, 47°N) was created in 1990 and is located in Bavaria, southeastern Germany (Fig. 29.2). It covers a surface of 840 km^2. Its core and main buffer zone include the mountainous Berchtesgaden National Park within the northern Alps, which has a long history of conservation (Lintzmeyer and Zierl 2009/10). The transition zone is mainly rural and covers 15 small municipalities, home to about 100,000 inhabitants. The overall landscape has a broad diversity of ecosystems, such as alpine and subalpine forests, rock fields, rivers, lakes, glaciers, and lowland meadows, harboring vulnerable fauna like golden eagles and endangered butterflies. People earn income from tourism, small-scale agriculture, forestry, and salt mining. A specialty of the biosphere reserve is its successful marketing of regional farm products (UNESCO 2021).

Outstanding Trans-European Education Initiative

The trans-European Project ELENA (Experiential Learning and Education for Nature Awareness) is a new approach to teach children sensitivity toward nature. Since Richard Louv's groundbreaking book *Last Child in the Woods* (2005), in which he expresses severe concern about the consequences of an increasingly digitized and urbanized society where children disconnect from the natural world, the importance of integrating learning environments with nature has received more attention in policy. For example, only 1% of German pupils recognized twelve of the most common native birds, while 8% did not know any of the species (Zahner et al. 2007). In ELENA, through personal experiences with living animals in classrooms and extracurricular environmental education, pupils develop interest, knowledge, and values (Fig. 29.4). As "heart openers," living animals lead to emotional experiences and learning about how life and its safekeeping works that can even stimulate behavioral changes (Sturm et al. 2014:2). Project ELENA (2013–2016), funded by the European Commission and implemented by schools in Georgia, Germany, Hungary, and Romania, was inspired by the German project "Animals Live" (Berthold and Sturm 2010) developed in response to the Bavarian Biodiversity Strategy. Modules about a variety of animals, such as bees, earthworms, snails, chickens, butterflies, and dogs can be downloaded from ELENA's homepage (http://elena-project.eu).

Applied Research on Dead Wood

With the creation of a chair for Ecosystem Dynamics and Forest Management at the Technische Universität München in 2018, research on the disturbance and resilience of mountain ecosystems and their implications for sustainable forest management has been strengthened (Seidl et al. 2017, 2018). Often neglected in forest management is the ecological importance of dead wood and dying trees. Researchers at Berchtesgaden and their colleagues draw attention to the largely unnoticed threat of forest degradation through the

(continued)

Box 29.2 (continued)
removal of dead wood (Thorn et al. 2020), accelerated by the extraction of biomass-based fuels (Jonsell 2007) and the predominantly negative perceptions of dead wood by the general public (e.g., Arnberger et al. 2018). This research has yielded concrete recommendations for managers, such as active and passive enrichment strategies like leaving unharvested tree parts in the forest or the creation of ecosystem-specific benchmarks for dead trees. Moreover, environmental education strategies help to raise awareness on the ecological importance of dead wood as habitats for invertebrates (Fig. 29.4) or charismatic bird species such as woodpeckers.

Fig. 29.4 Examples of similar education activities for children focusing on under-perceived and under-valued biota from the BLBR, Germany, and the CHBR, Chile. (**a**) Caterpillar touched by children in the trans-European project ELENA (photograph under a free Pixabay License) and (**b**) "ecotourism with a hand lens," an approach to perceiving lichens and other organisms in the miniature forests of Cape Horn (photograph by Adam Wilson in Rozzi et al. 2012). (**c**) Children's book on cryptic beetles that have been rediscovered in old-growth forests of the Bavarian Forest National Park (Michler and Zuda 2020) and (**d**) book illustrated with underwater photographs of hardly perceived river invertebrates in Cape Horn (photograph by Gonzalo Arriagada in Contador and Kennedy 2014)

29.6.3 Sweden

There are seven biosphere reserves in Sweden, protecting 27,000 km^2 of land with a total population of 433,000 inhabitants. According to their national environmental regulation, the biosphere reserves in Sweden compile, connect, and integrate already existing local, regional, national, and international policies and plans (Swedish EPA 2021). The Lima Action Plan for Biosphere Reserves, the UN 2030 Agenda, and the Sustainable Development Goals, are used as guidelines for socio-economic and ecological development in Swedish biosphere reserves (Heinrup and Schultz 2017). Four Swedish biosphere reserves received special recognition from the International Coordinating Council of the Man and the Biosphere (ICC/MaB) Programme regarding the implementation of good practices (Ruoss 2013): Kristianstads Vattenrike BR for the process of consultation and coordination/quality of the participatory process (Box 29.3); Blekinge Archipelago BR and Nedre Dalälven River Landscape BR for the high quality of the nomination process; and East Vättern Scarp Landscape BR for the holistic environmental policy carried out by the Biosphere Office. The biosphere offices for Sweden's biosphere reserves are usually located in municipal and/or county administrations and are responsible for coordinating the biosphere's action plan (e.g., KVBR 2020). In Sweden, the Biosphere Office has no official authority. Instead, actions are carried out through a continual dialogue among public services, governmental representatives, landowners, and other relevant stakeholders during meetings of the biosphere reserve's Consultation Group, which meets regularly as an advisory group (Wettemark et al. 2016).

Box 29.3 Kristianstads Vattenrike Biosphere Reserve (Sweden)
Kristianstad Municipality is the responsible authority for the oldest of Sweden's biosphere reserves, the Kristianstads Vattenrike Biosphere Reserve (KVBR), created in 2005. The protected area includes terrestrial (900 km^2) and maritime (140 km^2) zones and is home to over 75,000 habitants. The KVBR has the largest wet grassland landscape in Sweden, which has been historically used for hay-making and grazing (Schultz 2009). Its main attraction is the Helge å River, fringed by wetlands and seasonally flooded grasslands; there are also leafy forests and dry sandy arable lands. In fact, almost all natural habitats in Sweden are represented by the reserve. The protected area in the biosphere reserve represents less than 0.2% of the total country's area, but it is a refuge for at least 20% of the nation's red-listed species (KVBR 2020). This biosphere reserve has implemented an adaptive co-management strategy, meaning that management is conducted in cooperation with the relevant

(continued)

Box 29.3 (continued)
stakeholders. This approach includes the identification of synergies, transfer of knowledge to the community, constant evaluation, and communication to the public, scientists, and decision-makers. The Biosphere Office and the visitor center belong to the Municipal Executive Office and have 13 full-time officers who are responsible for the activities taking place in the biosphere reserve.

Selective Fishing to Reduce Invasive Fish

A 3-year biomanipulation project, funded by a local company (Local Water Management), started in 2019 in Lake Råbelövssjön and Lake Oppmannasjön by selective fishing to improve angling in fishery areas of the reserve. The Kristianstad Vattenrike region has a long history of agricultural activities which require annual flooding to be sustained (Schultz 2009). These activities provoked algal blooms, alterations in the fish communities, and the disappearance of underwater vegetation in the surrounding aquatic ecosystems. Consequently, ecosystem services such as fishing and swimming have diminished. Local fishing companies now remove invasive white fish, which are converted into biogas. That way, the water in eutrophic lakes is clarified allowing plant and algae growth at the bottom, which in turn fosters population growth in demersal fauna including predatory fish such as pike, perch, and zander (Fig. 29.5). Demersal predators are larger species with higher value in sport fishing and are good bioindicators of freshwater ecosystem health (Bernes et al. 2015).

Architecture Inspired by Nature

The Naturum Vattenriket is a visitor center in the middle of the KVBR, reached by a wooden bridge from the city of Kristianstad. It is situated on the banks of the river Helge and was designed as a "retreat among the reeds," like a bird's nest (Naturum Vattenriket 2021). Built in 2010, it offers a gallery space, restaurants, education, and conference rooms for around 80,000 visitors each year who can observe the restored wetlands that were once polluted and inaccessible. The building is constructed 4 m above the ground to overcome natural fluctuating water levels, thereby allowing native flora to grow wildly in the surroundings without disturbing waterfowl (Fig. 29.5). Timber boardwalks provide passages through the grasses, bushes, and reeds. The material is predominantly wood which will weather in natural silver-grey as time passes.

Fig. 29.5 Examples of similar activities focusing on invasive fish in the KVBR, Sweden, and the CHBR, Chile. (**a**) Local fishery workers remove white fish in order to reduce their populations and avoid eutrophication of lakes in KVBR (photograph in Vattenriket archive) and (**b**) students capturing exotic fish species of Salmonidae using electric fishing in Robalo River, Omora Park, CHBR (photograph by Javier Rendoll). (**c**) The architecture of the Naturum Vattenriket Visitor Center is inspired by the shape of nests in the reeds (photograph by Patrik Olofsson) and (**d**) Cape Horn Sub-Antarctic Center in the CHBR, imitating landscape elements such as mountain peaks and autumn leaves and designed using environmentally friendly principles

29.7 Discussion

In this chapter, we identified differences and similarities in the temperate zones of South America and Europe with respect to their biophysical, symbolic-linguistic-cultural, and institutional-socio-political domains, and focused on what this means for the management of biosphere reserves. We further presented examples of biocultural conservation initiatives in Chile, Estonia, Germany, and Sweden for mutual learning.

At latitudes 45–60°, contexts for biodiversity conservation in South America and Europe are prominently different. In this latitudinal range of biophysical space, countries of the European Union have 4.2 times more land than South America, a proportion that highlights the uniqueness of temperate South American ecosystems (Rozzi et al. 2012). South American ecosystems are better conserved than Europe's woodlands (Sabatini et al. 2018). However, are they well protected? The institutional-socio-political context shows that South America has fewer but larger protected areas (UNEP-WCMC 2021) compared to Europe, and this is particularly the case for southern Chile. Unfortunately, only 12 out of 145 protected areas are effectively managed in Chile (Petit et al. 2018). In contrast, Europe has more funding and administrative personnel allocated to the conservation of biodiversity and cultural landscapes. This seems, though, not to be sufficiently reflected in the preservation of linguistic diversity: South America and Europe have in common preventing the loss of languages. Thus, we identify a common need to strengthen the conservation of the linguistic-cultural and spiritual space, until now rarely considered in protected areas generally (Kothari et al. 2014; Bridgewater and Rotherham 2019).

Temperate biosphere reserves in the European Union work under good conditions. The countries of our three examples have placed biosphere reserves at the center of their biodiversity strategies, which are embedded in the international agenda on sustainability. Regarding the implementation of biosphere reserves, Estonia, Germany, and Sweden have biosphere reserve offices and staff on-site, where they can interact with local stakeholders in a participative way. Indeed, joint decision-making and bottom-up conservation policies are necessary in order to incorporate local knowledge and interrelationships (Harrison and Burgess 2000). The presence of local biosphere reserve offices is essential, not only to identify and interact with the many stakeholders, but also to improve the learning capacity of bureaucratic institutions (Stoll-Kleemann and Welp 2008; Schliep and Stoll-Kleemann 2010). In contrast, in Chile, a legal framework, local staff, and a financial budget are absent. Consequently, biosphere reserve initiatives largely depend on the commitment of universities and the goodwill of political authorities.

In Estonia, Germany, and Sweden, local businesses play a major role in the effective implementation of biosphere reserves. The German BR Rhön, for example, was identified as the most successful biosphere reserve among 90 biosphere reserves in 49 countries by Cuong et al. (2017) due to its participatory governance and strong fusion of conservation and sustainable development activities based on partnerships with local producers and enterprises. This also implied using the biosphere brand for marketing. A narrow focus on nature and landscape protection will also cause some stakeholders, such as farmers or fishers, not to recognize the biosphere reserve as a legitimate neutral partner (Stoll-Kleemann and Welp 2008). Hence, a participative co-management framework seems to be an important future direction for progress in Chilean biosphere reserves.

Despite the differences in domains and the implementation of biosphere reserves in South America and Europe, our three examples of conservation initiatives in Germany, Sweden, and Estonia had some similarities with biocultural education and conservation approaches in the CHBR, Chile.

First, we found that education approaches were similar to the field environmental philosophy (FEP) implemented in the CHBR with regard to the focus on under-perceived and under-valued biota. The ecotourism with a hand lens of cryptogamic organisms (bryophytes and lichens) (Rozzi et al. 2020) resembles the education project about learning experiences with living animals (Project ELENA, BRBA, Germany) which highlights "unspectacular," invisible biodiversity, such as earthworms and snails. This promotes "face-to-face" experiences with local fauna, a parallel to the CHBR ethical birding activity and freshwater invertebrate encounters with a hand lens (Rozzi et al. 2010; Contador et al. 2018).

Second, research contained important education components. The translation of research results into attractive education material such as children's books (BLBR, Germany) or professional photography books (Fig. 29.4, CHBR, Chile) acknowledges that research is a matter of the whole society and its outreach is essential for the conservation science toolbox. This is particularly true for research results that give hope for overcoming environmental problems (Swaisgood and Sheppard 2010), like the rediscovery of species believed to have disappeared (Fig. 29.4). Attractive visitor centers harmonizing with the surrounding landscape (Fig. 29.5) are a true holistic contribution to biodiversity conservation as they integrate education with sustainable landscape architecture, which emphasizes the "unity of man with nature" (Chen and Wu 2009).

Third, conservation actions such as the control of overpopulated species (selective fishing in the KVBR, Sweden) are essential to restore or protect locally valued biodiversity; however, they require funding that is not always available in countries with poor conservation priorities. In the CHBR for example, although some efforts to control invasive mammal species like American mink (*Neovison vison*) have been undertaken, they have been discontinuous and spatially limited (Schüttler et al. 2019). Hence, this conservation challenge largely remains in the realm of research (Fig. 29.5).

Fourth, our example from Estonia shows how the cultural dimension can be integrated into the biosphere reserve agenda through fostering traditional grazing practices in restored alvars or gastronomy courses. Gastronomy is also practiced in the CHBR to promote seaweeds as part of a healthy, local diet (Fig. 29.3). Some biosphere reserves support the preservation of cultural identities or indigenous peoples. For example, Sorbian names for watercourses are used in the Spreewald Biosphere Reserve, Germany (SBR 2012), ecotourism based on the Sámi cultural heritage is realized in the Vindelälven-Juhttátahkka BR in the north of Sweden (VJBR 2021), and artisanal Yagan basketry is included in university courses in the CHBR (Fig. 29.3). This important dimension needs to be further addressed in national agendas beyond the MaB Programme.

Finally, the use of metaphors to enhance conservation awareness is a central step in the FEP methodology. For example, the "Miniature Forests of Cape Horn" is used to describe the ecological relationships in the community of mosses, lichens, mushrooms, and other biota (Fig. 29.4). Apparently, a focus on metaphors has been less frequent in our examples from Europe and might be a promising tool to further develop (but see Taylor and Dewsbury 2018).

29.8 Conclusion

Lessons from our comparison of biocultural conservation initiatives in temperate biosphere reserves in the northern and southern hemispheres are as follows: (1) holistic conservation can work in the absence of great funding schemes when approaches are simple, integrative, and locally centered; (2) countries highly committed to the biosphere reserve concept allow implementing sustainability in a wider array; and (3) similar biomes have similar conservation challenges and hence, conservation stakeholders could learn from each other, highlighting the necessity of partnerships among biosphere reserves globally. Although MaB has ecosystem-specific networks on, for example, drylands, mangroves, mountains, and tropical forests, temperate forests remain absent. The addition of this biome into MaB networks would catalyze south–north inter-hemispheric collaborations. Despite the power of holistic conservation approaches, particularly in the areas of education and research, we think that economic sustainability can only be achieved when biosphere reserves are firmly anchored in policies. Such a commitment is still needed in Chile.

Acknowledgments We are grateful to Meriem Boumrane, Johanna MacTaggart, Daniela Nedelcheva, Lia Rosenberg, and Sebastian Seibold for providing us with local contacts and information about the projects of the biosphere reserves presented here. We also appreciate the suggestions and corrections of three reviewers that improved an earlier version of this chapter. Roy Mackenzie was funded by the Chilean National Agency for Research and Development (ANID) PAI/2017—79170119 and MPG/2019—190029. Authors Mackenzie and Schüttler receive basal support from the Cape Horn International Center (CHIC)—ANID PIA/BASAL (PFB210018).

References

Aguilar G, Lafosse S, Rojas H, Steward R (2010) South/North Exchange of 2009 - the constitutional recognition of indigenous peoples in Latin America. Pace Int Law Rev. Online Companion: 44

Arnberger A, Ebenberger M, Schneider IE et al (2018) Visitor preferences for visual changes in bark beetle-impacted forest recreation settings in the United States and Germany. Environ Manag 61:209–223

Arroyo MTK, Cavieres L, Peñaloza A et al (1996) Relaciones fitogeográficas y patrones regionales de riqueza de especies en la flora del bosque lluvioso templado de Sudamérica. In: Armesto JJ, Villagrán C, Kalin Arroyo MTK (eds) Ecología de los bosques nativos de Chile. Editorial Universitaria, Santiago de Chile, pp 71–92

Berentsen WH, East WG, Poulsen TM et al (2021) Europe. Encyclopedia Britannica. https://www.britannica.com/place/Europe. Accessed 4 Feb 2021

Bernes C, Carpenter SR, Gårdmark A et al (2015) What is the influence of a reduction of planktivorous and benthivorous fish on water quality in temperate eutrophic lakes? A systematic review. Environ Evid 4:1–28

Berthold T, Sturm P (2010) Die Bedeutung der biologischen Vielfalt – Eine Aufgabe der Umweltbildung am Beispiel des Projektes "Tiere live". Anliegen Natur 34:41–48

BNatSchG (Bundesnaturschutzgesetz) § 25 Abs. 1 Biosphärenreservate. https://dejure.org/gesetze/BNatSchG/25.html. Accessed 8 Nov 2021

Borrero LA, Martin FM, Barberena R (2011) Visits, "Fuegians" and information networks. In: Whallon R, Lovis WA, Hitchcock RK (eds) Information and its role in hunter-gatherer bands. The Cotsen Institute of Archaeological Press, Los Angeles, CA, pp 249–265

Bridgewater P, Rotherham ID (2019) A critical perspective on the concept of biocultural diversity and its emerging role in nature and heritage conservation. People Nat 1:291–304

Chen X, Wu J (2009) Sustainable landscape architecture: implications of the Chinese philosophy of "unity of man with nature" and beyond. Landsc Ecol 24:1015–1026

Contador T, Kennedy J (2014) Habitantes sumergidos bajo los ríos de Cabo de Hornos. Universidad de Magallanes, Punta Arenas

Contador T, Rozzi R, Kennedy J et al (2018) Sumergidos con lupa en los ríos del Cabo de Hornos: Valoración ética de los ecosistemas dulceacuícolas y sus habitantes. Magallania 46:183–206

CPI (2019) Global landscape of climate finance 2019. Climate Policy Initiative, London. https://climatepolicyinitiative.org/publication/global-climate-finance-2019/. Accessed 20 Jan 2021

Cuong CV, Dart P, Hockings M (2017) Biosphere reserves: attributes for success. J Environ Manag 188:9–17

Donghi TH, Eidt RC, Calvert PAR (2021) Argentina. Encyclopedia Britannica. https://www.britannica.com/place/Argentina. Accessed 4 Feb 2021

Elbakidze M, Hahn T, Mauerhofer V et al (2013) Legal framework for biosphere reserves as learning sites for sustainable development: a comparative analysis of Ukraine and Sweden. Ambio 42:174–187

Environmental Board (2014) Sustainable development programme for the West Estonian Archipelago Biosphere Programme Area 2014–2020. https://www.keskkonnaamet.ee. Accessed 27 Jan 2021

ESCCA (2019) Life to Alvars. The restoration and grazing reintroduction on 2500 hectares of Estonian alvar grasslands. Estonian Seminatural Communities Conservation Association. Layman's report on LIFE+ PROJECT 2014-2019. http://www.envir.ee/et/lifetoalvars. Accessed 27 Jan 2021

EUNIS (2017) EUNIS habitat classification of the European Environmental Agency. https://www.eea.europa.eu/data-and-maps/data/eunis-habitat-classification. Accessed 19 Jan 2021

European Commission (2016) Joint communication to the European Parliament and the Council. An integrated European Union policy for the Arctic. https://eeas.europa.eu. Accessed 19 Jan 2021

European Parliament (2015) Written declaration on the protection and promotion of regional and minority languages in the EU. https://www.europarl.europa.eu. Accessed 19 Jan 2021

European Union (2021) Official website of the European Union. 2021–2027 long-term EU budget. https://ec.europa.eu/info/strategy/eu-budget/long-term-eu-budget/eu-budget-2021-2027_en. Accessed 18 Jan 2021

Grote R (2006/07) On the fringes of Europe: Europe's largely forgotten indigenous peoples. Am Indian Law Rev 31:425–443

Guevara S, Laborde J (2008) The Landscape approach: Designing new reserves for protection of biological and cultural diversity in Latin America. Environ Ethics 30:251–262

Harrison C, Burgess J (2000) Valuing nature in context: the contribution of common-good approach. Biodivers Conserv 9:1115–1130

Heinrup M, Schultz L (2017) Swedish biosphere reserves as arenas for implementing the 2030 Agenda: analysis and practice. Swedish Environmental Protection Agency. http://www.swedishepa.se/Documents/publikationer6400/978-91-620-6742-7.pdf?pid=19696. Accessed 6 April 2021

Heyl K, Ekardt F, Roos P, Stubenrauch J, Garske B (2021) Free trade, environment, agriculture, and plurilateral treaties: the ambivalent example of Mercosur, CETA, and the EU-Vietnam Free Trade Agreement. Sustainability 13:3153

Jacobson AP, Riggio J, Tait MA et al (2019) Global areas of low human impact ('Low Impact Areas') and fragmentation of the natural world. Sci Rep 9:14179

Jallon N (2019) Life to alvars project: Socio-economic analysis report (2019). https://life.envir.ee/sites/default/files/pictures/Life%20to%20Alvars%20Project%20-%20Socio-economic%20analysis%20report%20II.pdf. Accessed 5 April 2021

Janssen JAM, Rodwell JS, García Criado M et al (2016) European Red List of habitats. Part 2. Terrestrial and freshwater habitats. Publications Office of the European Union, Luxembourg

Jonsell M (2007) Effects on biodiversity of forest fuel extraction, governed by processes working on a large scale. Biomass Bioenergy 31:726–732

Kothari A, Demaria F, Acosta A (2014) Buen vivir, degrowth and ecological Swaraj: alternatives to sustainable development and the Green Economy. Development 57:362–375
KVBR (2020) Kristianstads Vattenrike Biosphere Reserve activities 2019. https://vattenriket.kristianstad.se/wp-content/uploads/2020/04/2020_01_VB2019-ENG.pdf. Accessed 22 Jan 2021
Lintzmeyer K, Zierl H (2009/10) 100 Jahre Schutzgebiet Berchtesgaden. Wegbereiter alpiner Schutzgebiete. Jahrbuch des Vereins zum Schutz der Bergwelt (München), 74./75. Jahrgang: 317–346
Louv R (2005) Last child in the woods. Algonquin Books, New York
Massardo F, Rozzi R (2006) The world's southernmost ethnoecology: Yahgan craftmanship and traditional knowledge/La etnoecología más austral del planeta: artesanía y conocimiento ecológico tradicional Yagán. Programa aplicación y diversificación del ecoturismo, Cabo de Hornos. Sernatur, Punta Arenas
Michler T, Zuda S (2020) Die wilden 16. Die neuen Abenteuer der Urwaldkäfer im Bayerischen Wald. Edition Lichtland, Freyung
Ministry of Planification and Cooperation (1993) Ley 19253. Establece normal sobre protección, foment y Desarrollo de los indígenas, y crea la Corporación Nacional de Desarrollo Indígena. https://www.bcn.cl/leychile/navegar?idNorma=30620. Accessed 5 Nov 2021
Naturum Vattenriket (2021) Naturum Vattenriket visitor center, Sweden. https://www.world-architects.com/es/white-arkitekter-goteborg/project/naturum-vattenriket. Accessed 6 April 2021
Orquera L, Piana E (2009) Sea nomads of the Beagle Channel in Southernmost South America: over six thousand years of coastal adaptation and stability. J Island Coast Archaeol 4:61–81
Pärtel M, Rein K, Zobel M et al (1999) Alvar grasslands in Estonia: variation in species composition and community structure. J Veg Sci 10:561–570
Petit IJ, Campoy AN, Hevia MJ et al (2018) Protected areas in Chile: are we managing them? Rev Chil Hist Nat 91:1
Plumptre AJ, Baisero D, Belote RT et al (2021) Where might we find ecologically intact communities? Front For Glob Change 4:626635
Rozzi R (2013) Biocultural ethics: from biocultural homogenization toward biocultural conservation. In: Rozzi R, Pickett STA, Palmer C et al (eds) Linking ecology and ethics for a changing world: values, philosophy, and action, ecology and ethics. Springer, Dordrecht, pp 9–32
Rozzi R (2016) Bioética global y ética biocultural. Cuadernos Bioética 27:339–355
Rozzi R, Schüttler E (2015) Primera década de investigación y educación en la Reserva de la Biosfera Cabo de Hornos: el enfoque biocultural del Parque Etnobotánico Omora. An Inst Patagon 43:19–43
Rozzi R, Anderson CB, Pizarro JC et al (2010) Field environmental philosophy and biocultural conservation at the Omora Ethnobotanical Park: methodological approaches to broaden the ways of integrating the social component ("S") in Long-Term Socio- Ecological Research (LTSER) sites. Rev Chil Hist Nat 83:27–68
Rozzi R, Armesto JJ, Gutiérrez JR et al (2012) Integrating ecology and environmental ethics: earth stewardship in the southern end of the Americas. Bioscience 62:226–236
Rozzi R, La Valle MT, Russell S et al (2020) Ecotourism with a Hand-Lens: a field environmental philosophy experience from the South of the world. In: Brister E, Frodeman R (eds) Philosophy for the real world: an introduction to Field Philosophy. Taylor & Francis/Routledge, New York, pp 222–239
Ruoss E (2013) Biosphere reserves as model sites for sustainable development. In: Getzner M, Jungmeier M (eds) Protected areas in focus: analysis and evaluation. Verlag Johannes Heyn, Klagenfurt, pp 99–114
Sabatini FM, Burrascano S, Keeton WS et al (2018) Where are Europe's last primary forests? Divers Distrib 24:1426–1439

SBR (2012) Periodic review of the Spreewald Biosphere Reserve. http://www.unesco.org/new/fileadmin/MULTIMEDIA/HQ/SC/pdf/Spreewald_periodic_review_2013.pdf. Accessed 05 Feb 2021

Schliep R, Stoll-Kleemann S (2010) Assessing governance of biosphere reserves in Central Europe. Land Use Policy 27:917–927

Schultz L (2009) Nurturing resilience in social-ecological systems. Lessons learned from bridging organizations. Dissertation, Stockholm University

Schüttler E, Crego RD, Saavedra-Aracena L et al (2019) New records of invasive mammals from the sub-Antarctic Cape Horn Archipelago. Polar Biol 42:1093–1105

Seidl R, Thom D, Kautz M et al (2017) Forest disturbances under climate change. Nat Clim Chang 7:395–402

Seidl R, Klonner G, Rammer W et al (2018) Invasive alien pests threaten the carbon stored in Europe's forests. Nat Commun 9:1626

Stoll-Kleemann S, Welp M (2008) Participatory and integrated management of biosphere reserves. Lessons from case studies and a global survey. Gaia 17:161–168

Sturm P, Adelmann W, Berthold T et al (2014) Baseline study on biodiversity education using living animals in schools in Germany on the example of the state of Bavaria. ELENA Project Experiential learning and Education for Nature Awareness. Bayerische Akademie für Naturschutz und Landschaftspflege, Laufen

Swaisgood RR, Sheppard JK (2010) The culture of conservation biologists: show me the hope! Bioscience 60:626–630

Swedish EPA (2021) Swedish Environmental Protection Agency. http://www.swedishepa.se/Enjoying-nature/Protected-areas/Biosphere-reserves. Accessed 20 Jan 2021

Taylor C, Dewsbury BM (2018) On the problem and promise of metaphor use in science and science communication. J Microbiol Biol Educ 19:19.1.46

Thorn S, Seibold S, Leverkus AB et al (2020) The living dead: acknowledging life after tree death to stop forest degradation. Front Ecol Environ 18:505–512

UNEP-WCMC (2021) Protected Area Profiles for Latin America & Caribbean from the World Database of Protected Areas. Available at: www.protectedplanet.net. Accessed 20 Jan 2021

UNESCO (2017) A new Roadmap for the Man and the Biosphere (MaB) Programme and its World Network of Biosphere Reserves: MaB Strategy (2015–2025), Lima Action Plan (2016–2025) and Lima Declaration. UNESCO, Paris

UNESCO (2021) Man and the Biosphere (MaB) Programme. https://en.unesco.org/mab. Accessed 19 April 2021

Villagrán Moraga C (2020) Historia biogeográfica de las briófitas de Chile. Gayana Bot 77:73–114

VJBR (2021) The Sami Eco Lodge. http://samiecolodge.com/. Accessed 5 April 2021

WABPA (2013) Periodic review for the West Estonian Archipelago Biosphere Reserve. https://www.keskkonnaamet.ee/sites/default/files/periodic_review_2015_toim2017_0.pdf. Accessed 5 April 2021

Wettemark C, Magntorn K, Dahl J et al (2016) Periodic review for the Kristianstads Vattenrike Biosphere Reserve. https://vattenriket.kristianstad.se/wp-content/uploads/2020/01/periodic-review_eng.pdf. Accessed 5 April 2021

Zahner V, Blaschke S, Fehr P et al (2007) Eine Studie zur Artenkenntnis bei Vögeln, durchgeführt mit bayerischen Schülern. Vogelwelt 128:203–214

Chapter 30
Mindfulness and Reconnection with Freshwater Ecosystems at the Meadows Center Education Program

Rob Dussler, Justin Williams, and Samuel Massey

Abstract Opportunities for nature connection and equitable access to public green spaces are increasingly compromised and limited particularly within highly urbanized areas. Prevailing formal educational paradigms and pedagogies drive the biocultural homogenization of the mind as they privilege standardized, deterministic methodologies, and predetermined outcomes. The limiting of immersive and kinesthetic learning explorations and authentic meaning-making is problematic to biocultural conservation efforts and the formation of eco-consciousness, particularly among children. In this context, to cultivate an earth stewardship ethic, green public spaces open opportunities for eco-phenomenological, personally relevant experiences in nature and are integral to fostering ekistic understandings of physical, biocultural, and sociocultural environments. Sites for long-term research and education, such as the Meadows Center for Water and the Environment at Texas State University, are vital hubs for understanding socio-ecological systems through interdisciplinary perspectives and approaches with sensitivity and awareness. The center resides at one of the most prolific freshwater artesian springs in North America and is home to eight government-protected, threatened or endangered species. The educational program engages over 30,000 school children annually in outdoor education at the freshwater springs and nature preserve. This large educational interface creates rich opportunities for understanding the vital linkages of the inhabitants, habitats, and habits in an environmentally sensitive ecosystem. Mindfulness is an attentive lens for understanding and enhancing nature connection. Field environmental philosophy (FEP) provides a methodological approach to field experiences which allows students to examine themselves as co-inhabitants in an

R. Dussler (✉) · S. Massey
The Meadows Center for Water and the Environment, Texas State University, San Marcos, TX, USA
e-mail: robdussler@txstate.edu; sammassey@txstate.edu

J. Williams
Philosophy Department, Texas State University, San Marcos, TX, USA
e-mail: justin.williams@txstate.edu

R. Rozzi et al. (eds.), *Field Environmental Philosophy*, Ecology and Ethics 5,
https://doi.org/10.1007/978-3-031-23368-5_30

ecosystem, exploring ecological and ethical orientations and behavior. Current interdisciplinary research projects, curriculums, and strategies are explored.

Keywords Biocultural conservation · Contemplative pedagogy · Education · Field environmental philosophy · Mindfulness

30.1 The Meadows Center for Water and the Environment

The Meadows Center for Water and the Environment is one of Texas State University's six research centers. The University and the Meadows Center are located in central Texas in the city of San Marcos, along the innovation corridor between Austin and San Antonio (Fig. 30.1). San Marcos is one of the fastest-growing mid-sized cities in the United States with Austin being the country's fastest-growing large city in 2019 (KIUR 2015, 2019). Amazingly, amidst this urbanization boom, in a state whose land is 98% privately owned, is an almost unbelievable environmental oasis: the culturally and ecologically significant San Marcos Springs.

The springs emerge from limestone formations adjacent to a converted hotel that functions as the Meadows Center's base of operations since 2002. Hundreds of upwellings form a pool, known as Spring Lake, created by an impoundment 1 km downstream. The springs are entirely fed by the Edwards Aquifer, a biologically diverse groundwater system which provides primary water supply to over one million regional human inhabitants (Sharp et al. 2019) and to many economically and ecologically important Central Texas rivers and streams. In addition to their connection to this vital natural system making them a locus for understanding the interconnection of Central Texas ecoregions and hydrological systems, the San Marcos Springs are notable for their consistency and striking appearance. Most of the springs are small, putting out an amount of water comparable to an

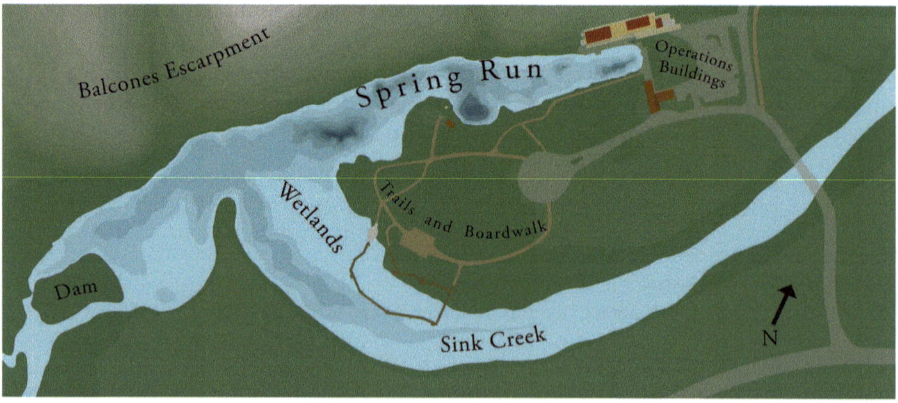

Fig. 30.1 Features of Spring Lake and Operations Area of The Meadows Center for Water and the Environment, Texas State University, San Marcos. Source: Texas Natural Resources Information System (https://data.tnris.org/collection/af1ca25e-b38b-4203-90b8-d90f881963ae)

upward-turned garden hose, displacing sediment with a characteristic boiling appearance. Some sediment disturbances may be 30 cm in diameter or more. By one engineering firm's estimate, most discharge less than 1 L of water a second during average flow conditions, although they are innumerable and account for the majority of the water in the pool (LBG-Guyton Associates et al. 2004).

Large and small springs are interspersed along the first 280 m of the pool and produce the majority of the water in the urban San Marcos River. Their consistency has kept the river flowing since at least the time of humans' first arrival in North America. Archeological evidence suggests persistent human interaction with the springs from the end of the last glacial period (Lohse 2013). For the past 171 years, an impoundment has increased the depth and residence time of the water at the beginning of the river. The springs have seen periods of drastically decreased output due to drought and groundwater extraction, but the quality and quantity of the spring flow continue to support a diversity of aquatic life including several endemic species that would be at risk of extinction if the springs ran dry.

The crystal blue, aquarium-like habitat created by the springs, seen in detail through the viewing wells of glass-bottom boats that have taken visitors on tours of the lake since the late 1940s, or the masks of snorkeling or SCUBA diving equipment, functions as an object of inquisitive curiosity, or as a striking focal point for meditative observation (Fig. 30.2). Using spring-viewing excursions as a central attraction, the Meadows Center engages over 120,000 people annually in environmental programming, 30,000 of whom are local school children and 50 percent of which originate from schools with high proportions of low-income students (Title I Schools according to Federal legislation).

30.2 Four Pillars of Action

The Meadows Center has the responsibility of being the key steward of the springs and the surrounding preserve. The Center fulfills its mission of inspiring research, innovation, and leadership that ensures clean, abundant water for the environment and all of humanity by integrating activities across four pillars of action:

(a) *Education*. The Meadows Center views environmental education as a chief conservation strategy. Children by their very nature are curious and when that curiosity is sparked, they often ask questions, engage fully in educational discussions, and grow immensely in short periods. The Center invests in environmental education knowing that children are happier, healthier, and smarter in the outdoors. Every time we invest in a child's understanding of the natural world, we imbue in them an appreciation for the world they live in. These children grow up to become adults who care deeply for and often advocate to protect the natural world.

(b) *Research*: Faculty, staff, and students conduct applied research to address real-world problems. Here researchers provide multidisciplinary expertise to advance

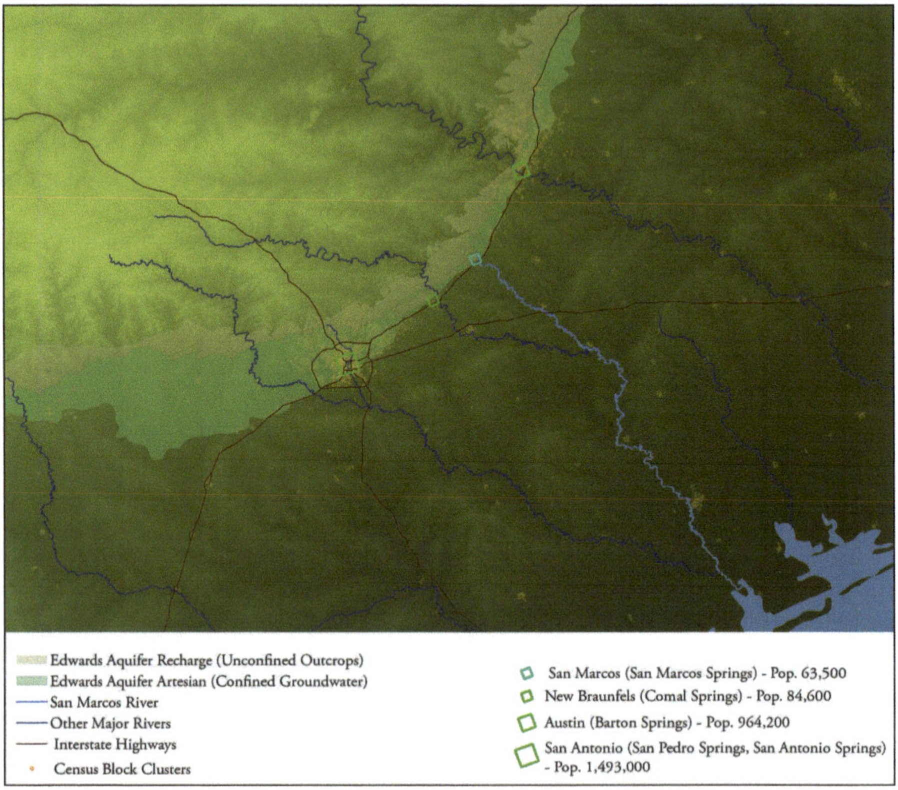

Fig. 30.2 The San Marcos Springs are part of a constellation of groundwater emergences from the Edwards Aquifer, which is in turn filled by rainfall to the north and west. Rhomboids indicate human inhabitation that are closely associated with the springs. All springs associated with the four urban areas shown contribute flow to Major Rivers and discharge to the Gulf of Mexico. River and Aquifer data adapted from Texas Water Development Board, 2020 (https://www.twdb.texas.gov/mapping/gisdata.asp), Census Block and Highway data adapted from US Census Bureau, 2019 (https://www.census.gov/cgi-bin/geo/shapefiles/index.php), Elevation data adapted from United States Geological Survey 3D Elevation Program (3DEP), 2020 (https://www.usgs.gov/core-science-systems/ngp/3dep/about-3dep-products-services). All in the public domain

science-based solutions for the most pressing water resource challenges facing Texas and the world beyond.

(c) *Stewardship*. The Meadows Center is active in the San Marcos community and beyond. It strives to strengthen those ties by connecting stewardship of water and natural resources to quality of life, and cultivates a stewardship ethic and practice through the community-driven management of Spring Lake as well as local watershed activities across Texas.

(d) *Leadership*. The Meadows Center supports responsible natural resource and water policy in Texas and convenes stakeholders to address critical water and natural resource concerns. Staff often serves with communities and organizations in voluntary leadership roles in diverse contexts to build the capacity of local communities for protecting and managing their own water and natural resources.

30.3 The Four-Step Cycle of Field Environmental Philosophy

The four pillars of the Meadows Center offer an approach that is complementary to the methodology of field environmental philosophy (FEP). FEP is centered on the interrelations between the ways of knowing and researching nature and the ways of dwelling (Rozzi 1997). Ecological sciences can play a crucial role in motivating ethical changes in the ways contemporary society relate to the biosphere. Sciences can change the ways humans perceive, know, and understand ecosystems and their place in them. FEP offers a methodology that is appropriate to integrate science and ethics at sites that conduct long-term socio-ecological research, such as the Meadows Center (Fig. 30.2).

FEP integrates ecological research and environmental ethics into biocultural education and conservation through an interrelated four-step cycle: (1) interdisciplinary ecological and philosophical research; (2) composition of metaphors and communication through simple narratives; (3) design of guided field experiences with an ecological and ethical orientation; (4) implementation of in situ conservation areas (Rozzi 2010, 2011). At the Meadows Center, the Four Pillars of Action intersect with FEP's 4-step cycle. The *education pillar* reflects FEP-Step 3. Experiences of "direct encounters" with ecosystems and different organisms in their habitats through field activities such as the glass-bottom boats, snorkeling and SCUBA, "bug picking," and other mindfulness activities transform knowledge about biodiversity. At the same time, these activities also promote the awareness of, and ethical attitudes toward, co-inhabiting with diverse living beings. The Meadows Center's *research pillar* directly corresponds to FEP-Step 1. Interdisciplinary research is based on endangered flora and fauna, such as the Texas Wild Rice (*Zizania texana*) and the Texas Blind Salamander *(Eurycea rathbuni*). The Meadow's *stewardship pillar* bridges FEP-Step 1 and Step 3. The *leadership pillar* converges with FEP-Step 4 in the spirit of developing multi-generational teams of volunteers to continue water conservation in Texas for years to come.

The mindfulness approaches being developed at the Meadows Center are directly linked to FEP-Step 2: communication of discoveries through the composition and use of metaphors and FEP—Step 3: that facilitate a deep understanding of our relationship to this highly diverse community. The four pillars of the Meadows Center show the significant value beyond quantitative sciences. The more we recognize the socio-ecological significance of a site, the more we are drawn into a stewardship or biocultural ethic.

30.4 Reconnecting through Mindfulness

Mindfulness has origins in Buddhism, Greek, and existential philosophy and humanism (Brown et al. 2007). It is important to note the two dominant conceptions of mindfulness in theory and practice. First, *meditative mindfulness* or eastern mindfulness is characterized by being present in the current moment in an open and non-judgmental way. It was popularized by Kabat-Zinn (1994) drawing heavily on meditative practices. Today it is often therapeutically utilized for stress reduction. Second, *creative mindfulness*, or western mindfulness is characterized by having an open, curious orientation to the environment and the development of novel distinctions (Carmody 2014). It has been largely influenced by Ellen Langer (2000). It employs approaches to facilitate the creative state where individuals are actively noticing, restructuring their environments, and creating new categories that are "consciously controlled" (Moscardo 1999).

Mindfulness is a vital ingredient to durable and meaningful nature connection. Schultz (2002, p. 67) suggests that nature connection can be described by the "extent to which an individual includes nature within his/her cognitive representation of self." Powerful nature experiences and the formation and expression of a vital ecological selfhood can be a catalyst to care for natural spaces and the formation of an earth stewardship ethic.

The genesis for nature connectedness is being mindfully awake and attentive to one's surroundings and one's existence and interaction with nature, in the present moment. How easily we can fall into mindless routine! We are so overstimulated and hurried that our ability to slow down, notice, and be fully present is appreciably limited. This current reality, coupled with limited opportunities for access to nature and a compromised ability to "be here now," accentuates the value and opportunity for robust environmental education programs. This has motivated the implementation of mindfulness into education programs at the Meadows Center. Specifically, understanding how humans connect and engage with nature and the associated meanings and outcomes of informal science programs within urban contexts, are relationships which cannot be examined or resolved in the short term. The ways in which we come to understand ourselves in relation to ecological systems co-evolve with how we understand our connection and engagement with nature in general. This orientation aids us in incorporating the myriad of metaphorical lenses we live by.

30.4.1 Multiple Mindfulness Approaches

Efforts to bridge the fundamental gaps between formal and informal science programs at the Meadows Center are being resolved through mindfulness approaches and practices at multiple levels of the educational program. Dussler and Deringer (2020) explored the effects of mindfulness interventions with environmental interpreters at the Meadows Center and their subsequent descriptions of mindfulness,

Fig. 30.3 Glass-bottom Boat (left) and Snorkeler (right) in Spring Lake. The "lenses" that grant vision into the waters of the San Marcos Springs can become "conceptual lenses" for socio-ecological investigations. Photos courtesy of The Meadows Center

nature connection, and interpretive practice. Interpreters participated in a 4 week mindfulness program that included an introduction to foundational elements and origins of mindfulness. Activities and meetings occurred at the preserve surrounding Spring Lake and in the waters of the lake via kayaking and snorkeling (Fig. 30.3). Focus groups, journaling, and debriefing provided space for interpreters to process and share their stories and experiences. Accompanying mindful exercises and journal prompts were included between weekly meetings at the lake. Mindful practice included noticing walks, wandering, body radar, engaging senses, nature sketching and "zooming in" with hand lenses, listening and breathing meditations, nature journaling, found objects, and stories of the day.

In qualitative interviewing, and constructivist grounded theory coding and analysis, several prominent themes emerged which have shaped the Center's future programming and staff development initiatives.

Interpreters experienced the construct of mindfulness in a variety of ways. It enhanced personal experiences of nature. It also created more authentic interpretive experiences of nature. New connections and novel distinctions emerged in areas that interpreters had often frequented in their interpretive programs. Interpreters realized that without slowing down and taking the time to immerse and connect mindfully, it was easy to be in "zombie mode" and just be "going through the motions." Mindfulness was also seen as a way to connect more deeply with nature. As expressed by one of the interpreters, it "reduced the stress in my head and filtered overstimulation."

Two mindfulness practices were particularly instructive. First, the practice of *wandering* encourages interpreters to let go of preconceived notions and expectations. Second, the practice of *imagining* the body as a powerful and receptive radar scanning the surrounding landscape can mobilize embodied sense experiences as well as trust in our exploratory capacities.

30.4.2 Mindfulness and FEP's Metaphor of the "Hand Lens"

Mindfulness approaches at the Meadows Center are expanded by FEP educational methodology that enables seeing a place through multiple complementary metaphors or "conceptual lenses." This combination of methodologies provides us with a more holistic perspective of the co-evolving nature of our life habits, the habitats, and the community of co-inhabitants we dwell within.

The *glass-bottom boats* at the Meadows Center is the most iconic lens offered to visitors (Fig. 30.3). The glass breaks the tension between two co-evolving worlds: it shifts our gaze from the surface into the aquatic world of the lake. This aerial perspective is inaccessible without the lens provided by the glass-bottom boats. This lens helps us to see the aquifer, which otherwise remains "invisible" as only an abstract concept.

The metaphors that we use embody our conceptual lenses and inform our behaviors. In other words, our orientation about ourselves in relation to nature functions as poetry that we live by. Our orientation, metaphorically, represents "lenses" through which society explores ecosystems, biodiversity, and the relations that we humans have with them. To "change one's lens" implies action. It can move us beyond observation into conservation. Moreover, it can move us to a new awareness and attitude of co-inhabitation with biodiversity (Rozzi 2011). Through these metaphors, "objects" we observe can be perceived as co-inhabitants rather than mere "natural resources." When we change our metaphor or lens, we see the "invisible" and we hear previously unheard voices. When we change our perceptions about ourselves in relation to the ecosystem we are co-inhabiting, our orientation changes as well. With this lens, a place is perceived as a mosaic of co-evolving narratives with a diversity of names, stories, and heritages, a diversity of co-inhabitants that spans generations.

30.4.3 Bug Picking

A favorite educational activity for our visiting school children is "Bug Picking." It involves tactile examinations of aquatic vegetation, macroinvertebrates, and water taken from the Spring Lake. Students equipped with species identification cards and hand lenses learn about the effects of water quality on biodiversity in ecological systems (Fig. 30.4). They become detectives as they touch, feel, and smell the water. "Eewwww" turns into "cooool" and most encouragingly, an "ahhhhhh" as they begin to understand that what lives in the water will let them know about the health of the water and ecosystems. Further study reveals the biological and ecological requirements these species need in order to exist and to thrive. A deeper appreciation is instilled when these young scientists determine that the San Marcos Springs are very clean and healthy. How is that possible given all of the surrounding threats to

Fig. 30.4 "Bug Picking" activity at The Meadows Center. (**a**) Aquatic macroinvertebrate reflex viewer. (**b**) A damselfly, (Calopterygidae), nymph. (**c**) A mayfly, (Baetidae), nymph. (**d**) Childrens' aquatic macroinvertebrate sample bucket and tools. Photographs by Sam Massey

the ecosystem? What does "protect" mean to you? Why would you want to protect this place, this ecological system, this part of you?

These salient moments involve an integration of "emotions and concepts derived from philosophical, scientific, and vernacular knowledge into the observation of surprising and unexpected behaviors of invertebrates, students and other participants experientially apprehend the concept of co-inhabitants" (Rozzi 2019, p. 277). Engaging all of our senses through recognition of our co-inhabitants as well as an awareness of their anatomy and unique habits can set the stage for numerous discoveries.

In addition to bug picking, activities such as observing and donning owl eyes, deer ears, employing raccoon touch, and dog noses can create sensory pathways for discovery and appreciation which are limited through overreliance on our bipedal stance and movement. For example, discoveries were made of an entire realm of olfactory sensations while moving our "dog noses" close to the ground. The fragrant understory of the preserve cannot be experienced from the standing position; hence, it had remained previously undiscovered by interpreters who had walked these trails many times over several years. Adding to the delight and power of such discoveries is the ability to share our experiences with others. Describing our sense experiences with others, hearing how others qualify sensation, intuition, and lived experience is vital to eco-phenomenological engagement with bugs and other organisms.

30.4.4 Sacred Springs Powwow

Understanding the depth and extent of the area's biocultural significance and importance to Indigenous cultures is essential to the Meadow Center's vision. The 1983 documentary *River of Innocence* triggered biocultural awareness when it was aired periodically on Texas public television stations. The documentary filmed underwater archeologists at San Marcos Springs uncovering a remarkably large and diverse assemblage of prehistoric stone tools. This discovery showed the importance of the springs to its past and contemporary Indigenous inhabitants. The archeological discoveries stimulated the first "Sacred Springs Powwow" held at Aquarena Springs in 1994, which evolved into a large annual Native American cultural celebration (Fig. 30.5). The San Marcos Indigenous Cultures Institute (ICI), a group led by local members of the Miakan-Garza Band of the Coahuiltecan Tribe, organizes this twenty-first-century version of the Sacred Springs Powwow in open spaces formerly occupied by a golf course and gift shop. Among the many dances, songs, and stories shared by the various attending tribal groups is the Miakan-Garza Coahuiltecan creation story. This story cites the largest spring in San Marcos as the

Fig. 30.5 2015 Sacred Springs Powwow. Photograph by Sam Massey

place of emergence for the first people onto the surface of the Earth (Xica Media 2020). Smaller ritual and cultural services are held at the springs by members of the ICI periodically throughout the year.

30.5 Environmental Interpretation and Mindfulness

The National Association of Interpretation defines interpretation as "a mission-based communication process that forges emotional and intellectual connections between the interests of the audience and the meanings inherent in the resource" (National Association for Interpretation n.d.). The starting point for sound environmental interpretation is authentic, emotive, and personally meaningful experiences of nature. Tilden (1977, p. 17) encourages interpreters to provoke emotion and positions interpretation as "an educational activity which aims to reveal meanings and relationships through the use of original objects, by firsthand experience, and by illustrative media, rather than simply to communicate factual information." Provoking emotion through the use of rigid schedules, facts, and rules is highly unlikely. This holds true on both personal levels and within the context of interpretive programming and participation. The revelation of meanings and relationships is forged through "intellectual and emotional connections" (National Association for Interpretation n.d.). The efficacy and durability of environmental interpretation are located primarily within powerful personal experiences in nature. These experiences are cultivated through intentional mindful engagement and practice (Dussler and Deringer 2020). Interpreters are the critical conduit for promoting and fostering nature connection and meaning within the Meadow Center for Water and the Environment's environmental education programs and public interface. Interpretation can create a forum for a social learning experience and construction of meaning which challenges assumptions and overly simplified approaches to learning and behavioral change (Orams 1994).

30.6 A Philosophical and Pedagogical Shift from Prevailing Formal Education

Informal science programs serve a vital function in connecting a variety of learners (possessing an array of interests and learning styles) to scientific knowledge in personally meaningful contexts. Interpreters are essential "bridges," carrying meaning from quantitative empirical, measured, and evaluative worlds into qualitative worlds of being. At the Meadows Center, interpreters adopt an eco-phenomenological approach, which asserts the holistic presence of the mind and body interrelated and within natural systems. In this way, "our reciprocal, embedded and embodied experiences with and within nature are seen to be loci of value, a priori of theoretical and philosophical abstraction and examination" (LeVasseur 2011, as cited in Newman and Robbins 2011, p. 160). Herein lies the critical challenge and tension of many informal science programs such as the Meadows Center: intentionally creating space for sense experiences, personalized learning, and reflective descriptions within the strictures and constraints of a formal educational field trip originating from a standardized paradigm. Standard efficiency and categorization do not lend themselves to absorption and deep meaning that are necessary for ecological consciousness.

Linking the intellect with a phenomenological engagement with nature, moves environmental education beyond a strictly cognitive enterprise. Personal meaning and relevance of nature engagement should include first-hand explorations and discoveries. However, current prevailing pedagogies privilege linear, standardized competencies that often are markedly separate from individualized experiences. In this way, the content of learning and what is determined to be important to the learner are externally located and predetermined; hence, not "in me" (Ergas 2015).

A metaphor might help here: the sitar. This musical instrument from India has both *external* and *internal* strings. When the external strings are moved, a harmonic vibration and resonance cause the internal strings to vibrate, which in turn enlivens the internal "resonance chamber." The internal strings are known as the sympathetic strings which respond to the external movement creating a unique and holistic sound. The sound is further complemented and formed by structural components of gourd, bone, ivory, and aged teak, of which much homage and mystery exists regarding the elemental influence of origin and age on the voice of the instrument. James A. Milward alludes to the amplification and expansion of resonance noting:

> Players of plucked stringed instruments have always regretted the fact that the note of a plucked string starts to decay as soon as it is plucked, unlike a sung, blown, or bowed note on voice or other instruments. By putting a set of tarab strings on a sitar, Indian luthiers partially defeated this problem, providing notes that magically swell *after* they are played (Milward 2018, p. 222).

The metaphor captures the need in environmental education for internal resonances of felt nature connection which have enlivened or accentuated an inherent biophilic quality. The body is a receptor, a resonance chamber, wherein one's innermost self is moving and being moved (Fuchs 2017). It is an active process that also is sustained beyond the initial moment, with a sympathy, an active empathy, that would not be possible via external stimulus alone. The internal and external sensory experiences, which are lasting, form the foundation for advocacy and stewardship.

This sentient acknowledgment and first-hand experience of being "with and within" is the locus for the recognition of oneself as a co-inhabitant (Rozzi 2019). Alas, the opportunity and capacity to connect in meaningful ways assume an attentive consciousness and presence for the current moment, amidst a culture so prone to mindlessness and routine. Pulkkli et al. (2017) explored environmental education as a lived body experience and situated contemplative pedagogical practices, such as mindfulness, as powerful ways to be present in nature and enliven the inner self as a site for environmental education. They conclude that:

> [M]erely exposing children to nature is often not enough. Their abilities to calm down, to focus, to clear their consciousness, to compassionately notice and care are also in need of cultivation, as are the abilities for respect and intrinsic valuation of nature. Without such abilities whatever experiences there are run the risk of dissolving into the continuous flux of "one thing after another" that characterise so much of modern life (Pulkkli et al. 2017, p. 216).

Vital connections demand not only a biophilic attachment to nature but a distinct recognition of the ecological systems and communities that encompass, as Rozzi (2011) so aptly clarifies in the 3H's, the habitats, the co-inhabitants, and the habits of a community. To be vitally engaged in this way is to be engaged in "a relationship to the world that is characterized both by experiences of flow (enjoyed absorption) and by meaning (subjective significance)" (Nakamura and Csikszentmihalyi 2003, p.87).

30.7 Conclusion: Snowy Egrets Are Our Guides

During the course of the mindfulness research study with interpreters at Spring Lake, there were awkward moments for the researchers and participants alike as we collectively engaged in mindfulness activities wondering at times if we were "doing it right." In follow-up interviewing and researcher debriefing it was apparent that occasionally self-awareness and critique were most prevalent and nature absorption was in the background as we wondered at times what we were supposed to feel through the activities and whether we were missing the point. "I'm being mindful, now what?"

In one pronounced moment the group was engaging in a noticing walk at the lake. It was a solo experience, yet the group was gravitating back to our debriefing spot prior to the conclusion of the activity. The mosquitoes had corralled all of us to the peninsula, and in a moment of collective surrender, we waded into the lake for reprieve, the cool waters instantly providing soothing relief and sanctuary. As we stood waist deep in the water, hands sliding back and forth across the surface, here they came. A flock of egrets swooped ever close to us with a dazzling display of aerobatics as they lost elevation, seemingly out of control, and then opening up their wings suddenly to lite gently on the branches of a tall cypress tree. There they would roost for the night, brilliant white dots on a background of deep green, with placid black water underneath reflecting the most wonderful scene and capturing the orange hues of the setting sun.

No one said a word. We turned to look at each other and were so delighted to be here, together, now, in this very moment, all senses engaged, absorbed, and connected. In later reflections a participant shared this poem from Mary Oliver (1983), which we cherished:

> Where the path closed
>   down and over,
>   through the scumbled leaves,

fallen branches,
through the knotted catbrier,
I kept going. Finally
I could not
save my arms
from thorns; soon
the mosquitoes
smelled me, hot
and wounded, and came
wheeling and whining.
And that's how I came
to the edge of the pond:
black and empty
except for a spindle
of bleached reeds
at the far shore
which, as I looked,
wrinkled suddenly
into three egrets ---
a shower
of white fire!
Even half-asleep they had
such faith in the world
that had made them ---
tilting through the water,
unruffled, sure,
by the laws
of their faith not logic,
they opened their wings
softly and stepped
over every dark thing.

It is in these moments, when we have changed lenses, that we become teachable, mosquitoes

and egrets become our guides. In these moments, beyond our reflection, beyond our abstractions, and as co-inhabitants we experience being part of the community of the Springs—moments when birds can teach us if we are mindful enough to watch and listen.

References

Brown KW, Ryan RM, Creswell JD (2007) Mindfulness: theoretical foundations and evidence for its salutary effects. Psychol Inq 18(4):211–237. https://doi.org/10.1080/10478400701598298

Carmody JF (2014) Eastern and western approaches to mindfulness: similarities, differences, and clinical implications. In: Le A, Ngnoumen CT, Langer EJ (eds) The Wiley Blackwell handbook of mindfulness. https://doi.org/10.1002/9781118294895

Dussler R, Deringer SA (2020) Exploring the effects of interpreters' experiences of mindfulness interventions on their connection with nature and subsequent environmental interpretation. J Interp Res 25:26–45. https://doi.org/10.1177/1092587220963556

Ergas O (2015) The deeper teachings of mindfulness-based curricular interventions as a reconstruction of "education". J Philos Educ 49(2):203–220
Fuchs T (2017) Intercorporeality and interaffectivity. Phenomenol Mind 11:194–209
Kabat-Zinn J (1994) Wherever you go, there you are: mindfulness meditation in everyday life. Hyperion, New York
Kinder Institute for Urban Research (2015) The fastest growing city in America? You will never believe. https://kinder.rice.edu/2015/06/19/the-fastest-growing-city-in-america-you-will-never-believe. Accessed 23 Mar 2021
Kinder Institute for Urban Research (2019) Austin named America's fastest-growing large city in 2019, report says. https://kinder.rice.edu/urbanedge/2019/10/22/wallethub-austin-dallas-houston-fastest-growing-cities-2019. Accessed 23 Mar 2021
Langer EJ (2000) Mindful learning. Curr Dir Psychol Sci 9(6):220–223. https://doi.org/10.1111/1467-8721.00099
LBG-Guyton Associates, BIO-WEST, Inc., Espey Consultants, Inc. and URS Corporation (2004) Evaluation of Augmentation Methodologies in Support of In-situ Refugia at Comal and San Marcos Springs, Texas. https://www.edwardsaquifer.org/science_docs/evaluation-of-augmentation-methodologies-in-support-of-in-situ-refugia-at-comal-and-san-marcos-springs-texas/. Accessed 23 Mar 2021
LeVasseur T (2011) Ecophenomenology. In: Newman J, Robbins P (eds) Green ethics and philosophy: an a-z guide. Sage, California, pp 158–161
Lohse JC (2013) Underwater archaeology at 41HY147, the Terrace Locality at Spring Lake. Center for Archaeological Studies, Texas State University, San Marcos, pp 53–57
Milward JA (2018) The silk road and the sitar: finding centuries of sociocultural exchange in the history of an instrument. J Soc Hist 52(2):206–233. https://doi.org/10.1093/jsh/shy050
Moscardo G (1999) Making visitors mindful: principles for creating sustainable visitor experiences through effective communication. Sagamore, Champaign
Nakamura J, Csikszmentmihalyi M (2003) The construction of meaning through vital engagement. In: Keyes CLM, Haidt J (eds) Flourishing: positive psychology and the life well-lived. American Psychological Association, pp 83–104
National association for interpretation (n.d.). https://www.interpnet.com/NAI/interp/About/nai/_About/ABOUT.aspx?hkey=6f63cbdc-2744-4b3f-9983-27a9e4f93d4a. Accessed 5 Mar 2020
Newman J, Robbins P (eds) (2011) Green ethics and philosophy: an A-Z guide. Sage, California
Oliver M (1983) American primitive. Back Bay Books, New York
Orams MB (1994) Creating effective interpretation for managing interaction between tourists and wildlife. Aust J Environ Educ 10:21–34
Pulkkli J, Dahlin B, Varri V-M (2017) Environmental education as a lived-body practice? A contemplative pedagogy perspective. J Philos Educ 51(1):214–229
Rozzi R (1997) Enseñar a los niños a ser ciudadanos, no meramente consumidores: El aporte de Hargrove a la educación ambiental en Chile. Ambiente Desarrollo XIII:52
Rozzi R (2010) Field environmental philosophy and biocultural conservation at the Omora Ethnobotanical Park: methodological approaches to broaden the ways of integrating the social component ("S") in long-term socio-ecological research (LTSER) site. Sociedad de Biologíade Chile, Santiago
Rozzi R (2011) Field environmental philosophy: regaining an understanding of the vital links between the inhabitants, their habits, and the regional habitats. Dialogue Universalism 1-2:85–109
Rozzi R (2019) Taxonomic chauvinism no more! Antidotes from Hume, Darwin and biocultural ethics. Environ Ethics 4:253–288

Schultz PW (2002) Inclusion with nature: understanding the psychology of human–nature interactions. In: Schmuck P, Schultz PW (eds) The psychology of sustainable development. Kluwer, New York, pp 61–78

Sharp J, Green R, Schindel G (2019) Introduction. The Edwards Aquifer: the past, present and future of a vital water resource. Geological Society of America, Colorado, pp 1–7

Tilden F (1977) Interpreting our heritage. University of North Carolina Press, Chapel Hill

Xica Media (2020) Coahuiltecan. https://youtu.be/-cMwWCp00Rc. Accessed 21 Mar 2021

Chapter 31
Nature, Humans, and Education: Ecohumanism as an Integrative Guiding Paradigm for Values Education and Teacher Training in Israel

Nimrod Aloni, Dafna Gan, Iris Alkaher, Nirit Assaf, Netta Baryosef-Paz, Adiv Gal, Naama Lev, Adva Margaliot, and Taly Segal

Abstract Ecohumanism is a timely topic for educational undertaking in our global reality amid the pressing climate crisis, mass migration, weakening commitment to social justice and liberal democracy, and the coronavirus pandemic that began in 2020. In this chapter, we explore the philosophical, theoretical, and practical approach of *ecohumanism* in the context of teacher training education, focusing on the interrelations between the ideas of field environmental philosophy, the biocultural conservation, and the social-ecological system frameworks. One avenue of exploration may focus on the pressing predicaments of environmental sustainability. Namely, Mother Earth sustains many cultures. Surely many cultures should be able to sustain Mother Nature. This is achieved by harnessing strengths from all cultures, with their different perspectives and traditions, to sustain Nature. Another avenue of exploration that has philosophical and practical elements, is to endorse and cultivate worldviews that stress the moral and civic virtue of respect for dignity, diversity, democracy, and harmony (social and natural). Along the lines of Neil Postman's book, *The End of Education*, we argue that *ecohumanism,* as a metanarrative or overarching goal for value education aiming for a good quality of life—for nature and humanity—defines the most appropriate and desirable common good for the twenty-first century. We include examples from the college's integrative and activist curriculum, as well as from the experiential and explorative teaching-learning methods, demonstrating how pre-service and in-service teacher training courses contribute to a new paradigm of ecohumanist education. Our approach considers issues of social-ecological systems, contemplative environmental education, multiculturalism, shared life, and biodiversity conservation.

N. Aloni (✉) · D. Gan · I. Alkaher · N. Assaf · N. Baryosef-Paz · A. Gal · N. Lev · A. Margaliot · T. Segal
Kibbutzim College of Education Technology and the Arts, Tel-Aviv, Israel
e-mail: Dafna.Gan@smkb.ac.il; adiv@tene-eco.co.il

R. Rozzi et al. (eds.), *Field Environmental Philosophy*, Ecology and Ethics 5,
https://doi.org/10.1007/978-3-031-23368-5_31

Keywords Contemplative pedagogy · social-ecological systems · multicultural education · interdisciplinary teacher-training

> Education is the process of forming fundamental dispositions, intellectual and emotional, towards nature and fellow men (Dewey 1966, p. 328)

31.1 Introduction

Ecohumanism is a timely topic for educational undertaking in our global reality, including today's pressing issues, such as climate and environmental change, millions of refugees and displaced people, weakening commitments to social justice and liberal democracy, and most recently, the coronavirus pandemic beginning in 2020. Ecohumanism consists of normative and integrative concern for sustaining the flourishing of both humanity and nature; hence its engagement with sustainability in the manifold realms of life—the natural environment, social justice and democratic regime, communal life, and cultural traditions—as well as its acknowledgment of the reciprocal influences among those realms and the emphasis on ethics of respect, harmony, solidarity, and moderation (Patterson 2008; also HSS 2021). Moreover, ecohumanism (seldom used overtly in this phrasing) has become a leading theme in social activism and educational practices that focus on coexistence and peace education, environmental education and ecopedagogy, biosphere environmental education, planetary sustainability, and contemplative pedagogy, as well as human rights, multicultural, and democratic citizenship education. As a worldview, ethical stance, and paradigm for academic research and social activism, ecohumanism also resembles the notion of field environmental philosophy (FEP), which integrates ecological research and environmental ethics into a social and educational agenda for biocultural conservation and the long-term study of social-ecological systems (Rozzi et al. 2010; Rozzi 2013; Avriel-Avni and Ben-Natan 2014; Gal and Gan 2017; Dunetz et al. 2017).

Notwithstanding the differences between ecohumanism and FEP in terminology and the relative weight given to the humanities and natural sciences, they both share an acute concern for the flourishing and sustainability of the social-ecological world, humanity, and nature, as one holistic and integrative ecosystem. Additionally, both have at their core a moral commitment to the values of respect, sensitivity, dignity, diversity, empathy, dialogue, well-being, coexistence, and moderation.

At stake here, is the development of a new notion of the common good. *Ecohumanism* is a regulative ideal that pursues an integrative and harmonious lived reality in which vital natural and social systems enjoy thriving through sustainable life habits. In an emphasis similar to Neil Postman's *The End of Education* (1995), in which he stresses the importance of an overarching end or metanarrative for sustaining the vitality of our educational enterprise, we consider ecohumanism to be the most appropriate and desirable vision of the common good for the twenty-first century and the regulative ideal for contemporary values education and teacher training.

At Kibbutzim College of Education—located in Tel Aviv, Israel, and serving about 6000 students in the bachelor and graduate programs—ecohumanism is new, yet old. It is *new* in the sense that only in 2020 did we set forth the educational credo "ecohumanism now so there will be a tomorrow." The curriculum at this school offers manifold courses that promote integrative sustainability to foster the quality of life of the individual, the community, and the natural environment. Ecohumanism is, however, *old* because the founders of Kibbutzim College, in 1939, held a progressive and naturalistic educational worldview, embodied in its holistic curriculum in which humanist psychology, social justice, unmediated experiential and explorative encounters with nature, and the arts were harmoniously interwoven.

It is also worth noting that like most others in the education profession, globally speaking, we have prior theoretical experience with the ecohumanist spirit via the great works of Jean Jacques Rousseau, the father of naturalistic education. Following his honed senses, more than 250 years ago Rousseau identified the enormous dangers looming due to the rise of the bourgeoisie—a class that even today worships artificial lifestyles, is devoted to wanton consumerism, and forms a culture that is offensive to both social justice and nature. In his own words,

> He [the bourgeois man] forces one soil to nourish the products of another, one tree to bear the fruit of another. He mixes and confuses the climates... He wants nothing as nature made it, not even man; for him, man must be trained like a school horse; man must be fashioned in keeping with his fancy like a tree in his garden (Rousseau 1979, p. 37).

These alarming words from the past echo strongly in the contention of present-day American psychologist and ecologist Peter Kahn, that due to our alienation from nature most of us on Earth "live in massive urban prisons" and are suffering from "underwhelming and dispiriting form of living" (Kahn 2018, p. 164).

To conclude this introduction, we argue that ecohumanistic values education offers a suitable response to today's human challenges. In light of the mass atrocities of the twentieth century (world wars, the holocaust, genocide, and totalitarian regimes), and the global risks to present day humanity (ecological, social, political, and cultural), it is not acceptable to perceive current crises and future challenges as separate arenas. Rather, these crises are intertwined and inter-influential. Our mission is to establish an ecohumanistic activist pedagogy, which follows the ethical-pedagogical vision of the late American educator Maxine Greene. She said that we should be attentive to "what the known demands" (1973, p. 21), to empower students to "recognize lack or deficiency and learn how to repair and transcend" (1978, p. 19) as well as to inspire their imagination "to look at things as if they could be otherwise" (2013, p. 19). It is an undertaking that aims to empower and instruct the younger generations how to successfully address the challenges of the twenty-first century, repair the world from the human-inflicted damages, and lay foundations for a thriving biocultural reality (Nussbaum 2002; Aloni 2008, 2017; Aloni and Weintrob 2017; Jenkins 2015; Rozzi et al. 2015).

We present an educational paradigm that cultivates sensitivities, sensibilities, capacities, and dispositions to enable future generations to enjoy environmental sustainability, humanistic democracy, social justice, physical and mental health, and to have a sense of belonging and cultural-educational affluence. To support this, we include examples from the college's integrative and activist curriculum for both undergraduate and graduate programs in various faculties, as well as from the experiential and explorative teaching-learning methods that we have been conducting in Israel. Our ecohumanist commitment merges the humanist "I and Thou" (Buber 1970) dialogical interpersonal relations (cherishing the humanity of every person as an end in itself) with the ecological consideration of nature not as a commodity to own and use egotistically but as a community to join harmoniously and respectfully (Cafaro 2001; Nodstrom 2008; Jordan 2016; Bonnett 2019; Hannis 2020). It amounts, in the words of John Dewey, to a pedagogical pursuit of forming desirable "dispositions, intellectual and emotional, towards nature and fellow men" (Dewey 1966, p. 328). It is the desire to create a peace within us, within our society, within our world. The notion of peace in Biblical Hebrew, *shalom*, as pointed out by Ravitzki (1993), is related to the words *whole*, *complete*, and *perfection*, and denotes a state of being characterized by security, wellness, tranquility, thriving, or existing in an intact mode of being. It refers to "blessedness and harmony in the manifold aspect of life—bodily and spiritually" (Ravitzki 1993, p.493). Or, as Ilana Avissar (2016, p. 451) states in *The Challenges of Sustainability* "we must learn to live with others and not at the expense of others."

31.2 Guiding Ideas

To be successful, education must pay attention to three basic elements: orientation, motivation, and competencies. Regarding orientation, we present 12 guidelines that provide a matrix for sound deliberation and action in our future engagement with ecohumanist education. Subsequently, we introduce and discuss integrative ecohumanist courses to address the two other basic elements: motivation and competencies.

31.2.1 Twelve Guidelines Regarding Pedagogical Orientation

Humanist and democratic ethical stances propose human dignity, equality, diversity, and growth as universal core values, beyond any particular values such as religious, ideological, or economic. It is a commitment to provide every man and woman with the rights and opportunities for social welfare, to enjoy self-realization, and participate meaningfully in the cultural, social, and political spheres of life (Aloni 2011, 2020).

Ecological approach requires considering life on the planet as ecosystems characterized by complex relationships and reciprocal influences among the diverse elements, natural and cultural, that need to be studied, managed, and protected for the sake of maintaining their health and vital functioning (Rozzi 2013).

Sustainability is an ethical stance committed to optimal long-term existence and preservation of societal and natural systems whose contribution to the thriving of humankind and/or nature is evident; i.e., social, democratic, cultural, and environmental sustainability. Providing long-term solutions, this approach aims at sustaining the biophysical and sociocultural components of the dynamic social-ecological systems as well as preventing investments and initiatives that are destructive of such systems (Bonnett 2019).

Quality of life is much more than economic growth and materialistic standard of living (Nussbaum 2000, 2010; Nussbaum and Sen 2003). The alternative, multidimensional and holistic, is looking into manifold elements of well-being and growth, very much along the lines of Maslow's psychological paradigm of self-actualization (pyramid of needs) (1954) and Nussbaum's capabilities approach (2000) (realization of the distinctively human capabilities). According to this holistic approach, we must evaluate our quality of living according to diverse standards, such as life expectancy and infant mortality, equal opportunity and gender equality, human rights and equality before the law, governmental transparency and trust in political institutions, environmental health and democratic participation, general education and quality of leisure, and community life and a sense of personal fulfillment.

Health signifies full and optimal function of an organism, of a societal organization, and of the natural environment, according to their innate nature and distinctive characteristics. The ecohumanist approach is holistic and aims towards healthy mind and body, healthy society and nature, health for people, community, and nature.

Dialogical culture replaces the win-lose approach of success at the expense of others, with that of attentive, respectful, and peaceful shared life and coexistence—hence a liaison between ecology and humanism, as well as between interpersonal and inter-cultural relationship. Resting on mutual trust, respect, and good will, dialogue rejects the modes of instrumental manipulation and control that in global society are so abundant in human behavior towards fellow humans and natural resources. In their stead, it lays the foundation for working together towards a common good through humane, reasonable, fair, and harmonistic dialogue (Aloni 2011, 2020).

Prevention approaches precede the repair and rehabilitation approach. Both morally (preventing suffering, destruction, and injustice) and economically (in utilitarian terms), it is preferable to prevent neglect, disease, poverty, and ignorance rather than enforcing, imprisoning, healing, and rehabilitating later on. For example, proper nutrition and a healthy lifestyle are more reasonable than endless hunger or dieting and daily intake of medication. Decreasing birthrate and reducing consumption in order to maintain and sustain our natural resources, are much more reasonable than coping with the detrimental outcomes of population explosion, air and water pollution, soil depletion, ocean-dead zones, and the extinction of species.

Glocalpatriotism is the caring, responsibility, and active involvement in favor of ecohumanistic welfare in our local and global arenas. It allows us to veer from the conventional clashes between political right and left, religious and secular, nationalism and cosmopolitanism, and one ethnic group confronting another. It enables us to unite efforts to address ecological, social, and political challenges with relative ease. In our current reality, different phenomena have implication beyond borders of place and time, such that collaborations promote the local and global common good (win-win situations) (Brooks and Normore 2010).

Academic life engaged with public problems rather than disciplinary scholarship is inspired by Maxwell's book *Global Philosophy* (2014). Maxwell urges us to identify the ultimate goal of academic life not to be the acquisition of knowledge but rather the development and mastery of wisdom in the art of living. Hence, in the tradition of Montaigne's educational thought, in our academic and educational undertaking we should connect the "tree of knowledge" with the "tree of life." Consequently, we should measure success not in the mastery of subject matter, but rather in sound and thriving life.

Slow-paced moderation grants more and harms less, because as explained in Ecclesiastes 5:10: "He that loveth silver shall not be satisfied with silver; nor he that loveth abundance with increase: this is also vanity." This same logic works for fast pace workaholism as well as for maximizing multi-tasking performances. When overloaded, pressured or over stressed, the result is illness, distorted perspective and conduct, and frequently mental collapse. In contrast, when we learn to take a break, have humble aspirations, and make the time for moral deliberation, spiritual meditation, and attentive dialogue, or when we are one with a work of art or experientially immersed in wilderness (in the spirit of FEP), we can achieve moments of slow-paced moderation and contemplation. These moments are cherished opportunities to regain our sense of beauty and trust in life, of personal growth, and meaningful life and harmony with others (Honoré 2009).

Endurance and resilience of the individual and the community alike are of crucial importance in times of volatility, uncertainty, complexity, and ambiguity. Endurance and resilience are "guardians of sanity" in a distinctively competitive, materialistic, workaholic, and consumerist era in which stress and depression are becoming prevalent mental illnesses. Hence the importance of attending to the classical dictums of Ecclesiastes 3: 1–13, "measure is best in everything" and "to everything there is a season and time to every purpose under the heaven." Endurance and resilience are not commodities to be bought in the marketplace. They should be developed and built from within the realms of personal autonomy, authentic and meaningful experiences, moral and spiritual internal core, intimacy with nature, trustworthy friendship, communal sense of belonging, and dialogical empathetic human relations (Taormina 2015). Adherence to the United Nations Sustainable Development Goals will support endurance and resilience as we are proposing. These goals include (1) eradication of poverty; (2) wiping out hunger; (3) good health and well-being; (4) quality education; (5) gender equality; (6) clean water and sanitation infrastructure; (7) affordable green energy; (8) decent work and economic growth; (9) industry, innovation, and infrastructure; (10) reduced inequality within and among countries; (11) sustainable cities and communities; (12) responsible consumption and production; (13) action against climate change; (14) conservation and sustainable use of marine life; (15) conservation and sustainable use of land; (16) peace, justice, and strong institutions; (17) partnerships to achieve these goals (Le Blanc 2015; Poole 2018; United Nations 2021).

31.2.2 From Theory to Practice: Ecohumanism in the Curriculum and Interdisciplinary Courses

The following section concentrates on specific examples of ecohumanist education and the 12 guidelines in relation to the dimensions illustrated in Fig. 31.1. This image exemplifies and clarifies the interrelated spheres of ecohumanist education manifested in three specific courses taught at Kibbutzim College of Education in

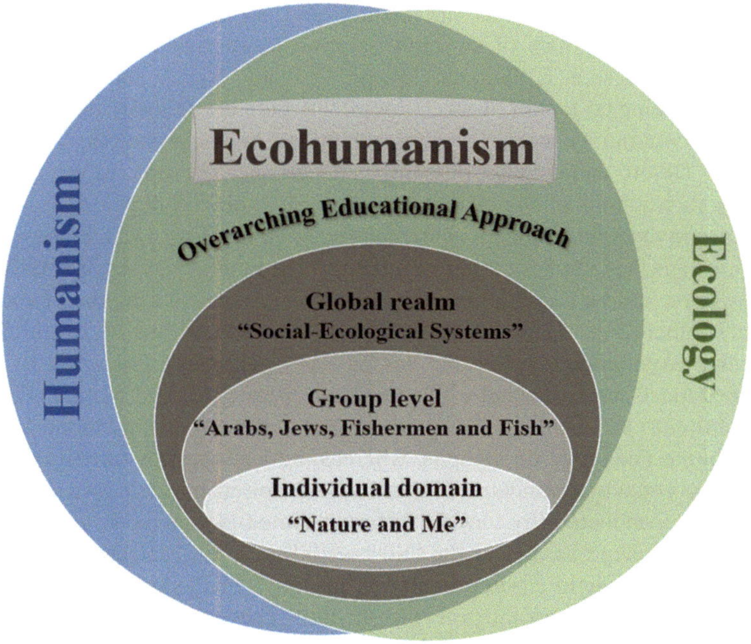

Fig. 31.1 Ecohumanism as an overarching educational approach at three scales: the individual domain, the group level, and the global realm. The illustration reflects the dynamic non-hierarchical direction of the educational processes encapsulated in three courses represented here by their titles in quotation marks. Once the learner is led to introspection s/he understands his/her surrounding social and ecological system and then to grasp the larger global scope. This process could work in different directions from the global inward or otherwise

2019–2020: “Nature and me” focused at the individual domain; “Arabs, Jews, Fishers and Fish” focused at the group level; and “Social-ecological Systems” focused at the global realm. The individual sphere involves an introspective process by means of experiencing the environment, and encourages better understanding of the self and the surroundings. The interpersonal sphere involves shared outdoor experience with a group and enables open and attentive dialogue both among group members and between group members and the environment. Finally, the third collective sphere facilitates understanding the system’s complexity. Students learn the interrelations between the system’s ecological and social components and the significance of their diversity, which provide essential foundations for the sustainable coexistence of humanity and nature.

In what follows we present three examples of courses, which demonstrate what we have portrayed in the earlier sections.

31.2.2.1 The Individual Domain: “Nature and Me”

“I wish to speak a word for Nature, for absolute Freedom and Wildness”, are the opening words of Henry David Thoreau’s renowned “Walking” (1851). Readings his fundamental text, as well as reading and writing contemporary Hebrew ecopoetry, were the pillars of an interdisciplinary experiential environmental and contemplative course at Kibbutzim College. Entitled “Nature and Me”, the course innovated by being co-taught by an ecologist and a literary scholar, and by being part of students’ community work. The course included 2 days of camping and hiking in the Negev Desert and 3 days on-campus meetings. The nine students enrolled (majoring in teaching of the Arts) read ecopoetry, practiced guided mindfulness meditations in the field, and kept contemplative journals (Fig. 31.2). During the time in the desert, each student took individual short explorative excursions studying the geology, flora, and fauna of the desert, taking photos, and sharing their experiences with each other (Fig. 31.3). Closely “tasting” the desert and its magnitude, the participants experienced sunset and sunrise, and personally interacted over open fire cooked meals, and reflected while speaking, meditating, and writing (Fig. 31.4).

In addition to this field excursion, the class held two meetings in the Center for Contemplative Pedagogy on campus, whereas the last one was on virtual platform due to the coronavirus pandemic. Afterwards, inspired by their own experiences, student teams conducted (by virtual platform) contemplative workshops with other classes in the college as part of the “Sustainability Week” celebrated annually on campus. Similar to field environmental philosophy (FEP) (Rozzi et al. 2010), we aimed at providing students integral “face-to-face” encounters with other human and other-than-human beings while also learning place-based awareness of ecology and environmental issues.

In addition to these, following the ecohumanist and holistic principles, the course objectives were to experience slowing down, contemplation, and dialogical culture (with oneself, with nature, and with others) combined with creative familiarity with the arts. Indeed, in their workshops the students independently and creatively

Fig. 31.2 A student practicing guided mindfulness meditations in the Negev Desert. Photograph by Nirit Assaf

designed and orchestrated a series of activities despite the pandemic restrictions. During a virtual meeting session, they asked their peers to take photos of the interiors of their physical settings and their reachable exteriors. They then led guided meditation followed by directed haiku readings (in Hebrew). As a product, the participants created their own haikus inspired by their interior-exterior sessions. This part of the course could be related to Step 2 of FEP's 4-step cycle: composition of metaphors and communication through simple narratives (Rozzi et al. 2010). Indeed, the students were asked to both decipher poetic metaphors and to create them. However, in contrast to FEP, our poetic endeavor is the means and ends of this workshop; that is, the literary experience (combined with the human interaction) is in itself the center of the lesson. Students do not create the metaphor in order to communicate an ecological discovery or to synthesize facts, values, or action but rather they immerse themselves in a creative and literary experience for the sake of experiencing the humanities. The haikus were thematically centered on nature, yet these poems were not subjugated to ecological scientific thought or agenda.

The tension between indoors and outdoors during quarantine (required as a precaution against the COVID-19 pandemic) was contemplated and translated into poetry. In another occasion the students allowed for a dialogical-contemplative moment when participants were asked, after having been partnered into breakout rooms, to silently share with each other the spaces in which they were enclosed, and

Fig. 31.3 Student sharing experiences and observations on the geology, flora, and fauna of the desert. Photograph by Nirit Assaf

then to share their experience with words. The personal habitat, poetry, and humanness were crystalized in these intimate moments of dialogue.

As demonstrated, the course practices ecohumanism in action: it acts out underlying perceptions that combine nature with human beings. Since the course includes an experiential journey to the desert, it does not "talk the talk" but rather "walks the walk." Merging the presence amid rocks, sand, sky, desert pancratium flowers, and birds with the reading of literary texts, and meditative writing, it exemplifies the non-dichotomous view of ecological systems, thus offering a holistic approach. The co-teaching by an ecologist and a literary scholar carries out a model of indiscrete and flowing interdisciplinary instruction, learning, and teacher training. Throughout the course, from planning to practice there are no divisions or categorizations, but rather climbing dissolves into meditating, which in turn, melts into bird watching, and then poetry reciting. Thus, an ecofocus unites with humanism.

My heart leaps up when I behold
A rainbow in the sky:
So was it when my life began;
So is it now I am a man.

This poetic enchantment with nature written in 1807 by the Romanticist poet William Wordsworth encapsulates the significance of the unmediated encounter with nature at a young age alongside its literary manifestation. Our educational

Fig. 31.4 Students admiring a sunrise in the Negev desert. Photograph by Nirit Assaf

endeavor wishes to train teachers whose hearts as well as their future students' would "leap up" beholding nature.

31.2.2.2 The Group Level: "Princesses, Sailors, Fishers, And Fish in Jaffa Beach"

In Tel Aviv, Jaffa's coastline ties together flowing streams, jellyfish swarms, local pollution, and ancient mythologies. The course entitled "Princesses, Sailors, Fishermen and Fish in Jaffa Shore" was designated for biology education undergraduate students at Kibbutzim College of Education in the 2019–2020 academic year. Led and designed by a literary scholar, its main objective was to identify and investigate some of the dilemmas embodied in the Jaffa coastline.

Located in the metropolitan center of Israel, this ancient port city is well known for its history from the Biblical Jonah who fled through its port, and from the different Muslim and Christian conquerors who left their mark on this city. Presently, Jaffa is an urban sphere in which escalating processes of gentrification, poverty, and crime reflect sociological, cultural, and national challenges Israel is currently confronting.

The course introduced the students to the tensions between ecological and cultural conservation in Jaffa port, that is, the cultural value of preserving the traditional heritage of Palestinian artisan fishers set against the ecological value of safeguarding the marine environment regulated by the Israel Nature and Park Authority. This conflict, existing in the heart of Israel, involves complex aspects such as cultural-political (Palestinian fishing local tradition in the Mediterranean Sea in a modern Jewish State), socio-economic (local hardworking fishers confronted by governmental regulation), and ecological (Western environmental standardization enforced upon Middle-eastern conservative community).

Visiting Jaffa port, the students met with Saado Zeenab, a Palestinian fisherman of half a century. He presented to them the local fishing tradition and heritage of the generations-old-community of Jaffa's fishers, their daily lives, and routines. The students gained specific knowledge about the traditional artisan fishing materials, tools, and methods (such as different fishing nets and various fishing crafts). Additionally, the students met with Yigal Ben-Ari, the marine inspector in Israel Nature and Parks Authority and learned about the governmental regulations enforced on overfishing in the Mediterranean Sea and unsustainable fishing methods, and their protective actions to save endangered marine species, according to European standards.

Directed at students who soon would become educators, the course promoted a holistic understanding of the term "environmental education," understood not only as biological-ecological but also culturally centered education, encompassing deep human issues such as heritage, identity, and history. The course's core was the tension between biological and cultural conservation. However, early in the process, ignorant of the fishers' tradition, the students tended to acknowledge merely the value of the biological conservation, claiming "…so, they can stop fishing!" Similarly to FEP, we aimed at developing varied methodologies to integrate an environmental understanding encompassing a wide ecological perception in which cultural diversity is considered a fundamental value (Rozzi et al. 2010). Getting to know Saado, Yigal, and others, the students gained nuanced-rich awareness of the fishers' tradition. Having internalized the conflict's complexity, the students further appreciated the sociocultural values embodied by the fishers' community and its right to conserve the fishing culture. They also understood that the sea is a fundamental source of income for this community.

As much as it may appear local and specific to Jaffa's coast, the Mediterranean and its fauna, this dilemma may serve as a metonymy for several substantial issues Israel is currently facing: the Israeli-Palestinian conflict, progress versus tradition, environmental versus cultural conservation, and European regulations versus native ethos. The ongoing dialogue between these sea-loving men (traditionally only men set out to sea in this community) who also deeply respect each other, Saado, and Yigal, encapsulates the hope for a sustainable peace for Israel.

The course pursued forming an ecohumanist worldview and pedagogical thinking which are not enclosed within dichotomous binaries but rather create a complex episteme of the eco-human-system, rich with details, balanced, and multilayered. It entails a sensitive awareness and harmonious dialogue between humans and the environment.

31.2.2.3 The Global Realm: "Social-Ecological Systems"

> "The complexity of social-ecological systems is hard to grasp ... I'm afraid of complications... I prefer simplicity... but this is difficult... what I have learnt... in the program I use as an inspiration for my whole life." (Students' reflection at the end of the social-ecological systems course).

Social and environmental injustice weakens social-ecological systems. Therefore, preserving such systems must include preventing and minimizing injustice. With this goal in mind, we created a "social-ecological system" course for the Kibbutzim College's master's degree in Education program specialized in environmental education. Reflecting the power relations in the Israeli society, most of the students in the course were Jewish and only a few Arabs. Based on a series of field trips, the course addresses the social-ecological systems in which the multicultural conflict is intertwined. Grounded in the ecohumanist approach, it is designed to promote a better understanding of this complexity through intergroup collaboration and to acquaint students with the voices of individuals who often go unheard. Furthermore, the course is based on the idea that the Arab-Jewish conflict impacts social-ecological systems and therefore must not be avoided.

Aimed at recognizing diverse cultures, encouraging openness to different perceptions, and inspiring a desire to act, the course enables open dialogue and discussions about both social-ecological systems and the power relations between the students in the group as part of the Jewish-Arab conflict. These goals are implemented in the co-led field trips to generate cognitive, emotional, and social experiences thus enacting diverse pedagogies. Each field trip examines a different social-ecological system in Israel (north, center, and south) and in different scales (small, medium, and large), and enhancing the students' familiarity with the characteristics and complexity of major social-environmental issues at local, regional, and international scales (Fig. 31.5). Focusing on knowledge construction, and emotional learning, the understanding of social-ecological systems included meeting key community figures, engaging in joint discussions, and sharing authentic food (Fig. 31.6). At the end of each study tour, the graduate students physically drew the social-ecological systems according to their perspectives and shared their thoughts (Fig. 31.7).

The entire learning process is both theoretical and practical, creating an open and democratic atmosphere, dialogical, empowering, and caring culture of ethical, political, and pedagogical elements that are part and parcel of the ecohumanist outlook. One of the students summarized her learning process by saying, "I learned about social and environmental aspects of the crisis... I also learned about the complex relationships... how the social problems lead to ecological degradation, which interrelated with the entire society... I learned that I can promote social change with my students by implementing pro-environmental behavior." In conclusion, the holistic process that students were engaged in will enable them not only to understand the principles of ecohumanism but also to develop their skills for environmental education implementation in their future educational work.

Fig. 31.5 Student sketching Northern Israel's social-ecological systems at multiple scales. Photograph by Adiv Gal

Fig. 31.6 Multicultural sharing food activity. Photograph by Adiv Gal

Fig. 31.7 Group of students in the field synthesizing through drawings their observations, experiences, and new understanding of social-ecological systems. Photograph by Adiv Gal

31.3 Concluding Remarks

From our perspective as activist pedagogues and educators, concerned with repairing the world and promoting the thriving of human culture and the natural environment, it is unacceptable to address humanity's pressing predicaments and challenges solely in terms of mastering some disciplinary knowledge and proficiencies. As portrayed and explained above, this awareness brought us to form and promote new curriculum and teaching methods that are characterized by interdisciplinary groups of scholars and activists. They come from the humanities, social studies, the natural sciences, and environmental studies. They all work together to develop insights, sensitivities, and competencies to address the current global risks and to act together for the sake of world betterment.

Another source of inspiration, orientation, and edification in developing our educational paradigm of ecohumanism was the encounter with the notions of biocultural conservation, field environmental philosophy, and social-ecological systems. They enabled us to give better sense, improve the logic, and make our activist narrative more coherent: connecting the various spheres of environmental sustainability, culturally sustainable pedagogy, and education for shared life, social equity, and human rights. The resulting product, as argued in this chapter, is the educational agenda of "Ecohumanism Now so there will be a Tomorrow".

Finally, whatever the motivation and vantage point for concern, the ethical and pedagogical paradigm of ecohumanism seeks to extend the notion of moral imperative to the non-human world. It amounts to adopting a caring, benevolent, and generous disposition towards nature and fellow humans, seeking to facilitate others to reach their best by actualizing their distinct nature while harmonizing with others: *to flourish with others and not at the expense of others*. Ecohumanism serves best as a framework and paradigm for such transformative education: integrating elements from environmental sustainability with social, cultural, and democratic sustainability for the sake of establishing and securing sustainable well-being, peace, and harmony for humanity and nature.

References

Aloni N (2008) The fundamental commitments of educators. Ethics Educ 3(2):149–159

Aloni N (2011) Humanistic education: from theory to practice. In: Education and humanism: linking autonomy and humanity. Sense Publishers, Rotterdam, pp 35–46

Aloni N (2017) Educators worthy of the name: intellectuals, generous, master dialogicians. In: Beyond bystanders: educational leadership for a humane culture in a globalizing reality. Sense Publishers, Rotterdam, pp 61–74

Aloni N (2020) Locally grounded, universally binding: the benefit of incorporating traditional care ethics, east and west, into current moral education. Educ Philos Theory 1:98–105

Aloni N, Weintrob L (eds) (2017) Beyond bystanders: educational leadership for a humane culture in a globalizing reality. Springer, Dordrecht

Avissar I (2016) The challenges of sustainability. Kibbutz HaMeuchad, Tel Aviv

Avriel-Avni N, Ben-Natan G (2014) Social-ecological coupled systems as a conceptual framework for environmental education. Negev Dead Sea Arava Stud 6(3):63–73

Bonnett M (2019) Towards an ecologization of education. J Environ Educ 50(4–6):251–258

Brooks JS, Normore AH (2010) Educational leadership and globalization: literacy for a glocal perspective. Educ Policy 24(1):52–82

Buber M (1970) I and Thou. Charles Scribner's Sons, New York

Cafaro P (2001) Thoreau, Leopold, and Carson: toward an environmental virtue ethics. Environ Ethics 23(1):3–17

Dewey J (1966) Democracy and education. Macmillan Co., New York

Dunetz D, Avissar I, Gan D (2017) Education for sustainability: using the society/environment and global/local binariesas catalysts for pedagogical renewal. In: Beyond bystanders: educational leadership for a humane culture in a globalizing reality. Sense Publishers, Rotterdam, pp 61–74

Gal A, Gan D (2017) The perceptions of college students from diverse cultural backgrounds towards the concept of socio-ecological systems within the context of the Jewish-Arab conflict. Negev Dead Sea Arava Stud 9(2):33–48

Greene M (1973) Teacher as stranger: educational philosophy for the modern age. Wadsworth, Belmont

Greene M (1978) Landscapes of learning. Teachers College Press, New York

Greene M (2013) Releasing the imagination: essays on education, the arts, and social change. Jossey-Bass, San Francisco

Hannis M (2020) Why virtue is good for you: the politics of ecological eudaimonism. In: A book of ecological virtue: living well in the anthropocene. University of Regina Press, Saskatchewan, pp 1–21

Honoré C (2009) In praise of slow: how a worldwide movement is challenging the cult of speed. Vintage

HSS (2021) Ecohumanism, humanist society of scotland. Available via: https://www.humanism.scot/what-we-do/policy-campaigns/ecohumanism/
Jenkins A (2015) Critical conversations: dialogue in global citizenship education. Widen Particip Lifelong Learn 17(2):30–47
Jordan K (2016) Sustainability, virtue ethics, and the virtue of harmony with nature. Environ Educ Res 23(9):1205–1229
Kahn P (2018) Technological nature and human well-being. In: Oxford textbook of nature and public health. Oxford, pp 160–164
Le Blanc D (2015) Towards integration at last? The sustainable development goals as a network of targets. Sustain Dev 23(3):176–187
Maslow AH (1954) Motivation and personality. Harper and Brothers, New York
Maxwell N (2014) Global philosophy: what philosophy ought to be. Imprint Academic, Upton Pyne
Nodstrom H (2008) Environmental education and multicultural education – too close to be separate? Int Res Geogr Environ Educ 17(2):131–145
Nussbaum M (2000) Women and human development: the capabilities approach. Cambridge University Press, Cambridge
Nussbaum M (2002) Education for citizenship in an era of global connection. Stud Philos Educ 21(4–5):289–303
Nussbaum M (2010) Not for profit: why democracy needs the humanities. Princeton University Press, Princeton
Nussbaum M, Sen A (2003) The quality of life. Oxford Scholarship Online, Oxford
Patterson W (2008) Ecohumanism: principles and practice. Essays Philos Human 16(2):71–88
Poole A (2018) The UN sustainable development goals and the biocultural heritage Lacuna: where is goal number 18? In: From biocultural; homogenization to biocultural conservation. Knopf, Toronto, pp 315–331
Ravitzki A (1993) Peace. In: Contemporary Jewish religious thought. Am Oved, Tel Aviv, pp 658–702
Rousseau JJ (1979) Emile, or on education. Translated by Allan Bloom. Basic Books, New York
Rozzi R (2013) Biocultural ethics: from biocultural homogenization toward biocultural conservation. In: Rozzi R, Pickett STA, Palmer C, Armesto JJ, Callicott JB (eds) Linking ecology and ethics for a changing world. Ecology and ethics, vol 1. Springer, Dordrecht, pp 9–32
Rozzi R et al (2010) Environmental philosophy and biocultural conservation at the Omora Ethnobotanical Park: methodological approaches to broaden the ways of integrating the social. Rev Chil Hist Nat 83(S):27–68
Rozzi R, Chapin FS III, Callicott JB, Pickett STA, Power ME, Armesto JJ, May RH Jr (eds) (2015) Earth stewardship: linking ecology and ethics in theory and practice. Ecology and ethics, vol 2. Springer, Dordrecht
Taormina RJ (2015) Adult personal resilience: a new theory, new measure, and practical implications. Psychol Thought 8(1):35–46
Thoreau HD (1851) Walking. Henry David Thoreau online. Available via: Walking (thoreau-online.org)
United Nations (2021) Transforming our world: The 2030 Agenda for Sustainable Development. Available via: https://sdgs.un.org/2030agenda

Chapter 32
The H.J. Andrews Experimental Forest Long-Term Ecological Research Program, Oregon, USA: A Historical Biocultural Perspective

Frederick J. Swanson

Abstract The H.J. Andrews Experimental Forest and Long-Term Ecological Research (LTER) Program, Oregon, USA, has a history of influence in biocultural conservation through basic and applied ecology and forestry research, close partnership with managers of public forest lands, and a growing arts/humanities program. Studies of old-growth forests, the northern spotted owl flagship species, and watersheds over several decades underpinned a major shift in federal forest lands management policies in the early 1990s as the public sense for native forests shifted from their utilitarian values to their intrinsic value. Since the 2000s, a program engaging arts and humanities, including creative writers and philosophers, has richly expressed the profound beauty, wonder, and mystery of the forest, perhaps foreshadowing a new era of forest stewardship. This transformation parallels merging of environmental science, arts, and humanities at other USA LTER sites and similar programs internationally, notably the Omora Ethnobotanical Park Long-Term Socio-Ecological Research program in the Cape Horn Biosphere Reserve, Chile.

Keywords Biocultural conservation · Environmental arts and humanities · Environmental education · Long-term ecological research · Old-growth forest

32.1 Introduction

Recognizing that conservation issues have become acute globally, sites and programs of long-term ecological science and other forms of inquiry have intensified efforts to advance biocultural conservation (Rozzi 2013). These programs are emerging at long-term socio-ecological research sites, biological field stations, marine laboratories, botanical gardens, and sites rooted in arts/humanities residency

F. J. Swanson (✉)
US Forest Service, Pacific Northwest Research Station, Corvallis, OR, USA
e-mail: fred.swanson@oregonstate.edu

R. Rozzi et al. (eds.), *Field Environmental Philosophy*, Ecology and Ethics 5,
https://doi.org/10.1007/978-3-031-23368-5_32

programs (Swanson 2015; Zhu 2022). The common objective has been to encourage human-nature connections in their bioregions.

The most explicit conceptualization and practice of these biocultural conservation objectives has been associated with the Omora Ethnobotanical Park and Long-Term Socio-Ecological Research site at the southern tip of South America (Rozzi et al. 2006; Rozzi 2013). Tauro et al. (2021), for example, advance Field Environmental Philosophy (FEP) as "a novel educational philosophy" to embody a "biocultural ethic that values the vital links among the life *Habits* of *co-in-Habitants* (human and other-than-humans) who share a common *Habitat*"—the "3Hs" model (Rozzi et al. 2008). Rozzi et al. (2006) outline ten principles, such as presence of a "flagship species," necessary to advance biocultural conservation at Omora. These principles can pertain to programs anywhere with adjustments to local socio-ecological conditions.

This chapter describes the history and influences of the H.J. Andrews Experimental Forest and Long-Term Ecological Research (LTER) program (Oregon, USA) in terms of fostering connections of people with nature. The different socio-ecological setting of the Andrews Forest stands as an interesting point of comparison with Omora Park and its FEP model; there is striking resonance, despite the differences. Work that began at the Andrews Forest with science to support forest exploitation evolved to explicitly embrace synergies among environmental sciences, humanities, and arts with the objective to "find new ways to understand and re-imagine our place in the natural world" in the words of philosophy professor and nature writer Kathleen Dean Moore (Panel 1 in Swanson et al. 2008).

The northern spotted owl (*Strix occidentalis caurina*) flagship species and the ancient rainforest flagship ecosystem of Andrews Forest speak to those who visit and work there, and to the broader public who learn about them through media communications and the storytelling of visual arts, music, and the spoken and written word (Fig. 32.1). This case study traces phases of development of the Andrews Forest program, the nature of the science work, the contexts of the values of the participants and the public in which the work takes place, and the methods employed to encourage connection between people and nature.

32.2 Setting the Stage

32.2.1 Andrews Forest: The Place, the Institutions

The Andrews Forest encompasses the 6400 ha Lookout Creek watershed in the Oregon Cascades Range. It ranges in elevation from 380 m to over 1600 m. The watershed is cloaked with old, native conifer forest up to 500 years in age and forest plantations created by clearcutting mainly in the 1950s and 1960s, and laced with cold, fast-running streams. It is an experimental forest in the US Forest Service system of such properties and is managed jointly by the Pacific Northwest Research Station and Willamette National Forest. Oregon State University (OSU) also

Fig. 32.1 Old-growth forest in the HJ Andrews Experimental Forest, Oregon Cascade Range, USA. The small aluminum tags on the two 500-year-old Douglas-fir (*Pseudotsuga menziesii*) trees indicate that this is in a long-term research plot. This photograph, created using a painting-with-light technique by artist-in-residence David Paul Bayles, is one of his "Outside of time/Forest landscapes" set of images (Bayles 2019)

participates in the management by virtue of sustained funding since 1980 from the National Science Foundation under the LTER program. Close collaboration among academic and agency scientists and federal land managers has proven essential for exchanging ideas with the public. Many of the humanities and arts activities associated with Andrews Forest since 2002 are conducted through a collaboration of the science community with the privately-funded Spring Creek Project for Ideas, Nature, and the Written Word based in OSU. While sometimes difficult to navigate, this institutional complexity also offers benefits, such as the opportunity to reach different audiences.

32.2.2 Facts and Values

The emergence of a biocultural ethic for human engagement with public lands, such as those where Andrews Forest is located, had to transcend the "command and control" management model of government control (Holling and Meffe 1996) in which decisions were to be made based on the "best science" without explicit consideration of the values dimension of decision making. Kathleen Dean Moore and fellow philosopher Michael Paul Nelson, who have long been associated with Andrews Forest, have written:

> Any argument that leads to a conclusion about how we ought to act or what policies we ought to adopt must have two premises. The first premise is empirical, based on observation and experiment, often grounded in science. This is the way the world is, this is the way the world may soon be. The second premise is normative, based on cultural values and ethical norms. Here is the collected human wisdom about what is of value, an affirmation of what is worthy and worth doing. From this combination of facts and values, but from neither alone, we reach conclusions about what we ought to do. Collaborations between science and humanities are, in this respect, a logical as well as a practical necessity. (Moore and Nelson 2010, xvii–xviii).

Consequently, connection of humans with nature involves both the clear distinction and the alignment of facts and values. Science is a source of "facts," and arts and humanities are expressions of "values." Scientists do their science and are custodians of processes of science (e.g., reviewers of proposals and publications); artists/ humanists do their work exploring and expressing values, while philosophers are custodians of moral reasoning to bring the facts and values together to reach a decision.

32.2.3 *A History of Andrews Forest Through the Lens of the Biocultural Ethic*

The history of the Andrews Forest in terms of human connections with the land is manifest in shifting societal expectations for public forest lands, motivations of the science community, and conditions of the land itself in the form of legacies of land use, such as clearcutting and roads. We trace the history of the Forest in terms of the mission of the science program in the context of the dominant value systems of the time. We mention the relation between the science and land management communities, because these are federal public lands subject to values expressed in federal legislation. The land management system is a channel for connections among science, society, and the forest. Furthermore, influences on federal policy can have great geographic reach.

This story begins at the end of World War II as the Timber Era of the US Forest Service commenced in the Pacific Northwest with the objective of converting native forest, much of it old-growth greater than 200 years in age established after wildfires, to tree plantations (Hirt 1994). Native forests are old-growth forests or secondary forests reestablished by natural processes after disturbances such as wildfires. The post-war housing boom, depletion of timber on private lands, and government incentives fueled this conversion of federal forests, which were widespread across the western USA. As part of this process, the research arm of the US Forest Service established the Blue River Experimental Forest in 1948. In 1953, it was renamed to honor Horace Justin Andrews, a prominent Forest Service leader who was instrumental in selecting the site. Management of the experimental forest aimed to develop and test methods of logging, road systems, and forest regeneration, while minimizing negative impacts on soil, streamflow, water quality, and plants and animals. Comprehensive histories of the Andrews Forest are told by historians Max Geier (2007) and William Robbins (2018, 2020).

32.2.3.1 1950s–1960s: Convert Old Growth to Plantations—"Domesticated Science"

Conversion of native forests on Forest Service lands in the Pacific Northwest was based on an instrumental/utilitarian view of the value of the forests—they are valuable only for human use as wood products. The native forest was being erased (Fig. 32.2), imposing a disconnect between humans and the natural world. In this era science was conducted at the Andrews Forest by a small cadre of Forest Service researchers mainly to support utilization, while minimizing environmental impacts. Communications efforts targeted scientists and land manager audiences, but interaction with land managers was otherwise minimal. In a sense, this was "domesticated science" in support of the management paradigm of the day (Franklin 1999). The Andrews program had no direct engagement with the notion of non-utilitarian values of the forest.

Fig. 32.2 Dispersed-patch clearcutting of native forest on the Willamette National Forest in the Mona Creek watershed adjacent to Andrews Forest in the 1980s (US Forest Service photo)

32.2.3.2 1970s–1980s: Ecosystem Science of Native Forests—"Wild Science"

Between 1950 and 1970, the old-growth liquidation policy and practice continued, even as the US Congress passed laws placing value on the well-being of species and ecosystems, such as the:

- National Environmental Policy Act of 1969
- Endangered Species Act of 1973
- National Forest Management Act of 1976

These acts express the values of native ecosystems, which aligned with the rise of the environmental movement. These values are also manifest in the annual Earth Day celebrations beginning in 1970, as well as in arts and creative writing telling ecosystem stories and extoling the value of old-growth forests and other native ecosystems. Also, beginning in 1970, ecosystem science in the International Biological Program was launched at Andrews Forest as a large group of academic scientists funded by the National Science Foundation (NSF) joined Forest Service researchers in studies of old, native forest, the northern spotted owl, watershed processes, and other topics. The NSF support for fundamental research created independence from the management system of the day—in this sense it was "wild science" free of commercial constraints and expectations (Franklin 1999, 2018). This science revealed the incredibly complex nature of wild, old forest systems (Franklin et al. 1981)—e.g., large standing live and dead trees and downed logs with a profusion of epiphyte, invertebrate, and other species (Fig. 32.1)—in sharp contrast with the simplicity of intensively managed plantations. The implications of these findings for forest management were explored intensively with local land managers, who incorporated attributes of natural forests into their practices, such as retention of large, dead wood on land and in streams. Realization of the distinctive properties of ancient forests also proved pivotal in environmentalists' efforts to terminate old-growth logging on public land.

During the 1970s–1980s, the Andrews Forest did not have an arts and humanities program to express values of native ecosystems, but many Andrews scientists, especially JF Franklin, spoke and wrote in many forums about the complexities and wonder of old-growth forests. These science findings and the scientists themselves became subjects of public communications through many media. Franklin promoted "New Forestry," later framed as "ecological forestry," as middle ground between clearcutting and no cutting (Franklin et al. 2018). Hundreds of field tours and media accounts reached broad audiences, including elected officials. The sense of the intrinsic value of native forests was clear in the scientists' passion for the subject matter. Environmentalists pushed for policy change to protect native species and ecosystems through many means, including arson on Forest Service facilities, occupation of tree tops, and lawsuits and timber sale appeals, based in part on science findings.

32.2.3.3 Late 1980s-Early 1990s: The Forest Wars—Injunction, the President's Forest Summit

By 1990 a national sense of commitment to environmental stewardship was widespread, and the science of old-growth forests and the declining populations of northern spotted owl, both prominent topics of Andrews Forest research, proved critical in the battles over the future of federal forest lands. A judge's injunction stopping logging on almost ten million hectares of federal land encompassing the range of the northern spotted owl brought the matter into sharp focus. The battle was often framed as demanding "over protection" of native forests and biodiversity versus "logging" to support jobs and local communities. Of course, it was much more complex than that. A defining moment of political drama was the convening of the President's Forest Summit by President Bill Clinton and Vice President Al Gore in April 1993 in Portland, Oregon. This aimed to establish an innovative forum for sharing *all points of view* to set the stage for the development of the Northwest Forest Plan as a path forward. This plan was envisioned to be a statement of *public values* expressed in the laws.

Leading up to and following the Summit, Andrews Forest scientists partnered intensively with land managers to demonstrate and host critical, public, field discussions of alternative forest practices (Fig. 32.3). Their innovative research and management experiences made the Andrews Forest a go-to place for such conversations, which were widely covered in the media, including articles in the *New York Times*. In addition to these public roles, Andrews scientists were deeply involved in the behind-the-scenes processes (i.e., the Forest Ecosystem Management Assessment Team, FEMAT 1993) that crafted what became the Northwest Forest Plan to guide federal forest land management emphasizing biodiversity protection. Andrews science was central to society's newly framed connection with the native forests on the public lands. That science revealed the complexity of forest ecosystems and interactions with stream systems, and knowledge of that complexity was translated into new practices, such as incorporation of dead wood in forest and stream management. Still, the arts and humanities dimension did not yet reside within the Andrews program.

32.2.3.4 2000–2020+: Humanities and Arts Emerge in the Andrews Forest

By 2000, federal forest ecosystems in the Pacific Northwest and elsewhere were managed with a high consideration for their intrinsic value, especially biological diversity and watershed integrity. Society was deepening its connections with the natural world, and interest was growing globally in art-literature-science synergies to promote conservation. Science was expanding its scope of inquiry and its tool kit with greater attention to societal issues, including, in the 2000s, social justice, especially for historically disenfranchised peoples, and the effects of climate change.

Fig. 32.3 Public tour circa 1991 to a site of recent logging with retention of living trees and dead wood on the ground as an example of practices intermediate between clear-cut and no-cut (US Forest Service photo)

As described in the next section, Andrews Forest programs in arts/humanities inquiry and communications began in earnest in this period (Fig. 32.4).

32.3 Andrews Forest Based Humanities/Arts Programs and Outcomes

32.3.1 Origin of the Long-Term Ecological Reflections Program

Explicit engagement with the humanities at Andrews Forest began in 2002 through a partnership with the privately-funded Spring Creek Project led by founding director Kathleen Dean Moore. Forest Service research administrator and former Andrews Forest scientist JR Sedell instigated the first activities with Forest Service funding,

Fig. 32.4 A gathering of creative writers, philosophers, scientists, and others convened by the Spring Creek Project in the Andrews Forest (US Forest Service photo)

and Andrews scientist FJ Swanson was Moore's counterpart on the science side. With her combination of perspectives of a philosopher, strong relationships in the world of creative writers, exceptional communication skills, and passion for the natural world, Moore provided critical leadership. In her words, the mission of Spring Creek is: "To bring together the wisdom of the environmental sciences, the clarity of philosophical analysis, and the creative, expressive power of the written word and the arts to find new ways to understand and re-imagine our place in the natural world" (Panel 1 in Swanson et al. 2008). This humanities-science partnership has come to be called the Long-Term Ecological Reflections program (Long-Term Ecological Reflections 2021), modeled after the Long-Term Ecological Research program, both based at the Andrews Forest.

32.3.2 *Field Symposia*

The Long-Term Ecological Reflections program began with a series of field symposia generally involving a weekend retreat in the forest, followed by a public event with community members who could critique the emerging ideas, contribute to them, and carry the work into the local community. In several cases, published statements of our ethical obligations to the well-being of the planet reached much further. The retreats involved around 20 creative writers, scientists, philosophers, and others relevant to the central themes. The first gathering, held in 2002, was entitled *New Metaphors for Restoration of Forests and Watersheds*, which prompted realization of the need for "re-story-ation"—the need for new stories to articulate the relationship we wish to have with the land. Science alone cannot give us complete stories of connection, so we need to engage a diversity of thinkers and accomplished storytellers.

Subsequent field symposia included *Catastrophe and Renewal* held in the volcanic eruption zone of Mount St. Helens in 2005, *The Meaning of Watershed Health* hosted at Andrews Forest in 2006, and the *Salmon Nation Environmental Philosophers Gathering* at Andrews Forest in September 2007, which brought environmental ethicists and philosophers together for a weekend retreat on the topic "what is our work in a 'wounded world'?" This led organizers philosophers Kathleen Dean Moore and Michael Paul Nelson to assemble the book *Moral Ground: Ethical Action for a Planet in Peril* (2010) with powerful testimonies by over 80 visionaries from around the globe stating our moral responsibility to the planet. The short essays are grouped into 14 affirmative responses to the question: "Do we have a moral obligation to take action to protect the future of a planet in peril?" The first two responses are "Yes, for the survival of humankind" and "Yes, for the sake of the children." Moore and Nelson then took this set of moral arguments to more than 60 cities to share the stories and encourage local communities to address climate change.

In 2011, a subsequent field symposium, *The Eye of the Storm: Re-imagining Ethics for a Changing Planet*, led to crafting the "Blue River Declaration: An Ethic of the Earth" (Blue River Quorum 2011). This gathering of philosophers, activists, theologians, writers, and scientists composed the Declaration written as a philosopher's argument addressing a series of points, which align with the 3H's model in Tauro et al. (2021). To quote the biocultural ethic from the Declaration:

> What is the world? Who are we humans? How, then, shall we live? Given that life on Earth is interconnected, that humanity is inescapably dependent on the Earth for gifts both material and spiritual, that the Earth's resources and resilience are finite, that life on Earth is resilient, we arrive at the necessity of achieving a concordance between ecological and moral principles, and the new ethic born of this necessity (Blue River Quorum 2011).

32.3.3 Residencies, Residents, and Their Works in the Forest Log

The Long-Term Ecological Reflections program began a writer-in-residence initiative in 2004, which grew to include visual and sound artists and visiting scholars. Over 100 writers, artists, and scholars have produced many works documented in The Forest Log on the Spring Creek website (Long-Term Ecological Reflections 2021). Articles have been published in *Orion*, *The Atlantic*, terrain.org, and the book *Forest Under Story: Creative Inquiry in an Old-Growth Forest* (Brodie et al. 2016). In addition, artworks have been exhibited in galleries, the halls of NSF, and as permanent installation art pieces; written works have been presented in public readings and other performances. As the diversity of views, "lenses," and voices has grown, synergies among writers, visual and sound artists, and scientists have also grown. For example, theologian Vince Miller found David Paul Bayles' artistic photography deeply moving, so he invited him to illustrate his essay *A Cathedral Not Made by Hands* (Miller 2020) and the cover of the journal in which it appears (Fig. 32.5). Visual artist Leah Wilson teamed with tree canopy scientists for inspiration, technical assistance, and artistic fidelity with the forest as she climbed a tree taller than 50 m in each season. With this art-action foundation, she created a large installation art piece entitled Listening to the Forest, inspired by the title and content of the essay Listening to Water by Grignon and Kimmerer (2019), based on Kimmerer's Andrews Forest residency.

A schematic overview of some of the participants in the residency program outlines a sample of the wide range of perceptions and representations of the natural world and the human presence (Fig. 32.6). Artists, writers, traditional knowledge holders, theologians, philosophers, and scientists all experience the complex, highly interconnected ecosystem, which is constantly changing through ecological succession and periodic disturbance by fire, wind, forest practices, scientific experimentation, and other processes. These residents are encouraged to visit several of the same places, noted as Reflections Plots, including a 200-year log decomposition experiment initiated in 1985. Despite this common ground, their perceptions and expressions of their responses vary greatly. Regardless of their formally designated "discipline," they all seem to share awe and wonder in the presence of the mystery of the ancient forest. They may also share senses of practice: humility, empathy, and "serene attentiveness" in the words of Pope Francis in his encyclical Laudato si' (Francis 2015). Together, these perceptions of the natural world may catalyze public commitment to stewardship of ecosystems—of nature.

32.3.4 Additional Andrews Forest Programs Advancing Biocultural Ethics

Philosopher and Andrews Forest LTER principal investigator Michael Paul Nelson leads a program in Conservation Ethics (Andrews Forest 2021). In his essay, *Ground Rules for Ethical Ecology* Nelson argues that all our major challenges—climate change, zoonotic disease pandemics, pollution, food insecurity, biodiversity loss—are intertwined scientific, economic, technological, and ethical problems (Nelson 2021). Using the book *Moral Ground* (Moore and Nelson 2010) as a model, Nelson makes a call to action to address such major issues as moral crises. The Conservation Ethics program has produced a series of articles addressing the conceptualization of the topic generally.

More specifically, Nelson's Conservation Ethics program seeks "to understand how ideas work at the interface of science and management" by examining the basis

Fig. 32.5 A synergy between an artist and a writer generated the cover of the journal *Commonweal* in which Vincent Miller's essay "A Cathedral Not Made by Hands" appears (Miller 2020). David Paul Bayles used a "painting-with-light" technique to create this image of a mossy log on the floor of an old-growth stand in the Andrews Forest

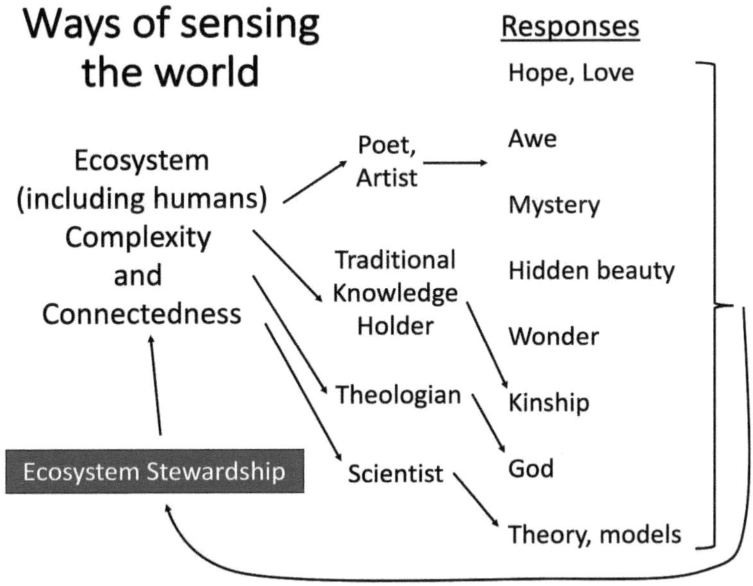

Fig. 32.6 Diagram of some of the many ways that people conduct inquiry of complex, highly-interconnected native ecosystems and the variety of reactions and expressions they have from those inquiries in the Andrews Forest. Writers in residence have included traditional knowledge holder Robin Wall Kimmerer and theologian Vincent Miller. The resulting communications feedback to catalyze a greater public sense for stewardship of ecosystems

for management decisions using philosophical methods such as argument analysis. While much of this work is not Andrews-based in the field in the sense of FEP, it has tested the rigor of science and ethical reasoning of some management recommendations with roots in Andrews Forest science. Batavia and Nelson (2016), for example, critique the proposal to use "ecological forestry" in management of federal forest lands (Franklin et al. 2018), and find significant inadequacies in the arguments because "facts" and "values" are not fully differentiated and disentangled.

Education programs for middle schoolers through graduate students are also an essential component of the Andrews Forest program, and several field experiences are designed specifically to engender a connection with nature and an environmental ethic in the students. For several years, Professor Moore taught a field philosophy course bringing landmark nature writings together with field ecology. A field program along the Discovery Trail in an old-growth forest near the Andrews Forest headquarters provides middle and high school students an arts-humanities-science field experience with prompts for reflection and creative work. Students use a digital curriculum on their tablets to explore the forests (Goralnik et al. 2020; Markiewicz et al. 2022). This experience includes readings and commentary by the writers in residence at Andrews in an effort to help students connect with the natural world. In studies of several high school student groups, investigators were encouraged because they found that most students reported that they connected with the place.

Remarkably, "more than 25% [of the students] demonstrated ethical growth by expressing agency for the care of the place" (Goralnik et al. 2020, p. 10).

Additional educational activities with roots in the Andrews Forest include the award-winning illustrated children's book *Ellie's Log: In the Forest where the Great Tree Fell* (Li and Herring 2013), which encourages young learners to engage closely with the natural world, observing, drawing, and keeping notes. At the advanced undergraduate and graduate student levels, numerous field courses in both the humanities and sciences have tapped creative writing from the Long-Term Ecological Reflections programs to prompt reflection and connection with the natural world.

32.4 Relations with Other Sites and Programs

Sharing perspectives concerning biocultural dimensions among sites and programs is highly valued because there is no grand plan and programs are evolving independently. We learn from one another, including learning about our own programs when juxtaposed with others. The Andrews Forest program aligns well with many features outlined for the practical development of Omora Park (Rozzi et al. 2008), principles for conduct of a biocultural conservation (Rozzi et al. 2006), and in the "3Hs" conceptual framework of the biocultural ethic (Rozzi 2012). The common ground is seeking to connect humans with nature in ways that respect co-inhabitants, their life habits, and the habits we share with them. This "3Hs" model (Rozzi 2012), and many of the ten principles outlined by Rozzi et al. (2006) are shared by Andrews Forest. More specifically, regarding the four-step cycle of Field Environmental Philosophy (Rozzi et al. 2012), the Omora Park and Andrews Forest Reflections Program converge in the integration ecological and philosophical views, practice of communication through metaphors in narrative compositions, "grounding" the connection with nature in the field, and work toward in-place implementation of biocultural conservation.

Significant points of difference between Andrews Forest and Omora Park programs result from differences in local circumstances. These include the engagement of local Indigenous communities essential in biocultural conservation at Omora Park. This engagement is very limited in the western Oregon setting of Andrews Forest where introduced diseases, displacement, and other factors nearly eliminated native people from the landscape early in the nineteenth century (Beckham 1977). Their past presence in the vicinity of Andrews Forest is clear in obsidian debitage. We endeavor to have the Andrews Forest community come into relationship with the land informed by the perspectives of Indigenous people, as articulated by Robin Kimmerer in *Braiding Sweetgrass* (2013) and consistent with the "3Hs" of the biocultural ethics (Rozzi 2012, 2013). In terms of communications with audiences from far away, ecotourism is not a factor at Andrews Forest, but it provides important opportunities for communication in the case of Omora Park.

Both programs converge in the strong influence of science on environmental policies. The Omora Park research team, for example, has led the creation of 250,000 km^2 of protected terrestrial and marine ecosystems, including the Cape Horn Biosphere Reserve and the Diego Ramirez Islands—Drake Passage Marine Park (Tauro et al. 2021). For its part, science findings from the Andrews Forest have had a strong record of influence on US federal land-use policies. In a broad sense, both programs are committed to Earth Stewardship (Rozzi et al. 2012; Andrews Forest 2021).

In another form of outreach, members of the Andrews Forest community foster arts-humanities-science collaborations at other sites of long-term ecological inquiry through a variety of activities. The Spring Creek Project manages a webpage to share information about programs at 20+ sites (Ecological Reflections 2021), and has collaborated in organizing multi-site workshops and art exhibits, including two in the NSF building in Washington, DC. Publications document the overall thrust of the programs and highlight examples of many LTER sites (Goralnik et al. 2015, 2016; Swanson 2015; Leigh et al. 2021). Spring Creek has also teamed with the online journal Terrain.Org to publish profiles of arts/humanities programs at several LTER sites.

Two studies of programs across the 28-site LTER network reveal that a strong majority has had significant engagements of the arts and humanities (Goralnik et al. 2015, 2016; Leigh et al. 2021). Surveys found that 19 of 24 sites surveyed agreed or strongly agreed that arts/humanities efforts were important and relevant to their mission, including fostering outreach and inspiring creative thought. This work engenders a sense of empathy, promoting earth stewardship, and enabling pro-environment attitudes and behaviors.

32.5 Final Points: The Path Forward

In its first 50 years, the Andrews Forest research findings revealed the incredible complexity of native ecosystems and helped society connect with ancient forests, streams, and the vast array of species therein. Through the 1970s and 1980s, basic research on these ecosystems rose from obscurity in the context of their conversion to tree plantations to become pivotal in social and environmental controversies that terminated the policy. Public values for federal forests shifted from exploitation for wood products to valuing biodiversity and other ecosystem attributes. Arts and humanities inquiry did not become a formal part of the Andrews Forest program until the 2000s, and the wealth of works and synergies among artists, writers, and scientists has grown steadily.

The outcomes of this work have never been formally assessed, but its enthusiastic persistence and growth, and the proliferation of similar programs at other field sites, attest to their value. The experience that Andrews Forest basic research on old growth and northern spotted owls took 20 years to go from irrelevant to hyper-relevant signals the importance of patience—the time for deeper societal relevance of the Reflections Program will come. A test of the relevance of this values inquiry may appear when federal forest land planning takes place in the coming few years—how will values be considered in that context?

The Andrews Forest community has important work to do in conceptualizing, articulating, and acting on our biocultural ethic, especially in relation with Indigenous people. Our plan for the future is to continue this public humanities work, which is consistent with the Omora Park mission to connect people with nature and to capitalize on the power of place. These two programs face exciting challenges and opportunities in times when a strong and broad sense of biocultural ethics is desperately needed. Only through shared understanding rooted in humanities, traditional knowledge, the arts, and environmental sciences can societies begin to address the social and environmental challenges here and now—and that lie ahead.

Acknowledgments The HJ Andrews Experimental Forest and Long-Term Ecological Research (LTER) program is administered cooperatively by the US Forest Service Pacific Northwest Research Station, Oregon State University, and the Willamette National Forest. This was supported by the National Science Foundation under LTER grants DEB-2025755 and DEB-1440409 and the Spring Creek Project in the School of History, Philosophy, and Religion, Oregon State University, from the Shotpouch Foundation, US Forest Service, and private donors. I greatly appreciate contributions over the years from KD Moore, C Goodrich, MP Nelson, C Lettero, W Robbins, R Rozzi, and many others.

References

Andrews Forest (2021) H.J. Andrews Experimental Forest. Available via: https://andrewsforest.oregonstate.edu/

Batavia C, Nelson MP (2016) Conceptual ambiguities and practical challenges of ecological forestry: a critical review. J For 114(5):572–581. https://doi.org/10.5849/jof.15-103

Bayles DP (2019) Outside of time|Forest landscapes. [online] Terrain.org https://www.terrain.org/2019/arterrain/david-paul-bayles/

Beckham SD (1977) The Indians of western Oregon: this land was theirs. Arago Books, Coos Bay

Blue River Quorum (2011) The Blue River declaration: an ethic of the earth. Minding Nat 4(3):146. [online] https://www.humansandnature.org/the-blue-river-declaration-an-ethic-of-the-earth

Brodie N, Goodrich C, Swanson F (eds) (2016) Forest under story: creative inquiry in an old-growth forest. Oregon State University Press, Corvallis

Ecological Reflections (2021) Ecological reflections network. Available via: http://www.ecologicalreflections.com

FEMAT (1993) Forest ecosystem management: an ecological, economic, and social assessment. Report of the Forest Ecosystem Management Assessment Team (FEMAT) 1993-793-071. US Government Printing Office, Washington DC

Francis (2015) Laudato Si'. On care for our common home. Encyclical letter. Our Sunday Visitor Publishing Division, Huntington

Franklin JF (1999) Foreword. In: Johnson KN, Swanson F, Herring M, Greene S (eds) Bioregional assessments: science at the crossroads of management and policy. Island Press, Washington DC

Franklin JF (2018) Foreword. In: Rozzi R, May RH Jr, Chapin FS III, Massardo F, Gavin M, Klaver I, Pauchard A, Nuñez MA, Simberloff D (eds) From biocultural homogenization to biocultural conservation. Ecology and ethics, vol 3. Springer, Dordrecht, pp v–vii

Franklin JF, Cromack K Jr, Denison W, McKee WA, Maser C, Sedell J, Swanson F, Juday G (1981) Ecological characteristics of old-growth Douglas-fir forests. Gen Tech Rep PNW-118. U.-S. Department of Agriculture, Forest Service, Pacific Northwest Forest and Range Experiment Station, Portland

Franklin JF, Johnson KN, Johnson DL (2018) Ecological forest management. Waveland Press, Long Grove

Geier MG (2007) Necessary work: discovering old forests, new outlooks, and community on the H.J. Andrews experimental forest, 1948–2000. Gen. Tech. Rep. PNW-GTR-687. U.-S. Department of Agriculture, Forest Service, Pacific Northwest Research Station, Portland

Goralnik L, Kelly SM, O'Connell KA, Nelson MP, Schulze M (2020) Forest discovery: place relationships on an environmental science, arts and humanities (eSAH) field trip. Aust J Environ Educ 1:1–12. https://doi.org/10.1017/aee.2020.28

Goralnik L, Nelson MP, Gosnell H, Leigh MB (2016) Arts and humanities inquiry in the long-term ecological research network: empathy, relationships, and interdisciplinary collaborations. J Environ Stud Sci 7:1–13. https://doi.org/10.1007/s13412-016-0415-4

Goralnik L, Nelson MP, Ryan L, Gosnell H (2015) Arts and humanities efforts in the US long-term ecological research (LTER) network: understanding perceived values and challenges. In: Rozzi R, Chapin FSIII, Callicott JB, Pickett STA (eds) Earth stewardship, linking ecology and ethics in theory and practice. Ecology and ethics, vol 2. Springer, Dordrecht, pp 249–268

Grignon J, Kimmerer RW (2019) Listening to the forest. In: Van Horn G, Hausdoerffer J (eds) Wildness: relations of people and place. University of Chicago Press, Chicago

Hirt PW (1994) A conspiracy of optimism: management of the National Forests since World War Two. University of Nebraska Press, Lincoln

Holling CS, Meffe GK (1996) Command and control and the pathology of natural resource management. Converv Biol 10(2):328–337

Kimmerer RW (2013) Braiding sweetgrass: indigenous wisdom, scientific knowledge and the teachings of plants. Milkweed Editions, Minneapolis

Leigh MB, Nelson MP, Goralnik L, Swanson F (2021) Integration of the arts and humanities with environmental science in the LTER network. In: Waide RB, Kingsland SE (eds) The challenges of long term ecological research: a historical analysis. Springer, Dordrecht, pp 315–334

Li JL, Herring ML (2013) Ellie's log: exploring the forest where the great tree fell. Oregon State University Press, Corvallis

Long-Term Ecological Reflections (2021) Available via: https://liberalarts.oregonstate.edu/centers-and-initiatives/spring-creek-project/programs-and-residencies/long-term-ecological-reflections

Markiewicz J, Goralnik L, O'Connell K (2022) Screens on trails: digital environmental science, arts, and humanities learning for biocultural conservation. In: Rozzi R, Tauro A, Wright T, Avriel-Avni N, May RH Jr (eds) Field environmental philosophy: education for biocultural conservation. Ecology and ethics, vol 5. Springer, Dordrecht, p XXX

Miller V (2020) A cathedral not made by hands. Commonweal 147(1):20–25

Moore KD, Nelson MP (eds) (2010) Moral ground: ethical action for a planet in peril. Trinity University Press, San Antonio

Nelson MP (2021) Ground rules for ethical ecology. Am Sci 109:246–249

Robbins WG (2018) The H.J. Andrews Experimental Forest: seventy years of pathbreaking forest research. Or Hist Q 119(4):454–485

Robbins WR (2020) A place for inquiry, a place for wonder: the HJ Andrews Experimental Forest. Oregon State University Press, Corvallis

Rozzi R (2012) Biocultural ethics: the vital links between the inhabitants, their habits and regional habitats. Environ Ethics 34:27–50

Rozzi R (2013) Biocultural ethics: from biocultural homogenization toward biocultural conservation. In: Rozzi R, Pickett STA, Palmer C, Armesto JJ, Callicott JB (eds) Linking ecology and ethics for a changing world: values, philosophy, and action. Ecology and ethics, vol 1. Springer, Dordrecht, pp 9–32

Rozzi R, Arango X, Massardo F, Anderson C, Heidinger K, Moses K (2008) Field environmental philosophy and biocultural conservation: the Omora Ethnobotanical Park education program. Environ Ethics 30:325–336

Rozzi R, Armesto JJ, Gutiérrez J, Massardo F, Likens G, Anderson CB, Poole A, Moses K, Hargrove G, Mansilla A, Kennedy JH, Willson M, Jax K, Jones C, Callicott JB, Kalin MT (2012) Integrating ecology and environmental ethics: earth stewardship in the southern end of the Americas. BioScience 62(3):226–236

Rozzi R, Massardo F, Anderson C, Heidinger K, Silander J (2006) Ten principles for biocultural conservation at the southern tip of the Americas: the approach of the Omora Ethnobotanical Park. Ecol Soc 11(1):43. [online] URL: http://www.ecologyandsociety.org/vol11/iss1/art43/

Swanson FJ (2015) Confluence of arts, humanities, and science at sites of long-term ecological inquiry. Ecosphere 6(8):1–23. https://doi.org/10.1890/ES15-00139.1

Swanson FJ, Goodrich C, Moore KD (2008) Bridging boundaries: scientists, creative writers, and the long view of the forest. Front Ecol Environ 6(9):499–504

Tauro A, Ojeda J, Caviness T, Mosses K, Moreno R, Wright T, Zhu D, Poole A, Massardo F, Rozzi R (2021) Field environmental philosophy: a biocultural ethic approach to education and ecotourism for sustainability. Sustainability 13:4526. https://doi.org/10.3390/su13084526

Zhu D (2022) Bridge the channel, enhance the inclusivity: a comparison between flagship species-centered and moss-centered conservation in Chile and China. In: Rozzi R, Tauro A, Wright T, Avriel-Avni N, May RH Jr (eds) Field environmental philosophy: education for biocultural conservation. Ecology and ethics, vol 5. Springer, Dordrecht, p nnn

Chapter 33
Screens on Trails: Digital Environmental Science, Arts, and Humanities Learning for Biocultural Conservation

Julie Markiewicz, Lissy Goralnik, and Kari O'Connell

Abstract The H.J. Andrews Experimental Forest (HJA), a Long-Term Ecological Research site in the Oregon Cascades, hosts long-term ecological research on the old growth forest and a long-running artist- and writers-residency program. This interdisciplinary investigation of place is featured on the Discovery Trail, an interpreted learning experience in the HJA for middle and high school students. The environmental science, arts, and the humanities (eSAH) inquiry-based curriculum on the Discovery Trail aims to facilitate engagement with both content and place, while providing learners opportunities to develop their own relationships with place through sensory interaction. The content is delivered through portable tablet computers (iPads), which allow for place engagement across time and season through audio and video media. While research on digital technology use in the classroom demonstrates a host of benefits, technology in field-based contexts is much less studied. We conducted an exploratory mixed methods study to investigate the impacts of technology on the Discovery Trail on learner engagement with place and content. Results show both benefits and challenges to learner engagement when using the tablets. In this chapter, we describe our research with 108 middle school learners from central Oregon on the Discovery Trail, as well as provide suggestions for content delivery and group preparation that can address potential challenges of employing this innovative pedagogy for biocultural conservation.

Keywords Affective learning · Digital technology · Engagement · Environmental education · Place-based relationships

---

J. Markiewicz · L. Goralnik (✉)
Michigan State University, East Lansing, MI, USA
e-mail: Markie11@msu.edu; Goralnik@msu.edu

K. O'Connell
Oregon State University, Corvallis, OR, USA
e-mail: Kari.OConnell@oregonstate.edu

R. Rozzi et al. (eds.), *Field Environmental Philosophy*, Ecology and Ethics 5,
https://doi.org/10.1007/978-3-031-23368-5_33

33.1 Introduction

Deep valleys and high peaks, towering Douglas firs, a floor of moss, and swift, clear mountain streams all characterize the H.J. Andrews Experimental Forest (HJA), one of 28 sites in the US Long-Term Ecological Research (LTER) network (LTER 2017). The LTER network is a collection of field sites across a range of biomes that are supported by the National Science Foundation for long-term study of place (HJA 2017a). Located in the Cascade Mountain range of central Oregon, the HJA's landscape is dominated by old growth forests blanketed in moss, the canopy of which is home to many charismatic species like the northern Spotted Owl (HJA 2017a).

In the HJA, environmental science, art, the humanities, and education intersect with the study of long-term change in the forest (see Swanson [Chap. 32] in this book for more of this story). This is the heart of the Discovery Trail (DT), an environmental science, arts, and humanities (eSAH) educational experience for middle and high school field trips in the HJA. Developed in 2015, the DT interpretive learning experience was designed to introduce learners to the landscape, forest management, and place-based ecology, as well as foster personal reflection about one's own place relationships and associated responsibilities. Through the integration of place-based ecological research, arts, and humanities, the interdisciplinary DT curriculum aims to foster biocultural conservation by facilitating the development of both knowledge about the ecosystem, as well as care for the natural world through interaction, reflection, curiosity, and wonder.

During the DT experience, student groups are guided through a series of stops along a half-mile forest trail. The inquiry-based curriculum is formatted on portable tablet computers (iPads) which are shared by groups of two to three students. The use of digital technology allows for the inclusion of multimedia elements in the curriculum, including videos and audio clips. This format enables a place experience across time and season, providing learners access to critters, places, and events they typically would not see on a forest visit, including the high canopy and the creek at flood volume. The digital platform accompanies learners' personal interactions, self- and group-reflections, and sensory explorations of the forest. The tablets also provide a non-invasive way to archive student assessment data, while facilitating increased student on-trail hours in a site with limited staff capacity.

Digital technology use in STEM (science, technology, engineering, and mathematics) learning environments has numerous benefits (Lakshminarayanan and McBride 2015). However, the use of digital technology in field settings is less studied, and likely has some limitations, including the simple intrusion of screens as a mediator of an immersed nature experience. To explore the use of digital technology as a delivery medium for an innovative field-based eSAH curriculum, we have been researching the impacts of the tablets on learner engagement with place and content on the DT. Content engagement is important for overall learning, particularly for encouraging persistence and maintaining motivation during learning experiences (Fredricks et al. 2004; Kuh et al. 2008; Reeve and Tseng 2011; Henrie et al. 2015). Place engagement, as we use the term, describes the depth with which a learner interacts with the storied built and natural landscape, a critical element of building relationships with places, and thus a central element of biocultural conservation. While scholarship on technology use in the field is limited, there is some hesitation in the literature about the use of screens on the trail because of the potential for distraction or a philosophical tension with taking technology outdoors to mediate the place experience (Anderson et al. 2015). If this is the case, then the learning medium on the DT—tablets—would prevent our learning goal for the trail experience, i.e., place-based learning, connections to the natural world, and biocultural conservation.

In 2018, we conducted an exploratory study on the DT to investigate the impacts of digital technology use on the learner experience. In this study, we asked the question: *What are the impacts of a digital curriculum on learner engagement with place and content?* Our results reveal both benefits and challenges to tablet use on the trail, with the positive impacts to biocultural conservation outweighing the logistical hurdles associated with using technology on the trail. In this chapter, we will describe our research with 108 middle school learners from central Oregon on the DT, as well as provide suggestions for content delivery and group preparation that can address potential challenges of employing this innovative pedagogy for biocultural conservation.

33.2 Background

33.2.1 The H.J. Andrews Experimental Forest

The US LTER network includes 28 sites with the shared mission to "provide the scientific community, policy makers, and society with the knowledge and predictive understanding necessary to conserve, protect, and manage the nation's ecosystems, their biodiversity, and the services they provide" (LTER 2017). Inquiry across the network is primarily ecological, though a number of sites (21 of 24 total sites in 2014) also host social science and/or arts and humanities inquiry (Goralnik et al. 2015). All sites are committed to education and outreach as central goals and some, including the HJA, also conduct research on these activities. The HJA in particular aims to "support research on forests, streams, and watersheds, and to foster strong collaboration among ecosystem science, education, natural resource management, and the humanities" (HJA 2017a). The HJA comprises 15,800 acres of mountainous old growth and managed forested terrain within the Lookout Creek watershed (Fig. 33.1). The forest is co-administered by the USDA Forest Service's Pacific Northwest Research Station, Oregon State University, and the Willamette National Forest. Foundational research at the HJA includes important studies on forest and stream ecology and forest management; ongoing long-term data collection includes climate, water quality, and forest succession.

Fig. 33.1 The HJA is home to towering Pacific Northwest old-growth trees such as Douglas fir, cedar, and hemlock. These iconic giants have grown uninterrupted for hundreds of years with some of them now exceeding 250 feet (75 m) (HJA 2017a)

The HJA also hosts a long-running arts and humanities residency program in collaboration with the Spring Creek Project for Ideas, Nature, and the Written Word at Oregon State University. Visual artists, writers, philosophers, and musicians spend 2 weeks in the forest to reflect on place, interact with scientists, and conduct their own scholarly and creative work. Since 2004, over 90 artists, writers, and humanities scholars have participated as residents in the HJA.

33.2.2 The Discovery Trail

In 2012, the DT loop was developed near the HJA headquarters to provide an opportunity for visitors to observe the characteristic ecosystems of the HJA forest (Fig. 33.2). The HJA hosted close to 1800 total visitors in 2018, and because much of the landscape is difficult to navigate due to steep terrain and active scientific studies, the DT provides these visitors an accessible outlet for exploration. In 2015, an eSAH digital interpretation of this trail was created specifically for middle and high school field trip groups.

In 2017, an additional quarter-mile trail extension was added to the original DT. This un-interpreted part of the trail is used for a silent sensory walk by field trip groups, whereby learners walk quietly and spaced apart to observe the forest with their senses. This experience effectively primes students for the DT learning experience by focusing their attention on place and their somatic experience.

The 5-h typical DT field trip experience begins with a brief orientation by HJA staff, who share a poem about the relationship between science and art, often part of "Art of Science" by Vicki Graham, a 2006 HJA writer-in-residence (Box 33.1).

Fig. 33.2 The trailhead for the Discovery Trail loop

Box 33.1 The final three stanzas from "Art of Science" (Brodie et al. 2016, p. 71–73) by Vicki Graham, a 2006 HJA writer-in-residence

Begin with love, a composite of science, art, imagination, and the pure world of the senses–with the things the hand can touch: The peeling papery bark of a yew, the curve of a snail shell, the beard lichen's wiry hair–with color and taste and smell and the call of a thrush at dusk, the velvet glide of a spotted owl	Begin with love and the questions the heart asks: Why stones arrange themselves in lapped patterns on a gravel bar. How old growth trees resist the quickly mutating pathogens that attack them. Where the orchid lacking chlorophyll gets its sugars. Let the body then the heart learn the forest and remember: Data collection, computer analysis, digitized imaging begin with hand and eye, tongue, nose, and ear	And while the pencil hovers Over the page or the hand grips A water gauge, the scientist Has time to stroke the willow leaf's silk, Breathe in the lemon scent Of chanterelles, follow the arc Of a swallow's flight. The artist, too, has time to taste and touch.

Table 33.1 The 11 learning objectives and activities of the DT, including four Next Generation Science Standards (NGSS) crosscutting concepts, and an example of an activity that fulfills that objective

Learning objective/ activity	Example activity
Place reflection and observation	Observing tree species
Creative activity	Drawing plant leaves
Personal reflection	Reflecting on how the growth that occurs in a forest after a disturbance is similar to how one changes after a disturbance in their own life
Values reflection	Reflecting on what the learner appreciates about forests
Mindfulness activity	Reflecting on what life is like for animals that live in HJA, meditation while listening to forest sounds
Creative writing	Reading aloud a short story about the relationship between a native American tribe and salmon, then interpreting the meaning of the story
Graph or diagram	Interpreting graphs of average rainfall, reading a diagram of the carbon cycle
NGSS: Stability and change	Learning about how logjams and flooding impact normally stable water bodies
NGSS: Cause and effect	Learning about how a forest disturbance changes plant communities
NGSS: Patterns	Comparing graphs of average rainfall to time lapse photos of Lookout Creek
NGSS: Systems	Learning about nutrient cycling and forest food webs

The students are then invited to participate on the solitary sensory walk on the un-interpreted trail. After a short debrief at the DT trailhead, student groups of two or three are taken to their first inquiry station and given their iPads. Each student group visits three of the ten interpretive stops, with each grouping of three stops designed to provide equivalent learning activities and objectives (Table 33.1). The stops include content related to HJA science, forest management, and forest dynamics, as well as reflective activities and creative content related to work from the HJA artists- and writers-in-residence (HJA 2017b). In addition to interviews with scientists and content from the long-running ecological inquiry at the HJA, the DT curriculum also includes writers reading poems and essay excerpts written during forest residencies and artist interviews about their creative process and HJA-related work. The intention is to provide opportunities to get to "know" the HJA through multiple modes of inquiry, as well as scales of perception: near and far; past, present, and future; ground and sky; interpreted and personal; human and animal; science and art. Ultimately, we hope that attention to both cognitive and affective learning outcomes—both learning and feeling—will lead learners to deeper understanding and potentially care and responsibility for the forest.

33.2.3 The Discovery Trail and Field Philosophy

The field philosophy (Goralnik et al. 2012, 2014, 2015, 2016; Goralnik and Nelson 2014, 2017) framework of our study is rooted in relational and care-based ethics, driven by the notion that right action derives from acting in ways that support the maintenance and wellbeing of our relationships (Warren 1990, 2000; Noddings 1984, 2002, 2005). As learners come to understand themselves to be in relationship with place(s), or to view place as a relational other worthy of moral consideration, they may also start to shift their actions in ways that reflect their responsibility within that relationship (Goralnik et al. 2020) (Fig. 33.3). Our approach aligns with Field Environmental Philosophy (FEP) (Tauro et al. 2021) in the shared focus of addressing the 'extinction of experience' common in contemporary educational settings (Anderson et al. 2015). The DT experience bridges the divide between mediated knowledge—the scientific, artistic, and literary narratives shared in the curriculum on the tablets—and direct encounters with the natural world, through guided meditations, sensory activities, reflection, observation, imaginings, and experiences in the forest (see Rozzi et al. 2006, 2008). Similar to FEP, we also hope to facilitate deeper connections between humans—in our case middle and high school learners—and the natural world, or rather place, which includes the

Fig. 33.3 At *Station 6: Lookout Creek,* students learn about the relationships between the forest, river ecology, and Indigenous storytelling. The entirety of the HJA, all 15,800-acres, belongs to the Lookout Creek watershed (HJA 2017a)

intersection of the natural world with layers of human relationships with the landscape over time, e.g., scientific, artistic, storytelling, land management, subsistence, political, recreational, psychological, emotional, and spiritual.

We have designed the DT learning experience around an environmental pedagogy of care (Goralnik et al. 2012), which bridges care-based ethics with scholarship in experiential and place-based learning (Gruenewald 2003; Knapp 2005), environmental education (Hungerford and Volk 1990; Goralnik and Nelson 2011), and emotional engagement (McCuen and Shah 2007; Skinner et al. 2009). This pedagogical foundation can lead to critical thinking and moral imagination in longer-term field philosophy experiences, as documented by Goralnik and colleagues in their work in Isle Royale National Park (Goralnik et al. 2014; Goralnik and Nelson 2014, 2015, 2017). An environmental pedagogy of care, as well as programming guided by this framework, is rooted in other examples of humanities-based field learning (Algona and Simon 2010; Johnson and Frederickson 2000), and foundational scholarship in environmental, place-based, and experiential learning. It uses empirical tools to describe learning and moral awareness in response to integrated field-based eSAH curriculum.

33.2.4 Digital Technology Use in the Learning Environment

Today's middle and high school students have been called "digital natives" due to their constant exposure to digital technologies (Prensky 2001). This fluency with technology has led many educators to use digital learning techniques in their classrooms to capitalize on student interests and strengths (Manuguerra and Petocz 2011), making digital learning the most rapidly developed method of learning in the past decade (Sebastian et al. 2012). A 2012 meta-analysis about the relationship between digital technologies and student academic achievement found that positive impacts of technologies have been consistently identified by researchers in many different learning contexts (Higgins et al. 2012; Lakshminarayanan and McBride 2015). Examples include the ability of digital technology to: foster deeper learning through guided critical thinking and problem solving (Kim et al. 2008); support student autonomy by allowing students to learn at their own pace and take ownership of the inquiry process, which can increase interest in learning (Kim et al. 2008; Lakshminarayanan and McBride 2015); and increase understanding of science and mathematics (Higgins et al. 2012). Digital technology can also improve learning facilitation, because it is portable and easily integrates multimedia elements like videos (Kukulska-Hulme et al. 2007).

Research on digital technology use in outdoor education contexts is limited, but while we might assume that many of the same benefits of technology in the indoor classroom apply, there are also reasons to proceed with caution when embracing technology in the field. Research—not specifically focused on environmental learning or field-based experiences—suggests that student dependence on technology can contribute to "alienation from nature" (Ruchter et al. 2010), the opposite of what we hope to accomplish with field-based environmental education. Richard Louv, author of *The Nature Principle: Reconnecting with Life in a Virtual Age*, echoes Leopold (1949) when he writes: "we cannot protect something we do not love, we cannot love what we do not know, and we cannot know what we do not see. Or hear. Or sense." (Louv 2012, p. 104). In other words, a relationship with nature is critical to the development of the motivation to care or protect it. Do screens inhibit these relationships in field-based learning? Or, is it possible to use digital technologies in the outdoor classroom in ways that do not *distract* from nature, but rather *enhance* connection-building? Our work on the DT is a step toward answering these questions.

33.2.5 Digital Tools in the Outdoor Learning Environment

Since the experience of being "digital natives" has changed the way learners think and process information (Prensky 2001; Selingo 2018), incorporating digital technology into the curriculum represents an innovation that accommodates the strengths and preferences of some learners (Prensky 2001). Learning according to one's preferred learning style improves learners' motivation to learn and to understand content (Wolf 2003; Kacoroski et al. 2016). While students are comfortable with technology use in general, the use of digital technologies in the field environment is likely an unfamiliar experience for most learners. Novel experiences like this—using a familiar tool in an unfamiliar context—can create excitement and motivate students to learn (Anderson et al. 2015). However, there are limited studies on the use of digital technology in field-based contexts like the DT to assess how it might support or detract from the outdoor learning experience (Kacoroski et al. 2016; McClain and Zimmerman 2016). Since learners today are increasingly alienated from nature (Ruchter et al. 2010), it is crucial to find ways to integrate digital tools in the field learning experience that balance the learning benefits and potential distractions.

Recent scholarship does suggest that it is, indeed, possible for learners to be engaged both with place and content in the digitally-mediated outdoor education classroom. In describing their study of a tablet-guided nature trail experience for students aged 8-11, McClain and Zimmerman (2016) offer suggestions for thoughtful digital field-based curriculum development, including prompts that require "heads-up" or "tactile investigation" to provide opportunities for students to engage directly with nature. This kind of activity encourages learners to engage with the landscape beyond the screen, instead of solely engaging with the technology (McClain and Zimmerman 2016). Being thoughtful about the integration of these activities can elevate the benefits of technology on the trail while addressing limitations.

33.2.6 Affective Learning

Research shows that learning requires cognitive, behavioral, and affective engagement (Reeve and Tseng 2011). The affective component refers to students' beliefs, feelings, and attitudes, including their emotional reactions to the learning experience (e.g., curious, excited, bored, or frustrated). Positive emotions increase affective engagement, while negative emotions decrease it (Fredricks et al. 2004; Reeve and Tseng 2011). Affective engagement is critical to the development of responsible environmental behaviors, one of the main goals of environmental education (Hungerford and Volk 1990; Littledyke 2008). This is because of its role in fostering a positive attitude toward the environment (Iozzi 1989) and its ability to cultivate a connection and a sense of respect, awe, and understanding of our place in the world (Littledyke 1996, 2008), as well an internal locus of control, or the feeling that one's actions can have an impact. On the DT, one way we engage the affective domain is through the integration of arts and humanities in the curriculum, which invites observations and expressions of feeling and emotion in the learning process more than science alone often encourages (Davis 2008; see Goralnik et al. 2020).

Recent scholarship on eSAH learning experiences shows that this kind of deeply interdisciplinary learning can contribute to changed attitudes about the natural world, a better understanding of complex socio-ecological systems, and emotional engagement with environmental issues, which can in turn foster pro-environmental intentions or action (Goralnik et al. 2016, 2020). Additionally, eSAH integration can increase learner capacity for creative and innovative thinking and decision-making, which are important skills for STEM-based problem solving and applying scientific concepts in the real world (Daugherty 2013). On the DT, art and humanities content—including poems, literary essays, visual art, Indigenous stories, values discussions, history, reflection and drawing—is integrated alongside place-based science curriculum, so that feeling and imagining about the natural world is rooted in concrete landscape dynamics. "Writers see the research work itself, such as long-term field experiments, in emotive terms the scientists themselves seldom use, such as faith, empathy, and love" explains Fred Swanson, an HJA scientist and eSAH collaborator (Swanson 2015, p. 18). This distinction is important, as Robin Kimmerer, an HJA writer-in-residence explains, because "The data may change our minds, but we need poetry to change our hearts." (Kimmerer 2004). Together, these creative perspectives can facilitate affective engagement in ways science, often presented as objective, cannot (Davis 2008), which in turn can foster meaningful learning (York 2014). The idea at the heart of the DT is that weaving environmental science, arts, and humanities together nests cognitive and affective learning on the trail. This is important for broad learner engagement, including content and modes of learning that appeal to all kinds of learners, and for facilitating feeling (and doing) in the process of learning.

33.2.7 Learning and Place Engagement

Engagement is a widely studied educational construct that can be applied across learning contexts (Connell 1990). It consists of three primary components: cognition, emotion, and behavior (Reeve and Tseng 2011) (Table 33.2). How a learner engages with content depends on many conditions, including the difficulty or simplicity of the material, the educator's or student's attitude, and the motivation to learn associated with perceived control and relatedness (Reeve and Tseng 2011). When a learner is engaged to the fullest extent in all three areas, many positive outcomes such as increased persistence and motivation have been documented (Fredricks et al. 2004; Henrie et al. 2015; Kuh et al. 2008).

In our work in environmental education, we also consider engagement a metric to describe the depth with which a learner interacts with a place. We call this "place engagement." It might also include interactions that take place *within* a learner, the internal experience of forest immersion, including the emotional connections learners make during creative or reflective activities (Fig. 33.4). The DT experience aims to strike a balance between student engagement with both content and place, with the goal of both imparting ecological knowledge as well as fostering feeling and emotional connections to the forest through the development of place relationships (Goralnik et al. 2020).

33.3 Methods

To explore the impacts of the digital curriculum on learner engagement with place and content, we compared two learner experiences on the Discovery Trail: (1) learners who used the digital curriculum on iPads, and (2) learners who used a printed curriculum workbook. Participants included the students and teachers from four eighth-grade classes from a single middle school in Eugene, Oregon. The two teachers both taught all of the students, but in different subjects (science and language arts). We used an exploratory, mixed methods approach, which enabled us to work with a large sample size, while at the same time provide rich context about the learner experience.

Four groups of students visited HJA on four separate days during May of 2019. Two control classes used the standard tablets and two experimental groups used an adapted equivalent paper version of the curriculum that we developed for the study. Translating the curriculum between a digital medium that featured multimedia elements to a paper version required some content and format modifications, including replacing open-ended questions with multiple choice, streamlining and condensing text, replacing videos with screenshot images and text, and replacing audio clips with text transcripts. Some content was adapted specifically for the different medium, so that the paper copy was not just a less dynamic version of the digital version, but a purposeful representation of the curriculum in its own right. The

Table 33.2 Descriptions for each of the three constructs of engagement (cognitive, emotional, behavioral) and place engagement in an environmental education context, including definitions and examples of what each component looks like in an education setting. These components are collective, not summative, and the DT aims to engage all components often at the same time throughout the experience. The example of Station 9 shows how all four engagement components are woven together in practice on the DT. Definitions and descriptions are borrowed from Fredricks et al. (2004), Dienno and Hilton (2005), and Reeve and Tseng (2011)

Engagement type	Component	Definition	Descriptions	Example from the Discovery Trail
Content engagement	Cognitive **(a)**	Being thoughtful and willing to invest the psychological effort required to comprehend material and master skills	Active self-regulation to sustain engagement, the use of strategic learning strategies to organize, monitor, and evaluate progress	*At Station 9: Sounds Maps* students listen to an interview recording of an HJA researcher explaining how she uses sound in her research on birds in the forest. Students are then asked inquiry-based questions about the content of the interview **(a)** before being prompted to set the tablet down and quietly listen to the forest **(d)**. After a few minutes, the voice narration instructs students to pay attention to the variety, direction, and strength of sounds around them **(c, d)**, then to map the sounds on paper using images, arrows, text, or symbols **(a, c)**. Afterward, students are asked to reflect with their group on how they felt while listening to the forest **(b, d)**, and then consider other times in their lives they have felt similarly **(b)**. The station wraps by asking learners to notice any connections between the scientist's bird research and their own sound mapping activity **(a)**
	Emotional **(b)**	The affective reactions to learning, such as how pleasurable and enjoyable learners find an experience, and their identification and sense of belonging with place	Learners are enthusiastic, anxious, sad, happy, curious, feel like they belong and are valued	
	Behavioral **(c)**	The amount of active participation and effort demonstrated by learners in learning tasks and in the classroom	On-task, paying attention, persistence at tasks, following directions, no conduct issues, asking questions, contributing to discussions	
Place engagement	Holistic (multi-sensory) **(d)**	The depth to which a learner interacts, considers, and connects with a place as a built and natural storied environment	Tactile investigations of landscape elements, spiritual connections to or an expressed feeling of ease in place, pointing out objects to peers, immersing oneself in a sensory experience in the landscape, or remarking on landscape features or observations	

Fig. 33.4 Students looking up at the tree canopy to observe patterns in how different species of trees are distributed in the stand around them during *Station 4: Forest Succession*

learning objectives for each station remained the same in the paper version, so while the two experiences were not identical, they covered the same concepts and material, making them equivalent versions of the same curriculum.

We used a pre- and post-survey consisting mostly of five-point scale Likert-type questions to measure (1) affective attitudes; (2) connection to nature; (3) engagement; and (4) content knowledge. The affective attitude section asked students to what degree they agree or disagree with prompts asking about their feelings about four prominent themes within the DT experience: science, nature, art, and digital technology. The connection to nature section combined items from the Connectedness to Nature Scale, a commonly used scale that was developed to measure an adult's emotional connection to the natural world (Mayer and Frantz 2004), and a modification of this scale that was adapted for fourth-grade participants (Cheng 2008). The combination of items designed for adults (Mayer and Frantz 2004) and fourth-grade students (Cheng 2008) was chosen to tailor our study for middle and high school learners.

The engagement section, on the post-survey only, included Likert prompts designed for this study and guided by the definitions for each engagement component (adapted from Cheng and Wu 2015) about how students perceived their own engagement during the experience. Items included statements like "the lessons on the trail were enjoyable" and "I often found my mind wandering from what I was supposed to be focusing on." The content knowledge section was designed to assess student understanding of four Next Generation Science Standards crosscutting concepts that are integrated into the DT curriculum: stability and change, cause and effect, patterns, and systems (NGSS 2013). The pre-surveys also collected basic demographic information and contained one open-ended question asking the students what they hoped to learn on the DT. The post-surveys included an additional section asking about specific activities on the DT, and three open-ended questions about what students learned, what they enjoyed, and what they disliked during the field trip. Illegible responses were not included in the analysis.

Additionally, we conducted participant observation and recorded extensive field notes during the trail experience. One researcher stationed herself at several stops along the trail—enough out of the way so as not to distract the students yet within hearing range of student conversations—and observed groups as they engaged with the inquiry and content at that stop. Field notes focused on engagement-specific behaviors such as: (a) "heads-up observations," when the learner responded to the curriculum by observing their natural surroundings; (b) "pointing," which indicates the learner is connecting a concept learned in the program to their natural surroundings, and (c) "tactile investigation," which indicates a learner is using touch to investigate a natural object that is referenced in the curriculum (McClain and Zimmerman 2016). Observations also noted distraction, socialization about off-task topics, and non-curriculum focused behaviors. We also interviewed both teachers before and after the field trips. These interactions contributed insights about the DT experience as a whole, while also capturing additional information about the teachers' expectations and experience with the tablets, expectations for and observations of student learning and engagement during the experience, observations of field trip impacts in the classroom, and overall satisfaction and perceived usefulness of the DT experience.

Statistical analysis of the Likert survey data was completed using the statistical software SPSS (IBM 2019). The shift in mean between the pre- and post-surveys was calculated for each component. Unpaired t-tests using the change of mean between the pre- and post-surveys were used to identify statistical differences in pre- to post-survey shifts depending on medium, i.e. tablets or paper books (Table 33.3). Paired t-tests were used to find differences between the pre- and post-surveys, which allowed us to see whether any of the measured constructs

Table 33.3 Statistical results from unpaired t-tests conducted on the mean pre- to post-survey shifts and on mean post-survey engagement scores between tablet and paper book-users. Threshold for significance is a p-value < 0.05

Construct	*N*	Mean shift in tablet-users (SD)	Mean shift in paper book-users (SD)	*p*-Value
Attitude toward *science*	84	0.131 (0.530)	−0.026 (0.566)	0.183
Attitude toward *nature*	84	0.103 (0.347)	0.038 (0.473)	0.471
Attitude toward *art*	84	−0.048 (0.386)	0.0179 (0.447)	0.474
Attitude toward *digital technology*	84	0.0159 (0.016)	−0.0519 (0.495)	0.531
Connection to nature	84	0.170 (0.358)	0.0003 (0.601)	0.170
Content knowledge	64	0.353 (0.884)	0.133 (1.008)	0.357
Construct	*N*	*Post-survey mean in tablet-users (SD)*	*Post-survey mean in paper-book users (SD)*	*Sig (2-tailed)*
Engagement	85	3.362 (0.492)	3.334 (0.569)	0.809

changed before and after the DT experience regardless of the instructional medium used.

Short-answer survey responses, interviews, and field notes were coded using an emergent thematic analysis (Hesse-Biber and Leavy 2008). Transcripts were read multiple times, noting themes in the margins. Themes were then condensed into codes to eliminate redundancy and a codebook was created for each data source. All data were then re-coded with the associated codebook, and codebooks were adapted if/when new themes arose. Data were then deductively analyzed with the final codebooks when no new themes arose. All results reflect categories of related codes that emerged in this final stage of the analysis.

33.4 Results

33.4.1 Survey Questions

For each of the constructs we measured (attitudes, connection to nature, engagement, and content knowledge), Likert scores were aggregated into a single mean and checked for reliability with a Cronbach's alpha test. Knowledge questions, each worth one point, were scored right or wrong out of five. Only participants that responded to at least 75% of the questions within each section were included in the analysis for that section.

Results from the paired t-tests showed a significant increase in content knowledge after the DT experience across all students. Before the DT experience, the average score out of five questions was 67%, whereas 73% answered correctly after the experience. These results indicate that both versions of the curriculum are effective at facilitating content engagement, at least enough that students are learning concepts. However, we did not find any statistically significant differences between the students that used tablets ($n = 51$) versus the students that used paper books ($n = 43$).

However, the survey likely did not capture the whole picture. We found no statistically significant differences between the pre- to post-survey shifts that occurred between the tablet and paper book-users (Table 33.3). This result shows that neither medium seems to be better or worse than the other, at least statistically. The qualitative participant observation data shows a more nuanced and different result than the survey, and thus provides a richer understanding of student engagement.

33.4.2 Participant Observation

Participant observation data showed that there were indeed barriers to engaging with content and place on the DT, though different barriers were associated with the tablets and the paper books. For example, groups using the tablets expressed frustration when experiencing technology issues, such as slow Wifi. Negative emotions, like frustration due to technology not working properly, can negatively impact a student's affective engagement (Reeve and Tseng 2011). Although students using paper books did not experience technology-related frustration, we did observe them skipping entire sections of the lesson, shuffling past some pages of instruction, a clear sign of being disengaged from the content. This is a challenge specifically associated with the paper books because the tablets require responses on each page before advancing the screen. Therefore, tablet-users needed to participate with the content on each page at some level.

There were also barriers observed *across* the tablet and paper book groups. Regardless of medium, students frequently expressed frustration with the number of questions embedded in the lessons and the difficulty of the questions and content. When a lesson is too challenging for a student, they often disengage (Turner et al. 1998). Students were also observed disproportionately engaged with the content rather than the place. Examples of place engagement, such as tactile investigation of objects in the forest or pointing out forest features to group members (McClain and Zimmerman 2016), were rarely observed in either group. Students were frequently observed spending a majority of their time at each station looking down at the tablet or paper book. Since this was seen across all student groups, the curriculum itself could be a barrier to place engagement, not the medium of transmission. While this detail was not reflected in the survey data, it is an important result.

33.4.3 Teacher Interviews

The pre- and post-interviews with the two teachers gave further insight into the DT experience, teacher perceptions of student engagement, and teacher thoughts on tablet-usage in general. The teachers stated convenience, ease of use, and multimedia capabilities as the main benefits to using tablets on the DT. They did not perceive any significant limitations. When asked to identify key differences between how students engaged with the tablet versus paper curriculum, neither teacher observed major differences in the level of commitment students were putting forth to complete the lessons. This result is notable because psychological investment to learn a lesson is one sign of cognitive engagement (Fredricks et al. 2004). One teacher said that regardless of medium, the students "were engaged with [the experience]. They were curious. They paid attention." Both teachers also expressed positive feelings toward using the tablets and thought the multimedia elements were exciting and novel for the students, feelings that can increase engagement (Fredricks et al. 2004).

Although the teachers did not perceive any significant content engagement limitations with the tablets, they had concerns about the difficulty of the curriculum for middle school-aged students. This finding also was reflected in the participant observation field notes. To mediate this concern, one teacher suggested two different versions of the DT curriculum, one for middle school learners and another for high school learners. They also had insightful ideas about how to modify the curriculum to increase engagement specifically for middle school learners and thereby improve the student experience. We plan to partner with the teachers in the future as we revise the curriculum.

33.4.4 Open-Ended Survey Questions

The four short open-ended questions on the surveys also provided deeper insight about the student experience. Two primary themes emerged during the qualitative analysis. The first identified a disconnect between what the students wanted to learn about and what they actually learned on the DT. For example, "forest animals" was the most frequent answer to the question "what do you hope to learn about at H.J. Andrews?" (36%). The second most common response was "climate change" and what they can do in their personal lives to mitigate environmental damage (24%). Currently, neither of these subjects are prominent in the curriculum. Students are more likely to be cognitively engaged in the content when they are interested in the subject matter, so adding more material in line with student expectations and interests, or better preparing learners for the experience and what to expect, could increase their overall engagement (Fredricks et al. 2004).

The second theme that emerged was that the experience was too much work and too little fun. When asked what their least favorite part of the DT experience was, 33 of 73 responses (45%) indicated *doing the work on the DT*. More specifically, students noted that there were too many questions, the videos and readings were too long, and the stations included too many activities.

Notably, when asked about their favorite part of the experience, the highest number of learners mentioned the silent sensory walk (23%). Other common responses described being present in the forest, for example: being outside, smelling and hearing the forest, walking on the trails, and being in a peaceful and quiet environment. Having a positive emotional reaction to the experience is a critical component of being affectively engaged, so if the experience featured a more satisfactory balance between focused learning and free exploration in the forest, the students might not only learn more, but also have a more positive experience (Fredricks et al. 2004). When students are cognitively engaged in work, but not provided sufficient opportunities to fully engage with the forest to facilitate other kinds of engagement, their overall engagement level will be sub-optimal. The challenge with responding to this critique is that teachers cannot justify a forest field trip if the experience does not satisfy classroom learning objectives. At the same time, biocultural conservation goals are not accomplished if the students are disengaged in the experience. We will have to negotiate this balance moving forward. Longer field trips would offer one solution, allowing us to provide equal time doing structured learning activities and unstructured exploring, though the remoteness of the site makes longer trips challenging unless they are overnight.

33.5 Discussion

Upon triangulating the results from the surveys, observation notes, and teacher interviews, we draw three main conclusions. First, some differences exist in engagement between students that use tablets and students that use paper on the DT, but neither greatly outweighs the other in contributing to engaged/disengaged behaviors. Second, on the DT, engagement with content and place depend more on the curriculum itself rather than the medium that is used to deliver the curriculum. Third, tablet-led experiences can be modified to increase both content and place engagement to facilitate an overall more enjoyable and beneficial experience, thereby supporting broader learning goals. Making thoughtful modifications to the curriculum can contribute to a more enjoyable experience for the students while also increasing their engagement. This effort will allow a better balance between place and content engagement, and ultimately positively impact place relationships, affective engagement with the forest experience, and biocultural conservation outcomes.

The study described here applies our previous field philosophy work in a new direction by exploring the impacts of short-term experiences in place and with younger learners, who are less steeped in the theoretical and popular culture dialogues about environmental issues, responsibility, and history. Recent scholarship on the DT demonstrates that these types of short-term, ethically grounded approaches to field-based environmental science, arts, and humanities (eSAH) learning can catalyze both passive and active place relationships (Goralnik et al. 2020), whereby learners either feel cared for by the forest, or even develop the motivation to care for the forest, as a result of their eSAH trail experience. This is a novel finding, as most place relationship literature focuses on repeated interactions with place over time. While longer-term immersive experiences in place are likely preferable for a host of reasons (Ardoin 2006; Kudryavtsev et al. 2012; Semken et al. 2009; Wattchow and Brown 2011), they are not always feasible for broad audiences, and may not even be appropriate for learners with little experience in the natural world, who need to acclimate to the sensory experience and conditions of learning beyond the classroom. The present study narrows the focus of our previous work on the DT to begin to understand more concretely why and how students engage during the eSAH trail experience. This work is important because engagement, especially emotional engagement, is integral to the learning process, and also a key ingredient in the development of care for the natural world. Therefore it is a necessary step toward biocultural conservation.

33.5.1 Limitations

All participants were middle school students belonging to a single middle school. While our results are relevant for this age group, our sample is small and the research is exploratory. More research is necessary to understand engagement across audiences, including high school learners, who are a target audience for the DT. We also experienced some incomplete data due to survey logistics and miscommunication between the teachers and the research team. In future studies on the trail, we will mediate these challenges with more streamlined and focused data collection tools and processes. Finally, our results are rooted in the context of a specific eSAH curriculum created for the DT experience. More research is necessary to understand this comparison across contexts.

33.5.2 Suggestions for Best Practices and Next Steps

While our results are exploratory, they do address novel and relevant questions related to innovative field philosophy methods. By identifying barriers to content and place engagement, we can revise the curriculum to provide a more meaningful, personal forest experience that engages and immerses students in place and with the forest ecology, art, and humanities content. Our sense is that there are more subtle differences between the tablet and paper curriculum experiences that are not reflected in our data, therefore we intend to further develop the qualitative data protocol (perhaps including interviews with student participants) to better capture the student experience. This insight will help us revise the curriculum in a way that it increases both content and place engagement by including more "heads up" prompts. These activities will reduce the time students spend looking at the tablet screen and encourage deeper engagement with place. In practice, these prompts can involve asking students to look for and observe specific objects in the forest, as well as encouraging them to engage in more tactile or sensory investigations. We can also incorporate additional audio narration into the curriculum that invites students to set down their tablets and fully sense their surroundings without the screen. Currently there are several activities like this in the DT curriculum, including the sound mapping activity, a color observation activity, and a drawing activity. This kind of passive tablet use simultaneously facilitates both content and place engagement and offers an effective way to balance the benefits and limitations of technology use in field learning environments for biocultural conservation.

Acknowledgments We conducted this research at H.J. Andrews Experimental Forest, which is funded by the US Forest Service, Pacific Northwest Research Station, and the National Science Foundation LTER Grants: LTER8 DEB-2025755 and LTER7 DEB-1440409. We would also like to acknowledge Adam Kennedy, whose work with the digital curriculum and database management made this project possible, and Ginny White, Chris Smith and Matt Viehdorfer, who developed the application and database for the digital curriculum. We appreciate the support of an anonymous donor to the Discovery Trail and the NSF Graduate Research Fellowship Program, which supported the lead author's contributions.

References

Algona PS, Simon GL (2010) The role of field study in humanistic and interdisciplinary environmental education. J Exp Educ 32(3):191–206

Anderson CL, Miller BG, Bradley EK et al (2015) Exploring techniques for integrating mobile technology into field-based environmental education. Electron J Sci Educ 19(6):1–19

Ardoin N (2006) Toward an interdisciplinary understanding of place. Can J Environ Educ 11:112–126

Brodie N, Goodrich C, Swanson FJ (2016) Forest under story: creative inquiry in an old growth forest. University of Washington Press

Cheng JC (2008) Children, teachers and nature: an analysis of an environmental education program. Dissertation, University of Florida

Cheng TM, Wu HC (2015) How do environmental knowledge, environmental sensitivity, and place attachment affect environmentally responsible behavior? An integrated approach for sustainable island tourism. J Sustain Tour 23(4):557–576

Connell JP (1990) Context, self, and action: a motivational analysis of self-system processes across the life-span. In: Cicchetti D (ed) The self in transition: infancy to childhood. University of Chicago Press, Chicago, pp 61–97

Daugherty MK (2013) The prospect of an "A" in STEM education. J STEM Educ 14(2):10–15

Davis JH (2008) Why our schools need the arts. Teachers College Press, New York

Dienno CM, Hilton SC (2005) High school students' knowledge, attitudes, and levels of enjoyment of an environmental education unit on nonnative plants. J Environ Educ 37(1):13–25

Fredricks JA, Blumenfeld PC, Paris AH (2004) School engagement: potential of the concept, state of the evidence. Rev Educ Res 74(1):59–109

Goralnik L, Dobson T, Nelson MP (2014) Place-based care ethics: a field philosophy experience. Can J Environ Educ 19:180–196

Goralnik L, Kelly S, O'Connell KB et al (2020) Forest discovery: active and passive place relationships on an environmental science, arts, and humanities (eSAH) field trip. Aust J Environ Educ 37:1–12. https://doi.org/10.1017/aee.2020.28

Goralnik L, Millenbah K, Nelson MP et al (2012) An environmental pedagogy of care: emotion, relationships, and experience in higher education. J Exp Educ 35(3):412–428

Goralnik L, Nelson MP (2011) Framing a philosophy of environmental action: Aldo Leopold, John Muir, and the importance of community. J Environ Educ 42(3):181–192

Goralnik L, Nelson MP (2014) Field philosophy: from dualism to complexity through the borderland. Dialect Anthropol 38(4):447–463. https://doi.org/10.1007/s10624-014-9346-1

Goralnik L, Nelson MP (2015) Empathy and agency in the Isle Royale field philosophy experience. J Sustain Educ 10:3. http://www.jsedimensions.org/wordpress/content/empathy-and-agency-in-the-isle-royale-field-philosophy-experience_2015_12/. Accessed 9 May 2021

Goralnik L, Nelson MP (2017) Field philosophy: environmental learning and moral development in Isle Royale National Park. Environ Educ Res 23(5):687–707. https://doi.org/10.1080/13504622.2015.1074661

Goralnik L, Nelson MP, Gosnell H et al (2016) Arts and humanities inquiry in the long-term ecological research network: empathy, relationships, and interdisciplinary collaborations. J Environ Stud Sci 7(2):361–373

Goralnik L, Nelson MP, Ryan L et al (2015) Arts and humanities efforts in the US long-term ecological research (LTER) network: understanding perceived values and challenges. In: Rozzi R et al (eds) Earth stewardship, ecology and ethics, 2nd edn. Springer International, Cham, pp 249–268

Gruenewald D (2003) Foundations of place: a multidisciplinary framework for place-conscious education. Am Educ Res J 40(3):619–654

H.J. Andrews Experimental Forest (2017a) About the Andrews Forest. https://andrewsforest.oregonstate.edu/about. Accessed 9 May 2021

H.J. Andrews Experimental Forest (2017b) Arts and humanities. https://andrewsforest.oregonstate.edu/outreach/arts-and-humanities. Accessed 9 May 2021

Henrie CR, Halverson LR, Graham CR (2015) Measuring student engagement in technology-mediated learning: a review. Comput Educ 90:36–53

Hesse-Biber SN, Leavy P (eds) (2008) Handbook of emergent methods. Guilford Press, NY

Higgins S, Xiao Z, Katsipataki M (2012) The impact of digital technology on learning: a summary for the Education Endowment Foundation. Education Endowment Foundation, pp 1–52

Hungerford HR, Volk TL (1990) Changing learner behavior. J Environ Educ 21(3):8–21

IBM Corp. Released 2019. IBM SPSS statistics for windows, version 26.0. IBM Corp, New York

Iozzi LA (1989) What research says to the educator. Part one: environmental education and the affective domain. J Environ Educ 20:3–9

Johnson BL, Frederickson LM (2000) 'What's in a good life?' Searching for ethical wisdom in the wilderness. J Exp Educ 23(1):43–50

Kacoroski J, Liddicoat KR, Kerlin S (2016) Children's use of iPads in outdoor environmental education programs. Appl Environ Educ Commun 15(4):301–311

Kim MC, Hannafin MJ, Bryan LA (2008) Technology-enhanced inquiry tools in science education: an emerging pedagogical framework for classroom practice. Sci Educ 91:1010–1030

Kimmerer R (2004) Interview with a watershed. Available via https://liberalarts.oregonstate.edu/sites/liberalarts.oregonstate.edu/files/kimmerer1.pdf. Accessed 9 May 2021

Knapp CE (2005) The "I-thou" relationship, place-based education, and Aldo Leopold. J Exp Educ 27(3):277–285

Kudryavtsev A, Stedman RC, Krasny ME (2012) Sense of place in environmental education. Environ Educ Res 18(2):229–250

Kuh GD, Cruce TM, Shoup R et al (2008) Unmasking the effects of student engagement on first-year college grades and persistence. J High Educ 79:540–563

Kukulska-Hulme A, Traxler J, Pettit J (2007) Designed and user-generated activity in the mobile age. J Learn Des 2(1):52–65

Lakshminarayanan V, McBride AC (2015) The use of high technology in STEM education. Educ Train Opt Photon 9793:1–12. https://doi.org/10.1117/12.2223062

Leopold A (1949) A sand county almanac. Oxford University Press, New York

Littledyke M (1996) Science education for environmental awareness in a postmodern world. Environ Educ Res 2:197–214

Littledyke M (2008) Science education for environmental awareness: approaches to integrating cognitive and affective domains. Environ Educ Res 14(1):1–17

Louv R (2012) The nature principle: reconnecting with life in a virtual age. Algonquin Books, Chapel Hill

LTER (2017) Long-term ecological research (LTER) network vision, mission, and goals. Lternet.edu/vision-mission. Accessed 9 May 2021

Manuguerra M, Petocz P (2011) Promoting student engagement by integrating new technology into tertiary education: the role of the iPad. Asian Soc Sci 7(11):61–65

Mayer FS, Frantz CM (2004) The connectedness to nature scale: a measure of individuals' feeling in community with nature. J Environ Psychol 24:503–515

McClain LR, Zimmerman HT (2016) Technology-mediated engagement with nature: sensory and social engagement with the outdoors supported through an e-Trailguide. Int J Sci Educ B 6(4): 385–399

McCuen RH, Shah G (2007) Education of recent neuroscience research on emotions. J Leadersh Stud 1(3):44–56

Next Generation Science Standards (NGSS) (2013) Appendix G – crosscutting concepts. https://www.nextgenscience.org. Accessed 10 Dec 2019

Noddings N (1984) Caring: a feminine approach to ethics and moral education. University of California Press, Berkeley

Noddings N (2002) Educating moral people. Columbia University Teachers College Press, New York

Noddings N (2005) The challenge to care in schools. Columbia University Teacher's College Press, New York
Prensky M (2001) Digital natives, digital immigrants part 1. On Horizon 9(5):1–6
Reeve J, Tseng C (2011) Agency as a fourth aspect of students' engagement during learning activities. Contemp Educ Psychol 36(4):257–267
Rozzi R, Arango X, Massardo F et al (2008) Field environmental philosophy and biocultural conservation: the Omora Ethnobotanical Park educational program. Environ Ethics 30(3): 325–336
Rozzi R, Massardo F, Anderson C et al (2006) Ten principles for biocultural conservation at the southern tip of the Americas: the approach of the Omora Ethnobotanical Park. Dig Ecol Soc 11(1):43. http://www.ecologyandsociety.org/vol11/iss1/art43
Ruchter M, Klar B, Geiger W (2010) Comparing the effects of mobile computers and traditional approaches in environmental education. Comput Educ 54(4):1054–1067
Sebastian D, Ali S, Ivo B et al (2012) Determinants of physicians' technology acceptance for e-health in ambulatory care. Int J Med Inform 81(11):746–760
Selingo JJ (2018) The new generation of students: how colleges can recruit, teach, and serve gen Z. The Chronicle of Higher Education, pp 1–48
Semken S, Freeman CB, Watts NB et al (2009) Factors that influence sense of place as a learning outcome and assessment measure of place-based geoscience teaching. Int Electron J Math Educ 13(2):1–24
Skinner EA, Kindermann TA, Furrer CJ (2009) A motivational perspective on engagement and disaffection: conceptualization and assessment of children's behavioral and emotional participation in academic activities in the classroom. Educ Psychol Meas 69(3):493–525
Swanson FJ (2015) Confluence of arts, humanities, and science at sites of long-term ecological inquiry. Ecosphere 6(8):1–23
Tauro A, Ojeda J, Caviness T et al (2021) Field environmental philosophy: a biocultural ethic approach to education and ecotourism for sustainability. Sustainability 13(8):4526
Turner JC, Thorpe PK, Meyer DK (1998) Students' reports of motivation and negative affect: a theoretical and empirical analysis. J Educ Psychol 90(4):758–771. https://doi.org/10.1037/0022-0663.90.4.758
Warren KJ (1990) The power and promise of ecological feminism. Environ Ethics 12(2):125–146
Warren KJ (2000) Ecofeminist philosophy. Rowman & Littlefield Publishers, Inc., Lanham, MD
Wattchow B, Brown M (2011) A pedagogy of place: outdoor education for a changing world. Monash University Press, Victoria
Wolf CJ (2003) Technology in environmental education, vol 9. New Horiz Learn Online J, pp 1–6
York RA (2014). Re-connecting with nature: transformative environmental education through the arts, Dissertation. University of Toronto. Retrieved from: https://tspace.library.utoronto.ca/handle/1807/68404

Chapter 34
Reframing Sense of Place: A Goal for Environmental Education and an Indicator Supporting Social-Ecological Resilience

Noa Avriel-Avni, Miri Lavi-Neeman, and Jennifer M. Holzer

Abstract Education for sustainability has become a main thrust of environmental education since the early 1990s. In this context, the development of ecological and environmental literacy has been favored as a key tool for cultivating environmental concern and responsible human behavior towards ecosystems. In recent years, the concept of *sense of place* (SoP) has added an affective dimension to environmental education. In this chapter, we challenge prevailing understandings of SoP. By drawing on critical geography, we propose an alternative framing for place and *sense of place* that capture the relationship between social and ecological systems. Taken together the concepts of *place* and *sense of place* complement each other, and amplify the perception of *place* as co-produced by humans and nonhumans, in long-term processes and through different multiple spatial scales. Blurring the conceptual lines between the social and the ecological, and between the local and the global, this theoretical framework opens a new direction for environmental education, which corresponds better with the realities in the era of capitalist globalization and climate change. We give an example of the use of such a perception of *place* and *a sense of place* in environmental education, and we also discuss the use of the new concept of *place* for monitoring the resilience of social-ecological systems in Long-Term Social-Ecological Research platforms. This requires a change in the usual measures of *sense of place* and therefore calls for further research.

Keywords Metabolic rift · Place · Sense of place · Social-Ecological system

---

N. Avriel-Avni (✉)
Dead Sea and Arava Science Center, Mitzpe Ramon, Israel
e-mail: noa@adssc.org

M. Lavi-Neeman
Arava Institute, Ketura, Israel

J. M. Holzer
Brock University, St. Catharines, ON, Canada

R. Rozzi et al. (eds.), *Field Environmental Philosophy*, Ecology and Ethics 5,
https://doi.org/10.1007/978-3-031-23368-5_34

34.1 Introduction

The flash flood that occurred on January 18, 2010, in the valleys of the Negev Highlands, Israel, left astonishingly heavy damage. Olive trees and grapevines were uprooted and swept away, dragging many other trees with them. The flow caused intense soil erosion, forming deep ditches between the rows of grapevines. Farmers, who planted their vineyards and olive groves a few years before, were stunned by this extreme weather event. It was indeed exceptional; a flood like this usually only occurs once every 10 years. Modern agricultural activity on these arid lands (average annual precipitation of 80 mm) has been encouraged by the Israeli government since the 1990s, as part of its population dispersal policy. These new farms were connected to the national pipeline system, which carries water from subtropical regions in northern Israel to the arid zone of the country. This steady water supply has provided a sense of certainty that has allowed farmers to manage their lands according to external market forces while ignoring their impact on local and remote ecosystems. For example, farmers tend to use large quantities of imported water, which are required because of the high level of evaporation and salinity in the desert. At the same time, most newly arrived farmers invested in removing floodwaters from their lands, while disregarding ancient agricultural terraces and runoff collection ditches. These archeological remains from the Byzantine period (1300 years ago) are still visible throughout the watershed. They preserve ancient knowledge for moderating soil erosion while still allowing the use of flood waters for irrigation (Avriel-Avni et al. 2019).

The farmers' story illustrates a case of modern land management practices that lacks attachment and awareness of the fragility of social-ecological systems at multiple scales, and ignores long-term climate cycles and traditional local practices for mitigating extreme environmental events. Large cities in industrialized countries are often extreme examples of communities living a conceptual rift (Berkes 2012; Leopold 2004; Moore 2011; Orr 1992; Rozzi 2012) between the social system and the ecosystems that sustain them (Kissinger and Stossel 2019). Yet, rural residents may develop such conceptual rifts as well (Costanza et al. 2014), and may actually act without environmental responsibility anywhere (Mathevet et al. 2018).

This chapter addresses the factors that contribute to the development of responsible behavior towards ecosystems. This goal forms the basis of most of the world's environmental education programs (Aarnio-Linnanvuori 2019; Carleton-Hug and Hug 2010; Disinger and Roth 1992; McBeth and Volk 2009). The definition of the term *sense of place* (SoP) that has entered environmental education in recent years is the attachment and meaning that people or groups attribute to a space and thereby make it a *place* for them (Masterson et al. 2017). We propose that expanding the meaning of *place* from *a meaningful "space"* bounded to particular locale, to a relational concept which sees place as produced by multiple geographies at multiple scales, may further expand the concept of SoP and transform it into a relevant tool for environmental education, thereby strengthening the resilience of social-ecological systems at different scales. Our case is based on the environmental education program of the Negev Highlands Long-Term Social-Ecological Research (LTSER) platform at the Arava and Dead Sea Science Center in Mitzpe-Ramon, Israel. Located in the southern desert of Israel, this area hosts a rich biocultural history, traditional agriculture, goat and sheep grazing, and tourism. It is characterized by its hyper-arid climate and a diversity of geological processes. We use the LTSER networks as platforms to demonstrate our ideas, but our insights could be valid for other platforms dealing with human-ecosystem relations, such as botanical gardens or protected areas.

34.2 The Role of LTSER Platforms for Developing Environmental Responsibility

The international Long-Term Ecological Research (LTER) network was established in the early 1990s following the growing awareness by ecologists for the need of long-term and regional research to understand ecological processes (Anderson et al. 2008; Mirtl et al. 2018). LTSER platforms were proposed to better integrate humans as ecosystem components in landscapes that include settlements, agricultural areas, industry, and commerce, alongside remote areas (Redman and Miller 2015; Maass et al. 2016; Dick et al. 2018). A main goal of LTSER platforms is to support the resilience of social-ecological systems (Mirtl et al. 2018).

To meet this goal, LTSER platforms have adopted an approach based on the Social-Ecological System (SES) model (Turner et al. 2008). SES model portrays humans as part of the ecosystem, conceptualizing the social component as a web of interactions among different stakeholders in the ecosystem (Fig. 34.1). This creates specific (though dynamic) social, economic, and political settings (Ostrom 2010). Social systems depend on local and remote ecosystems that provide vital ecosystem-services (e.g., provision, regulation, culture) (Assessment, Millennium Ecosystem

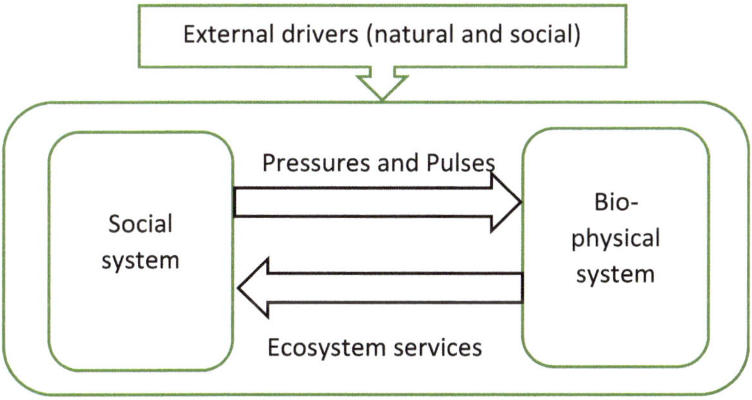

Fig. 34.1 A simple scheme of the Social-Ecological Model (after Collins et al. 2011)

2005). At the same time, the behavior of different stakeholders within the social system creates pressure and pulses that affect the structure and function of their life-support systems and their ability to provide ecological services (Folke 2006; Holling 2001).

Researchers on LTSER platforms take a transdisciplinary approach (Holzer et al. 2018) to maintain long-term ecological and social monitoring and knowledge production that sustain the resilience of SES in terms of time and space (Dick et al. 2018). Understanding the interdependence between the social and the ecological, LTSER researchers position responsible behavior towards life-supporting ecosystems as an essential condition for the SES resilience (Costanza et al. 2014; Folke et al. 2010; Ostrom 2010; van der Hel 2018). The fact that LTSER platforms operate as an international network can help them in this task (Maass and Equihua 2015; Angelstam et al. 2018).

Our definition of "resilience" follows Canadian ecologist Crawford Stanley Holling's (2001) focus on the ability of the SES to tolerate unknown or unforeseen shocks caused by natural forces (e.g., climate change, floods, droughts, earthquakes) and by human society (e.g., fluctuations in the world market, demographic growth). SES resilience is reflected in its ability to absorb, accommodate, or implement changes (adaptation), or even by fundamentally reorganizing in response to challenges that surpass a current SES state or regime (transformation) (Folke et al. 2010). As mentioned above, a key condition for SES resilience is the responsible behavior of humans within the SES.

According to the Cambridge Dictionary, responsible persons have good judgment and the ability to act correctly and make decisions on their own. French ecologist Raphaël Mathevet and collaborators have defined environmental stewardship as:

> a form of responsible management of human activities according to their impacts on the natural environment in order to ensure the conservation of biodiversity, natural resources and their values in terms of use and non-use for future generations of humans and non-humans (Mathevet et al. 2018, p. 3)

Complementarily, based on a biocultural conservation perspective, Chilean ecologist and philosopher Ricardo Rozzi and collaborators have emphasized that:

> Earth stewardship emerge from praxis, a praxis that is ancient and contemporary, collective and individual... Ethical values, citizenship traditions, metaphors and poetic creativity, contemporary and traditional vernacular ecological knowledge, political ecology, institutional networks, local communities, and exemplary lives complement the scientific perspectives of Earth Stewardship initiative... The hybridization of disciplines and traditions will stimulate . . . a new transdisciplinary stage of long-term socio-ecological research at a global scale, and a biocultural approach that includes all beings with whom humans co-inhabit the biosphere. (Rozzi et al. 2015, p. 13)

Finally, in the context of SES, responsible behavior means a mode of action that regulates and reduces pressures and pulses on life-supporting ecosystems (Wu and Wu 2013). To foster responsible behavior, many environmental education programs are based on a simple model proposed by Lucas (1980). That is, environmental education should include three components: education *about, in*, and *for* the environment. Over the years, the component of learning *about* the environment has become central and the development of "environmental literacy" has become the main focus of environmental education (Carleton-Hug and Hug 2010; Disinger and Roth 1992; NAAEE 2010). Recent trends on environmental literacy have given a grooving emphasis to the development of knowledge of local and remote ecosystems, understanding that we humans are part of these systems and depend on their integrity, and the awareness of socio-ecological problems that exist in them (Berkowitz et al. 2005; McBride et al. 2013).

The relationship between environmental literacy and responsible environmental behavior is, however, not strong enough (Goldman et al. 2006; Negev et al. 2008). It is much more complex than linear relationships (Kollmuss and Agyeman 2002). Accordingly, there are calls to foster the emotional dimensions (psychological, social, and moral motives) alongside the cognitive dimension (Bamberg and Möser 2007; Carmi et al. 2015).

34.3 Fostering Sense of Place for Environmental Responsibility

One way to address the challenge of environmental responsibility is to foster a Sense of Place (SoP). As indicated above, SoP refers to the attachment and the meaning that a person or a group attributes to a geographical site thereby making it a *place* for them (Relph 1976; Tuan 1977). Place attachment is described as consisting of place-dependency—a sense of ability to express his/her essence in this geographical location, and of place-identity—a positive integration of the place in individuals' or groups' self-identity. Place meaning refers to the meaning that residents attach to the geographical site. Place attachment and place meaning are in dialectical relationships (Fig. 34.2). Creating a positive "meaning" towards a geographical location influences "place attachment" and vice versa (Kudryavtsev et al. 2012).

Close connections among SoP, environmental concern, and pro-environmental behaviors have been revealed (Kudryavtsev et al. 2012; Masterson et al. 2017). Studying rural residents in Norway, cultural geographers Vorkinn and Riese (2001) found that place attachment predicts the attitudes towards specific environmental changes and concern for environmental issues. Similarly, studying rural communities in the United States, sociologist Joan Brehm and collaborators (2006) found that natural-environment attachment is a predictor of local environmental concern. Both the social and the environmental aspects of SoP are necessary to predict local environmental concern (Armstrong and Stedman 2019). Social ties regarding environmental concern also has been found in Beijing, China (Song and Soopramanien 2019). Based on a literature review, environmental education researcher Alex Kudryavtsev and collaborators (2012) proposed integrating SoP into environmental education to achieve pro-environmental behavior. In the same line, sustainability science researcher Vanessa Masterson and collaborators (2017) suggested using SoP

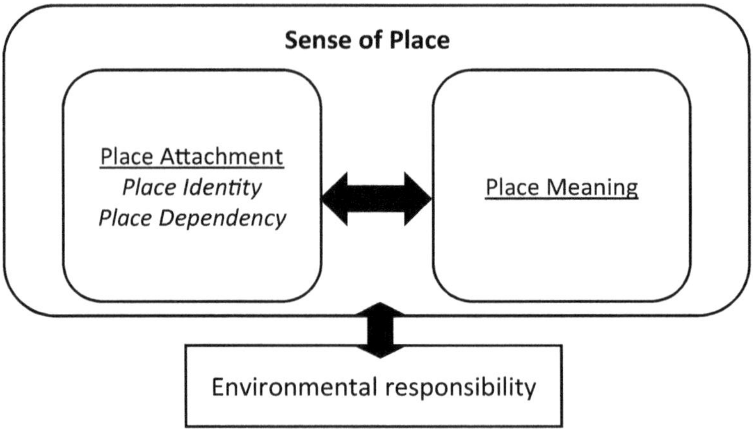

Fig. 34.2 Sense of Place and its relationship with environmental responsibility (after Masterson et al. 2017)

as a tool for research and for developing stewardship of ecosystem services and caring for social-ecological systems.

Despite the potential for strengthening SESs' resilience that lies in fostering an SoP, there are also some weaknesses in this concept. One major problem is that the concept of place is local, whereas modern and semi-modern societies rely on remote ecosystems for their ecosystem services supply (Costanza et al. 2014; Kissinger and Stossel 2019). Consequently, a local SoP does not necessarily contribute to responsible behavior towards these remote ecosystems nor to their resilience (Devine-Wright and Batel 2017; Ryfield et al. 2019).

Recalling our opening story about the desert farmers, it may indicate that the newcomers lacked an SoP towards the specific local *and* the remote life-supporting ecosystems. The desert farmers who rely on a regular supply of water imported from the subtropical zone, through their behavior, express a lack of awareness about the significance of the local water regime; e.g., the distribution of precipitation throughout the year and flash floods that characterize the area. Following this perception, they also do not attach much importance to the traditional knowledge of ancient Byzantine farmers (1300 years ago) who maintained prosperous agriculture in those valleys for 600 years and whose remnants of sophisticated agricultural systems are evident in contemporary vineyards (Avriel-Avni et al. 2018). Instead of restoring and preserving ancient agricultural terraces designed from long-term environmental learning to prevent soil erosion and make optimal use of floods to irrigate vineyards, some contemporary farmers plow and destroy the terraces and consequently must rely heavily on resources from remote ecosystems (Avriel-Avni et al. 2019).

Another widespread problem is the sharp distinction between the social and the environmental and between the urban and the rural, where frameworks of SoP are put to work (see Armstrong and Stedman 2019; Eaton et al. 2019). This distinction might lead to misrepresentation of the complexity and the co-productive nature characterizing the relationship between social systems and ecosystems in the creation of a *place* (Moore 2011), and may establish a reductive linear understanding of mutual "impacts."

A broad spatial and long-term understanding of the environment challenges whether SoP offers an adequate conceptual and educational framework to address the questions that LTSER platforms seek to respond to.

34.4 An Alternative Concept of "Place"

Prevailing conceptualizations of *place* have been criticized in geography since the mid-1990s. Critics called for a conceptualization of *place* that is less bounded, less static, and broader (Sayer 1991; Massey 1991). However, responsible environmentalism and healthier ecology were not initially addressed in new conceptualizations which focused more on epistemology. In the twenty-first century, problems resulting from capitalists society's reliance on remote SES and particularly the implication for environmental responsibility have been addressed in critical scholarship as *"metabolic rift"* (see Foster 2000).

The theory of metabolic rift describes global capitalism as an ecological regime that disrupts forms of *social metabolism*—the transformation of the biophysical environment derived from the flows of materials and energy between nature and society—that alienate humans from the natural environment (Foster 1999; Moore 2000, 2011). Humans' relationships with life-supporting systems have been reorganized following capitalism in ways harmful to both. Social-ecological disruption and individual alienation are part of the processes and its results (McClintock 2010). As the story at the beginning of our chapter demonstrates, a "rift" can be detrimental for ecological, social, and individual well-being, and can be responsible for a type of conduct resulting from alienation. This problem is what education for SoP has been trying to address at the local level. SoP serves as a way to understand in a grounded and differentiated way these "rifts" produced by global and local social economic processes. SoP also serves to generate forms of environmental concern to mend the rifts.

We suggest that to develop environmental responsibility towards SESs, a different concept of *place* (and thereby SoP) should be put into practice. Critical human geographers, particularly British Marxist and feminist cultural geographer Doreen Massey (1991, 2004), have proposed to redefine, and re-conceptualize, *place* by rethinking space relationally. Rather than "the bounded," "the local," "the concrete," and "the grounded" place, standing in contrast to "the global," place should be understood as a spatial identity, forged and formed through interactions at multiple scales—national, regional, local, and more (Massey 2004). Consequently, *place* is never defined a priori, and is essentially unboundable in any absolute sense. It is also "internally complex, and inevitably, historically changing" (Massey 2004 p. 7).

Thinking about space and place relationally addresses concerns of "rift theory" and makes sense from a social-ecological perspective. However, it was seldom incorporated into social-ecological thinking (Baldwin 2012). As we show in our opening story and the examples below, defining place as relational historically and geographically may help change the widely accepted understanding of SoP in SES as nurturing responsibility towards the nearest, the smallest, and the immediate. This understanding, we argue, is limiting from an ecological perspective, and as Massey described, it is "utterly territorial" (2004, p. 6).

Defining place and SoP in the expanded geographical and historical way Massey describes, leads to changing these concepts in educational and community projects. It also expands social and environmental responsibility in terms of time and space. Understanding how our past and how other geographies (and SES) continue to shape our present, may help us understand our responsibility for the past we carry with us, "not because of what we have done but because of what we are" (Massey, p. 12), socially and ecologically. Responsibility, in other words, derives from those relations through which identity (or *place*) is constructed.

SoP provides a complementary methodology with which we can build an understanding of, and engagement with, wider relationships nesting and structuring a specific geography, but spreading beyond and into broader temporal, spatial, ecological, and social systems. SoP may develop and lead to more responsible behavior towards the environment of both local and distant SES, as illustrated by the example in the next section.

34.4.1 "Place": From Theory to Practice

The environmental education curriculum "People of the Desert" (Avriel-Avni 2004) was developed using the participatory action research method (Ballard and Belsky 2013; see also Zent and Zent 2021 this volume). It was developed together with four teachers and 25 14- and 15-year-old students working in the remote desert town of Mitzpe Ramon, Israel with a population of about 5000 residents. In the past, environmental education programs in this town were only taught to children up to the age of 12, arguing that high school students should focus on closing gaps with children studying in the cities.

Interviews conducted with the students before the program began indicated a weak connection of the students with their extensive landscapes surrounding the town. Although most of the students were born and raised locally, they could not identify landscape features nearby and could not recollect significant experiences of time spent in the surrounding environment. Some students expressed fear of the desert that surrounds the town, and for most of them, it was obvious that they only perceive the town as a place and feel safe only within its built-up boundaries.

Some students expressed negative feelings towards the town and even anger towards their parents who chose to raise them in this remote location. One of the girls put it bluntly: "This settlement is like a greenhouse, but the opposite. Instead of thriving we are withering here." Another student pondered: "What are my chances of becoming an influential person in the country, if I grow up here?"

Most students believed that this town lowers their chance to achieve their potential (negative place dependence). They considered it a negative contributor to their place-identity. When describing the significance of the place to them, students generally used negative terms and metaphors such as "hole in the map" or "nothingness." Based on this perception, the overall goal of the curriculum was therefore three-dimensional: (a) to foster a positive SoP, (b) to develop a systemic and dynamic perception of the environment connecting social and ecological aspects at different spatiotemporal scales, and (c) subsequently cultivates a sense of responsibility towards the environment.

34.4.2 *Changing the Perception of Place*

The program was launched with a 3-day hiking tour near the town (Fig. 34.3). It aimed to dispel the intimidation of the desert and to stimulate interest and place attachment in the local socio-ecosystem. The students were constantly within a radius of 2–7 km from the town. This enabled them to observe and experience their surroundings from unfamiliar angles, and to expand the realm considered by them as their *place*.

The journey included encounters with various people who live and work in the area. They shared their personal concept of place with the students. Camels, belonging to the Bedouin semi-nomadic natives who live near the settlement, carried the students' personal effects throughout the journey. Bedouin guides led part of the journey. This created an opportunity for students to form a new perception of the Bedouin, thus transforming the fear and segregation to which they were accustomed.

Among other things, students explored the remains of Byzantine agricultural systems in the desert valleys (Fig. 34.4). They were excited to see these ancient systems and to learn that people prospered here, relying on local water, 1300 years ago.

Fig. 34.3 Students in the journey around their town. (**a**) View of the students walking towards the industrial area of the town. (**b**) Girl observing a limestone quarry near the town, while her friends tour around the quarry. Photographs by Noa Avriel-Avni, taken in November 2000

Fig. 34.4 Students study the Byzantine terraces for irrigation and flood control. Photograph by Gadi Laybrok, taken in December 2000

During the journey, learning was a multi-sensory experience that required two consecutive sessions to process and conceptualize it. This phase of the program deepened and expanded the students' familiarity with and knowledge of their surroundings, in terms of culture and time. This encouraged students to accept the place as a dynamic process, rather than as a permanent reality. The analysis of the content of the meetings with the students after the journey showed a significant change in their perception of place. One of the students summed up the experience: "This was the first time I understood what tourists are excited about when they come here."

34.4.3 Local System's Connections with Remote SESs

During the next phase, a new journey with students further expanded the space dimension of *place* by following the pipeline that supplies water to the town. Throughout a 4-day trek, the students encircled the Sea of Galilee that supplies this water. This freshwater lake is 320 km from their hometown. Students learned about this socio-ecosystem from both scientists and fishermen living in the villages around the lake. Experiential learning taught the students how water consumption in their town affects the lake's socio-ecosystem, and highlighted the town's dependence on remote ecosystems. In a discussion about the town's dependence on remote lake water, students raised various ideas for reducing water use, as illustrated by the following dialogue.

Student A: "I can't manage with less water."

Student B: "Why not? You can simply supply water according to the number of people in the house."

Student C: "But if you stop the water supply for someone, then [the resident] becomes aware. He will finally understand that water is an important and finite resource and he will use it [in a more economical way]..."

Student D: [cuts off Student C] "No. Because if they do this to someone, he will take them to court for stopping the water. And he could have died."

Student C: "No. Let's say when he washes dishes, he'll think twice: Okay, if I'm not exactly rinsing [the soap from the dishes], then I'll turn off the water for a second. Little things like that. He will think of them more when he doesn't have water."

Besides developing the awareness of dependence on distant socio-ecosystems, this conversation also demonstrates the evolution of a sense of ability and environmental responsibility.

34.4.4 Nurturing Environmental Responsibility Through Action

In the third phase of the program, students chose to participate in a project that revolves around challenges in managing their local socio-ecosystem. Two examples are: (1) disseminating the advantages of local desert gardening compared to vegetation that needs frequent irrigation; and (2) designing a home that efficiently uses solar energy (which is very abundant in the desert community). A project that was popular among many students was the restoration of ancient Byzantine structures to establish an orchard and demonstrate the wise use of local floodwaters. An expert advisor joined each of the students' teams. These experts supervised the project and provided knowledge on the subject. For example, the team of students that studied the runoff-based orchard visited well-preserved Byzantine farms. Team members learned the principles that guided ancient farmers for collecting floodwaters and slowing their flow in channels (Fig. 34.4).

The students later built a model orchard irrigated by floodwaters. Students made a mistake in planning the drainage route; however, they recalculated it and worked to achieve a better result (Fig. 34.5). The last rain fell late that year so the students were lucky to see the ditches that they had dug channel the water from the hillsides and flood the orchard they had built. The students were overjoyed with their success and spontaneously began dancing in the water.

This small-scale project conducted with the participatory action research method demanded that participants met regularly to plan their research and actions. These meetings also served as a reflection group (McNiff 1988). The students became aware of the connections among projects and began to develop a systemic and dynamic perception of the environment (Avriel-Avni 2004).

Throughout the program, teachers and students held reflective talks, in addition to the reflection meetings of the teaching staff. These conversations were recorded, transcribed, and analyzed using the semiotic evolution method (Taborsky 1997),

Fig. 34.5 Students working in the field to correct the drainage route. Photograph by Noa Avriel Avni, taken in May 2000

which tracks changes in discourse and perception of place. This analysis follows how the meaning of key concepts develops over time. For instance, the term "responsible behavior" was used by teachers and students from the beginning of the program. In the first stage, however, it meant "carrying out the teachers' instructions independently." One of the teachers said:

> We clearly stated the required assignments, the students brought notebooks, and in the end, did not do much with them. The assignment was given as homework. We said that on Sunday they would submit it, but they forgot about it and today I spoke to them again and told them that as a result, everything was postponed. It all depends on you. [I made it clear to them] The responsibility [to perform the tasks] is yours and not only ours. (Recorded Avriel-Avni February 7, 2000)

However, teachers perceived themselves as responsible for the whole process of learning as illustrated by the following statement by one of the teachers,

> Since here we are working with small groups, it will be our [the teachers'] personal responsibility, each with his own personal group... What I want is for my group to be the best. (Recorded Avriel-Avni on February 7, 2000)

Similarly, students also perceived themselves as responsible only for doing what they were told to do. They introduced themselves during a conversation after the opening trip as follows: "We are a group of students who want to be told how to improve this place."

As the program progressed, students became equal partners in conversations about environmental responsibility and became leaders in action. The concept "responsibility" gained the meaning of taking an initiative to improve the environment. In the teachers' reflection after the tour, one said: "Guiding the students to take responsibility for their lives is the main goal. That's why it matters that things are done in the town; it also gives them [the students] the feeling that they have an impact."

In the summary talk of the program, after completion of their projects, it became evident that this meaning of "environmental responsibility" had become accepted by the students as well. One student suggested that students who demonstrated responsibility towards the environment "*beyond what is expected of them*" should score bonus points in class. This idea was met by a flood of negative reactions from his schoolmates who claimed that: "Responsibility for the environment was our goal in the program and therefore it does not make sense to give it a bonus in the grade." Environmental responsibility became a built-in value for many of the students.

34.5 Discussion

By changing the perception of *place*, the concept *sense of place* (SoP) may become relevant in fostering the resilience of social-ecological systems. Educational institutions, as well as other platforms such as LTSER can use this concept as a foundation for environmental education to encourage long-term responsible behavior that conserves regional life-supporting ecosystems. SoP also can be used as a tool to monitor the development of environmental responsibility, although the measures—as detailed below—vary slightly compared to the accepted measures of place attachment and meaning.

Based on Massey (2004), we have introduced a new definition of place that deviates from compartmentalized distinctions between natural environment (ecosystems unaffected by human presence, such as parks and nature reserves) and human-made environment. This new definition of SoP acknowledges the mutual existence and co-production of these systems and relates to a *place* as the product of relationships between dynamic Social-Ecological Systems (SES), which evolve through interactions between humans and their geographical environment (Ryfield et al. 2019). The social-cultural-economic system cannot be understood correctly disconnected from its ecosystem. Each social or economic system must also be perceived as a particular ecosystem at different spatiotemporal scales. By formulating the concept *place* in this way, the concept *sense of place* may expand the realm of environmental responsibility.

The "People of the Desert" curriculum developed in a small and isolated town in the Negev Desert in Israel offers guidelines for changing approaches to environmental education by incorporating expanded conceptualizations of *place* and *sense of place*. Similar to the first step in the four-step cycle of the field environmental philosophy (FEP) methodology developed at the LTSER site of Omora Park in southern Chile (Rozzi et al. 2010; Tauro et al. 2021), the "People of the Desert" curriculum in the Negev Desert began by broadening the students' familiarity with their geographical environment. Various local residents (e.g., Jews, Bedouins, farmers, tour-guides, quarrymen) shared their viewpoints with the students. Learning about ancient agricultural practices added a dimension of time, emphasizing that a place is dynamic over time. This exposure and holistic learning experience occurred throughout the seemingly familiar area surrounding the town.

The reflection meetings held after the tour allowed the students to discuss their experience and change the way they perceived a *place*, like Step 2 in the FEP cycle. Composition of metaphors and communication through narratives (Rozzi et al. 2010). Although metaphors are not formally used as a tool in the People of the Desert curriculum, they emerged organically from the reflective conversations with the students and indicated a change in the perception of *place* following the opening tour. The numerous research questions proposed by the students, and their motivation to improve the social-ecological system, illustrated their renewed interest in the socio-ecosystem in which they live.

The journey to the remote lake that provides water to the community demonstrated that water is a regional and global system, simultaneously artificial and natural, which sustains local SES, including local human consumption. Students' awareness significantly increased by showing how they are an emplacement of a remote system and structures, and how their environmental culture and behaviors (consumption, forms of development, knowledge, and more) influence distant SESs. The broader geographical perception of *place* was expressed in the students' research projects to reduce the town's pressures on remote ecosystems. Our study shows that it is highly effective to incorporate such "journeys" in environmental education. Where physical journeys are impossible, encounters between students of distant SESs that are linked (e.g., providers of ecosystem services) can be organized through remote communication. Because LTSER platforms operate as international networks, they may facilitate the establishment of such connections. Ties between students of unique SESs are likely to be very significant for the reformulation of *place* and for the development of environmental responsibility.

The action phase of the educational program was a major contributor to the students' sense of competence and to the development of their environmental responsibility. This step parallels steps 3 and 4 of FEP's cycle. FEP's Step 3 involves field activities guided with an ecological and ethical orientation, while Step 4 calls for implementation of in situ biocultural conservation (Rozzi et al. 2010). Rozzi and collaborators have proposed that:

> Under an FEP approach, Earth stewardship is intended to maintain not only human welfare but also the welfare of the whole community of life. Earth stewardship requires that global phenomena and regional biocultural heterogeneity be linked. To fulfill this goal, LTSER networks should aim to forge appropriate "conceptual lenses" as much as these networks aim to forge appropriate technological sensors to research and monitor socioecological systems. In this mission, environmental philosophy and ethics undertake a task that is as relevant as the one undertaken by environmental engineering and the environmental sciences, to better integrate local and global forms of ecological knowledge to promote a stewardship ethics on a bioculturally diverse planet (Rozzi et al. 2012, p. 235).

Consistent with and complementarily to FEP, the People of the Desert curriculum focused on encouraging students to become actively involved in managing their SES, a necessity for the dynamic definition of *place*. Understanding how ancient cultures survived in the desert by using natural water sources for irrigation contributed to: (1) expanding the time dimension of the perception of *place*; (2) acknowledging the connections between place and culture; (3) comprehending the role of the past in shaping the present *place*.

Throughout the program the meaning of "responsibility" transformed. This change indicates the effectiveness of the action phase and is consistent with other studies that show a change in students' SoP, responsibility and competence following active involvement in environmental management (Avriel-Avni et al. 2010). Like educational institutions, LTSER platforms can catalyze changes in the perception of place among older stakeholders, such as the desert farmers mentioned at the beginning of the chapter, LTSER platforms may consequently elevate the SoP and the responsibility of stakeholders towards the resilience of SESs at various spatio-temporal scales. Transdisciplinary research approaches, such as FEP, used in LTSER platforms expose stakeholders to multicultural local knowledge (Holzer et al. 2018). Collaboration between stakeholders in LTSER platforms throughout the globe can help expand the space dimension of *place* (Angelstam et al. 2018).

Environmental responsibility indicates a change in behavior, which may be directly measurable. Long-term monitoring of changes in the SoP may provide environmental educators and scientists of LTSER platforms with vital information on the impact of educational activities and transdisciplinary research, on the resilience of social-ecological systems. Such monitoring is required to deviate from the accepted measures of place attachment and place meaning, and from the early definitions of the geographical space that is considered a *place*. Our relational definition of *place* is dialectic in terms of time and space, integrates natural and social aspects, and focuses on their interactions within the socio-ecological system. The theoretical framework we have laid out in this chapter can serve as a basis for collaborative research action and educational efforts at LTSER sites.

References

Aarnio-Linnanvuori E (2019) How do teachers perceive environmental responsibility? Environ Educ Res 25(1):46–61

Anderson CB, Likens GE, Rozzi R, Gutiérrez JR, Armesto JJ, Poole A (2008) Integrating science and society through long-term socio-ecological research. Environ Ethics 30(3):295–312

Angelstam P, Manton M, Elbakidze M, Sijtsma F, Adamescu CM, Avriel-Avni N, Beja P, Bezak P, Zyablikova I, Cruz F, Bretagnolle V, Díaz-Delgado R, Ens B, Fedoriak M, Flaim G, Gingrich S, Lavi-Neeman M, Medinets S, Viesturs Melecis Muñoz-Rojas J, Schäckerman J, Stocker-Kiss A, Setälä H, Stryamets N, Taka M, Tallec G, Tappeiner U, Törnblom J, Yamelynets T (2018) LTSER platforms as a place-based transdisciplinary research infrastructure: learning landscape approach through evaluation. Landsc Ecol 34(7):1–24

Armstrong A, Stedman RC (2019) Understanding local environmental concern: the importance of place. Rural Sociol 84(1):93–122

Assessment, Millennium Ecosystem (2005) Ecosystems and human well-being: wetlands and water, vol 5. Island Press, p 563
Avriel-Avni N (2004) Ecological thinking as a base for environmental education. Unpublished doctoral dissertation. Ben Gurion University, Bear Sheva, Israel Print
Avriel-Avni N, Spektor-Levi O, Zion M, Levi NR (2010) Defining how urban children perceive their environment as home – implications for aspiring environmental citizens. Int Res Geogr Environ Educ 19(3):241–259
Avriel-Avni N, Holzer J, Shachak M, Orenstein D, Groner E (2018) Using transdisciplinary action research. Toward sustainable management in the Negev highlands. In: Mapotse TA (ed) Cross-disciplinary approaches to action research and action learning. IGI Publisher
Avriel-Avni N, Avni Y, Babad A, Meroz A (2019) Wisdom dwells in places: what can modern farmers learn from ancient agricultural systems in the desert of the Southern Levant? J Arid Environ 163:86–98
Baldwin J (2012) Putting Massey's relational sense of place to practice: labour and the constitution of jolly beach, Antigua, West Indies. Geogr Ann B Hum Geogr 94(3):207–221
Ballard HL, Belsky JM (2013) Participatory action research and environmental learning: implications for resilient forests and communities. In: Krasny ME, Lundholm C, Plummer R (eds) Resilience in social-ecological systems: the role of learning and education. Routledge, New York, pp 152–168
Bamberg S, Möser G (2007) Twenty years after Hines, Hungerford, and Tomera: a new meta-analysis of psycho-social determinants of pro-environmental behavior. J Environ Psychol 27: 14–25
Berkes F (2012) Implementing ecosystem-based management: evolution or revolution? Fish Fish 13(4):465–476
Berkowitz AR, Ford ME, Brewer CA (2005) A framework for integrating ecological literacy, civics literacy, and environmental citizenship in environmental education. In: Johnson EA, Mappin MJ (eds) Environmental education and advocacy or advocacy: perspectives of ecology and education in environmental education. Cambridge University Press, New York, pp 227–266
Brehm JM, Eisenhauer BW, Krannich RS (2006) Community attachments as predictors of local environmental concern: the case for multiple dimensions of attachment. Am Behav Sci 50(2): 142–165
Carleton-Hug A, Hug JW (2010) Challenges and opportunities for evaluating environmental education programs. Eval Program Plann 33:159–164
Carmi N, Arnon S, Orion N (2015) Transforming environmental knowledge into behavior: the mediating role of environmental emotions. J Environ Educ 46(3):183–201
Collins SL, Carpenter SR, Swinton SM, Orenstein DE, Childers DL, Gragson TL, Whitmer AC (2011) An integrated conceptual framework for long-term social–ecological research. Front Ecol Environ 9(6):351–357
Costanza R, de Groot R, Sutton P, van der Ploeg S, Anderson SJ, Kubiszewski I, Farber S, Turner RK (2014) Changes in the global value of ecosystem services. Glob Environ Chang 26(1): 152–158
Devine-Wright P, Batel S (2017) My neighbourhood, my country or my planet? The influence of multiple place attachments and climate change concern on social acceptance of energy infrastructure. Glob Environ Chang 47:110–120
Dick J, Orenstein DE, Holzer JM, Wohner C, Achard AL, Andrews C, Avriel-Avni N, Beja P, Blond N, Cabello J, Chen C, Díaz-Delgado R, Giannakis G, Gingrich S, Izakovicova Z, Krauze K, Lamouroux N, Leca S, Melecis V, Miklós K, Mimikou M, Niedrist G, Piscart C, Postolache C, Psomas A, Santos-Reis M, Tappeiner U, Vanderbilt K, Van Ryckegem G, Chen C (2018) What is socio-ecological research delivering? A literature survey across 25 international LTSER platforms. Sci Total Environ 622:1225–1240
Disinger JF, Roth CE (1992) Environmental literacy. Environmentalist 12(3):68–165
Eaton WM, Eanes FR, Ulrich-Schad JD, Burnham M, Church SP, Arbuckle JG, Cross JE (2019) Trouble with sense of place in working landscapes. Soc Nat Resour 32(7):827–840

Folke C (2006) Resilience: the emergence of a perspective for social–ecological systems analyses. Glob Environ Chang 16:253–267
Folke C, Carpenter S, Walker B, Scheffer M, Chapin T, Rockström J (2010) Resilience thinking: integrating resilience, adaptability and transformability. Ecol Soc 15(4):20–29
Foster JB (1999) Marx's theory of metabolic rift: classical foundations for environmental sociology. Am J Sociol 105(2):366–405
Foster JB (2000) Marx ecology: materialism and nature. Monthly Review Press, New York
Goldman D, Yavetz B, Pe'er S (2006) Environmental literacy in teacher training in Israel: environmental behavior of new students. J Environ Educ 38(1):3–22
Holling CS (2001) Understanding the complexity of economic, ecological, and social systems. Ecosystems 4:390–405
Holzer JM, Adamescu MC, Bonet-García FJ, Díaz-Delgado R, Dick J, Grove JM et al (2018) Negotiating local versus global needs in the International Long Term Ecological Research Network's socio-ecological research agenda. Environ Res Lett 13(10):105003
Kissinger M, Stossel Z (2019) Towards an interspatial urban metabolism analysis in an interconnected world. Ecol Indic 101:1077–1085
Kollmuss A, Agyeman J (2002) Mind the gap: why do people act environmentally and what are the barriers to pro-environmental behavior? Environ Educ Res 8(3):239–260
Kudryavtsev A, Stedman RC, Krasny ME (2012) Sense of place in environmental education. Environ Educ Res 18(2):229–250
Leopold C (2004) Living with the land ethic. Bioscience 54(2):149–154
Lucas AM (1980) Science and environmental education: pious hopes, self-praise and disciplinary chauvinism. Stud Sci Educ 7:1–26
Maass M, Equihua M (2015) Earth stewardship, socioecosystems, the need for a transdisciplinary approach and the role of the International Long Term Ecological Research Network (ILTER). In: Earth stewardship. Springer, Cham, pp 217–233
Maass M, Balvanera P, Bourgeron P, Equihua M, Baudry J, Dick J et al (2016) Changes in biodiversity and trade-offs among ecosystem services, stakeholders, and components of well-being: the contribution of the International Long-Term Ecological Research network (ILTER) to Programme on Ecosystem Change and Society (PECS). Ecol Soc 21(3):14
Massey D (1991) The political place of locality studies. Environ Plan A 23(2):267–281
Massey D (2004) Geographies of responsibility. Geogr Ann B Hum Geogr 86(1):5–18
Masterson VA, Stedman RC, Enqvist J, Tengö M, Giusti M, Wahl D, Svedin U (2017) The contribution of sense of place to social-ecological systems research: a review and research agenda. Ecol Soc 22(1):14
Mathevet R, Bousquet F, Larrère C, Larrère R (2018) Environmental stewardship and ecological solidarity: rethinking social-ecological interdependency and responsibility. J Agric Environ Ethics 31(5):605–623
McBeth W, Volk TL (2009) The national environmental literacy project: a baseline study of middle grade students in the United States. J Environ Educ 41(1):55–67
McBride BB, Brewer CA, Berkowitz AR, Borrie WT (2013) Environmental literacy, ecological literacy, ecoliteracy: what do we mean and how did we get here? Ecosphere 4(5):1–20
McClintock N (2010) Why farm the city? Theorizing urban agriculture through a lens of metabolic rift. Camb J Reg Econ Soc 3(2):191–207
McNiff J (1988) Action research. Principles and practice. Routledge, New York
Mirtl M, Borer ET, Djukic I, Forsius M, Haubold H, Hugo W, Jourdane J, Lindenmayer D, McDowell HM, Muraoka H, Orenstein DE, Pauw JC, Peterseil J, Shibata H, Wohner C, Yu X, Haaseel P (2018) Genesis, goals and achievements of long-term ecological research at the global scale: a critical review of ILTER and future directions. Sci Total Environ 626:1439–1462
Moore JW (2000) Environmental crises and the metabolic rift in world-historical perspective. Organ Environ 13(2):123–157

Moore JW (2011) Transcending the metabolic rift: a theory of crises in the capitalist world-ecology. J Peasant Stud 38(1):1–46

Negev M, Sagy G, Garb Y, Salzberg A, Tal A (2008) Evaluating the environmental literacy of israeli elementary and high school students. J Environ Educ 39(2):3–20

North American Association for Environmental Education (NAAEE) (2010) Excellence in environmental education: guidelines for learning (K-12). Author, Washington, DC

Orr DW (1992) Ecological literacy: education and the transition to a postmodern world. State University of New York Press, New York

Ostrom E (2010) Polycentric systems for coping with collective action and global environmental change. Glob Environ Chang 20(4):550–557

Redman CL, Miller TR (2015) The technosphere and earth stewardship. In: Rozzi R, Chapin FS III, Callicott JB, Pickett STA, Power ME, Armesto JJ, May RH Jr (eds) Earth stewardship: linking ecology and ethics in theory and practice. Ecology and ethics, vol 2. Springer, Dordrecht, pp 269–279

Relph E (1976) Place and placelessness, vol 1. Pion, London

Rozzi R (2012) Biocultural ethics: recovering the vital links between the inhabitants, their habits, and habitats. Environ Ethics 34(1):27–50

Rozzi R, Anderson CB, Pizarro JC, Massardo F, Medina Y, Mansilla A et al (2010) Field environmental philosophy and biocultural conservation at the Omora Ethnobotanical Park: methodological approaches to broaden the ways of integrating the social component ("S") in long-term socio-ecological research (LTSER) sites. Rev Chil Hist Nat 83(1):27–68

Rozzi R, Armesto JJ, Gutiérrez J, Massardo F, Likens G, Anderson CB, Poole A, Moses K, Hargrove G, Mansilla A, Kennedy JH, Willson M, Jax K, Jones C, Callicott JB, Kalin MT (2012) Integrating ecology and environmental ethics: earth stewardship in the southern end of the Americas. Bioscience 62(3):226–236

Rozzi R, Chapin FS III, Callicott JB, Pickett STA, Power ME, Armesto JJ, May RH Jr (2015) Introduction: linking ecology and ethics for an interregional and intercultural earth stewardship. In: Rozzi R, Chapin FS III, Callicott JB, Pickett STA, Power ME, Armesto JJ, May RH Jr (eds) Earth stewardship: linking ecology and ethics in theory and practice. Ecology and ethics, vol 2. Springer, Dordrecht, pp 1–14

Ryfield F, Cabana D, Brannigan J, Crowe T (2019) Conceptualizing 'sense of place' in cultural ecosystem services: a framework for interdisciplinary research. Ecosyst Serv 36:100907

Sayer A (1991) Behind the locality debate: deconstructing geography's dualisms. Environ Plan A 23(2):283–308

Song Z, Soopramanien D (2019) Types of place attachment and pro-environmental behaviors of urban residents in Beijing. Cities 84:112–120

Taborsky E (1997) The textual society. University of Toronto Press, Toronto

Tauro A, Ojeda J, Caviness T, Mosses K, Moreno R, Wright T, Zhu D, Poole A, Massardo F, Rozzi R (2021) Field environmental philosophy: a biocultural ethic approach to education and ecotourism for sustainability. Sustainability 13(8):4526. https://doi.org/10.3390/su13084526

Tuan YF (1977) Space and place: the perspective of experience. University of Minnesota Press

Turner RK, Georgiou S, Fisher B (2008) Valuing ecosystem services: the case of multi-functional wetlands. Earthscan, London

van der Hel S (2018) Science for change: a survey on the normative and political dimensions of global sustainability research. Glob Environ Chang 52:248–258

Vorkinn M, Riese H (2001) Environmental concern in a local context: the significance of place attachment. Environ Behav 33(2):249–263

Wu J, Wu T (2013) Ecological resilience as a foundation for urban design and sustainability. In: Pickett STA et al (eds) Resilience in ecology and urban design: linking theory and practice for sustainable cities. Future city 3. Springer, Dordrecht, pp 211–229

Index

R. Rozzi et al. (eds.), *Field Environmental Philosophy*, Ecology and Ethics 5,
https://doi.org/10.1007/978-3-031-23368-5

F

G

H

I

J

K

L

M

U

V

W

MIX
Papier aus verantwortungsvollen Quellen
Paper from responsible sources
FSC® C105338

Printed by Books on Demand, Germany